The Privileged Planet

THE PRIVILEGED PLANET

HOW OUR PLACE IN THE COSMOS IS DESIGNED FOR DISCOVERY

Guillermo Gonzalez and Jay W. Richards

Library of Congress Cataloging-in-Publication Data

Gonzalez, Guillermo.
The privileged planet : how our place in the cosmos is designed for discovery / Guillermo Gonzalez and Jay W. Richards.
p. cm.
ISBN 0-89526-065-4
1. Solar systems. 2. Earth. 3. Planets. 4. Cosmology. 5. Discoveries in science. I. Richards, Jay Wesley, 1967– II. Title.
QB501.G66 2004
523.2—dc22

2004000421

Published in the United States by
Regnery Publishing, Inc.
An Eagle Publishing Company
One Massachusetts Avenue, NW
Washington, DC 20001

Visit us at www.regnery.com

Distributed to the trade by
National Book Network
4720-A Boston Way
Lanham, MD 20706

Printed on acid-free paper
Manufactured in the United States of America

1

uantity for promotional or premium use. Write to Director of
Publishing, Inc., One Massachusetts Avenue, NW, Washing-
rmation on discounts and terms or call (202) 216-0600.

In memory of

GUILLERMO J. GONZALEZ

and

JOSIAH WESLEY RICHARDS

Table of Contents

Introduction

Section 1. Our Local Environment

Chapter 1: Wonderful Eclipses 1
Chapter 2: At Home on a Data Recorder 21
Chapter 3: Peering Down 45
Chapter 4: Peering Up 65
Chapter 5: The Pale Blue Dot in Relief 81
Chapter 6: Our Helpful Neighbors 103

Section 2. The Broader Universe

Chapter 7: Star Probes 119
Chapter 8: Our Galactic Habitat 143
Chapter 9: Our Place in Cosmic Time 169
Chapter 10: A Universe Fine-Tuned for Life and Discovery 195

Section 3. Implications

Chapter 11: The Revisionist History of the Copernican Revolution 221
Chapter 12: The Copernican Principle 247
Chapter 13: The Anthropic Disclaimer 259
Chapter 14: SETI and the Unraveling of the Copernican Principle 275
Chapter 15: A Universe Designed for Discovery 293
Chapter 16: The Skeptical Rejoinder 313

Conclusion: Reading the Book of Nature 331

Appendix A: The Revised Drake Equation 337

Appendix B: What about Panspermia? 343

Notes 347

Acknowledgments 417

Figure Credits 419

Index 421

Introduction

The Privileged Planet

Discovery is seeing what everyone else saw and thinking what no one thought.
—Albert von Szent-Györgyi[1]

On Christmas Eve, 1968, the Apollo 8 astronauts—Frank Borman, James Lovell, and William Anders—became the first human beings to see the far side of the Moon.[2] The moment was as historic as it was perilous: they had been wrested from Earth's gravity and hurled into space by the massive, barely tested Saturn V rocket. Although one of their primary tasks was to take pictures of the Moon in search of future landing sites—the first lunar landing would take place just seven months later—many associate their mission with a different photograph, commonly known as *Earthrise*. (See Plate 1.)

Emerging from the Moon's far side during their fourth orbit, the astronauts were suddenly transfixed by their vision of Earth, a delicate, gleaming swirl of blue and white, contrasting with the monochromatic, barren lunar horizon.[3] Earth had never appeared so small to human eyes, yet was never more the center of attention.

To mark the event's significance and its occurrence on Christmas Eve, the crew had decided, after much deliberation, to read the opening words of Genesis: "In the beginning, God created the heavens and the Earth. . . ." The reading, and the reverent silence that followed, went out over a live telecast to an estimated one billion viewers, the largest single audience in television history.

In his recent book about the Apollo 8 mission, Robert Zimmerman notes that the astronauts had not chosen the words as parochial religious expression but rather "to include the feelings and beliefs of as many people as possible."[4] Indeed, when the majority of Earth's citizens look out at the wonders of nature or Apollo 8's awe-inspiring *Earthrise* image, they see the majesty of a grand design. But a very different opinion holds that our Earthly existence is not only rather ordinary but in fact insignificant and purposeless. In his book *Pale Blue Dot*, the late astronomer Carl Sagan typifies this view while reflecting on another image of Earth (see Plate 2.), this one taken by Voyager 1 in 1990 from some four billion miles away:

> Because of the reflection of sunlight... Earth seems to be sitting in a beam of light, as if there were some special significance to this small world. But it's just an accident of geometry and optics.... Our posturings, our imagined self-importance, the delusion that we have some privileged position in the Universe, are challenged by this point of pale light. Our planet is a lonely speck in the great enveloping cosmic dark. In our obscurity, in all this vastness, there is no hint that help will come from elsewhere to save us from ourselves.[5]

But perhaps this melancholy assumption, despite its heroic pretense, is mistaken. Perhaps the unprecedented scientific knowledge acquired in the last century, enabled by equally unprecedented technological achievements, should, when properly interpreted, contribute to a deeper appreciation of our place in the cosmos. In the following pages we hope to substantiate that possibility by means of a striking feature of the natural world, one as widely grounded in the evidence of nature as it is wide-ranging in its implications. Simply stated, the conditions allowing for intelligent life on Earth also make our planet strangely well suited for viewing and analyzing the universe.

The fact that our atmosphere is clear; that our moon is just the right size and distance from Earth, and that its gravity stabilizes Earth's rotation; that our position in our galaxy is just so; that our sun is its precise mass and composition—all of these facts and many more not only are necessary for Earth's habitability but also have been surprisingly crucial to the discovery and measurement of the universe by scientists. Mankind is unusually well positioned to decipher the cosmos. Were we merely lucky in this regard?

Scrutinize the universe with the best tools of modern science and you'll find that a place with the proper conditions for intelligent life will also afford its inhabitants an exceptionally clear view of the universe. Such so-called habitable zones are rare in the universe, and even these may be devoid of life. But if there is another civilization out there, it will also enjoy a clear vantage point for searching the cosmos, and maybe even for finding us.

To put it both more technically and more generally, "measurability" seems to correlate with "habitability."[6] Is this correlation simply a strange coincidence? And even if it has some explanation, is it significant? We think it is, not least because this evidence contradicts a popular idea called the Copernican Principle, or the Principle of Mediocrity. This principle is far more than the simple observation that the cosmos doesn't literally revolve around Earth. For many, it is a metaphysical extension of that claim. According to this principle, modern science since Copernicus has persistently displaced human beings from the "center" of the cosmos, and demonstrated that life and the conditions required for it are unremarkable and certainly unintended. In short, it requires scientists to assume that our location, both physical and metaphysical, is unexceptional. And it usually expresses what philosophers call naturalism or materialism—the view that the material world is "all that is, or ever was, or ever will be," as Carl Sagan famously put it.[7]

Following the Copernican Principle, most scientists have supposed that our Solar System is ordinary and that the emergence of life in some form somewhere other than Earth must be quite likely, given the vast size and great age of the universe. Accordingly, most have assumed that the universe is probably teeming with life. For example, in the early 1960s, astronomer Frank Drake proposed what later became known as the Drake Equation, in which he attempted to list the factors necessary for the existence of extraterrestrial civilizations that could use radio signals to communicate. Three of those factors were astronomical, two were biological, and two were social. They ranged from the rate of star formation to the likely age of civilizations prone to communicating with civilizations on other planets.[8] Though highly speculative, the Drake Equation has helped focus the debate, and has become a part of every learned discussion about the possibility of extraterrestrial life. Ten years later, using the Drake Equation, Drake's colleague Carl Sagan optimistically conjectured that our Milky Way galaxy alone might contain as many as one million advanced civilizations.

This optimism found its practical expression in the Search for Extraterrestrial Intelligence, or SETI, a project that scans the skies for radio transmissions containing the "signatures" of extraterrestrial intelligence. SETI seeks real evidence, which, if detected, would persuade most open-minded people of the existence of extraterrestrial intelligence. In contrast, some advocates (and critics) of extraterrestrial intelligence rely primarily on speculative calculations. For instance, probability theorist Amir Aczel recently argued that intelligent life elsewhere in the universe is a virtual certainty. He is so sure, in fact, that he titled his book *Probability One: Why There Must Be Intelligent Life in the Universe*.[9]

Although attractive to those of us nurtured on *Star Trek* and other fascinating interstellar science fiction, such certainty is misplaced. Recent discoveries from a variety of fields and from the new discipline of astrobiology have undermined this sanguine enthusiasm for extraterrestrials. Mounting evidence suggests that the conditions necessary for complex life are exceedingly rare, and that the probability of them all converging at the same place and time is minute. A few scientists have begun to take these facts seriously. For instance, in 1998 Australian planetary scientist Stuart Ross Taylor challenged the popular view that complex life was common in the universe. He emphasized the importance of the rare, chance events that formed our Solar System, with Earth nestled fortuitously in its narrow habitable zone.[10] Contrary to the expectations of most astronomers, he argued that we should not assume that other planetary systems are basically like ours.

Similarly, in their important book *Rare Earth: Why Complex Life Is Uncommon in the Universe*,[11] paleontologist Peter Ward and astronomer Donald Brownlee, both of the University of Washington, have moved the discussion of these facts from the narrow confines of astrobiology to the wider educated public.[12] Ward and Brownlee focus on the many improbable astronomical and geological factors that united to give complex life a chance on Earth.

These views clearly challenge the Copernican Principle. But while challenging the letter of the principle, Taylor, Ward, and Brownlee have followed its spirit. They still assume, for instance, that the origin of life is basically a matter of getting liquid water in one place for a few million years. As a consequence, they continue to expect "simple" microbial life to be common in the universe. More significant, they all keep faith with the broader perspective that undergirds the Copernican Principle in its most expansive form. They argue that although Earth's complex life and the rare conditions that allow for it are highly improbable, perhaps even unique, these conditions are still nothing more than an unintended fluke.[13] In a lecture after the publication of *Rare Earth*, Peter Ward remarked, "We are just incredibly lucky. Somebody had to win the big lottery, and we were it."

But we believe there is a better explanation. To see this, we have to consider these recent insights about habitability—the conditions necessary for complex life—in tandem with those concerning measurability. Measurability refers to those features of the universe as a whole, and especially to our particular location in it—in both space and time—that allow us to detect, observe, discover, and determine the size, age, history, laws, and other properties of the physical universe. It's what makes scientific discovery possible. Although scientists don't often discuss it, the degree to which we can "measure" the wider universe—not just our immediate surroundings—is surprising. Most scientists presuppose the measurability of the physical realm: it's measurable because scientists have found ways to measure it. Read any book on the history of scientific discovery and you'll find magnificent tales of human ingenuity, persistence, and dumb luck. What you probably won't see is any discussion of the conditions necessary for such feats, conditions so improbably fine-tuned to allow scientific discoveries that they beg for a better explanation than mere chance.

Our argument is subtle, however, and requires a bit of explanation. First, we aren't arguing that every condition for measurability is uniquely and *individually* optimized on Earth's surface. Nor are we saying that it's always easy to measure and make scientific discoveries. Our claim is that Earth's conditions allow for a stunning diversity of measurements, from cosmology and galactic astronomy to stellar astrophysics and geophysics; they allow for this rich diversity of measurement much more so than if Earth were ideally suited for, say, just one of these sorts of measurement.

For instance, intergalactic space, far removed from any star, might be a better spot for measuring certain distant astronomical phenomena than the surface of any planet with an atmosphere, since it would contain less light and atmosphere pollution. But its value for learning about the details of star formation and stellar structure, or for discovering the laws of celestial mechanics, would be virtually worthless. Likewise, a planet in a giant molecular cloud in a spiral arm might be a great place to learn about star formation and interstellar chemistry, but observers there would find the distant universe to be hidden from view. In contrast, Earth offers surprisingly good views of the distant *and* nearby universe while providing an effective platform for discovering the laws of physics.

When we say that habitable locations are "optimal" for making scientific discoveries, we have in mind an optimal balance of competing conditions. Engineer and historian Henry Petroski calls this constrained optimization in his illuminating book *Invention by Design*: "All design involves conflicting objectives and hence compromise, and the best designs will always be those that come up with the best compromise."[14] To take a familiar example, think of the laptop computer. Computer engineers seek to design laptops that have the best overall compromise among various conflicting factors. Large screens and keyboards, all things being equal, are preferable to small ones. But in a laptop, all things aren't equal. The engineer has to compromise between such matters as CPU speed, hard drive capacity, peripherals, size, weight, screen resolution, cost, aesthetics, durability, ease of production, and the like. The best design will be the best compromise. (See Figure 0.1) Similarly, if we are to make discoveries in a variety of fields from geology to cosmology, our physical environment must

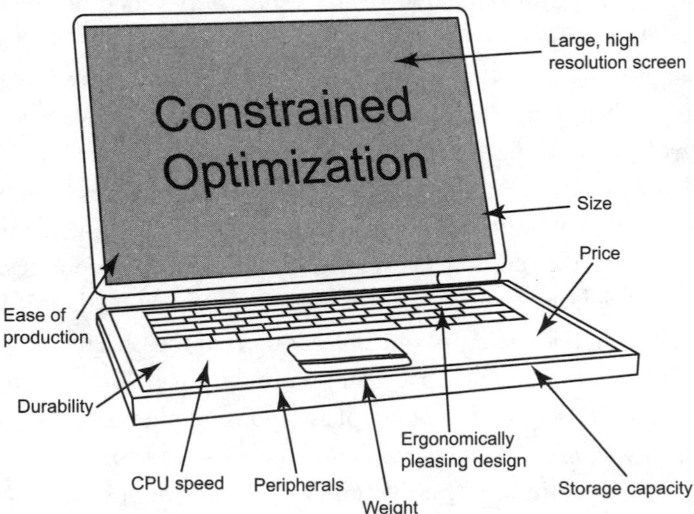

Figure 0.1: A laptop computer, like many well-designed objects, exhibits "constrained optimization." The optimal or best-designed laptop computer is the one that is the best balance and compromise of multiple competing factors.

be a good compromise of competing factors, an environment where a whole host of "thresholds" for discovery are met or exceeded.

For instance, a threshold must be met for detecting the cosmic background radiation that permeates the universe as a result of the Big Bang. (Detecting something is, of course, a necessary condition for measuring it.) If our atmosphere or Solar System blocked this radiation, or if we lived at a future time when the background radiation had completely disappeared, our environment would not reach the threshold needed to discover and measure it. As it is, however, our planetary environment meets this requirement. At the same time, intergalactic space might give us a slightly better "view" of the cosmic background radiation, but the improvement would be drastically offset by the loss of other phenomena that can't be measured from deep space, such as the information-rich layering processes on the surface of a terrestrial planet. An optimal location for measurability, then, will be one that meets a large and diverse number of such thresholds for measurability, and which combines a large and diverse number of items that need measuring. This is the sense in which we think our local environment is optimal for making scientific discoveries.[15] In a very real sense the cosmos, our Solar System, and our exceptional planet are themselves a laboratory, and Earth is the best bench in the lab.

Even more mysterious than the fact that our location is so congenial to diverse measurement and discovery is that these same conditions appear to correlate with habitability. This is strange, because there's no obvious reason to assume that the very same rare properties that allow for our existence would also provide the best overall setting to make discoveries about the world around us. We don't think this is merely coincidental. It cries out for another explanation, an explanation that suggests there's more to the cosmos than we have been willing to entertain or even imagine.

Section 1

Our Local Environment

Chapter 1

Wonderful Eclipses

Perhaps that was the necessary condition of planetary life:
Your Sun must fit your Moon.
—Martin Amis[1]

INSPIRED

October 24, 1995: the date I had long awaited.* I awoke at 5 A.M., along with several other astronomers in our group. It was a cool, clear morning in Neem Ka Thana, a small town in the dry region of Rajasthan, India, a great place for an eclipse. By 6 A.M. I had staked my claim within a roped-off compound in a local schoolyard and was setting up my scientific instruments. Half a dozen other experimental setups were scattered around me in the compound, each with its own team of astronomers. Some had mounted their experiments on stable concrete piers built weeks before. Around the compound were TV and radio news crews and hundreds of curious onlookers, staring at us as if we were rare zoo exhibits. I had joined the expedition at the invitation of the Indian Institute of Astrophysics in Bangalore. Although the eclipse was not the main purpose of my trip to India, I couldn't pass up this rare opportunity.

Strictly speaking, like snowflakes, no two solar eclipses are exactly alike, but astronomers sort these events into three types: partial, annular, and total. In a partial eclipse, the Moon fails to completely cover the Sun's bright photosphere.[2] In an annular eclipse, although their centers may pass

*In this section, Guillermo is speaking in the first person.

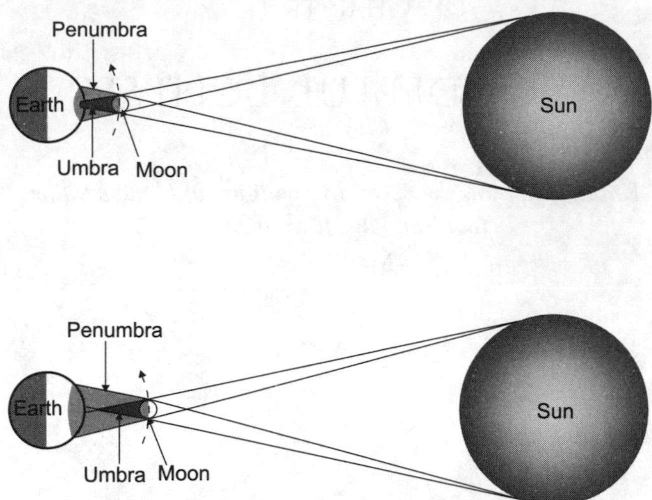

Figure 1.1: A total solar eclipse (above) compared to an annular eclipse (below). In a total eclipse, viewers within the Moon's umbra will see the Moon block the Sun's entire photosphere. Those within the penumbra will see a partial eclipse. During an annular eclipse, however, the Moon's shadow cone converges above Earth's surface, leaving a bright ring of the Sun's photosphere visible even for the best-placed viewers. Sizes and separations are not drawn to scale.

very close to each other, the Moon's disk is too small to cover the Sun's photosphere. To qualify as a total eclipse, the Moon's disk must completely cover the bright solar disk as seen from Earth's surface. These are the eclipses everyone wants to see. Observers far from the eclipse "centerline" see only a partial eclipse. Only total and very close annular eclipses noticeably darken the sky, while only total eclipses allow us to view the eerie pink chromosphere and silvery-white corona. Under such conditions, the chromosphere looks like a fragile, jagged crown, with pink flames protruding around it like a ring of fire. The corona is the outermost part of the Sun's atmosphere, extending several degrees farther out from the chromosphere.

I had witnessed a number of partial solar eclipses—including two annular ones in 1984 and 1994—but this was to be my first (and, to date, only) experience of a total solar eclipse. My experiment was simple: to measure the changing atmospheric conditions of temperature, pressure, and humidity, and to photograph the event with my 35 mm camera and a telephoto lens.

It was a complete success. The perfect weather both that day and the previous day allowed me to compare the meteorological changes occurring during the eclipse.[3] I managed to shoot thirty frames during the fifty-one seconds of totality, the period when the Moon fully eclipses the Sun. The long coronal streamers were plainly visible to the naked eye. (See Plate 3.) Unfortunately, I was so busy snapping photos that I had only a brief glimpse of the eclipsed Sun with my naked eyes. My best view was through the camera's viewfinder—a common complaint of eclipse watchers.

To experience a total solar eclipse is much more than simply to see it. The event summons all the senses. The dramatic drop in temperature was just as much a part of it as the blocked Sun and the "oohs" and "aahs" from the crowd. Just after the total phase ended, many burst into spontaneous applause, as if rewarding a choreographer for a well-executed ballet.

This was only the fourth total solar eclipse visible from India in the twentieth century. Still, I was surprised at the Indians' interest in this eclipse. National television covered the event, with crews set up at three or four locations spread across the eclipse path. One of them shared our site. Prior to departing India, I received a videotaped copy of the TV coverage from a colleague. A number of scholars were interviewed on the scientific aspects of solar eclipses; others discussed Indian eclipse mythology and superstitions. The TV producers, it seemed, were trying to show the world that India had finally discarded religious superstition and entered the era of scientific enlightenment. But the widespread superstitious practices in evidence during this eclipse, such as people—especially pregnant women—remaining indoors, suggest they were not quite successful.

Finally, there were the amateur astronomers and eclipse chasers, people who try to see as many total solar eclipses as they can fit into a lifetime. Eclipse chaser Serge Brunier explains in his book *Glorious Eclipses: Their Past, Present, and Future,* what drives them:

> Passionately interested in astronomy ever since the age of twelve, for me eclipses remained, for a long time, simple dates in the ephemerides, and I had to wait until I was thirty-three before witnessing, for professional reasons, my first total eclipse, that of 11 July 1991, from the Hawaiian Observatory on top of Mauna Kea volcano.
>
> It would be an understatement to say that I immediately became passionate about celestial events, which I have followed ever since, over the course of the years and the lunations, more or less all over the planet. Each time, there is the same astonishment and, each time, the feeling has grown that eclipses are not just

> astronomical events, that they are more than that, and that the emotion, the real internal upheaval, that they produce—a mixture of respect and also empathy with nature—far exceeds the purely aesthetic shock to one's system.[4]

Brunier describes his first total solar eclipse experience:

> The sight is so staggering, so ethereal, and so enchanting that tears come to everyone's eyes. It is not really night. A soft twilight bathes the Mauna Kea volcano. Along the ridge, the silvery domes, like ghostly silhouettes of a temple to the heavens, stand rigidly beneath the Moon. The solar corona, which spreads its diaphanous silken veil around the dark pit that is the Moon, glows with an other-worldly light. It is a perfect moment.[5]

Amateur astronomers who have traveled abroad to watch solar eclipses have told me that responses are always the same. The locals and the visiting astronomers are equally in awe and often in tears. Being able to predict the circumstances of total solar eclipses to within a second of time anywhere on Earth has not quenched our deepest emotional responses to them; neither has it stopped a modern astronomer like Brunier from describing this most physical of phenomena as ethereal, as spiritual. Is there something more to total solar eclipses than just the mechanics of the Earth-Moon-Sun system? Is there some deep connection, perhaps, between observing them and conscious life on Earth? We believe there is.

THE PHYSICS OF THE MOON

First, consider a little-known fact: A large moon stabilizes the rotation axis of its host planet, yielding a more stable, life-friendly climate. Our Moon keeps Earth's axial tilt, or obliquity—the angle between its rotation axis and an imaginary axis perpendicular to the plane in which it orbits the Sun—from varying over a large range.[6] A larger tilt would cause larger climate fluctuations.[7] At present, Earth tilts 23.5 degrees, and it varies from 22.1 to 24.5 degrees over several thousand years. To stabilize effectively, the Moon's mass must be a substantial fraction of Earth's mass. Small bodies like the two potato-shaped moons of Mars, Phobos and Deimos, won't suffice. If our Moon were as small as these Martian moons, Earth's tilt would vary not 3

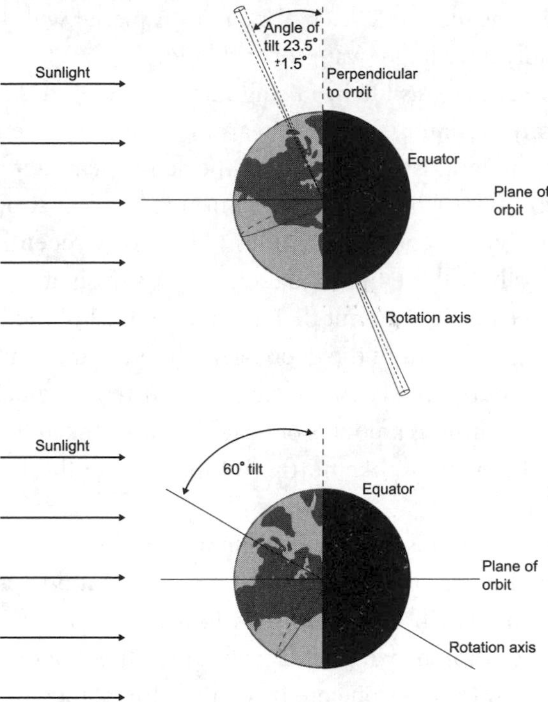

Figure 1.2: Earth's axis currently tilts 23.5 degrees from a line perpendicular to the plane formed by the Earth's orbit around the Sun, and varies a modest 2.5 degrees over thousands of years. Such stability is due to the action of the Moon's gravity on Earth. Without a large Moon, Earth's tilt could vary by 30 degrees or more, even 60 degrees, which would make Earth less habitable.

degrees but more than 30 degrees. That might not sound like anything to fuss over, but tell that to someone trying to survive on an Earth with a 60-degree tilt. When the North Pole was leaning sunward through the middle of the summer half of the year, most of the Northern Hemisphere would experience months of perpetually scorching daylight. High northern latitudes would be subjected to searing heat, hot enough to make Death Valley in July feel like a shady spring picnic. Any survivors would suffer viciously cold months of perpetual night during the other half of the year.

But it's not just a large axial tilt that causes problems for life. On Earth, a small tilt might lead to very mild seasons, but it would also prevent the wide distribution of rain so hospitable to surface life. With a 23.5-degree axial tilt, Earth's wind patterns change throughout the year, bringing seasonal monsoons to areas that would otherwise remain parched. Because of this, most regions receive at least some rain. A planet with little or no tilt would probably have large swaths of arid land.

The Moon also assists life by raising Earth's ocean tides. The tides mix nutrients from the land with the oceans, creating the fecund intertidal zone, where the land is periodically immersed in seawater. (Without the Moon, Earth's tides would be only about one-third as strong; we would experience only the regular solar tides.) Until very recently, oceanographers thought that all the lunar tidal energy was dissipated in the shallow areas of the oceans. It turns out that about one-third of the tidal energy is spent along rugged areas of the deep ocean floor, and this may be a main driver of ocean currents.[8] These strong ocean currents regulate the climate by circulating enormous amounts of heat.[9] If Earth lacked such lunar tides, Seattle would look more like northern Siberia than the lush, temperate "Emerald City."

The Moon's origin is also an important part of the story of life. At the present time, the most popular scenario for its formation posits a glancing blow to the proto-Earth by a body a few times more massive than Mars.[10] That violent collision may have indirectly aided life. For example, it probably helped form Earth's iron core by melting the planet and allowing the liquid iron to sink to the center more completely.[11] This, in turn, may have been needed to create a strong planetary magnetic field, a protector of life that we'll discuss later. In addition, had more iron remained in the crust, it would have taken longer for the atmosphere to be oxygenated, since any iron exposed on the surface would consume the free oxygen in the atmosphere. The collision is also believed to have removed some of Earth's original crust. If it hadn't, the thick crust might have prevented plate tectonics, still another essential ingredient for a habitable planet. In short, if Earth had no Moon, we wouldn't be here.[12]

Of course, with eclipses it takes three to tango: a star, a planet, and its moon. As long as they are the right relative sizes and distances apart, a total eclipse can happen with a larger or smaller moon or star. But two factors vary considerably: the life-support potential of the host planet and the usefulness of the eclipse for science. Let's start with the former.

Habitability varies dramatically, depending on the sizes of a planet and its host star and their separation. There are good reasons to believe that a star similar to the Sun is necessary for complex life.[13] A more massive star has a shorter lifetime and brightens more rapidly. A less massive star radiates less energy, so a planet must orbit closer in to keep liquid water on its surface. (The band around a star wherein a terrestrial planet must orbit to maintain liquid water on its surface is called the Circumstellar Habitable Zone.) Orbiting too close to the host star, however, leads to rapid tidal locking, or "rotational synchronization," in which one side of the planet perpetually faces its host star. (The Moon, incidentally, is so synchronized in its orbit around Earth.) This leads to brutal temperature differences between the day and night sides of a planet. Even if the thin boundary between day and night, called the terminator, were habitable, a host of other problems attend life around a less massive star (more on this in Chapter Seven).

If a planet's moon were farther away, it would need to be bigger than our Moon to generate similar tidal energy and properly stabilize the planet.[14] Since the Moon is already anomalously large compared with Earth, a bigger moon is even less likely. A smaller moon would have to be closer, but then it would probably be less round, creating other problems.

As for the host planet, it needs to be about Earth's size to maintain plate tectonics, to keep some land above the oceans, and to retain an atmosphere (more on these requirements in Chapter Three). To maintain a stable planetary tilt, a planet needs a minimum tidal force from a moon. A larger planet would require a larger moon. So indirectly, even the size of Earth itself is relevant to the geometry of the Earth-Sun-Moon system and its contribution to Earth's habitability. In short, the requirements for complex life on a terrestrial planet strongly overlap the requirements for observing total solar eclipses.

SUPER-ECLIPSES AND PERFECT ECLIPSES

What if the Moon were much closer to Earth, as it was in the distant past? About 2.5 billion years ago, the Moon was, on average, about 13 percent closer than it is now.[15] Such total eclipses of the Sun, what we will call super-eclipses, would then have been more common and visible over a wider region of Earth's surface. During a super-eclipse, the pink chromosphere and parts of the innermost corona are visible briefly only near the start and end of totality. Today we can observe the entire chromosphere throughout much of the total phase of an eclipse.

In eclipses like the one on October 24, 1995, when the Moon's black disk just barely covered the Sun's bright photosphere,[16] the Sun's extended atmosphere was fully visible for almost a minute. We'll refer to an eclipse

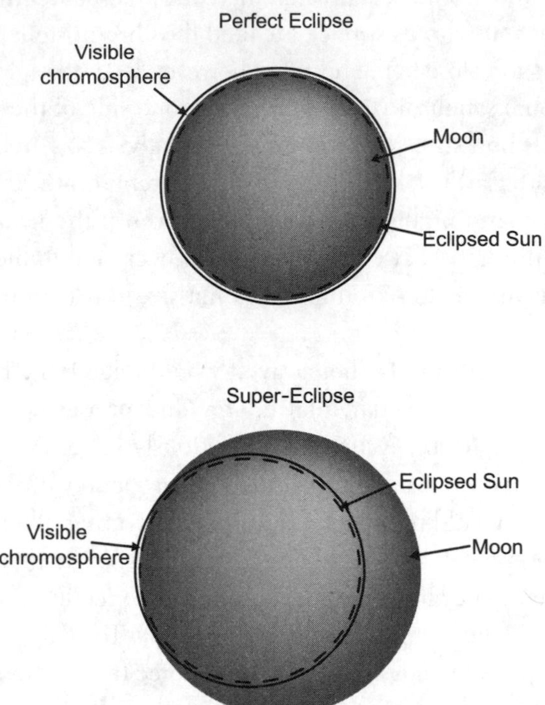

Figure 1.3: A perfect solar eclipse compared to a super-eclipse. For scientific discovery, perfect eclipses are better than super eclipses. In a perfect eclipse the Moon just covers the Sun's bright photosphere, revealing the Sun's thin chromosphere. In contrast, a super-eclipse would reveal only a small sickle of the scientifically valuable chromosphere, and then only at the beginning and end of totality. The thickness of the chromosphere has been exaggerated for clarity; in reality, its thickness is about one three-hundredth the radius of the Sun.

of this type as a "perfect eclipse," because it lasts long enough for an observer to take it in. The Moon is just large enough to block the bright photosphere but not so large that it obscures the colorful chromosphere. A briefer total eclipse leaves a brighter sky, with less time for our eyes to adapt to the darkness, making the faint outer corona harder to see. A slightly larger moon would provide longer eclipses but block more of the scientifically revealing chromosphere.

If the Moon were "less round," we would still enjoy solar eclipses (if the minor axis of a squashed moon appeared larger than the Sun). But such eclipses would be "less perfect," since the chromosphere would be obscured along the major axis during mid-totality. The Moon and the Sun, as it happens, are two of the roundest measured bodies in the Solar System. Neither is precisely a geometric sphere, of course, but the Sun comes closer than just about any natural object known to science.[17] Because the Moon is rocky, its roundness is a bit surprising. In contrast, the moons in the outer Solar System are a mixture of rock and ice, which leads to a rounder shape, as ice is less resistant to stress than rock. Although the Moon has virtually no ice, its profile is quite round. This is probably the result of the peculiar way it formed, as compared with the moons in the outer Solar System. After the Moon formed as a result of a giant impact with the proto-Earth, the ejected material quickly coalesced while some of it was still partially molten; the remaining material accreted onto the Moon soon thereafter.[18]

What if the Moon had an atmosphere? Total lunar eclipses provide some clues. The Moon turns deep red during the central phase of a total lunar eclipse, because sunlight refracts through Earth's atmosphere on its way to the Moon. The light looks red for the same reasons the Sun looks red at sunrise and sunset. An observer on the Moon would be bathed in deep red light, and he would see a bright red ring encircling Earth. We would also see such a ring around the Moon, if it had an atmosphere, during a total solar eclipse. It would completely obscure the pink chromosphere and much, if not all, of the corona.[19]

Finally, what if we were living on another planet in the Solar System? Figure 1.4 shows how big a given moon looks to an observer on its host planet compared with the Sun.[20] The apparent size of a moon is what an observer at the equator of the parent planet would observe; for the gas giants, imagine the observer floating above the cloud tops in a research balloon. This figure illustrates an astonishing fact: Of the more than sixty-four moons in our Solar System, ours yields the best match to the Sun as viewed from a planet's surface, and this is only possible during a fairly narrow window of Earth's history, encompassing the present. The Sun is some four hundred times farther than the Moon, but it is also four hundred times larger. As a result, both bodies appear the same size in our sky.

The so-called Galilean Moons cast large shadows on the cloud tops of Jupiter, which are familiar to amateurs who have spent any time observing them. (Had they more closely matched the apparent disk of the Sun, their shadows would probably not be visible in amateur telescopes.) In general, the Sun looks smaller and total eclipses become more common as one goes outward from the Sun. Total solar eclipses are much more difficult to pull off when the Sun looms close and large.

In fact, if your only goal were mere total solar eclipses, you might wish to relocate to a planet farther from the Sun. But for scientific purposes, Earth's eclipses are the best available, since in general the farther a planet is from the Sun, the briefer its eclipses. Because the Sun looks smaller on those outer planets, all other things being equal, an average moon orbiting one of them passes over the Sun's disk more quickly. All other things aren't equal, however, and those other things just make matters worse for our intrepid outer-planet eclipse chaser. Moons orbit the giant planets much faster than our Moon orbits Earth, because the giant planets are more massive. Moreover, only four moons in the Solar System are larger than the Moon. As a result, the typical total solar eclipse seen on the outermost planets lasts only a few seconds.

Of the sixty-four moons plotted in Figure 1.4, only two appear the same size (on average) as the Sun from their host planets—our Moon and Prometheus, a small, potato-shaped moon of Saturn. But Prometheus produces eclipses lasting less than one second as it whips around Saturn. Moreover, its highly elongated shape compromises the view of the chromosphere. As the figure shows, a typical moon appears larger than the Sun in the outer Solar System. The average ratio is near one at Saturn, so it's not so surprising that a Saturnian moon most closely matches the Sun among the other planets. But can chance also account for the Moon's match to the Sun? The Moon bucks this trend. We think an additional explanation is called for.

In fact, compared with the other moons in the Solar System, the Moon gives us eclipses that are "more than perfect," since the Sun appears larger from Earth than from any other planet with a moon. So an Earth-bound observer can discern finer details in the Sun's chromosphere and corona than from any other planet.

REVEALING ECLIPSES

Besides their intrinsic beauty, perfect solar eclipses have played an important role in scientific discovery. In particular, they have helped reveal the nature of stars, provided a natural experiment for testing Einstein's General Theory of Relativity, and allowed us to measure the slowdown of Earth's rotation.

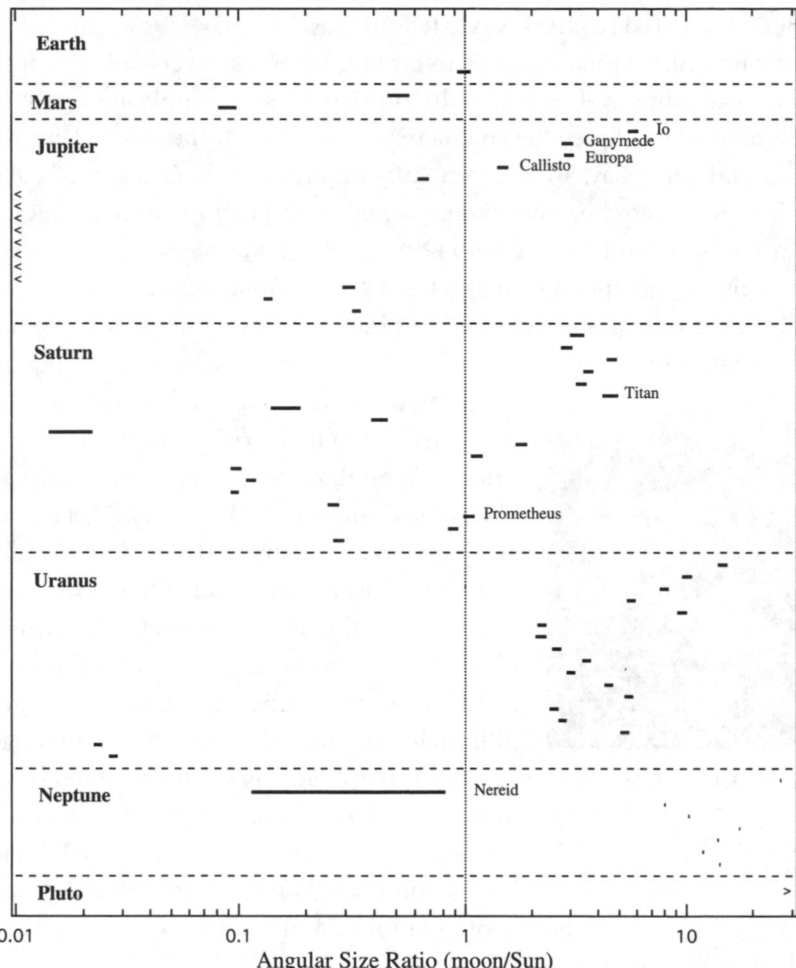

Figure 1.4: Comparison of the average angular size ratios of sixty-four moons to the Sun from the surfaces of their host planets (many smaller, recently discovered, moons around the giant planets are not included). The ratios are plotted on a logarithmic scale on the horizontal axis, so the main tick marks represent multiples of ten. If a moon is non-spherical, then its smallest dimension is used for calculating its apparent size. If a moon has a ratio of one, it is a perfect match for the Sun from the surface of its host planet. There are only two such matches in our Solar System: Earth's Moon, and Prometheus, a small potato-shaped moon of Saturn. Unlike our Moon, however, Prometheus produces eclipses lasting less than one second. Notice that there is a range of angular size ratios of the moons, so a line rather than a point represents them. This is because the orbits of the planets and moons are not perfectly circular. As a result, the angular sizes of the Sun and the moons vary from the respective planetary surfaces. Nereid, one of Neptune's moons, has a quite eccentric orbit. "<" represents moons too small to appear on the chart. ">" represents the one moon too large for the chart—Pluto's Charon.

SPECTRA AND THE SUN'S ATMOSPHERE

The Sun's full corona is visible to ground-based observers only during a total solar eclipse.[21] It is one of the primary reasons people are drawn to view total solar eclipses: the corona never looks exactly the same at any two eclipses. Even today, astronomers still conduct experiments at total solar eclipses to discern how the corona can be heated to millions of degrees.

More important was the help perfect solar eclipses gave to early spectroscopists for interpreting the spectra of stars. Astronomers use instruments called spectroscopes to separate light into its constituent colors. The different colors in the light spectrum, we now know, correspond to different wavelengths of electromagnetic radiation. The traditional colors of the visible spectrum are the stripes of a rainbow: red, orange, yellow, green, blue, indigo, and violet. Wavelengths get longer as we go from the blue to the red end of the spectrum. (Visible light is actually an extremely tiny part of the electromagnetic spectrum, which extends from radio waves on the long end to X-ray and gamma rays on the short end.) Although scientists since a bit before Isaac Newton (1666) had known that sunlight splits into all the colors of the spectrum when passed through a prism, it was not until 1811 that Joseph von Fraunhofer first described the dark gaps that intersperse the smooth continuum of the solar spectrum, often called Fraunhofer lines.

Figure 1.5: In 1811, Joseph von Fraunhofer (1787–1826) first described the dark gaps that cross the solar spectrum.

Over the following decades, laboratory experiments revealed that atoms and molecules both emit and absorb light at characteristic points on the spectrum, called emission and absorption lines. When a gas is heated to a certain temperature, it emits light unique to its composition. Such a gas absorbs light when illuminated from behind, producing absorption lines in the spectrum like a bar code superimposed on a rainbow. Each element impresses its own unique fingerprint on the spectrum. As a result of these laboratory experiments, astronomers were eventually able to identify many of the Fraunhofer lines in the Sun's spectrum with emission lines produced by specific elements.

But astronomers did not know where the Sun's Fraunhofer lines formed or understand the properties of the gas absorbing the light until two notable eclipses in the latter half of the nineteenth century. During the eclipse of August 18, 1868, the French astronomer Pierre Jules César Janssen pointed his spectroscope at prominences—plumes of gas that surge out from the photosphere into the corona—during the few minutes of totality, revealing a spectrum of bright emission lines. Most were quickly identified as hydrogen by comparing them with laboratory spectra.[22] The brightness of the emission lines motivated Janssen to search for prominences the following day, when there was no eclipse. He succeeded, and soon thereafter invented the spectroheliosope, which produces an image of the Sun in the light of one spectral line; this allows astronomers to study the gas motions in the Sun's atmosphere in great detail. Observations of prominences and the chromosphere against the backdrop of dark space during an eclipse demonstrated that they are made of hot, low-density gas, like the gas-filled glass tubes excited by an electric current in laboratories. In fact, the color of such a tube filled with hydrogen is similar to that of the chromosphere and the prominences. (See Plate 4.)

These discoveries helped confirm the conjecture of Jesuit priest Angelo Secchi and John Herschel in 1864 that the Sun is a ball of hot gas. Today, this seems obvious, but it was not so to early-nineteenth-century astronomers. George Airy was the first astronomer to describe what we now know as the chromosphere, during the July 28, 1851, total solar eclipse, the first one to be photographed. He had called it the sierra, mistaking it for a range of mountains on the Sun.

The English astronomer Joseph Norman Lockyer independently recorded spectra of prominences without the benefit of an eclipse to guide and inspire him, though he was certainly aware of the results from successful solar eclipse expeditions. Both Janssen and Lockyer independently discovered a bright emission line in the yellow part of the Sun's emission line spectrum, which Lockyer identified with a new element he named helium, after the Greek word for the Sun, *helios*. (Helium was not isolated in the laboratory until 1895.) Helium doesn't have any spectral features in the absorption spectrum of the Sun, so its discovery would have been greatly delayed had astronomers continued to focus their attention only on its absorption spectrum. Today we know that helium makes up about 28 percent of the Sun's mass; it's the second most abundant element in the universe. It's very unlikely that either Janssen or Lockyer would have thought of obtaining spectra of prominences if previous solar eclipse observers had not described them.

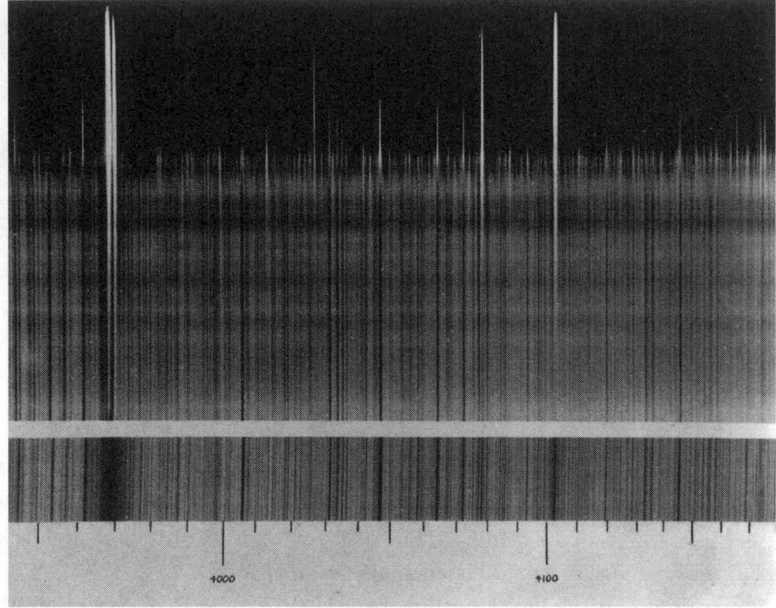

Figure 1.6: Ultraviolet region of the solar spectrum obtained by W. W. Campbell during the total eclipse of August 30, 1905. Campbell used a clever "moving plate" method to record the changing solar spectrum as the Moon's limb covered the last bit of the Sun's photosphere. Wavelength runs horizontally and time vertically on the photo. Note how the spectrum changes from absorption to emission. The Sun's photospheric spectrum is shown on the bottom panel for comparison.

During the total solar eclipse of December 22, 1870, American astronomer and one-time missionary Charles A. Young noticed that the Sun's spectrum changed from its usual appearance of sharp, dark lines superimposed on a bright continuum to emission lines just as totality began. In Young's own words:

> As the Moon advances, making narrower and narrower the remaining sickle of the solar disk, the dark lines of the spectrum for the most part remain sensibly unchanged, though becoming somewhat more intense. A few, however, begin to fade out, and some even begin to turn palely bright a minute or two before totality begins. But the moment the Sun is hidden, through the whole length of the spectrum—in the red, the green, the violet—the bright lines flash out by the hundreds and thousands almost

> startlingly, for the whole thing is over in two or three seconds. The layer seems to be only something under a thousand miles in thickness, and the Moon's motion covers it very quickly.[23]

Hence, this thin region came to be called the reversing layer, which today we know is part of the chromosphere. Young's observation first demonstrated the location and state of the gas producing the absorption lines in the out-of-eclipse solar spectrum.[24] By applying Gustav Kirchhoff's laws of spectroscopy (see Plate 5),[25] Young realized that the reversing layer is made of cooler gas than the underlying photosphere. It is only during a total solar eclipse that the bright photosphere is conveniently blocked. Had the Moon loomed larger, Young's experiment would have been possible only over a short segment of the Sun's limb (the apparent "edge" of the photosphere).

Since Young's historic observations of the 1870 eclipse, the so-called flash spectrum of the chromosphere has been photographed several times. In essence, the Moon acts as a giant slit, allowing only a thin sliver of light from the chromosphere to reach the observer during the first and last few seconds of totality. If there were a rainbow during a total solar eclipse, and one had sensitive video equipment, one could see it change from a continuous spectrum to an emission line spectrum for a few brief seconds.[26] In effect, Earth, the Moon, and the Sun form the primary components of a giant spectroscope. All that remains is for an observer to hold a prism to his eye.

It's hard to exaggerate the significance of the insights afforded by the 1868 and 1870 eclipses for developing stellar astrophysics later in the nineteenth and twentieth centuries. Only because we understand how absorption lines form in the Sun's atmosphere can we interpret the spectra of distant stars, and thereby determine their chemical makeup, all without leaving our tiny planet. Such knowledge is the linchpin for modern astrophysics and cosmology.

Eddington and Einstein's Theory of General Relativity

Arthur Eddington was a famous theoretical astrophysicist of the early twentieth century, but today most know him for his observations of a total solar eclipse that confirmed a prediction of Einstein's General Theory of Relativity—namely, that gravity bends light. On May 29, 1919, two teams, one led by Eddington and Edwin Cottingham on Principle Island off the coast

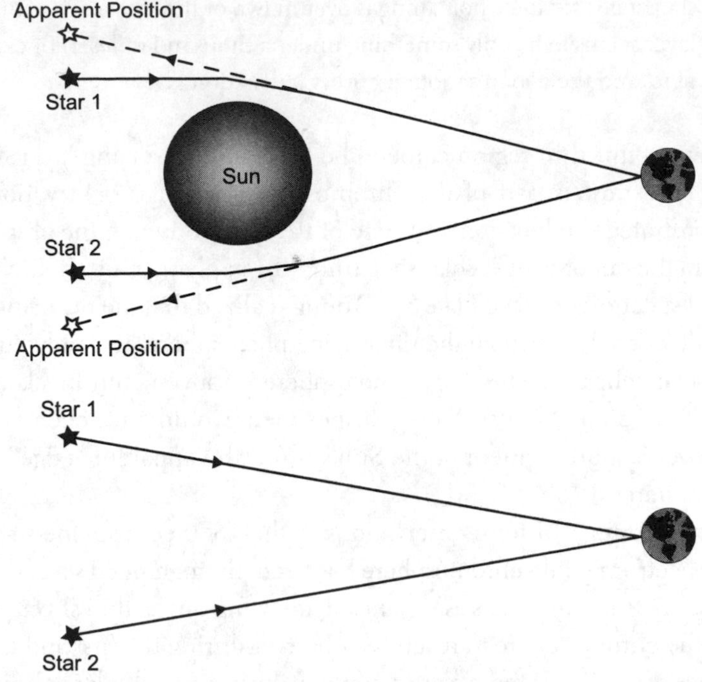

Figure 1.7: According to Einstein's General Theory of Relativity, gravity should cause starlight passing near the Sun's limb to "bend." A perfect total solar eclipse creates the best natural experiment for testing this prediction. Sizes and separations of bodies are not drawn to scale; the amount of bending has been exaggerated for clarity.

of West Africa and the other led by Andrew Crommelin and Charles Davidson in Brazil, used a total solar eclipse to test Einstein's 1916 theory. Their goal was to measure the changes in the positions of stars near the Sun compared with their positions months later or before. Both teams succeeded in photographing the eclipse. Their results confirmed Einstein's predictions and won him immediate acclaim.

Astronomers have repeated the 1919 experiment at many eclipses since, generally agreeing with Einstein's predictions, although the first observed deflections tended to be a bit too large and displayed considerable scatter, this perhaps due to the less than ideal weather conditions.[27] The most carefully executed starlight deflection experiment was conducted during the June 30, 1973, solar eclipse, and the results again confirmed General Relativity.[28] Only a couple of years later, radioastronomers tested Einstein's predictions to much higher precision with observations made without an eclipse.[29] Other tests involving radio transmissions from space probes have also confirmed related aspects of General Relativity. Therefore, although more stringent tests of General Relativity have gone far beyond those requiring a solar eclipse, and although the British 1919 results were somewhat imprecise, solar eclipse experiments clearly played a crucial role in speeding the adoption of General Relativity.

Discerning the Past Rate of Earth's Rotation

Historical observations of total solar eclipses are by far the best known way to measure the change in Earth's rotation period over the last few thousand years.[30] Careful observations of stars show that Earth's rotation period is slowing at a rate of two milliseconds per day per century, due mostly to the action of the tides on Earth by the Sun and Moon.[31] However, such precise observations have only been possible for the last couple of centuries.

Since it casts a narrow shadow across Earth's surface, a total solar eclipse is visible only by the lucky or ardent few in its track. Variations in Earth's rotation period translate into errors in the placement of the predicted shadow track. By examining ancient accounts of total solar eclipses at known dates and places, astronomers can determine the error in the predicted longitude and translate it into an error in time. This kind of information has several uses. For example, knowing the precise variations of Earth's rotation period helps us to discern subtle changes in its shape over centuries and millennia, such as changes due to the retreat of the glaciers in the Northern Hemisphere. More importantly, total solar eclipse observations allow historians to translate the calendar systems of ancient civilizations into our modern system, permitting us to place events from different civilizations on a common timeline. We can then establish the configuration of the Sun, Moon, and planets on any place and any date on that calendar. Other types of astronomical phenomena, such as lunar eclipses and planetary conjunctions, are not as useful as total solar eclipses for historical studies, since they are visible over a much broader geographical area and/or last much longer.

Perfect solar eclipses are optimal for all three of these uses—discovering the nature of the Sun's atmosphere, testing General Relativity, and timing Earth's rotation. If we experienced super-eclipses instead, we would be able to observe the chromosphere only over a small fraction of the solar limb.[32] Also, we wouldn't be able to measure the deflection of starlight as closely to the solar limb.[33] Finally, the eclipse shadow on Earth would be larger, limiting its usefulness for studying Earth's rotation.

We would be even more deprived had the Moon's disk not covered the Sun's bright face, yielding only annular eclipses. The difference between an annular and a total eclipse is not just a matter of degree. To a casual observer, an annular solar eclipse is hardly different from a partial one. Since the chromosphere is wafer-thin, we would have learned far less about stellar atmospheres had the Moon's apparent size been only a little smaller.

The awe-inspiring beauty of total solar eclipses no doubt motivated astronomers in the last two centuries to travel great distances to observe them. This may seem trivial, but it's clear from their diaries and accounts that the experience of an eclipse was an important part of their interest. A number of important discoveries about the Sun were unplanned. If the beauty of total eclipses had not attracted astronomers to their narrow shadow tracks, some discoveries would have been delayed, perhaps indefinitely.

Today, observatories in space can image the Sun's outer corona, and mountaintop telescopes with coronagraphs can image the inner corona. But the occulting disks in the space-based coronagraphs cover everything within about two solar radii, and the spatial resolution is less than that of ground-based observations during total solar eclipses.[34] Since only total solar eclipses allow the full corona to be imaged, they still provide useful and inexpensive information about the corona.

There's a final, even more bizarre twist. Because of Moon-induced tides, the Moon is gradually receding from Earth at 3.82 centimeters per year.[35] In ten million years, the Moon will seem noticeably smaller. At the same time, the Sun's apparent girth has been swelling by six centimeters per year for ages, as is normal in stellar evolution. These two processes, working together, should end total solar eclipses in about 250 million years, a mere 5 percent of the age of Earth. This relatively small window of opportunity also happens to coincide with the existence of intelligent life.[36] Put another way, the most habitable place in the Solar System yields the best view of solar eclipses just when observers can best appreciate them.[37]

FROM AN ECLIPSE TO AN INSIGHT

Perfect solar eclipses are perhaps the most aesthetically striking example of the correlation between habitability and observability. Yet the correlation in this case is more encompassing, since it applies also to measurability and discoverability, both of which are subtly different from observability. Earth is ideal for observing perfect solar eclipses. Beyond this, perfect solar eclipses are optimal for measuring a range of important phenomena, such as the solar flash spectrum, prominences, starlight deflection, and Earth's rotation. But even more than this, perfect solar eclipses provide great opportunity for discoveries about the Sun. Finally, besides inspiring awe and allowing us to discover the nature of the Sun's atmosphere and the element helium—both unanticipated—perfect solar eclipses became the occasion for discovering the correlation between habitability and measurability itself, hardly an insignificant point.

Eclipses are but the tip of the iceberg. The universe is filled with similar evidence. Let's consider that evidence, working our way out in concentric spheres, beginning at the smallest scale—the surface of Earth—and then moving on to the sky, planets, and starry heavens, and ending with the cosmos itself.

Chapter 2
At Home on a Data Recorder

But now ask the beasts, and let them teach you;
And the birds of the heavens, and let them tell you.
Or speak to the earth, and let it teach you;
And let the fish of the sea declare to you.
—Job 12:7–8[1]

RHYTHMS

In the warmth of the summer of 1975, a young male fur seal waddled across the steep rocks of Signy Island, off the coast of continental Antarctica. He had swum and shuffled through ice and water from the cold and salty Stygian Cove, where the ocean meets the tiny island. Upon clearing the rocks, he found what he hoped to find—a freshwater lake. Although Sombre Lake was still mostly frozen, continuous sunlight had melted enough of its icy crust to give him all the water he could wish to drink in one sitting. After a few minutes in the fresh slushy ice, he headed back out to the cove to the scores of breeding fur seals. Without knowing it, he had left a trace of his visit: a few thick, black hairs.

In 1992, Dominic Hodgson and other members of the British Antarctic Survey submerged sediment traps in Sombre Lake, extracting cores from the dark, chilly bottom. The variation of seal hairs in the layers of extracted sediment revealed startling details. The researchers discerned the annual fur seal populations fairly directly. But with careful analysis, the hairs gave them detailed clues about the effects of seal and whale hunting on the various Antarctic flora and fauna over the past two hundred years.[2] Hodgson's team had tapped into a tiny sliver of the information stored on our planet as a result of regular sedimentary processes. Their ingenuity in deciphering such information is apparent. Less apparent but equally important are the rare conditions that preserve it. For this sedimentation process is just one of dozens of natural recording "devices" that preserve detailed information about the past.

NATURE'S DATA LOGGERS

We all learn as children that we can estimate the age of a tree by counting the concentric rings in a cross section of its trunk. But scientists can discern more than age from such rings. Comparing their thickness, for instance, tells us something about the changes in temperature and precipitation in a tree's vicinity while it was alive.

Tree rings, as well as many other geological layering and biological growth processes, are like scientists' data-recording devices. We're all familiar with one such device, if only from hokey TV dramas: the dreaded lie detector, with its black, spindly pens scribbling across a scroll of paper. The heart of a lie detector is a strip chart recorder, the basic parts of which are akin to many natural recorders.[3] Consider the layered polar snow/ice deposits. The ice is like the paper, the diurnal and annual ice layers are like the pens that mark time intervals, and any other substances in the ice sensitive to environmental changes are like the other pens that record signals from the transducers, which convert environmental data to electrical signals.

Of course, to be useful, the paper roll must be protected from damage. Moreover, new information must be added without disturbing previously recorded data. If the motor on a chart recorder breaks down and the paper fails to advance, the pens will keep writing over the same small area of paper, creating an indecipherable glob of ink. Likewise, in a snow deposit, if several feet of snow accumulate for a few years, and the topmost one or two years' accumulation erodes before the next year's snow falls, then the process won't be a good data recorder. A proper natural recorder preserves the sequential order of events as layers grow one at a time, new over old, accreting like memories not easily forgotten.

Of course, interpreting a record objectively requires that it be calibrated. To calibrate a recorder strip, a researcher must relate the ink markings on the paper to some environmental property known independently, such as temperature, pressure, or humidity. As we'll show, there are many ways to calibrate natural recorders as well.[4] Our planet contains vast libraries with immense quantities of time-stamped information that we're only now learning how to read. Let's open a few books in that library.

IN COLD STORAGE

Arguably, Earth's best natural data recorders are the snow/ice deposits in the polar regions.[5] Ice contains many things scientists employ as "proxies," which, as the name suggests, "stand in" for various facts about Earth's past. Such proxies include carbon dioxide, oxygen, and methane gas trapped as bubbles; dust; marine aerosols; ash and sulfuric acid from large volcanic eruptions; soot from forest fires; pollen; micrometeorites; and the ratios of isotopes.[6] Isotopes of common elements, such as hydrogen and oxygen, have the same number of protons but different numbers of neutrons in the nuclei of their atoms. Though isotopes of the same element behave the same way in most chemical reactions, environmental changes can still subtly affect their ratios in deposited layers. For instance, in a hotter climate, more water evaporates, leaving behind more of the heavier isotopes and changing the ratio of its isotopes in the water that remains and in the water that precipitates elsewhere. This water, if preserved in layered deposits, can leave telling clues about climate changes as far back as hundreds of thousands of years.[7]

Today scientists travel to the frigid polar regions to drill for "white gold." With equipment similar to oil exploration platforms, they drill deep into the ice sheets and carefully extract and store the ice cores for later analysis. Recently, scientists working in Antarctica finished extracting and measuring the Vostok station ice core, a 2.25-mile-deep record of snowfall on East Antarctica going back about 420,000 years. Though these records are from just one location, they provide more than only local information. An ice core contains data that are local, such as snowfall amount and temperature; regional, such as windblown dust, sea salt, volcanic dust, and other aerosols; and global, such as atmospheric levels of methane and carbon dioxide and total ice volume.[8]

From the Vostok cores, we've learned that the levels of carbon dioxide and methane have changed with temperature and ice volume. We've also learned that the cold glacial periods were dusty and windy. (Since the sea level was low enough to expose the continental shelves during a glacial period, strong winds swept what was once fine underwater sediment into

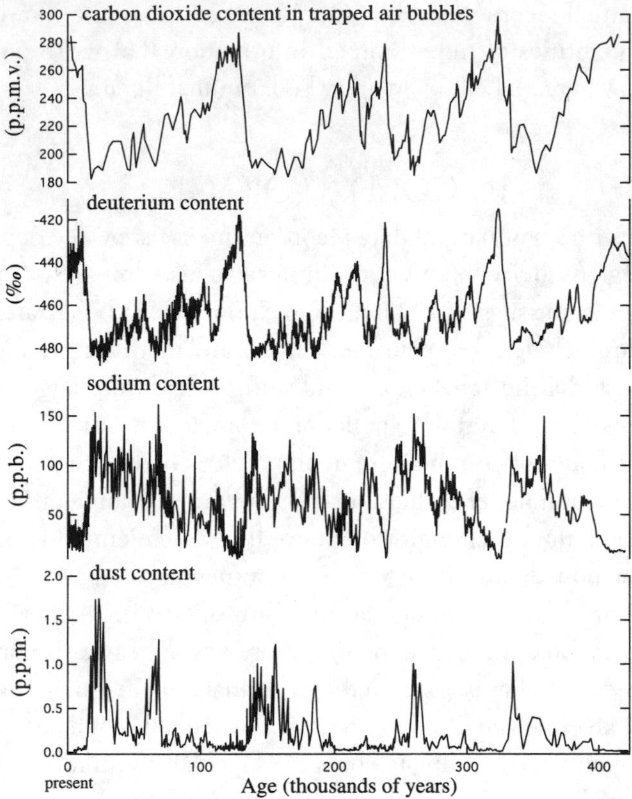

Figure 2.1: Environmental information derived from four of the many diverse proxies that can be traced back some 400,000 years in the Vostok ice core from Antarctica. Trapped air bubbles tell us about the composition of the atmosphere, including the carbon dioxide content. Deuterium content in the ice tells us about the local temperature. Sodium content tells us about wind speeds over the oceans surrounding Antarctica. And dust content tells us about global winds and the extent of deserts.

the atmosphere.) Perhaps most important, we now know that Earth's climate has been colder and less stable than the present, with brief warmings, for most of the last 420,000 years.[9]

Researchers have also obtained ice cores in Greenland, the forbidding island near the North Pole that was named by Norsemen when it was still green. Since snow accumulates there more rapidly than in Antarctica, these ice cores go back only about 100,000 years. But while shorter in total time span, the Greenland cores have a "higher resolution," providing greater detail than the Antarctic cores. This allows us to assign fairly accurate dates to short-lived events like volcanic eruptions and sudden climate shifts. Large volcanic eruptions appear in ice cores as increases in the acidity of the ice or even as ash layers. Historically documented eruptions that leave a signal in the ice—going back to Vesuvius in A.D. 79—serve as independent checks of ice core dating methods. Once scientists can identify known eruptions in a core, they can catalog other eruptions not well documented elsewhere.[10]

The Greenland cores also show us that dramatic global climate changes have been the rule over the past 100,000 years. The most recent one was the Younger Dryas, a glacial period lasting about a thousand years, which ended abruptly some twelve thousand years ago.

Because of their differences, the two sets of ice cores in Greenland and Antarctica complement each other, not only in their duration and resolution but also in their positions on nearly opposite sides of the planet. This allows cores from one to be a check on the other, providing a better reconstruction of the past global climate. What's more, if one ice core is carefully dated, then scientists can use it to date other cores by matching patterns of change in carbon dioxide and methane. Since these gases are well mixed in the atmosphere, measuring the trapped gas at one location effectively gives us the global value.

They have even revealed changes in Earth's magnetic field. For example, the ice cores recorded the Laschamp event, a temporary weakening of Earth's magnetic field, which occurred about forty thousand years ago.[11]

Ice cores (as well as tree rings) record extraterrestrial phenomena as well. As high-energy cosmic ray particles—mostly protons—strike atoms in the upper atmosphere, nuclear reactions produce the unstable isotopes carbon-14, beryllium-10, and chlorine-36. (The sunspot cycle regulates the rate of production of these in the atmosphere through the interaction of the Sun's extended magnetic field and solar wind with Earth's magnetic field and the background flux of galactic cosmic rays.) Fluctuations in the strength of the eleven-year sunspot cycles appear clearly as changing concentrations of beryllium-10 in the Greenland cores.[12] Evidence of other, longer-term changes in the Sun's energy output over the last twelve thousand years shows up in the cores as well.[13]

Finally, there is the evidence of much more distant events. In 1979, astronomers and ice core researchers jointly published a study tentatively correlating nitrate spikes in an Antarctic ice core with historical supernovae (exploding stars).[14] In 2000, another group of researchers suggested that a recently discovered supernova remnant caused the deepest nitrate spike noted in the original study, which had remained unmatched with a supernova.[15] These claims, if confirmed with additional deeper ice cores at the South Pole, may enable us to catalog all supernovae close to Earth occurring over the last few hundred thousand years, something otherwise impossible. Not only would astrophysics benefit from a better sampling of the nearby supernova rate, but we might also be able to discern the biological effects of these powerful events.[16]

BIOLOGICAL DATA RECORDERS

We tend to think of living organisms primarily in terms of their survival and reproduction. But biological processes also provide some of nature's most sensitive historical records. They may do this directly by producing growth layers useful for recording environmental information in a chronological sequence, or indirectly by enhancing inorganic deposits.[17] For example, marine sediments would still be informative in a lifeless world, but the skeletal shells of single-celled marine organisms embedded in these sediments make them far more so.[18]

Many organisms display a slight preference for certain isotopes, often in a way that is temperature-sensitive.[19] For instance, in 1946 Harold Urey discovered that the oxygen isotope ratios in the skeletons of planktonic foraminifera, or foram—microscopic organisms that live in shallow, warm waters—are sensitive to water temperature. Since then, researchers have confirmed this in the laboratory by culturing these species over a range of temperatures. Today, with such information, researchers can convert the measured isotope ratios in the marine sediments to changes in temperature. As we noted earlier, however, the oxygen isotope ratio in ocean water also depends on the global ice volume. Thus, the ratios measured in the planktonic foram skeletons will depend on both water temperature and global ice volume. There are at least two ways of separating out these effects.

One method uses another species of foram. Benthonic forams live near the ocean floor, where water temperatures are near freezing and aren't expected to change much over long periods. As a result, changes in the oxygen isotope ratios in the benthonic forams tell us mostly about changing global ice volume. By measuring the sediment-trapped skeletons of both species in the same general region, scientists can discern changes in both temperature and global ice volume.[20]

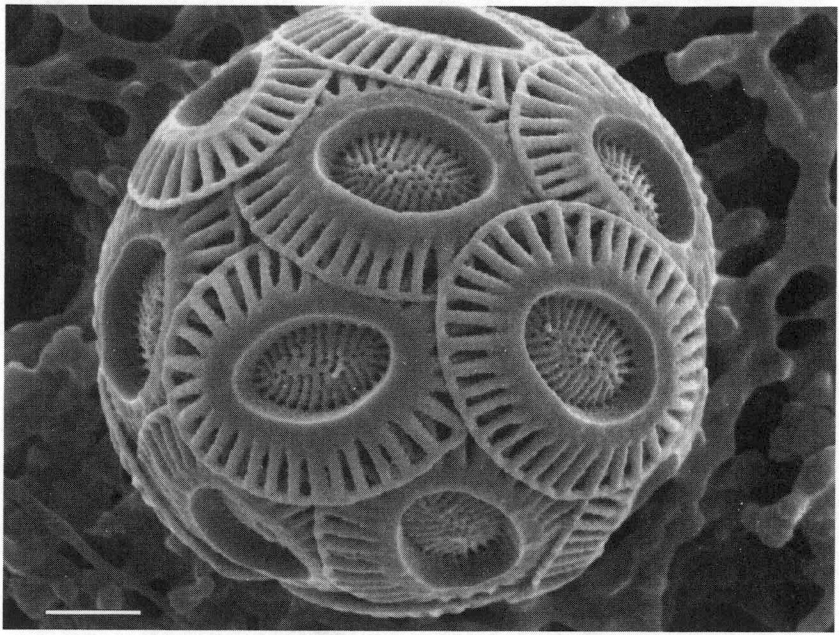

Figure 2.2: *Emiliana huxleyi* (type A), the most common species of phytoplankton. The platelets, called coccoliths, are made of calcium carbonate. These tiny organisms play two important roles in regulating the climate. First, when deposited on the ocean floor, their skeletons participate in an important part of the carbon cycle. Second, by emitting dimethyl sulfide into the atmosphere, they help form marine clouds. The scale bar on the lower left is equal to one micron.

Freshwater deposits are also good data recorders. Lake Baikal in Siberia provides the longest continuous sediment record in the continental interior of Asia. The high-resolution record retains information on temperature going back at least 800,000 years![21] While not as data-rich as ice cores, marine and lake sediment cores give us a reasonably detailed look at even more ancient climates.[22]

Other processes allow us to peer much further back in time. We can measure how quickly the Moon is receding from Earth by timing the round trip of laser beams reflected off the mirrors left on its surface by the Apollo astronauts.[23] Geologists, collaborating with astronomers, have estimated the Moon's orbital period and Earth's rotation period back to about 2.5 billion years ago.[24] This amazing reconstruction is possible because the tides—as established by the cycles of Earth's orbit, Earth's rotation, and the

Moon's orbit—leave repeating patterns in the growth layers of corals and mollusk shells.[25] To reconstruct these cycles, all we need is a good measure of the patterns of thickness in the preserved layers.[26]

Inorganic processes allow similar studies. For example, data from ancient tidalites—fossilized remains of sediments deposited during periodic coastal tidal flooding—tell us that 500 million years ago days were twenty hours long and months were 27.5 modern Earth days.[27] The Moon has been receding at a fairly constant rate over the last few hundred million years. By reading modern tidalites, astronomers can confirm the known lunar orbital period to better than one percent. It's hard to imagine a more elegant and precise way of measuring the orbital properties of the Earth-Moon system while preserving them so accessibly.

By providing much narrower and more recent information, trees rings complement more ancient climate data. As we noted above, these rings

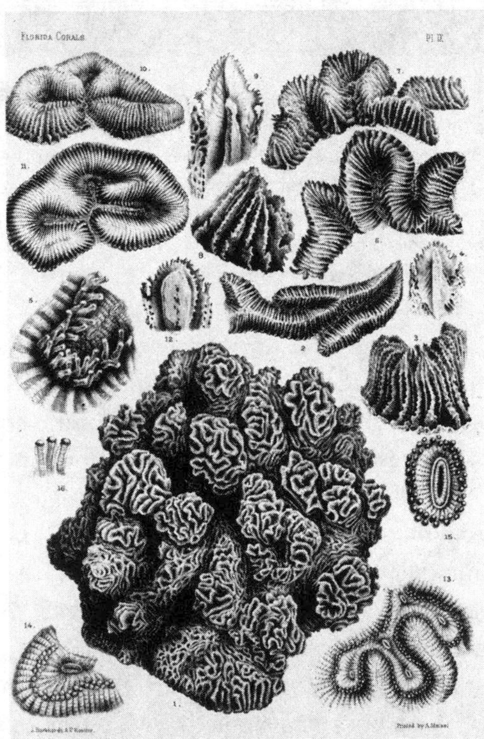

Figure 2.3: A variety of Florida corals (Louis Agassiz, 1880). Corals, like many other living things that produce growth rings or non-biological processes that produce layering, are remarkably high-fidelity data recorders of Earth's past climate.

allow us to reconstruct ancient local climates. Thicker tree rings imply warmer and wetter conditions, while thinner rings suggest the opposite. Those trees most sensitive to local climate fluctuations grow in mountain areas without direct access to steady ground water. For this reason, the mountainous western United States has been a prime field laboratory for the scientists, called dendrochronologists, who specialize in this work. As stationary temperature and rainfall recorders, trees give us detailed information about local climate changes, such as regional droughts and fires, missed in the global records.[28] Continued study of tree rings may reveal other ways in which they record their local environment.[29] The next time you walk in a forest, besides admiring its intrinsic beauty, think of the tall trees as tightly wound scrolls bearing recorded information waiting to be read.

Trees leave other useful records. The leaves of fossilized trees can serve as "palaeobarometers" of carbon dioxide in the atmosphere.[30] Trees exchange gas with the atmosphere through small openings on leaf surfaces called stomata. From laboratory experiments and historical herbarium data, botanists have found that the concentration of stomata on a leaf surface is sensitive to the concentration of carbon dioxide in the atmosphere. (Leaf fossils often reveal details fine enough to enable counting of the tiny stomata.) These studies, when compared with temperature reconstructions from oxygen isotopes in marine fossils, give us important clues about the long-term relationship between temperature and carbon dioxide.

Some tiny organisms not only are highly compact data recorders but are nearly ubiquitous on and under Earth's surface. Mites, for instance, occupy almost every habitat, like the clutter that fills the house of a pack rat. We even find mites entombed in dark caves, preserved in stalagmites and stalactites by mineral-rich water. Because each mite species is adapted to a specific climate, the dominance of certain species of mites can teach us about the environment in which they lived. One recent study catalogued twelve types of mites in stalagmites in a cave near Carlsbad Caverns dating back 3,200 years,[31] revealing that the southwestern United States was wetter and cooler from 3,200 to 800 years ago.

EARTH'S CLOCKS

The most basic way to date deposited or grown records is just to count layers. Tree ring counting has been particularly useful for calibrating the carbon-14 dating technique, since dendrochronologists can measure the carbon-14 in individual rings. This is necessary, because carbon-14 isn't produced in the atmosphere at a constant rate.[32]

The organic excretions of living things contain carbon-14, and any living thing will contain some carbon-14 when it dies. So any growth layers or organic materials in layered deposits, like pollens and tree rings, are potential tools for absolute dating. Pollens are especially valuable. Trees and plants prodigiously produce pollens specific to their species (say that five times, quickly). Charles Darwin famously complained that pollination seems extravagant and even wasteful, but that very extravagance allows us to use pollen granules to date layered deposits. Pollens, cast far and wide by the wind, have been dated in lake sediments with carbon-14 to sixty thousand years ago and in bogs to about twelve thousand years ago.

There are other fairly short-lived radioactive isotopes in the marine environment, including a number of intermediate decay products from long-lived radioactive isotopes. For example, uranium-234 and thorium-230 hide in corals and ocean sediments, allowing us to date them back a few tens of thousands of years.[33]

A very different sort of clock uses the polarity reversals in Earth's magnetic field (more on this in Chapter Three). Any magnetic minerals will align themselves with Earth's magnetic field when they are deposited as sediments and left undisturbed. Oceanographers have mapped magnetic variations back nearly 200 million years, providing another way to check other records independently.[34]

Finally, there are the Milankovitch cycles, probably the single most useful type of clock for layered deposits.[35] These are the various long-term dynamic rhythms, such as the changes in the angle and direction of Earth's axial tilt and the subtle changes in its orbit. These cycles affect anything that involves the condensation and evaporation of water, or which is sensitive to changes in temperature or sunshine. Many such cycles occur on Earth, with ranges from 19,000 to 400,000 years, and are catalogued in ice, marine, and deep lake sediment cores.[36]

Once records are properly calibrated, researchers can use one record as an independent check on others by comparing their common proxies.[37] (This means layered records are much more useful when more than one kind is available.) Earth's many distinct records, therefore, allow scientists to reconstruct much of the ancient global climate confidently, as these records preserve events from Earth's deep interior to the distant stars.

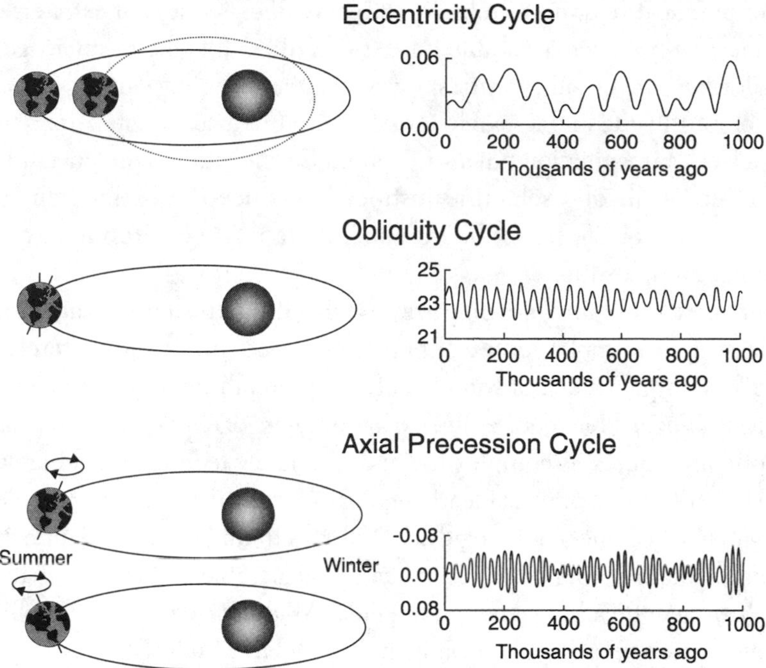

Figure 2.4: Milankovitch astronomical cycles for Earth going back one million years. The small variations in the orbital parameters of Earth cause climate changes that leave their mark on its surface. All three types of variation—eccentricity, obliquity (tilt), and precession—are observed in the geological record. The dominant cycles are eccentricity, 100,000 and 400,000 years; obliquity, 41,000 years; precession, 23,000 and 19,000 thousand years. The actual precession period is 26,000 years, but modulation by the eccentricity cycles results in the two periods seen in the geological record. Earth and the Sun are not drawn to scale.

There are many more examples, but this brief survey should reveal that Earth's surface is peppered with millions of natural data recorders, all patiently measuring a diverse range of phenomena.

WHAT DOES ALL THIS HAVE TO DO WITH HABITABILITY?

This may come as a surprise, but all the measurable aspects of our environment discussed above are closely related to its habitability. To see this, however, it's important to understand the nature of life and its basic requirements.

For practical reasons, astrobiologists generally assume that extraterrestrial life, if it exists, will resemble Earthly life. After all, if their primary goal is to discover life on other worlds, it is much easier if they can use knowledge of Earthly life in the search. Critics claim this assumption betrays not only a lack of imagination but also a certain Earth-centered provincialism. In fact, it has a solid scientific justification, since the basic chemical requirements for life prescribe the kind of planetary environment conducive to complex life.

Although the boundaries are fuzzy, we can distinguish three categories of life: simple, complex, and technological. Since simple life is a prerequisite for complex life, and complex life is a prerequisite for technological life, technological life requires the narrowest range of conditions.[38] Because our primary interest is complex life and ultimately technological life, we need the criterion that most clearly separates simple life from the complex life that may become technological.[39] With this in mind, we can define the minimum complex organism as a macroscopic aerobic metazoan—that is, largish, oxygen-breathing, and multicellular. Without oxygen, large mobile organisms aren't possible, especially those with large brains. This is a basic limitation resulting from simple chemistry and physiology.

LIFE'S CHEMISTRY

Science-fiction stories often describe alien life forms based on completely different chemistry from that of life on Earth.[40] But sci-fi writers use artistic license, which allows their imaginations to run unfettered. One popular notion is silicon-based life—attractive, no doubt, because of silicon's proximity to carbon in the periodic table.

CARBON

At its basic level, however, chemical life must be able to carry the instructions for the construction of its progeny from basic atomic building blocks.[41] These instructions, or "blueprints," require, among other important things, a complex molecule as the carrier. This molecule must be stable enough to withstand significant chemical and thermal perturbations, but not so stable that it won't react with other molecules at low temperatures. In other words, it must be *metastable*.[42] Also, to allow for diverse chemistry, it must have an affinity for many other kinds of atoms comparable with the affinity it has for itself. Carbon excels in this regard, but silicon falls far short. Other elements aren't even in the race.

There are other arguments in favor of carbon, such as the fact that it forms gases when combined with oxygen (to make carbon dioxide) or hydrogen (to make methane), and both gases allow free exchange with the atmosphere and oceans.[43] And most important, when other key atoms—hydrogen, nitrogen, oxygen, and phosphorus—are added to carbon, we get the informational backbones (DNA and RNA), and the building blocks (the amino acids and proteins) of life. Carbon gives these molecules an information-storage capacity vastly exceeding that of hypothetical alternatives.[44] In fact, the half-dozen or so key chemical requirements for life discussed in the literature are rare or absent in other elements but are all present in carbon. (And in case you think there's a loophole, it doesn't work to try to create a carbon equivalent by combining several kinds of atoms.[45])

WATER

Life also needs a solvent, which provides a medium for chemical reactions. The best possible solvent should dissolve many types of molecules, transporting them to reaction sites while preserving their integrity. It should be in the liquid state, since the solid state doesn't allow for mobility and the gaseous one doesn't allow for sufficiently frequent reactions. Further, the solvent should be liquid over the same range of temperatures where the basic molecules of life remain largely intact and in the liquid or gaseous state. Water, the most abundant chemical compound in the universe, exquisitely meets these requirements.[46]

In fact, water far exceeds these basic requirements for life chemistry. Harvard chemist Lawrence J. Henderson described the many ways that water and carbon are uniquely suited for life in his classic 1913 work, *The Fitness of the Environment*.[47] Our increased knowledge of chemistry over the past century has only reinforced his arguments.[48] The topic is too extensive to cover here in detail, except to note a few important examples. First, water is virtually unique in being denser as a liquid than as a solid (the element bismuth is another substance with this property). As a result, ice floats on water, insulating the water underneath from further loss of heat.[49] This simple fact also prevents lakes and oceans from freezing from the bottom up. It's very difficult, if not impossible, to alter such a situation once attained. If ice were to sink to the bottom, it would remain there, unable to melt, separated from the Sun's warmth. Surface ice also helps to regulate the climate by altering Earth's ability to absorb or reflect sunlight, as we'll discuss at length in Chapter Four.

Second, water has very high latent heats when changing from a solid to a liquid to a gas.[50] So more heat is needed to vaporize one gram of water than the same amount of *any* other known substance at ambient surface temperature (and higher than most others at any temperature).[51] This means that it takes an unusually large amount of heat to convert liquid water to vapor. Similarly, vapor releases the same amount of heat when it condenses back to liquid water. As a result, water helps moderate Earth's climate and helps larger organisms regulate their body temperatures. This characteristic also permits smallish bodies of water to exist on land; otherwise, ponds and lakes would evaporate more easily. In all three cases, if a gram of water evaporated with less heat, it would remove less heat from a surface. It's probably no coincidence that water is found in all three states at Earth's surface, and that the mean surface temperature is near the triple point of water—a unique combination of pressure and temperature where all three states can coexist. Not only does this provide a diverse set of surfaces (the relevance of which will be covered in Chapter Four), but it also best exploits water's anomalous properties for regulating the temperature.

Third, liquid water's surface tension, which is higher than that of almost all other liquids, gives it better capillary action in soils, trees, and circulatory systems, a greater ability to form discrete structures with membranes, and the power to speed up chemical reactions at its surface. Finally, water is probably essential for starting and maintaining Earth's plate tectonics, an important part of the climate regulation system.[52]

Frank H. Stillinger, an expert on water, observed, "It is striking that so many eccentricities should occur together in one substance."[53] While water has more properties that are valuable for life than nearly all other elements or compounds, each property also interacts with the others to yield a biologically useful end. Michael Denton describes one of these ends, the weathering of rock:

> Take, for example, the weathering of rocks and its end result, the distribution of vital minerals upon which life depends via rivers to the oceans and ultimately throughout the hydrosphere. It is the high surface tension of water which draws it into the crevices of the rock; it is its highly anomalous expansion on freezing which cracks the rock, producing additional crevices for further weathering and increasing the surface area available for the solvation action of water in leaching out the elements. On top of all this, ice

> possesses the appropriate viscosity and strength to form hard, grinding rivers or glaciers which reduce the rocks broken and fractured by repeated cycles of freezing and thawing to tiny particles of glacial silt. The low viscosity of water confers on it the ability to flow rapidly in rivers and mountain streams and to carry at high speed those tiny particles of rock and glacial silt which contribute further to the weathering process and the breaking down of the mountains. The chemical reactivity of water and its great solvation power also contribute to the weathering process, dissolving out the minerals and elements from the rocks and eventually distributing them throughout the hydrosphere.[54]

This chemical and mechanical distribution of vital elements is an important part of chemical weathering, which is also an important part of Earth's climate regulation system (a topic we will cover in Chapter Three).[55]

All Together Now

John Lewis, a planetary scientist at the University of Arizona, agrees that carbon and water have no equals. After considering possible alternatives, he concludes:

> Despite our best efforts to step aside from terrestrial chauvinism and to seek out other solvents and structural chemistries for life, we are forced to conclude that water is the best of all possible solvents, and carbon compounds are apparently the best of all possible carriers of complex information.[56]

Henderson was also struck by the overall fitness of carbon and water for life: "From the materialistic and the energetic standpoint alike, carbon, hydrogen, and oxygen, each by itself, and all taken together, possess unique and preeminent chemical fitness for the organic mechanism."[57] Water appears to be an ideal match for carbon-based chemistry. For starters, organic reactions are optimal over the same range of temperatures that water is liquid at Earth's surface.[58] At low temperatures, reactions become too slow, while at high temperatures, organic compounds become unstable.

Earth's ability to regulate its climate hinges on both water and carbon, not least because carbon dioxide and water vapor—and to a lesser extent, methane—are important atmospheric greenhouse gases. These life-essential vapors are freely exchanged among our planet's living creatures, atmosphere, oceans, and solid interior. Moreover, carbon dioxide is highly soluble in water. Together, they create a unified climate feedback system, and have kept Earth a lush planet for the past 500 million years. Indeed, it's hard to ignore the need for the planetary environment to be so closely linked to the chemistry of life.[59]

We're made from the dust of Earth and to dust we will return. Life is not just an ephemeral dross clinging to an inert surface. Life, rocks, and the atmosphere interact in a complex web of feedback loops reminiscent of the classic dilemma of the chicken and the egg: Life needs a habitable planet to exist, but simple organisms seem to be necessary ingredients for making a habitable planet. James Lovelock even goes so far as to call Earth's geophysical and biological processes a type of planetary physiology, because, like an animal's metabolism, our planetary environment remains relatively stable despite changing external conditions. But the analogy goes deeper, since both systems use the special properties of carbon and water.

Because carbon and water are so well suited for life at the scales of molecules, cells, organisms, and planets,[60] environments without enough carbon and water are very probably lifeless. Once we recognize the high degree of fitness of carbon and water chemistries for life, we must also accept the constraints this places on a habitable planet. A planet less flexible than Earth at regulating its climate with water and carbon will surely be less habitable.[61]

Of course, even simple life requires far more chemical elements than carbon, hydrogen, and oxygen. A tiny bacterium needs seventeen elements, and humans need twenty-seven.[62] In general, the larger and more complex the organism, the more diverse the proteins and enzymes it requires. While most of the essential elements are concentrated enough in seawater for life, the oceans aren't an adequate source of all elements. For example, the atmosphere is the primary source of nitrogen, and the continents are the primary source of several mineral nutrients, including molybdenum.[63] This suggests that planetary environments lacking a nitrogen-rich atmosphere and continents may not be able to support a robust biosphere.[64]

Apart from these essential elements, life requires a stable, long-term energy source. The basic sources are stellar radiation, geothermal heat,[65] and chemical energy. An environment needs enough energy to maintain liquid water, but even with liquid water, an environment with weak energy sources diluted over its surface won't be able to support a lush biosphere. For that, you need lots of energy.

Complex life also requires a certain minimum biological support system through the activity of autotrophs, organisms that synthesize organic molecules from simple inorganic matter.[66] For example, photosynthetic algae and some bacteria synthesize food from such inorganic materials as carbon dioxide, nitrogen, methane, hydrogen, and various minerals. These algae and bacteria, and their organic products, then become food for other organisms that require organic food—heterotrophs like us. Some environments might be able to support low-level microbial life, but if it lacks the energy to sustain an abundant autotroph population, it won't allow for larger, more complex organisms.

Life relies on chemical energy for its immediate metabolic needs, and chemical energy is all about the exchange of electrons. The most energy is released when elements located on opposite ends of the periodic table exchange electrons. Oxygen is second only to fluorine in the amount of chemical energy released when it combines with other elements.[67] Hydrogen, and carbon combined with hydrogen, or hydrocarbons, are the best substances to combine with oxygen. All complex life forms use such oxidation reactions (other common reactions yield far less chemical energy). And not incidentally, the products of oxidation are water and carbon dioxide, the nontoxic and essential components of the climate regulation system. So hydrogen, carbon, and oxygen, together, offer the best source of chemical energy. This remarkable fact was not lost on Henderson: "This is the last argument which I have to present, but it is one of the most potent. The very chemical changes, which for so many other reasons seem to be best fitted to become the process of physiology, turn out to be the very ones which can divert the greatest flood of energy into the stream of life."[68]

EXTREME ENVIRONMENTS

Recent studies of the organisms called extremophiles have made many astrobiologists more optimistic about the prospects for finding life of this sort on other planets.[69] We find these hearty critters in the severely salty Dead Sea and Great Salt Lake, the superheated water in deep-sea thermal vents, the stinky, gurgling Yellowstone springs, and the frigid ice fields of the Arctic and dry valleys of the Antarctic. But we may not find similar organisms in isolated extraterrestrial settings, because these Earthly organisms are not nearly as independent of other life as it appears. For example, the biological communities found around some deep-sea thermal vents

1 H																	2 He
3 Li	4 Be											5 B	6 C	7 N	8 O	9 F	10 Ne
11 Na	12 Mg											13 Al	14 Si	15 P	16 S	17 Cl	18 Ar
19 K	20 Ca	21 Sc	22 Ti	23 V	24 Cr	25 Mn	26 Fe	27 Co	28 Ni	29 Cu	30 Zn	31 Ga	32 Ge	33 As	34 Se	35 Br	36 Kr
37 Rb	38 Sr	39 Y	40 Zr	41 Nb	42 Mo	43 Tc	44 Ru	45 Rh	46 Pd	47 Ag	48 Cd	49 In	50 Sn	51 Sb	52 Te	53 I	54 Xe
55 Cs	56 * Ba	71 Lu	72 Hf	73 Ta	74 W	75 Re	76 Os	77 Ir	78 Pt	79 Au	80 Hg	81 Tl	82 Pb	83 Bi	84 Po	85 At	86 Rn
87 Fr	88 ** Ra	103 Lr	104 Rf	105 Db	106 Sg	107 Bh	108 Hs	109 Mt	110 Ds	111 Uuu	112 Uub	113 Uut	114 Uuq	115 Uup	116 Uuh	117 Uus	118 Uuo

*57 La	58 Ce	59 Pr	60 Nd	61 Pm	62 Sm	63 Eu	64 Gd	65 Tb	66 Dy	67 Ho	68 Er	69 Tm	70 Yb
**89 Ac	90 Th	91 Pa	92 U	93 Np	94 Pu	95 Am	96 Cm	97 Bk	98 Cf	99 Es	100 Fm	101 Mb	102 No

Elements necessary for *E. coli*

Additional elements necessary for Humans

Figure 2.5: The periodic table of elements. Hydrogen (H) is the lightest element, and so is number one on the periodic table. Helium (He) is two. Complex life requires more elements than "simple," single-celled life. A bacterium needs seventeen elements, while a human being needs twenty-seven. In addition, some of these essential elements, such as iron, along with radioactive elements located near the end of the table, are required for geological processes that produce a habitable planet.

include many creatures that require oxygen. Surface-dwelling organisms produce the oxygen by photosynthesis, which is then mixed into the deep ocean waters. There may be other, more direct ways that these vent communities are linked to present and past surface life.[70] For instance, these animals may even have migrated there from shallower waters.

Extremophiles also make a go of it off rock and hydrogen about a mile deep in the Columbia River basalt in eastern Washington and other subsurface spots around the globe. Typical subsurface life metabolizes very slowly compared with life at the surface, and the concentration of the cells is typically very dilute. And once again, it is probably dependent on surface life. There's increasing evidence that deep subsurface microbial communities feed off dissolved organic matter, either from fossil soils or from fresher organic material brought down from the surface.[71] As a result, thermal vent and deep subsurface communities may not be able to exist on a world that has never had abundant surface life or is far from a source of light energy.

Research by Abel Méndez, an astrobiologist at the University of Puerto Rico at Arecibo, suggests that most prokaryotes—"simple"[72] organisms without a nucleus—grow best between 70 to 126 degrees Fahrenheit, with optimum growth at 96.8 degrees.[73] This is important because the biodiversity in the tropical regions depends mostly on the growth rate of such organisms.[74] Complex life is even less tolerant to changes in temperature.[75] Temperatures quite different from this optimum would provide much less support for a complex biosphere.

Méndez also notes that life cannot survive at arbitrarily high pressures; the maximum limit is nearly one thousand times the pressure at Earth's surface. Several environments in the Solar System where liquid water may exist exceed this limit. Moreover, while various species of extremophiles can tolerate extremes in temperature, salt content, moisture, and pH, few can tolerate a very broad *range* of environmental conditions.[76] In fact, they're somewhat challenging to maintain in the laboratory.[77] So while we can certainly learn something of the extreme range of conditions in which life can exist by studying extremophiles, we shouldn't assume we will find them on planets with environments significantly different from ours.

Moreover, studying the other planets in the Solar System enhances our knowledge of the range of conditions required for life. And the more we can compare Earth with the other planets, the more we realize that Earth is an exceptional host for both simple and complex life (more on this in Chapter Five).

Nevertheless, we may yet find life elsewhere in the Solar System (see Appendix B on panspermia, 343–345). Astrobiologists now generally recognize that the terrestrial planetary bodies have been exchanging material, especially during the early history of the Solar System. Even today fairly intact pieces of Mars and the Moon are collected as meteorites on Earth. Similarly, Earth has probably contaminated most of the other planetary bodies in the Solar System with its microbes. On most bodies—such as Mercury, the Moon, Jupiter, and asteroids—Earthly life can't flourish. Mars, however, was probably wet for some time in its early history, and might have supported life. But today we find no evidence of life on its surface. Even with an early helping of our microbes, the harsh conditions on the other planetary bodies in the Solar System prevented them from surviving or transforming their host planets into more habitable environments.[78] One or two hardy species clinging to a few oases in a mostly barren world have no opportunity to regulate climate; they are completely at the mercy of their environment. This implies that simple life may not be as widespread in the universe as many astrobiologists believe, even given what we know about extremophiles on Earth. Nevertheless, if only because of contamination from Earth, we may find some microbes below the Martian surface. In fact, it would be more surprising to find none.

THE FEEDBACK FROM LIFE TO MEASURABILITY

In taking basic mineral elements and energy sources to produce organic compounds, autotrophs make their environment more habitable for all life. For example, marine organisms deposit carbonates—an important part of the carbon cycle—on the ocean floor (we'll discuss this cycle in Chapter Three). In addition, marine phytoplankton produce most of the oxygen in the atmosphere. We and our animal cohabitants also depend on simple life directly: as food sources, digestive aids, and decomposers. And simple life makes Earth a more measurable place—through tree rings, stomata on leaves, mites in cave stalagmites, foram skeletons in deep ocean sediments, and pollen in lake sediments, to name a few.

All the sedimentation and growth processes described in this chapter ultimately depend on the hydrological cycle. This cycle encompasses snowfall on Antarctica and Greenland; rainfall on the continents that replenishes rivers, lakes, and springs and nourishes trees and other living things; and mountain erosion that provides the life-essential minerals for lakes and oceans. But not just any water-cycle-in-a-bottle will do. A hydrological cycle must be fine-tuned to produce the high-quality natural recorders we find on Earth's surface. Too little water would result in the erosion of deposited sediments; too much water would leave too little land surface for stable ice sheets, trees, or corals. Both extremes lead to conditions less hospitable for life and discovery. Since this is easier to appreciate by comparing Earth with the other planets, we'll wait to explore this until Chapter Five.

FORETELLING THE FUTURE

Isn't it surprising that processes on Earth would encode such high-grade, accessible information as a mere accidental byproduct of cosmic evolution? It's equally surprising from the perspective of *biological* evolution, since this information conferred no survival advantage on living things throughout Earth's long past. After all, we've only recently noticed it and are still perfecting the technology required to recover and read it.

At the same time, Earth's capacity for recording data, especially in high-resolution ice cores, could confer survival advantages on an *advanced* civilization.[79] In particular, it could help us maintain Earth's present level of habitability long into the future by teaching us the proper relationship between temperature and atmospheric carbon dioxide. This requires a bit of explanation.

Ice cores have revealed that very large climate swings can occur over just a few years—at least they have in the past. Civilized humankind has not experienced such events. The last one, the Younger Dryas, occurred about twelve thousand years ago. Cooling our climate to glacial temperatures over just a few years would severely disrupt global food production and render cities far from the equator uninhabitable. The ice core record from central Greenland shows that events like the Younger Dryas were the norm for most of the last 100,000 years, while the time corresponding to human recorded history has been quite exceptional.[80] Extending the record further back in time with the less detailed Antarctic ice cores, it appears that the present warm period is the longest-lived one of the past 420,000 years. There's clearly something special about our time.

These records provide an objective test of computer simulations, which otherwise can be highly subjective. Climatologists can now develop long-term simulations of the global climate by adjusting their models to the present climate and testing them on the paleoclimate data derived from the diverse Earthly archives.[81] With this growing database, they'll continue to improve their ability to predict future climate changes. Long-term forecasting once seemed a dream, but the ice-filled pipes of ice cores, alongside other records, may one day make that dream a reality.

We still have much to learn about climate change, of course, but one surprising discovery from this work is that atmospheric carbon dioxide could help prevent glaciation in the future. Research by climatologists A. Berger and M. Loutre of the Institut d'Astronomie et de Geophysique in Belgium suggests that variations in the average amount of sunlight received recently by the Northern Hemisphere are quite exceptional.[82] They compared the near-term changes (due to the Milankovitch cycles), five thousand years back to sixty thousand years into the future, to those of the past

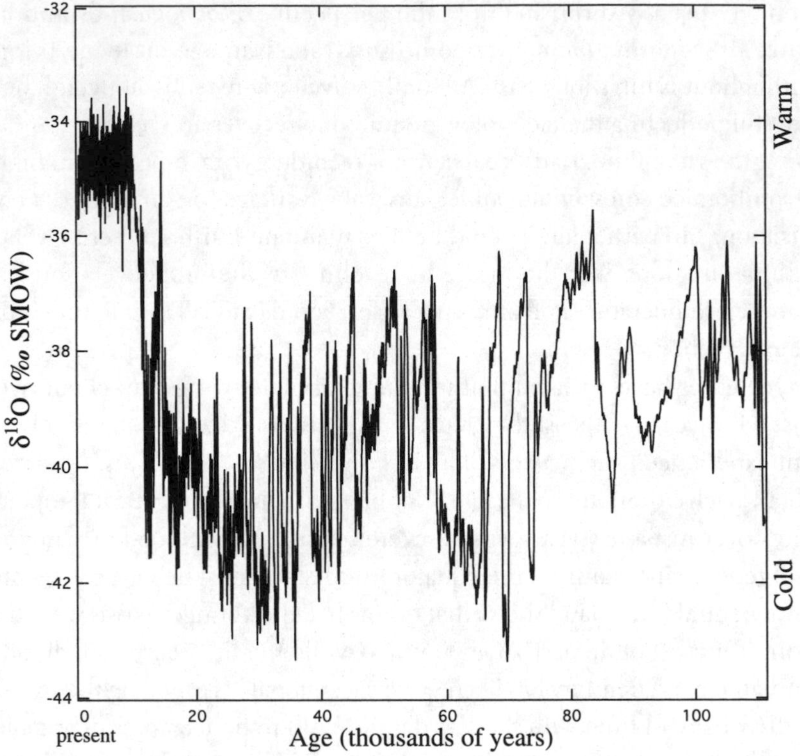

Figure 2.6: Variation in the isotope ratio of oxygen 18 to oxygen 16 in Greenland ice going back some 100,000 years. The oxygen isotope ratio is related to the local temperature. According to these data, the temperature over most of this time period has been marked by very large and rapid changes, and at points has been as much as 25 degrees C colder than the present. Moreover, the climate over the last 12,000 years has been anomalously warm and stable.

three million years. They found that only five intervals in the past three million years had variations as moderate as our present.[83]

Loutre and Berger also found that for up to 130,000 years into the future, the onset of a glaciation depends on the atmospheric level of carbon dioxide, with less carbon dioxide leading to more pronounced growth of ice sheets in the Northern Hemisphere.[84] So not only has the climate been anomalously warm, with fairly stable temperatures for the last twelve thousand years, but we could enjoy stability for at least a few tens of thousands of years into the future. That puts us at the beginning of a long-lived stable, warm period.

You're probably astonished to learn that a high carbon dioxide level could inoculate the planet, and us, against a near-term glaciation (as long as it's not too high, of course). The modern Industrial Revolution has maintained, and will continue to maintain, carbon dioxide levels well above the minimum threshold Loutre and Berger predict.[85]

We should be glad that the era since the last glacial period has lasted this long. You might not be reading this book had the next major glacial period started, say, one thousand years ago.[86] The Northern Hemisphere's climate would have been too severe for Europe to drag itself out of the so-called Dark Ages and to give us enough leisure time to make the web of scientific, philosophical, and artistic advancements that laid the groundwork for the Scientific and Industrial Revolutions. And without industrial man pouring his extra carbon dioxide into the atmosphere by burning fossil fuels, as he began to do some 150 years ago, the tendency toward increased glaciation might have continued unchecked, making it more and more difficult for civilization to progress. Carbon dioxide emissions are a natural consequence of the rise of civilization (in fact, concentration of carbon dioxide tracks closely with world population).

Human activity has always affected our planet locally, but our striving since the early Holocene has brought us to the point where we can hope to understand the global climate system just as we're beginning to have a significant impact on it.[87] If we're smart, the measurability of our environment can lead to improved habitability in the near term by allowing us to attune our behavior with the natural processes of global change.[88]

Chapter 3

Peering Down

Plate tectonics plays at least three crucial roles in maintaining animal life: It promotes biological productivity; it promotes diversity (the hedge against mass extinction); and it helps maintain equable temperatures, a necessary requirement for animal life. It may be that plate tectonics is the central requirement for life on a planet and that it is necessary for keeping a world supplied with water.

—Peter D. Ward and Donald Brownlee[1]

TREMORS

Just as we began to write this chapter, on February 28, 2001, we experienced the strongest earthquake in Seattle in the past five decades. We were both in tall buildings that rocked eerily, like trees in the wind. A friend in a ground-floor coffee shop saw a solid tile floor rippling like water, its solidity suddenly revealed as an illusion. Although it could be felt as far south as Salt Lake City and as far north as Vancouver, the earthquake's epicenter was thirty life-preserving miles underground. As a result, although the earthquake was terrifying and powerful—magnitude 6.8 on the Richter scale—it caused relatively minor damage.

Earthquakes destroy property and kill many people every year; nevertheless, they benefit both our planet's habitability and scientific discovery. Without earthquakes, we probably wouldn't even be here and, if somehow we were, we would know far less about Earth's interior structure.[2]

Like a hammer hitting a bell, a strong earthquake generates waves in the solid Earth that radiate in all directions and travel across its entire diameter. A seismograph anchored to the ground can detect waves from both strong, distant earthquakes and weak, local ones. Differences in density leave their marks on the wave paths, which then carry information about that part of Earth's interior. For instance, sharp discontinuities abruptly

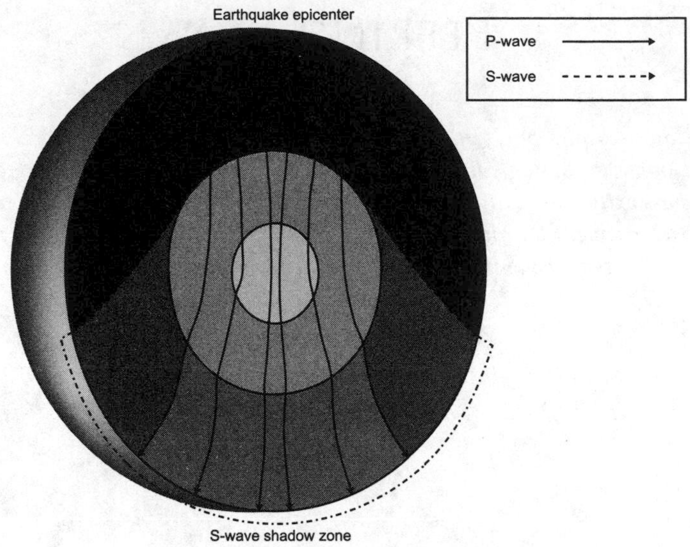

Figure 3.1: Cross section of Earth showing its internal structure inferred from the study of P and S waves generated by earthquakes. This figure shows a hypothetical earthquake and the P and S waves it propagates through the mantle, outer core, and inner core. Seismographs, which are widely distributed on the continents around Earth, are used to detect the waves. Note that only the P waves can traverse the liquid outer core.

deflect the waves, like light bending as it passes from air to a lens. Other wave features can reveal whether any of the traversed regions are liquid.[3]

But the real insight comes from collecting tracings from seismographs spread across Earth's surface. With thousands of earthquakes measured on thousands of seismographs over the past several decades, geophysicists can "invert" a vast database to produce a three-dimensional map of the structure of Earth's interior. The technique, called three-dimensional tomography, is like a geological CAT scan.

Earthquakes also help geophysicists probe small- and large-scale structures, such as an oceanic plate subducted beneath a continental plate. As two crustal plates slide past each other, sections from the opposing plates often catch and build up stress. At some point, a sudden jolt relieves the stress and generates earthquakes. The recent deep earthquake near Seattle was produced within the Juan de Fuca oceanic plate, which is subducting beneath

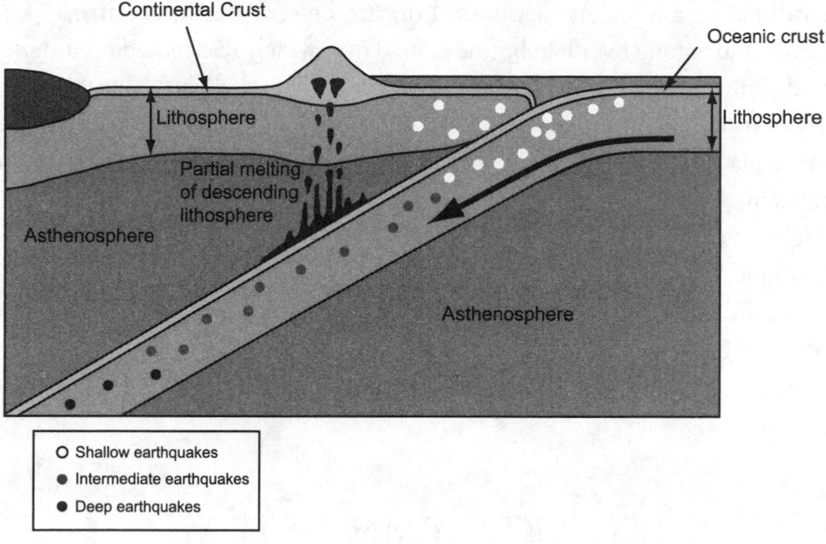

Figure 3.2: Subduction of oceanic plate beneath a continental plate as revealed by earthquakes (in this case near the Japan trench). Earthquakes trace the three dimensional shape of the sinking lithosphere (rigid part of Earth's outermost layer, which includes the crust); the asthenosphere is the part of the upper mantle that is easily deformed. As viewed from the top, the earthquakes display a clear pattern of increasing depth as one moves away from the trench, towards the continental plate.

the North American plate. By measuring such deep earthquakes over several decades, geophysicists have developed three-dimensional pictures of subduction zones around the globe. They've learned that deep earthquakes occur only at subduction zones. Earthquakes also trace the mid-ocean ridges, where fresh crust forms like hot icing on a dry cake. Together, the subduction zones and spreading ridges delineate the plate boundaries (more on this below). If we had only half a dozen well-separated seismograph stations, we could still produce a map of Earth's major plate boundaries.

Today seismographs are spread rather uniformly across Earth's surface. This would have been impossible earlier in Earth's history. About two hundred million years ago, only one supercontinent, called Pangaea, pierced the ocean surface of a watery, lopsided planet. Pangaea has since divided like a puzzle, with its pieces well distributed over Earth's face.

To map our planet's liquid outer and solid inner cores, seismographs need to be nearly opposite to an earthquake's epicenter. Like the continents, earthquakes are widely distributed on the present Earth, occurring predominantly along the plate boundaries. Thus, widely distributed continents and earthquakes together enable us to optimally probe Earth's structure.

Fortunately for science, earthquakes, unlike lightning, often strike the same place twice. If geophysicists can measure the seismic waves of such repeating quakes at stations nearly antipodal to them—that is, on the opposite side of Earth—they can uncover secrets about the inner core. For example, for three decades, researchers at a seismic station in Alaska measured the arrival times of waves generated by South Sandwich Island earthquakes.[4] Since the waves arrived at different intervals after the quake, the researchers concluded that the solid core rotates a little faster than the rest of Earth.

PLANETARY MAGNETISM

In the nineteenth century, physicists discovered that rotating a wire coil inside a set of magnets generates an electrical current in the wire. Today we use this dynamo process to convert mechanical energy into electricity. The basic ingredients of a dynamo are a conducting medium, a magnetic field, and motion. From studying earthquakes and other events, geophysicists know that Earth's outer core is made of piping-hot liquid iron, which probably exceeds 3000° C. Heat flowing through Earth's outer liquid core causes it to convect, like the heat-transferring convection on a hot afternoon that produces cumulus clouds. Because it conducts electricity, the outer core sets up a dynamo generator. But Earth's magnetic dynamo is even more demanding than the man-made variety. In an electrical dynamo, the magnets have permanent fields, but the geodynamo must regenerate its magnetic field. Otherwise, it would decay after only a few hundred years. Such a self-sustained planetary dynamo requires, among other things, the circulation provided by a planet rotating fast enough to produce eddies in the outer core.[5]

Geophysicists use the planetary magnetic field to reconstruct Earth's geologic past. As we noted in the previous chapter, Earth's magnetic field aligns ferrimagnetic minerals in grains as they sink to form undisturbed marine sediments. This isn't the only process that records Earth's magnetic field. Whenever basaltic lava cools below the Curie point—the temperature above which a rock loses its magnetism—it "freezes in" Earth's magnetic field. Such fossil magnetism, or paleomagnetism, remains preserved in a

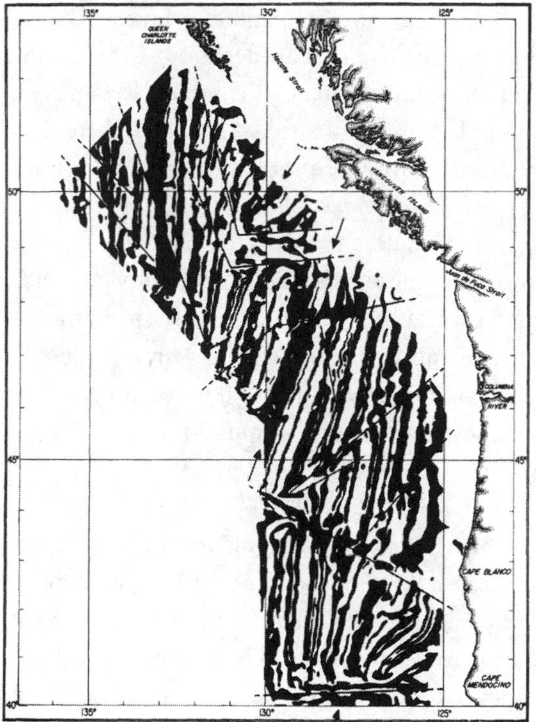

Figure 3.3: Geomagnetic reversals measured in the northeastern Pacific with a magnetometer towed by a research ship. The present magnetic field direction is shown as black, and the reversed direction is shown as white. These magnetic patterns are used to reconstruct the detailed history of seafloor spreading. The locations of spreading ridges as well as faults are apparent in the figure. This figure (first published in 1961) played a very significant role in the eventual adoption of plate tectonic theory.

rock as long as it's not heated above the Curie point. We now know from studies of dated lava flows on land that the planetary magnetic field has changed polarity many times in the past. Magnetic reversals aren't strictly periodic, but occur roughly every million years.[6] The last one occurred 780,000 years ago. Because they are global events, magnetic reversals serve as universal markers for geologists to match widely separated seams of rock.

In the late 1950s, oceanographers began towing devices called magnetometers behind research ships to map the ancient magnetic fields of the rocks forming the ocean floor. The purpose of the first surveys was to search for magnetic anomalies associated with large structures like volcanoes. But the resultant maps revealed an unexpected and remarkable pattern of magnetic polarity reversals running parallel to and symmetric on both sides of the mid-ocean ridges.[7] We now understand that these magnetic stripes result from reversals of Earth's magnetic field as fresh sea floor crust forms and spreads out on both sides of a ridge. In effect, the ocean floor acts as a giant magnetic tape recorder. To read it, all you need is a ship sporting a magnetometer. (See Plate 6.)

But access to these data shouldn't be taken for granted just because they're easy to read. As geophysicist David Sandwell notes, "Indeed, the ability to observe magnetic reversals from a magnetometer towed behind a ship relies on some rather incredible coincidences related to reversal rate, spreading rate, ocean depth, and Earth temperatures."[8] Sandwell goes on to describe how these four scales conspire to produce measurable fields at the ocean surface:

> Most of this magnetic field is recorded in the upper mile or two of the oceanic crust. If the thickness of this layer were too great, then as the plate cooled as it moved off the spreading ridge axis, the positive and negative reversals would be juxtaposed in dipping layers; this superposition would smear the pattern observed by a ship. On Earth, the temperatures are just right for creating a thin magnetized layer.[9]

The remnant field variations on the sea floor would look attenuated and smooth if measured from too far away. For the signal to remain strong at the ocean surface, the spacing of the magnetic stripes must have a typical spacing of about 6.3 times the ocean depth. The final two scales must be tuned as follows:

> Half-spreading rates on Earth vary from 6 to 50 miles (10 to 80 kilometers) per million years. This suggests that for the magnetic anomalies to be most visible on the ocean surface, the reversal rate should be between 2.5 and 0.3 million years. It is astonishing that this is the typical reversal rate observed in sequences of lava flows on land.... This lucky convergence of length and timescales makes it very unlikely that magnetic anomalies, due to crustal spreading, will ever be observed on another planet.[10]

This rare convergence is "lucky" because it creates a nearly optimal environment for accessing magnetic data inscribed in the ocean floor from

Bar code

Magnetic reversals

Tree rings

Figure 3.4: Aperiodic patterns convey much more information than regular or repeating patterns. Varying thickness of bars in the common UPC bar code convey a message to those equipped to decode it. This type of information is also present in nature, as in tree rings and geomagnetic polarity reversals.

the ocean's surface. Geologists can then compare these data with data gathered from treasured magnetic rock samples from ancient lava flows. The amounts of certain long-lived radioactive isotopes reveal a sample's age. And the spatial orientation of a sample's acquired, or remnant, magnetic field reveals the state of the local geomagnetic field when it solidified, leaving behind a detailed historical record. By comparing sea floor magnetic tracings to these dated lava flows around the world, geologists can reconstruct a detailed history of sea floor spreading in both space and time.[11] The pattern is similar for all the spreading ridges around the globe. If the field were steady and unchanging, it would be far less useful. Even a strictly periodic pattern would be inferior to the actual one, since each magnetic reversal cycle would look like all the others. This would make it impossible to match a magnetic pattern on the sea floor uniquely to the dated pattern on the land. But a semi-periodic cycle displays a unique, unambiguous pattern, like a fingerprint. For this same reason, scientists can combine growth ring patterns of different trees, whose lifetimes might only partially overlap, into one longer sequence.

Earth's magnetic field is surprisingly rich in information. For most of us, the compass is the only reminder that Earth even has a magnetic field. Flat compasses create the illusion that Earth's magnetic field is merely two-dimensional, since the needle is restricted to an imaginary flat plane like a pointer on a road map or board game. This simple image has allowed

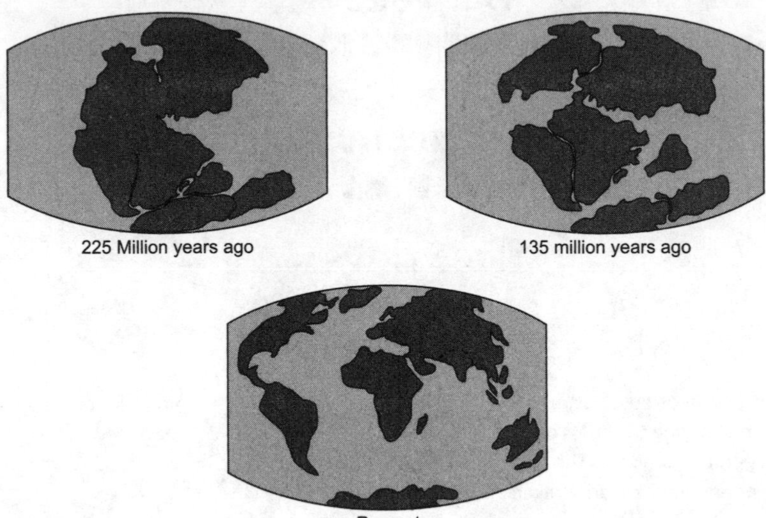

Figure 3.5: Continental drift from the Pangaea supercontinent to the present. These reconstructions of the relative positions of the continents are based primarily on measurements of the fossil magnetism in basaltic rocks. Earth's magnetic field serves as a kind of global positioning system.

humans to navigate Earth's surface since the Chinese invented the compass at least two thousand years ago. (Still, it's an imprecise tool, because it points to the magnetic north, not the geographic north. As a result, navigators must use charts of the difference between the two poles, or the declination, to stay on course.) But Earth's magnetic field can move a compass's needle in three dimensions, a fact that German cleric Georg Hartmann first discovered in 1544. If you place a compass on its side so that its needle can move vertically, it will assume an angle, called inclination, which is perpendicular to the more familiar horizontal plane on which we locate the directions north, south, east, and west.

Rock samples from all the continents retain this three-dimensional remnant magnetization. And more dimensions mean more information. With enough toil, these samples even reveal the locations of now-separated continents up to a few hundred million years in the past. These rock records work like global positioning system (GPS) receivers; Earth's magnetic field serves as the network of satellites to define the absolute coordinate system.

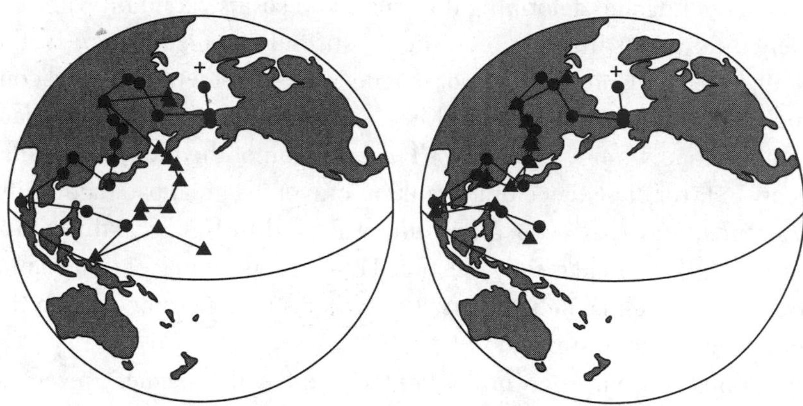

Figure 3.6 Apparent Polar Wander (APW) paths for the North American (circles) and European (triangles) continents superimposed on the modern globe. Geologists routinely measure declination and inclination values of the remnant magnetization of rock samples from all the continents. Since the inclination angle varies simply with magnetic latitude, it's fairly easy to determine the distance to the magnetic pole when a sample formed. Adding the sample's declination allows them to pinpoint the location of the magnetic pole at that time. They can also prepare an APW path from the ancient magnetic pole positions derived from volcanic lava samples of various ages at a given site. By comparing polar wander paths from samples collected on different continents, geologists can reconstruct their motions over long stretches of time. Such plots reveal the past locations of continents up to a few hundred million years in the past. The APW paths in this figure were determined from the fossil magnetism in rocks with accurately known ages. If the European APW path is rotated 38 degrees clockwise (as viewed from the North Pole), then the two paths overlap. This implies that the two continents were moving together over the time covered by the data (150 to 500 million years ago).

Unlike man-made GPS receivers, however, the natural variety retains geological data about the deep past.

This is the evidence that finally convinced skeptics of the theory of plate tectonics. The German meteorologist Alfred Wegener first proposed one of the main elements of plate tectonics theory, continental drift, in the 1920s. His theory of moving continents nicely accounted for the jigsaw-puzzle fit of the outlines of some continents and the similar rocks found on now widely separated lands. The geophysicists of his day who were too strongly wedded to the so-called geosynclinal theory rejected

Wegener's theory.[12] (Geosyncline theory was tied to the idea that Earth was contracting and deforming the crust. The basins would fill with sedimentary deposits from the continents, furthering subsidence.) In the 1960s, however, land and sea floor magnetic measurements provided convincing evidence for continental drift and sea floor spreading and for plate tectonics theory more generally.[13] The rejection of the old geosynclinal theory in favor of plate tectonics ranks as one of the great paradigm shifts in science, and Earth's measurability was vital in the breakthrough.[14] Earth's magnetic field greatly enhanced our understanding of its dynamic interior, and the magnetic reversals provided striking visual clues to the underlying mechanism.

We don't yet completely understand what drives the magnetic reversals, but Earth isn't unique in displaying them. For instance, the Sun also has a magnetic field and exhibits magnetic polarity reversals every twenty-two years on average, twice the length of the eleven-year sunspot cycle. Earth's magnetic field even tells us something about the ever-changing state of the Sun's magnetic field. The solar wind carries the Sun's field out among the planets. For this and other reasons, this interplanetary magnetic field is constantly changing, interacting with Earth's magnetic field, and causing small, rapid fluctuations. By monitoring these daily fluctuations[15] from ground magnetic stations since the early twentieth century, geophysicists have discovered that they vary with the sunspot cycle. Space physicists are now beginning to understand how the Sun's magnetic field affects the interplanetary magnetic field, which they can use to predict a sunspot cycle five to six years in advance.[16] This, in turn, helps NASA estimate the atmospheric drag of low-altitude satellites. Such drag determines the lifetime of such satellites. During sunspot maximum, the outermost region of Earth's atmosphere is more puffed up, increasing the drag.

An even more fruitful collaboration between astronomers and geophysicists has allowed them to measure continental drift in real time. Widely separated radio telescopes aimed at the same distant "fixed" quasar, an extremely luminous object near the edge of the observable universe, can detect changes in continents' distances from each other amounting to about one centimeter per year.[17] Astronomers use radio observations of several quasars over several years, with radio telescopes located on different continents, to determine the relative motions of the continents. Such an experiment is not possible with nearby stars, since their motions in the Milky Way would overwhelm the weak signature of continental drift. In this case, our access to the distant universe gives us knowledge about our local environment. (We'll learn more about the useful qualities of quasars in Chapter Nine.)

WHAT DOES ALL THIS HAVE TO DO WITH HABITABILITY?

Most of us associate earthquakes with death and destruction, but ironically, earthquakes are an inevitable outgrowth of geological forces that are highly advantageous to life.[18] Heat flowing outward from Earth's interior is the engine that drives mantle convection and, in turn, crustal motions. A tectonically active crust builds mountains, subducts old sea floor, and recycles the carbon dioxide in the atmosphere, all of which make Earth more habitable. But it's not obvious that a cold, rigid, floating chunk of crust should be subducted deep into Earth. The continuous presence of liquid water on Earth's surface may explain why it has maintained long-lasting plate tectonics.[19] Apparently, the chemical reactions of water with the minerals in the crust weaken it, providing lubrication that allows the crust to bend without breaking.

THE CARBON CYCLE

As already suggested, this means more than just headaches and high insurance fees for people living close to major fault lines. Plate tectonics makes possible the carbon cycle, which is essential to our planet's habitability. This cycle is actually composed of a number of organic and inorganic subcycles, all occurring on different timescales.[20] These cycles regulate the exchange of carbon-containing molecules among the atmosphere, ocean, and land. Photosynthesis, both by land plants and by phytoplankton near the ocean surface, is especially important, since its net effects are to draw carbon dioxide from the atmosphere and make organic matter. Zooplankton, such as the forams mentioned in Chapter Two, consume much of the organic matter produced in the sunlight-rich surface. The carbonate and silicate skeletons of the marine organisms settle obligingly on the ocean floor, to be eventually squirreled away beneath the continents.

Also central to the carbon cycle is the chemical weathering of silicate rocks on the continents.[21] This occurs when rainwater, made acidic with dissolved carbon dioxide from the atmosphere, dissolves minerals in exposed rock. The rivers eventually carry dissolved silica (SiO_2), calcium,

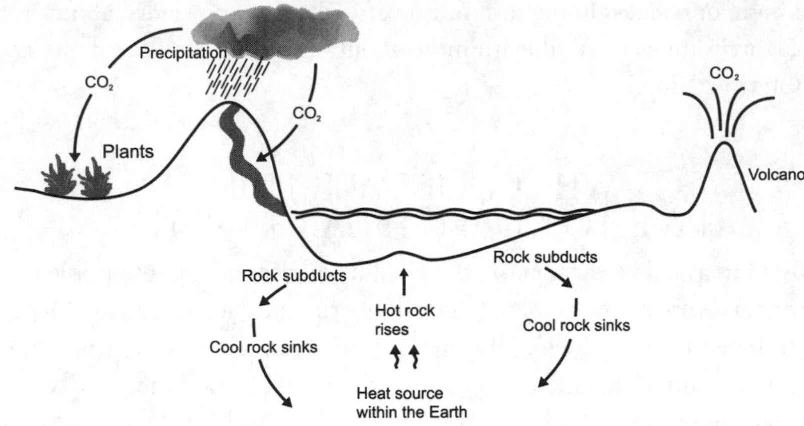

Figure 3.7: Plate tectonics and the carbon and water cycles work together to produce a planet that is hospitable to life. Heat from Earth's interior makes its way to the surface, setting the mantle in motion. The dynamic interior builds mountains and releases carbon dioxide into the atmosphere through volcanoes. Energy from the Sun evaporates water into clouds, which precipitates, returning water and CO_2 to plants, the soil, and the oceans. Transformed into other chemical forms, the carbon originally in the atmosphere makes its way to the seafloor. The carbon-enriched oceanic crust is then subducted into the mantle, and eventually reheated and returned to the surface as CO_2. The entire process keeps nutrients, water, and land available for life, and regulates the global temperature on timescales of millions of years.

and bicarbonate ions (derived from the carbon dioxide) to the oceans. Phytoplankton and zooplankton—and to a lesser extent, corals and shellfish—then remove these dissolved chemicals from the ocean to build their silicate and calcium carbonate skeletons. The carbon cycle is completed when the subducted carbonates are pressure-cooked deep in the crust, releasing carbon dioxide that eventually finds its way to the surface through volcanoes and springs.

Negative feedback loops maintain the whole cycle in balance. Perhaps the most important stabilizing feedback is the dependence of the rate of chemical weathering on temperature. Here's how it works: Suppose a prolonged period of large volcanic eruptions rapidly increased the amount of carbon dioxide in the atmosphere. Through the greenhouse effect, the upsurge in carbon dioxide would raise the global temperature. The higher temperature and carbon dioxide level, in turn, would speed up chemical weathering, and thus the removal of carbon dioxide from the atmosphere. Eventually, the carbon dioxide and temperature would return to their pre-eruption levels. Conversely, a drop in carbon dioxide would slow chemical weathering, allowing carbon dioxide to build up in the atmosphere. In either case, the loop comes full circle.

As we'll discuss in Chapter Four, a rise in atmospheric carbon dioxide increases plant growth. Rocks fuzzy with plants weather chemically about five times faster than bare rocks, allowing Earth to tuck carbon away under its surface all the more quickly. In other words, this accelerated weathering reinforces this stabilizing feedback loop, allowing Earth's climate system to respond much more effectively to perturbations than could a lifeless world. Thus, plate tectonics, together with plant life, makes a planet much more nurturing for all life.

Continents

A carbon cycle, at least the planetary variety, needs both continents and oceans. Continents serve as a mixing bowl for minerals and water at the surface, where energy-rich sunlight is available. The continents began to appear about one billion years after Earth formed. As they grew, they and the crust they ride on extracted potassium, thorium, and uranium from the mantle. Because these radioisotopes have been the primary sources of heat in Earth's interior over most of its history, siphoning them from the convecting mantle weakens tectonic activity. If the continents and crust had grown more rapidly, they would have drawn more heat-producing elements from the mantle. This would have slowed down mantle convection and tectonic activity in recent times, resulting in poorer climate feedback.

Plate tectonics plays another life-essential role: it maintains dry land in the face of constant erosion. A large rocky planet like Earth wants to be perfectly round, with erosion eventually wearing down the mountains and even the continents, creating a true "waterworld." Its interior must continuously supply energy to keep it from getting bowling-ball smooth. Without geological recycling, such a place would probably become lifeless, since it would lack a way to mix all the life-essential nutrients in its sunlight-drenched surface waters.[22]

Magnetic Fields

A terrestrial planet with plate tectonics is also more likely to have a strong magnetic field, since both depend on convective overturning of its interior.

And a strong magnetic field contributes mightily to a planet's habitability by creating a cavity called the magnetosphere, which shields a planet's atmosphere from direct interaction with the solar wind. If solar wind particles—consisting of protons and electrons—were to interact more directly with Earth's upper atmosphere, they would be much more effective at "sputtering" or stripping it away (especially the atoms of hydrogen and oxygen from water). For life, that would be bad news, since the water would be lost more quickly to space.

Just as *Star Trek*'s *Enterprise* uses a force field to protect it from incoming photon torpedoes, Earth's magnetic field serves as the next line of defense against galactic cosmic ray particles, after the Sun's magnetic field and solar wind deflect the lower-energy cosmic rays. These cosmic ray particles consist of high-energy protons and other nuclei, which, together with highly interacting subatomic particles called mesons, interact with nuclei in our atmosphere. These secondary particles can pass through our bodies, causing radiation damage and breaking up nuclei in our cells.

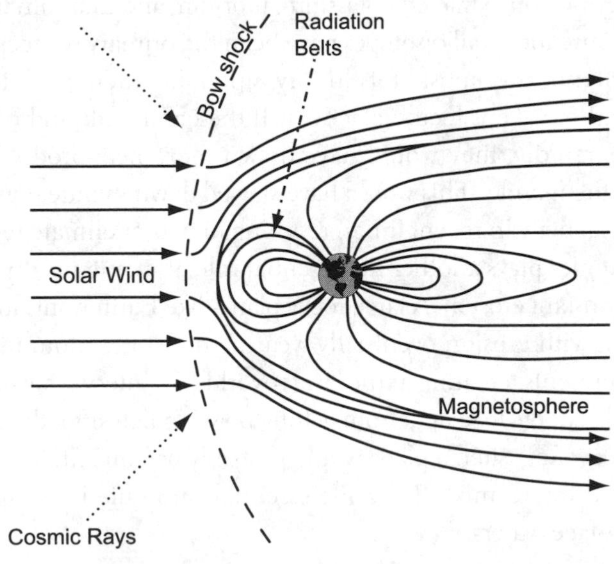

Figure 3.8: Like a giant magnet, Earth's geology conspires to create a magnetic "shield," which protects Earth's atmosphere from solar wind particles and low-energy cosmic ray particles.

Planetary Size and Mass

Also vitally important is a planet's mass.[23] A planet's habitability depends on its mass in many ways; terrestrial planets significantly smaller or larger than Earth are probably less habitable. Because its surface gravity is weaker, a less massive Earth twin would lose its atmosphere more quickly, and because of its larger surface-area-to-volume ratio, its interior might cool too much to generate a strong magnetic field. And as we will show in Chapter Five, smaller planets also tend to have more dangerously erratic orbits.

In contrast, without getting more habitable, a more massive Earth-twin would have a larger initial inventory of water[24] and other volatiles, such as methane and carbon dioxide, and would lose less of them over time. Such a planet might resemble the gas giant Jupiter rather than our terrestrial Earth. In fact, Earth may be almost as big as a terrestrial planet can get. While life needs an atmosphere, too much atmosphere can be bad. For example, high surface pressure would slow the evaporation of water and dry the interiors of continents. It would also increase the viscosity of the air at the surface, making it more difficult for big-brained, mobile creatures like us to breathe.

In addition, more surface gravity would create less surface relief, with smaller mountains and shallower seas. Even with more vigorous tectonic churning, rocks could not support mountains as high as those we enjoy.[25] The planet probably would be covered by oceans and too mineral-starved at the surface (and too salty throughout) to support life. Even a gilled Kevin Costner, cast as a lone mariner, would find such a waterworld unappealing.

To add insult to injury, the surface gravity of a terrestrial planet increases with mass more rapidly than you might guess.[26] Intense pressures compress the material deep inside,[27] so that a planet just twice the size of Earth would have about fourteen times its mass and 3.5 times its surface gravity.[28] This higher compression would probably result in a more differentiated planet; gases like water vapor, methane, and carbon dioxide would tend to end up in the atmosphere. Earth has kept dry land throughout its long history, in part, because some of its water has been sequestered in the mantle; in contrast, a more massive planet would probably have degassed more than Earth.

Maybe you're still pining away for some adventure on a sci-fi–inspired giant terrestrial planet, but there's another problem with larger planets—impact threats.[29] To put it simply, they're bigger targets. Asteroids and comets have a really hard time avoiding larger planets, so these planets suffer more frequent, high-speed collisions. While their bigger surfaces distribute the greater impact energy over more area, this doesn't compensate for the larger destructive energy, since surface area increases slowly with mass for terrestrial planets more massive than Earth.[30]

Not only would both smaller and larger terrestrial planets probably be less livable than Earth, but they would also offer poorer overall platforms for discovery. While smaller planets would have taller mountains, providing a better view of the stars, they would have fewer earthquakes, delaying discoveries in geophysics. Smaller planets also provide a smaller, less effective platform for VLBI (very long baseline interferometry) radio observations—which require distant telescopes on different continents—and a smaller "lowest rung on the distance ladder," which we'll discuss later. A planet larger than Earth would probably have more tectonic activity, but it would have smaller mountains and a thicker atmosphere, which would hinder astronomy. On Earth, the atmosphere is not an insurmountable problem. Mountain observatories equipped with large optics, like the Keck telescopes on Mauna Kea or the Very Large Telescope in South America, can achieve a spatial resolution rivaling that of the Hubble Space Telescope.[31] From their home on Earth's surface, scientists can learn about Earth's interior and the distant stars more efficiently than would observers on planets of quite different sizes.

TECHNOLOGICAL LIFE

To have advanced technology, you need life forms complex enough to develop it. That's technological life. Now, technological life requires a narrow set of conditions, narrower than the conditions for complex life, and much narrower than the conditions for simple life. But it turns out that life and technology are intimately yoked, since the many geological factors crucial for complex life also make high technology possible—the very technology that extends our knowledge of the universe. In fact, virtually all of the detailed knowledge discussed in this chapter hinges on advanced technology.

We could say a species is technological if, at the very least, it can plan several years ahead, control and shape the basic elements with fire[32] as a central energy source, and alter its environment enough to enhance its survival and prosperity. Controlling fire may be the single most important requirement for technology.[33] Early on, fire was used to make pottery and smelt metal ores.[34] More recently, fire has allowed our species to sustain the large population needed to develop high technology, which requires that plenty of individuals be free to focus their attention on matters not directly related to survival. A species that shapes metal to form tools to build large dams, bridges, and skyscrapers would therefore be technological, while one that spends all its time hunting wild animals, gathering wild fruits, and picking lice would not.

Some environments would prevent technological development altogether. A large aquatic animal lumbering around a waterworld, or Carl Sagan's giant "floaters," drifting like gossamer through Jupiter's atmosphere, could not develop high technology because fire would be impossible and electrical circuits impractical.[35] But even the more benign terrestrial planets must have enduring periods with a climate stable enough to maintain widespread agriculture, which is probably necessary for growing and sustaining a large stable population. As we showed in Chapter Two, the climate stability that has existed on Earth since civilization arose has been unusually long-lasting.

The Fuels of Discovery

Technological life needs simple and complex life directly as food sources and as beasts of burden, but it also needs them indirectly to regulate the climate and provide energy. Through glorious acts of creative imagination, the wood from trees and the vast accumulated deposits of ancient life were transformed into the fuels that drive the engines of modern research—coal, gas, and oil.

Most of the coal we burn today is probably inherited from the lush forests of the Carboniferous period (about three hundred million years ago); petroleum was formed mostly from marine plants over roughly the same period.[36] Both fuels made possible the modern Industrial Revolution (begun in the nineteenth century) and the wealth that has followed in its wake.[37] The early modern Industrial Revolution—with its fuel-hungry steam engines, electric generators, and iron smelters—would not have happened without wood or coal. If you doubt this, just visit a part of the world where people use dried dung as fuel because wood is scarce; they can typically do little more than boil a small pot of water.

The wealth of industrial nations has allowed them to devote enormous time and resources to scientific research and technological development, which have led to scientific discoveries unparalleled in human history. To sustain the present level of research, earthly life had to be highly productive for several hundred million years. A planet with feeble primary productivity would take much longer to build up similar levels of coal and petroleum reserves.

Interestingly, each new energy source has been abundant and long-lasting enough to hold us over while we developed the technology needed to reach the next level of energy generation. Burning wood kept us going for thousands of years. Without wood, inhabitants of, say, a rocky, lichen-covered world would be hard-pressed to find an abundant, concentrated source of long-lasting heat. On Earth, all the forests would have been cut down decades ago had we not learned to use coal. In fact, England lost its forests before coal became widely available as a substitute for wood. Eventually, petroleum and then uranium entered the mix, and the polluted, dark, lichen-covered trees of England returned to their natural, ashen color. With luck, these fuels will sustain us until we can develop the next grade of energy—probably fusion. Fortunately, each new grade of fuel demands less from the environment, since its energy is more concentrated. And the carbon dioxide we've been producing while developing new energy sources apparently has improved the overall habitability of Earth (for reasons we gave in Chapter Two).[38]

CONCENTRATED ORES

We would not have reached our present technological level so quickly if geological processes hadn't concentrated mineral ores in the crust. The cost of mining an ore depends a great deal on its concentration. Berkeley geologist George Brimhall writes:

> The creation of ores and their placement close to the Earth's surface are the result of much more than simple geologic chance. Only an exact series of physical and chemical events, occurring in the right environment and sequence and followed by certain climatic conditions, can give rise to a high concentration of these compounds so crucial to the development of civilization and technology.[39]

The same processes that make Earth a habitable planet, especially plate tectonics and the hydrological cycle, are also indispensable for concentrating mineral ores. It's easy to understand why ores are concentrated near the surface. The crust is the interface between land, water, and air. Hydrothermal mineral-rich solutions flow through cracks in fractured crustal rock, precipitating minerals as they come into contact with cool rock surfaces. But these dissolved minerals linger close to the surface because open fractures in rock are rare in the deep, highly compressed Earth, and the liquids cannot flow very far down. In fact, water has been involved, in one way or another, in concentrating nearly all ores in the crust. The rich cornucopia of Earth's mineral ores also owes much to the dissolving power of water.

Consider the important radioisotopes. Despite some popular claims to the contrary, most of Earth's uranium and thorium probably didn't sink into its core. It turns out that uranium, along with potassium and thorium, is most concentrated in the crust. They're lithophiles—lovers of rock. Uranium is nearly one hundred times more concentrated in the continental crust than in the mantle.[40] Only the much greater mass of the mantle makes it Earth's primary source of radioactive heating. The same forces that built the continents concentrated these elements in the crust; every overturning motion of the mantle deposits some radioisotopes there. Uranium mining would have been impractical had the quantities of these elements in the crust matched those found in the mantle. As with other mineral ores, the same small fraction of Earth's mass that contains the most easily mined uranium is also its most habitable part.

Life has also helped concentrate ores in the crust. A particularly unusual example of such "biomineralization" is the collection of uranium oxide left by microbial mats—layered communities of bacteria in salt evaporite lagoons—nearly two billion years ago in what is now Oklo in Gabon, Africa. The relatively short-lived isotope uranium-235 was then far more abundant in Earth's crust. In Oklo, so much uranium was concentrated that nuclear reactions started spontaneously, with water acting as a moderator.[41] The fission reactions ceased long ago; too little uranium-235 remains in natural concentrations for sustained reactions. Today the fossil Oklo reactors are the major source of uranium for the French nuclear industry. The uranium ores in Oklo were able to accumulate only because the atmosphere had become oxygen-rich, and the oxygen allowed uranium to exist in a water-soluble mineral form. Prior to the rise in atmospheric oxygen, organisms could not concentrate uranium in the microbial mats. Perhaps the rise of oxygen in the atmosphere had to await the decay of most of the uranium-235, lest it pose too great a threat to the life in the mats. In other words, this could turn out to be a process that set the timescale for the rise of oxygen in the atmosphere.

Brimhall also highlights the importance of atmospheric oxygen for distilling certain metallic ores, such as copper. So-called secondary enrichment of ores occurs because water can transport some metal ions more easily in an oxygen-rich atmosphere, an atmosphere vital to technological life in other ways. A planet needs both a long history of plant life for its inhabitants to fuel their technological development and an oxygen-rich atmosphere to burn their fuel.[42] So Michael Denton writes, "Another fascinating coincidence is that only atmospheres with between ten and twenty percent oxygen can support oxidative metabolism in a higher organism, and it is only within this range that fire—and hence metallurgy and technology—is possible."[43]

Although much more could be said, it should be clear that Earth's magnetic field and plate tectonics—as well as the associated carbon cycle, nutrient mixing, and continent building—are crucial for both life and scientific discoveries in fields as diverse as geophysics and astronomy. The development of technology, which is essential for decoding mysteries in many corners of the cosmos, also hinges on a number of key planetary processes, including the right kind of atmosphere. Luckily for us, we have just the right atmosphere.

Chapter 4

Peering Up

The combined circumstance that we live on Earth and are able to see stars—that the conditions necessary for life do not exclude those necessary for vision, and vice versa—is a remarkably improbable one.

This is because the medium in which we live is, on the one hand, just thick enough to enable us to breathe and to prevent us from being burned up by cosmic rays, while, on the other hand, it is not so opaque as to absorb entirely the light of the stars and block any view of the universe. What a fragile balance between the indispensable and the sublime.

—Hans Blumenberg[1]

CLEAR VISION

In the sci-fi classic *Nightfall*, Isaac Asimov tells the story of Lagash, an inhabited planet that orbits Alpha, one of six stars in a multiple star system. Because of their crowded sky, the inhabitants of Lagash, unlike Earthlings, know neither darkness nor night. The story opens just a few hours before one of Lagash's large moons is predicted to eclipse the one star still in the sky after the others had set. On Lagash, such an event happens only once every 2,049 years—just often enough to give rise to myths, but not knowledge, of stars. With only their hazy cultural memories, the people await the eclipse in fear of what darkness might bring. But despite their fear, they survive the event. As the eclipse comes to an end, the story wistfully concludes, "The long night had come again." The "long night" refers to the next 2,049 years, during which time they will again be ignorant of the stars.[2]

In a rare moment of insight, Emerson once said, "If the stars should appear one night in a thousand years, how would men believe and adore, and preserve for many generations the remembrance of the city of God!" As it is, we risk taking the starry hosts for granted in our night sky. These points of light have inspired all of the world's great cultures and religions.

But they have also fostered the curiosity and careful observations that gave rise to astronomy, a discipline that would have been obstructed without two features of our environment—dark nights and an atmosphere transparent to visible light. Without these features, we might be utterly ignorant of the cosmos beyond our tiny neighborhood. To truly appreciate our good fortune, at least once everyone should stand on a mountaintop under the open sky, on a clear, moonless night. The air is so clear, and the stars so vivid, that only your lungs will remind your eyes that you're on a planet with an atmosphere.

Consider the atmosphere's transparency, which is actually just part of the story. Our atmosphere participates in one of the most extraordinary coincidences known to science: an eerie harmony among the range of wavelengths of light emitted by the Sun, transmitted by Earth's atmosphere, converted by plants into chemical energy, and detected by the human eye. The human eye perceives light of different wavelengths as different colors, ranging from violet blue (the shortest wavelengths in the visible spectrum) to red (the longest). Looking at a diagram of the electromagnetic spectrum, we see the visible light emerging gracefully from the ultraviolet, differentiating into the familiar colors of the rainbow, and disappearing seamlessly into the warm, invisible infrared. The visible wavelengths rise and fall at the nano-scale, the distances from their peaks measured in ten-billionths of a meter, called Ångstroms (Å). We see blue at approximately 4,800 Å, green at 5,200 Å, yellow at 5,800 Å, and red at 6,600 Å. Earth's atmosphere is transparent to radiation between 3,100 to 9,500 Å and to the much longer radio wavelengths. The radiation we see is near the middle of that range, between 4,000 and 7,000 Å, the range in which the Sun emits 40 percent of its energy. Its spectrum peaks smack in the middle of this visible spectrum, at 5,500 Å. This is but a tiny sample of the entire range. The near-ultraviolet, visible, and near-infrared spectra—the light most useful to life and sight—are a razor-thin sliver of the universe's natural, electromagnetic emissions: about one part in 10^{25}. That is much smaller than one star out of all the stars in the entire visible universe: about 10^{22}. (See Plate 7.)

As it happens, our atmosphere strikes a nearly perfect balance, transmitting most of the radiation that is useful for life while blocking most of the lethal energy. Water vapor in the atmosphere is likewise accommodating,[3] a fact that even the fifteenth edition of the staid *Encyclopaedia Britannica* picks up on: "Considering the importance of visible sunlight for all aspects of terrestrial life, one cannot help being awed by the dramatically narrow window in the atmospheric absorption . . . and in the absorption spectrum of water."[4] The oceans transmit an even narrower window of the spectrum, mainly the blues and greens, while halting the other wavelengths near the surface, where they nourish the marine life that figures prominently in Earth's biosphere (we'll return to this point in Chapter Seven).

It seems plausible that this is merely an artifact of our eyes having evolved through natural selection to decipher just the spectrum of light that happens to get through the atmosphere. But this fact isn't so easily dismissed.[5] As George Greenstein notes:

> One might think that a certain adaptation has been at work here: the adaptation of plant life to the properties of sunlight. After all, if the Sun were a different temperature could not some other molecule, tuned to absorb light of a different color, take the place of chlorophyll? Remarkably enough the answer is no, for within broad limits all molecules absorb light of similar colors. The absorption of light is accomplished by the excitation of electrons in molecules to higher energy states, and the general scale of energy required to do this is the same no matter what molecule you are discussing. Furthermore, light is composed of photons, packets of energy, and photons of the wrong energy simply cannot be absorbed.[6]

In other words, because of the basic properties of matter, the typical energy involved in chemical reactions corresponds to the typical energy of optical light photons. Otherwise, photosynthetic life wouldn't be possible. Photons with too much energy would tear molecules apart, while those with too little energy could not trigger chemical reactions. Similar arguments hold for the wavelength range over which the atmosphere is transparent. So stars that don't emit the right sort of radiation in the right amounts won't qualify as useful energy sources for life.

The radiation a star emits and its peak emission depend on its temperature. Temperatures of normal stars range over a factor of about twenty. The Sun is a little below the mid-range, while exotic stars emit most of their energy on the far ends of the electromagnetic spectrum, as gamma rays, X-rays, or radio waves.[7] A star like our Sun, which produces mostly photons that can energize chemical reactions on Earth, will shine light that comes from just that region in its photosphere where atoms can combine to form stable molecules.[8] So we can't explain away one part of this coincidence in terms of the others. Life can't use just any type of light from any type of star. Our Sun, it turns out, is near the optimum for any plausible kind of chemical life.[9]

To see the stars, though, a translucent atmosphere that merely allows light to reach the surface won't cut it. Like a clean, clear pane of glass as compared to a frosted one, we need an atmosphere, like Earth's, that is transparent in order to see the stars and the wider cosmos.[10] (See Plate 8.) Of course, we enjoy these views because we have the good fortune of living on land. The view would not be nearly as good if we lived underwater.

But even dry land and a transparent atmosphere aren't enough, since they wouldn't be much use without the dark nights. Although we take them for granted, dark nights depend on several astronomical variables, some local and some not. A dark sky requires that our planet regularly rotate away from the intense direct light of the Sun.[11] If our day were the same length as our year, Earth would always keep the same face pointed toward the Sun, much as the Moon does Earth. The resulting large temperature difference between the day and night sides would be hostile to complex life (more on this in Chapter Seven). Any complex life, such as there could be, would stay on the day side. Unless the planet enjoyed total solar eclipses, like Lagash, it would never see dark skies.

We would suffer similar but less severe problems if we had several moons staring unblinkingly in the night sky like headlights on a busy highway. Our single moon does interfere with astronomers' ability to observe distant faint objects, but only when it's out. Moreover, visibility is better now than it was in the distant past, when the Moon was nearly three times closer to Earth, and thus nearly nine times brighter.

Of course, a translucent atmosphere might work for merely converting sunlight into chemical energy or hunting prey and avoiding predators. But because we are big-brained, mobile, surface-dwelling creatures, we need a certain type of atmosphere. It just so happens that that very atmosphere—predominantly nitrogen and oxygen with some carbon dioxide and water vapor—is mostly transparent to optical radiation. All four of these chemical constituents are essential for our biosphere. Water does partially hinder our view of the distant universe, but the valuable trade-offs it provides more than compensate for this hindrance.

For example, one particularly impressive consequence of atmospheric water is the rainbow. Rainbows, like total solar eclipses, seem equal parts whimsy and mystery, summoning the artist's creativity and the naturalist's curiosity.[12] The early attempts to explain their formation started the first of our many tutorials on the nature of light. The scientists who followed those clues eventually learned how to unweave the white light of the Sun. René Descartes and, later, Isaac Newton, who performed experiments with sunlight and prisms in 1666, groped toward the modern explanation for the rainbow and the modern science of spectroscopy. A rainbow is in effect a natural spectroscope as big as the sky. Once scientists learned how to use a prism to replicate a rainbow, it was only a matter of time before someone scrutinized the solar spectrum. But rainbows won't appear on just any planet. A good rainbow needs a partially cloudy atmosphere,[13] the golden mean between the uniformly cloudy and uniformly dry.

Earth's water results in skies that, on average, are about 68 percent cloudy.[14] Clouds help balance the global energy by contributing to the global albedo, that fraction of sunlight reflected back into space. Earth currently reflects about 30 percent of the sunlight that strikes it. Four major types of reflecting surfaces contribute to the global albedo: land, ice, oceans, and clouds, each with its own particular reflective properties.

FEEDBACKS AND A CLEAR VIEW

For regulating the climate, flexibility is essential. If we were building a habitable planet from scratch, we would want the various albedo types to be adaptable to changing conditions. If some part of Earth's climate system changes appreciably, its temperature will change, unless another part of the system can compensate. For example, say the Sun brightens significantly over a few million years. If no other part of the system counteracts this, Earth will heat up, possibly with disastrous consequences for life. The polar ice caps could melt, causing Earth to deflect less of the Sun's energy, leading to a spiraling cycle of overheating that over time could evaporate the oceans. But such a gloomy scenario ignores counterbalancing factors. The environment might compensate by increasing cloudiness and precipitation. These would reflect the excess solar energy back into space and encourage plant growth, thereby sequestering carbon dioxide, which might otherwise lead to a sweltering atmosphere. We call such stabilizing forces negative feedbacks.

Earth's four albedo types operate on different timescales. Clouds are the rapid responders. Trees and plants react more slowly, changing as quickly as a single season or as slowly as several decades. Changes in sea level may take hundreds or even thousands of years. And geological changes may take a million years or more. With such variety available, the climate can respond to changes lasting anywhere from hours to billions of years. This is vital, because the Sun does change its energy output over those timescales. A large sunspot group produces short-term changes against a backdrop of slower variations produced over the prominent eleven-year sunspot cycle and other, subtler cycles that last hundreds of years. Over billions of years the Sun has slowly brightened as its core has heated up.

While we still don't completely understand the mechanisms regulating cloud cover, negative feedbacks probably link it to other parts of the climate system. But for regulating the climate, a partly cloudy atmosphere is better than either a completely clouded or a cloud-free one. For either of these extremes, it's impossible to regulate climate merely by tweaking the cloud cover. Only an atmosphere in which cloud cover is quite different from 0 or 100 percent can provide negative feedbacks to help keep life comfortable. The same is true for the contributions of land, ocean, and ice. Earth has the most diverse collection of reflective surfaces in the Solar System, all providing intricate biological and nonbiological climate controls. Life thrives at the interfaces: atmosphere-land, atmosphere-ocean, land-ocean, and atmosphere-land-ocean.

The idea that Earth has feedback processes in which its biota interact with its nonliving parts to regulate the global climate is known as the Gaia hypothesis. James Lovelock and Lynn Margulis first proposed this in the 1970s, arguing that such interactions tend to make the environment more fit for life, mostly by taking the edge off climate change.[15] Lovelock has compared Earth's regulation system, which he calls geophysiology, to the metabolism of a mammal or a redwood.[16]

Lovelock and Andrew Watson illustrated the Gaia hypothesis with a mathematical "toy model" called Daisyworld.[17] Daisyworld has an orbit just like Earth's and a host star just like the Sun. On the Daisyworld planet, there are two species of daisies, one black and one white (or dark and light as compared with bare soil), growing in large, distinct patches. Both species grow best at the same optimum temperature, and they both have the same lower and upper temperature limits on their growth; the overall temperature dependence on growth is a simple parabolic function (looks like a

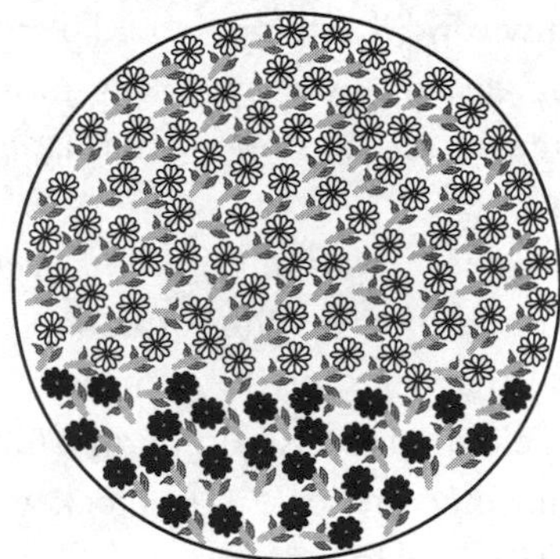
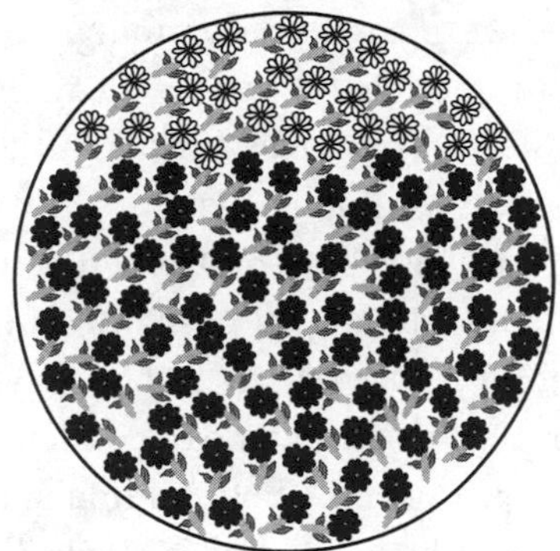

Figure 4.1: Two snapshots of a hypothetical Daisyworld planet. When the host star is faint, black daisies predominate; when it is bright, white daisies predominate. In this way the changing relative numbers of white to black daisies maintain a stable planetary temperature, even as the host star brightens over time.

helmet). Since the white daisies are more reflective than the black daisies, a patch of white daisies produces a cooler local environment than a patch of black daisies. The relative populations of black and white daisies determine the average global temperature. For a given brightness of the host star, the relative numbers of daisies will adjust to provide the best global temperature for growth of daisies.

Now, suppose the host star begins to brighten slowly. Increased light brings the white daisy patches closer to the temperature they prefer, which encourages them to multiply. The areas draped in black daisies also heat up, moving them away from the optimum temperature for growth, so their numbers diminish. Globally, then, the number of white daisies increases and the number of black ones decreases, so the planet reflects more light into space than it did before. This keeps Daisyworld's average temperature constant, even though its host star has brightened.[18]

As intriguing as the Daisyworld model is, it's just an illustration—Earth's biosphere is far more complex and far more awe-inspiring. Recall that a key part of the carbon cycle is the rate of chemical weathering of surface rocks by carbon dioxide dissolved in rainwater, which forms a weak acid. This reaction can occur without life, but plants, trees, and tiny marine life greatly accelerate it.[19] Since both rising temperature and the concentration of carbon dioxide increase chemical weathering, somewhat paradoxically, they speed up the removal of carbon dioxide from the atmosphere, which reduces greenhouse warming. And like Daisyworld, Earth's trees and plants alter the local albedo with their dark foliage and by evaporating more water, which cools Earth's surface by drawing the curtains—that is, by increasing cloud cover.

Scientists have only recently recognized other important links between life and the global climate. One such process involves the formation of cloud condensation nuclei (CCN), small particles in the atmosphere around which water can condense to form cloud droplets. Although CCN are produced by both natural and anthropogenic processes, Lovelock was among the first to link biologically produced CCN to the climate.[20] Phytoplankton, such as *Emiliana huxleyi*, produce dimethyl sulfide, the first step in a chemical chain to build the CCN. Phytoplankton respond to a warming ocean surface by producing more dimethyl sulfide, which concentrates more CCN and enhances the albedo of marine stratus clouds. Reflecting more light back into space cools the ocean below. Higher carbon dioxide levels also stimulate production of dimethyl sulfide and CCN.[21]

All of this contributes to an atmosphere that lets us see distant stars and galaxies from Earth's surface. But such clarity does not spring eternal. We know from ice and marine cores that the glacial periods were much windier and dustier than the present.[22] So even if continental interiors were less cloudy, the trade-off was a much dustier atmosphere.

Astronomers would prefer a few cloudy nights interspersed with crystal clear ones to continuously dusty and windy conditions. Dust blocks starlight and makes observations far more difficult to calibrate. It's also murder on optical instruments. Observatories have strict rules on the maximum tolerable dust and wind levels before they are forced to close; for example, astronomers on the Canary Islands frequently deal with the dust that blows in from the Sahara. When astronomers lose a few nights to clouds on an observing run, they should thank their lucky stars that they weren't born a mere twelve thousand years ago.

While its precise makeup still eludes us, Earth's early atmosphere was clearly quite different from the present one; in particular, it was more reducing, which means that it contained more hydrogen. Oxygen was a trace gas, and carbon dioxide and probably methane were much more abundant. This early Earthly atmosphere may have resembled a warmer version of the present atmosphere of Titan, Saturn's largest moon. The Sun's ultraviolet radiation would tend to form more complex hydrocarbon molecules in such a more reducing atmosphere, leaving behind a thick, oily, planet-wide haze.[23] That haze probably cleared off with the rise of oxygen about two billion years ago.

Not only does the great diversity of interfaces and negative feedbacks maximize life's diversity, it also fosters scientific discovery. Recall that virtually all the natural recording processes described in the previous chapter require the hydrological cycle. You might be tempted to think that clouds or an atmosphere of any sort is a detriment to measuring the universe. Wouldn't it be better to have no atmosphere, or at least a cloud-free one? Even if the dust problem, which is itself a trade-off with cloudiness and rainbows, is forgotten, this scenario ignores the added advantages of an atmosphere like Earth's. By providing a hydrological cycle while still being fairly transparent in the same part of the spectrum where most stars emit most of their light, Earth's atmosphere combines otherwise competing factors to strike the best overall compromise for measuring the universe.

Earth's present-day atmosphere, then, gives us exceptional access not only to the past but also to the wider universe, while maintaining a life-nurturing environment. And like the layering processes discussed previously, there's no reason to suppose that we are specially adapted to decode this information, since doing so gave us no survival advantage—that is, not until very recently.

MAKING LONG-TERM PLANS

Curiously, Earth's transparent atmosphere does provide survival advantages to a civilization advanced enough to use the stored information, but much too late in the process to be explained in Darwinian terms.[24] For example, the transparency of the atmosphere, especially with respect to high-precision astrometry (measuring position on the sky) of faint objects, allows astronomers to catalog the population of near earth objects, or NEOs, which include both near Earth asteroids and comets. NEOs have Earth-crossing orbits, which means they can hit us. Currently, about 1,250 NEOs are known, most discovered in the past few years.[25] The extent of damage caused by a NEO impact with Earth depends on its size. A NEO two kilometers in diameter is considered a "civilization ender," while one ten to fifteen kilometers in diameter is a "K/T event," like the one that probably killed off the dinosaurs sixty-five million years ago.[26] K/T events are believed to occur once every fifty to one hundred million years. There is about a 1 percent chance of a two-kilometer NEO impacting in a 10,000-year time period, and a 50 percent chance for a two-hundred-meter impact over the same time period. In the latter case, enormous tsunamis could cause severe damage by flooding coastal areas.

The 1908 Tunguska event in Siberia, thought to have resulted from the atmospheric explosion of a roughly 50-meter asteroid or comet fragment, is the only well-documented example of a destructive impact in recorded history. This 15-megaton explosion occurred over a mostly uninhabited region of Siberia. Were it not for this disturbing natural experiment (and the safely distant comet Shoemaker-Levy 9 impact with Jupiter in 1994), we might not take impact threats seriously.[27]

Comets seem to impact Earth less frequently than asteroids, but the typical comet impact may be more devastating because of comets' greater velocity. Hale-Bopp, the intrinsically brightest visible comet since the sixteenth century, inspired people worldwide in 1997. (See Plate 9.) It's understandable why such a sight both inspired and frightened ancient peoples. Today, we fear comets for a different reason. If Hale-Bopp had hit Earth, with an energy perhaps one hundred times greater than the K/T extinction event, it would have wiped out everything but the hardiest microbes.[28]

That's the bad news. The good news is that we can protect Earth from a devastating impact if we can discover a NEO headed for us with enough lead time. Observations over a few years allow astronomers to predict the precise position of a NEO a few decades into the future. Comets are a different matter, since astronomers usually discover them only a few months before they pass through the inner Solar System. But that would still give us some time to prepare, and as our technology progresses, so will our capacity to respond.

Had our atmosphere been completely cloud-covered or translucent, we would not know about impact threats until it was too late. Nor would we have developed a space program, which is a prerequisite for deflecting NEOs. Had our atmosphere been thicker but still partly transparent, the additional distortions of telescopic images would have severely limited our ability to catalog these faint objects. So once again we see life and discovery walking hand-in-hand.

It's obvious that if large comets and asteroids were now frequently bombarding Earth, our planet would be quite dreadful. But ironically, these rogue bodies in the Solar System were important early on, not only for future habitability but for present measurability as well. As valuable as it is for us to view extraterrestrial objects from a distance, sometimes a little bit of the sky comes down, allowing us the chance for direct inspection.

SPACE ROCKS

On January 18, 2000, a fiery ball streaked across the sky, its explosions breaking the early-morning silence of the frigid Yukon and waking its sleepy residents. Defense satellites had tracked it from space and confirmed its extraterrestrial origin. Its smoke trail lingered in the sky for a full day like the Cheshire cat's grin. This had been the largest bright meteor, called a bolide, detected over land in ten years. A few days later, a resident near Whitehorse, Yukon, searched the ice-covered Tagish Lake and found several fragments of the space visitor.[29] Today, the few Tagish Lake fragments on the market fetch a handsome sum. Scientists called meteoriticists are willing to pay top dollar for meteorite fragments, not only because they are rare and hard to locate, but also because they contain unique and important information.

Astronomers are fairly confident that they understand how most meteorites make their way to Earth. Meteorites are basically shards from large, broken-up parent bodies in the asteroid belt, located between the orbits of Mars and Jupiter. Its largest member is Ceres, a hefty one thousand kilometers in diameter. But there are still thousands of smaller asteroids in the belt, creating enough traffic that collisions are still fairly common. A single collision can produce thousands of rock-sized fragments. Some wander into unstable zones, which resonate with Jupiter's orbit. Jupiter's gravity will perturb a body in such a zone in the same part of its orbit every time it completes a circuit around the Sun.[30] Once perturbed into less circular—that is, more "eccentric"—orbits, the fragments begin to visit the inner Solar System. Those that enter Earth-crossing orbits can hit us.[31] When a fragment approaches Earth, its surface begins to heat from friction with the atoms in our atmosphere. We see it as a meteor. If it's big enough, it makes it to the ground as a meteorite.

The Tagish Lake fragments belong to a previously unknown type of meteorite. Astronomers can identify a class of meteorite with a particular asteroid or class of asteroids by comparing their broad spectroscopic signatures. If they have similar so-called reflectance spectra, researchers conclude that they have a common composition, and hence a common source. Based on such analysis, researchers have tentatively identified the Tagish Lake meteorites with D-type asteroids, which tend to lurk in the colder, outer reaches of the asteroid belt.[32] They preserve some of the most pristine material from the early Solar System.

So why are meteorites so important? For starters, they contain samples of the early Solar System. Such material is not available from the homogenized and processed planetary bodies. The larger the body, the more heated and hence the more dynamic its interior. Some meteorites come from asteroidal parent bodies that have clearly differentiated; the metals (mostly iron and nickel) sink to the center, leaving the stony material behind in the crust. These don't preserve information about conditions prior to the asteroid's formation. In contrast, since the most primitive types of meteorite, the carbonaceous chondrites, presumably come from undifferentiated parent bodies, they are the most treasured.

Until the early 1970s, most meteoriticists believed that the violent processes that formed the bodies in the Solar System would melt and homogenize any pre-solar interstellar grains beyond recognition. So they were surprised to discover that some micron-sized grains in primitive meteorites exhibit isotopic ratios inconsistent with an origin in a processed parent body. This means that some small bodies in our Solar System contain matter that predates the birth of the planets and even the Sun. These grains hold unique clues about the chemical history of the Milky Way galaxy itself, and their discovery has led to an entirely new field of astrophysics. The Tagish Lake meteorites are a fine example, since they contain more interstellar dust grains than any other specimens.

Interstellar grains are grouped according to their bulk mineral type and according to the ratios of various isotopes.[33] Not only can astronomers connect individual meteorites to specific asteroid types, but they can also connect individual interstellar grains to specific types of sources in the Milky Way. For instance, most of the silicon carbide grains seem to come from asymptotic giant branch (AGB) stars—very luminous, cool stars near the end of their lives—and most of the other types come from supernovae, AGB stars, and possibly some novae. Astronomers can know this, in part, because some of the isotope ratios they can measure in the interstellar grains can also be measured in stellar atmospheres.[34] While interstellar grains in meteorites came together at the time and place of the Solar System's birth, a mere pinpoint in the broader context of the Milky Way galaxy, they are believed to sample all the previous history of the galaxy and a fairly broad region of space. With isotope ratio and atomic abundance trends in hand, astronomers can determine more details about the buildup

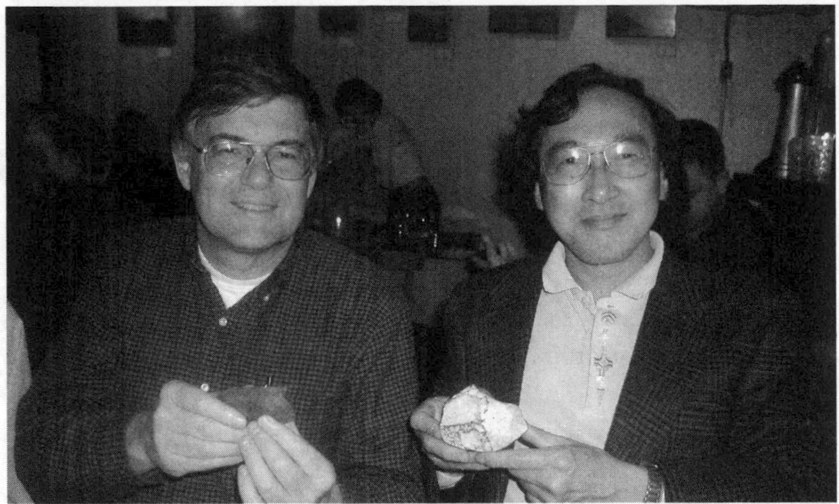

Figure 4.2: Astronomer Donald Brownlee (University of Washington) holding a 200 gram Mars meteorite, with Kuni Niishiumi (Berkeley) holding a 300 gram Moon meteorite, in the Asteroid Café in Seattle. Both rocks were found in North Africa within the last two years.

of elements within stars—called stellar nucleosynthesis—and galactic chemical evolution than they could with only one of these available. It's interesting that these two primary sources of information are so complementary to each other: interstellar grains provide us only with isotopic ratios, while stars give us mostly atomic abundances. What's more, they overlap much of the same time and space.

We still have a lot to learn about interstellar grains. Somehow, individual grains form in the winds of AGB stars or in the debris of supernova explosions, wander through space for millions or billions of years, merge with one another in the dense environment of a giant molecular cloud, survive the early trauma of planet formation, and eventually end up on a meteoriticist's shelf. As a result, we can hold in one hand a two- or three-pound rock containing thousands of original pre-solar grains, each possibly from a different star!

ASTEROIDS, COMETS, JUPITER, AND LIFE ON EARTH

Delivering meteorites to Earth's surface requires a fine-tuned chain of events. Apart from preserving pre-solar grains in the earliest stages of planet formation, which is not yet well understood, at least some small primitive bodies in the inner Solar System must *not* be incorporated into larger ones. Otherwise, their unique identities would be erased by a planet's hot, dynamic interior. Recent gravity simulations of the early Solar System indicate that the asteroid belt has lost about 99.9 percent of its original mass. Probably several Moon- and Mars-sized bodies once lurked in the region between Jupiter and Mars, helping to perturb most of the smaller bodies into the Jupiter and Saturn resonances.[35] Once there, the asteroids' orbits become more eccentric, allowing them to visit the terrestrial planet neighborhood or Jupiter's orbit. Jupiter will usually then fling them out of the Solar System.

But some of the scattered asteroidal debris ended up on Earth, blessing it with almost all of its life-essential volatile elements and simple organics.[36] Jupiter's gravity eventually removed the larger perturbing bodies from the region of the asteroid belt. The details of Jupiter's formation are critical in all this. On one hand, if it had formed a little earlier, or been a little more massive, or had a more oval-shaped, eccentric orbit, probably too few asteroids would have remained to provide enough carbon for Earth. The same problem would have occurred if Jupiter hadn't cleared the asteroid belt region of planets, for then perhaps these planets would have cleared away too many asteroids. On the other hand, if Jupiter had formed later or had a significantly smaller mass, too many asteroids might have remained, the larger ones pummeling Earth too frequently for life to take hold.[37]

Asteroids are a subtle contributor to Earth's habitability—we need them early on to deliver water and organics, but not too many later on, since they have an unfortunate tendency to extinguish life. This most habitable of circumstances left us a small remnant population of asteroid remains to study the formation of the Solar System and the preceding history of the production of elements within stars. Just enough meteorites are still delivered to Earth to make studying them a practical endeavor.[38]

While meteorites are the handiest source of encoded information about cosmic history, they're not the only source. Comets also deposit debris onto Earth. But unlike asteroids, comet fragments can't survive the journey to Earth's surface in an identifiable form. Instead, they fall to Earth as comet dust. So high-flying aircraft, like NASA's U2 research planes, must nab this comet dust before it mixes with Earth's dust. Although most planetary systems are probably accompanied by comets, probably few have just enough asteroids to keep meteoriticists happy with little presents without wiping them out.

Who would have guessed that the asteroid belt, which first seemed like a failed experiment in planet-building or a dangerous planetary junkyard, should play a role not only in Earth's habitability but in scientific discovery as well? As we learn more about the seemingly accidental features of our atmosphere and Solar System, we begin to recognize a trend: The Earth system offers not only a habitat but also a great viewing platform for its inhabitants. Because the processes that produce such a happy planetary state are intricate and interdependent, Earth is likely to be a very rare kind of place. But we can do better than guess. With the knowledge of the previous pages firmly in hand, we can begin to place Earth in its proper context by comparing it with other planets in the Solar System and even to those beyond it.

Chapter 5

The Pale Blue Dot in Relief

A very great part of the surface of Venus is no doubt covered with swamps.... The temperature on Venus is not so high as to prevent a luxuriant vegetation.... The organisms [at the poles] should have developed into higher forms than elsewhere, and progress and culture, if we may so express it, will gradually spread from the poles toward the equator. Later... perhaps not before life on the Earth has reverted to its simpler forms or has even become extinct, a flora and a fauna will appear, similar in kind to those which now delight the human eye, and Venus will then indeed be the "Heavenly Queen" of Babylonian fame, not because of her radiant lustre alone, but as the dwelling-place of the highest beings in the Solar System.

—Nobel laureate Svante Arrhenius, 1918[1]

ALL THOSE OTHER PLACES

Human beings have speculated about life on the other planets for as long as we've been aware that there were other planets. But such questions really picked up steam in 1543, when Copernicus suggested that Earth itself was a planet, which, like the others, revolved around the Sun. No longer were the planets just mysterious, wandering points of light. They were *places*. And places, of course, might have people.

Only recently, however, have we been able to compare those places with Earth. Starting with the Mariner probes in the 1960s and continuing to the present day, with probes in orbit around Mars and (soon) Saturn, our understanding of the other planets and their moons (save Pluto) has expanded exponentially. This new knowledge has altered our perception of Earth compared with the Sun's other children. It has also disappointed speculations that, in retrospect, look a bit fanciful. Nevertheless, what we've learned, especially combined with the material we've already discussed, provides some objectivity for comparing Earth with the other planets and moons in our Solar System, as well as with the newly discovered "extrasolar planets." In a sense, these other bodies serve as a control group for comparing—or better, contrasting—the remarkable features of our home planet. Despite all we've been told about this pale and insignificant blue dot, Earth is really quite an extraordinary hostess for both life and scientific discovery.

THE PLANETARY MENAGERIE

Astronomers group bodies in the Solar System into planets, their satellites (or moons), asteroids (also called minor planets), and comets. We can also put the planets into three groups: terrestrial planets (Mercury, Venus, Earth, and Mars); the gas giants (Jupiter, Saturn, Uranus, and Neptune); and Pluto, in a class of its own, possibly as a citizen of the Kuiper Belt. Some moons are comparable in size to terrestrial planets, and Titan, one of Saturn's moons, even has a thick atmosphere. Let's begin with a tour of Earth's closest neighbors.

Mars: Our Most Earth-like Neighbor

Mars's climate is more like Earth's than any other planetary body in the Solar System, but the differences are still profound. Its surface pressure is less than one-hundredth that of Earth's. And although some water condenses on its polar regions, the planet's surface is dry and dusty. The size of both polar caps varies dramatically over a Martian year. Its south polar ice cap is mostly frozen carbon dioxide—dry ice—while the larger, northern one has a large permanent water ice component. The northern cap has about half of the volume of Greenland's ice cap. If melted and spread over the entire planet, it would cover Mars in a red sea about ten to twenty feet deep.

Even with polar caps, however, Mars lacks the Earth-like features that make historical geology more than a wild goose chase, and this absence makes it a nasty place to live. High-resolution orbiter images suggest that wind erodes the entire surface of Mars, destroying important information in the process. (See Plate 10.) This dust-loaded wind erodes previously deposited layers. Deposits with no source of cohesion are particularly vulnerable. Over thousands of years, strong, persistent winds can turn a delicately layered plain into a chaotic dune field. Since Mars has precious little precipitation, the ice laid down on the northern polar cap over several years could easily be destroyed during a large dust storm or evaporate during a drier-than-average season. Even if ice is still accumulating in the Martian caps, it must do so very slowly, yielding very thin annual layers.

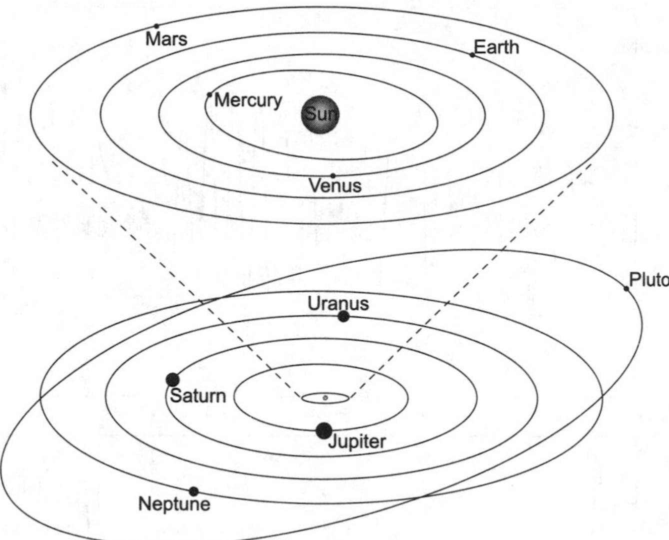

Figure 5.1: Our Solar System contains eight major planets in fairly circular, stable orbits in roughly the same plane, and one outlier, Pluto. Mercury, Venus, Earth, and Mars are rocky terrestrial planets. The asteroid belt lies between Mars and Jupiter. Jupiter and Saturn are gas giant planets. Uranus and Neptune are also mostly gas with some rock and ice. Comet-like Pluto is largely rock and ice. The orbits but not the planets are drawn to scale.

The dust deposited on the caps also reduces the ice's ability to reflect sunlight back into space, causing it to warm more than pure ice. While some layering is visible on high-resolution images of the Martian caps from orbiting probes, it almost surely fails to preserve information with a fidelity approaching Earth's polar ice deposits, which accumulate steadily from one year to the next.[2]

Coincidentally, the present Martian and Earthly rotation periods and axial tilts are quite similar. But Mars lacks a large moon, and as a result, its tilt wobbles widely over millions of years. Thus, while Earth's rotation axis is very stable, Mars's tilt has ranged from fifteen to forty-five degrees over the past ten million years.[3] Even at its currently favorable angle, the Martian polar caps are not very good data recorders. At higher tilt angles, the polar ice would melt or evaporate even more completely.[4] And we've already discussed the problems that such an unstable tilt poses for complex life.

Comparing the Martian climate with Earth's highlights the exquisite quality of Earth's ice deposits. Oceans surround Greenland and Antarctica,

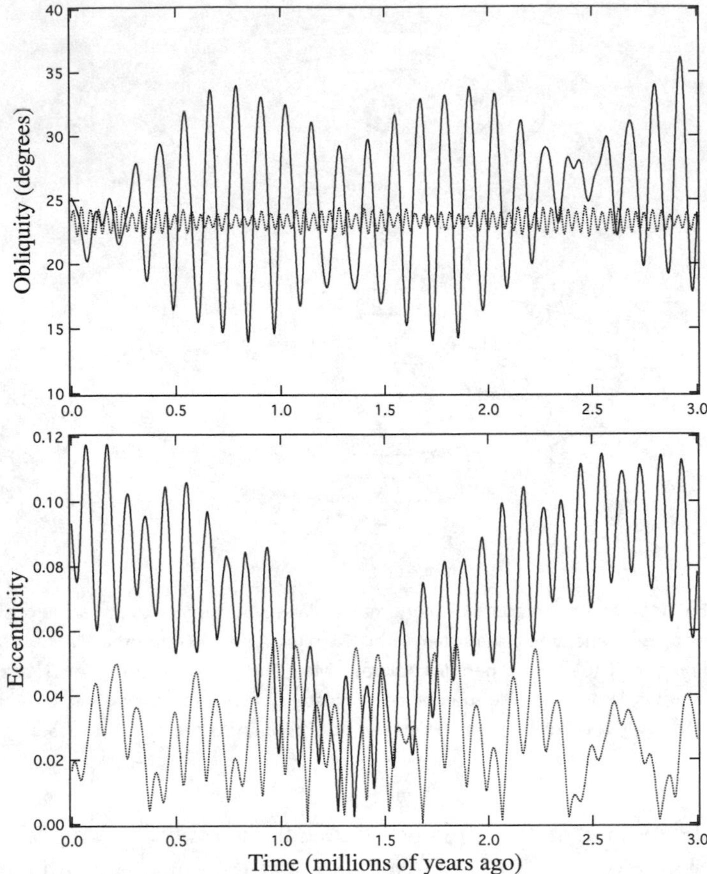

Figure 5.2: Variations of the obliquity and eccentricity of Earth (dotted) and Mars (solid) over the past three million years. Even this long time span does not fully capture the range of Mars's variability. In the past ten million years Mars's obliquity reached about 45 degrees, and over the past one billion years it probably reached 60 degrees. These large variations result in large climate swings.

providing a continuous source of moisture for ice deposits far into the interior of these giant islands. The parched Martian climate doesn't guarantee net ice growth from year to year, and Mars's vast oceans of sand erode its delicate polar caps, reversing the familiar Earthly scenario of water waves washing over a sand castle. Carbon dioxide, which changes directly from ice to gas at Mars's surface, is the most abundant part of the Martian atmosphere, and is a poor substitute for water as a stable matrix for layered deposits. Mars's large orbital eccentricity—its swings near and far in its annual course around the Sun—and the wide wobbles in its axial tilt make it less likely that annual ice layers will be deposited and preserved on its surface. In contrast, Earth's layered deposits record evidence of its small variations in eccentricity and axial tilt, a fact that itself testifies to those deposits' great sensitivity.[5]

Mars's geology also differs substantially from Earth's. Starting in the late 1990s, the Mars Global Surveyor magnetometer began to measure the Martian magnetic field. The data revealed a weak fossil magnetic field in the crustal rocks, and confirmed that Mars now lacks a dynamo-generated planetary field, that is, one generated by circulation of a fluid metallic core. The fossil field does point to a time in Mars's past when it had a planetary field, though it was probably short-lived.[6] Some of the remnant field patterns are similar to Earth's striped crustal fields. This has encouraged some to speculate that Mars may have briefly displayed plate tectonics.[7] Its small mass led to a rapid shutdown of Mars's planetary magnetic field and geologic activity. This caused it to quickly lose heat from its interior.[8] The shutdown was probably hastened because heat-producing radiogenic elements, such as potassium, uranium, and thorium, were sequestered in its crust within about a half billion years of the planet's formation.[9] Mars's crust is two to four times thicker than Earth's, and probably contains over 50 percent of the planet's heat-producing elements (compared with 30 to 40 percent in Earth's crust). Once removed from the mantle, the radiogenic elements can no longer contribute to its convection, unless the crust is subducted back down into the mantle. This problem is exacerbated on smaller planets, because their crust takes up a larger fraction of the planet's mass.

Mars probably lacks ore deposits as diverse and rich as those on Earth's surface, although some ores probably formed during Mars's early wet era and during its longer-lasting volcanic episodes. So Mars probably comes closer than any other planetary body in the Solar System to matching the diversity of Earth's ore deposits. Indeed, iron may be more abundant in Mars's crust than Earth's, since Mars probably has not undergone as much internal differentiation as Earth. In addition, the rain of meteorites (some of which are nearly pure nickel-iron) onto its surface over the past few billion years, made more common by its thinner atmosphere and closer proximity to the asteroid belt, provides a rich source of iron. Other metals are probably not so accessible. Particularly important would be copper, which was employed for tools on Earth before iron was. In any case, Mars's atmosphere lacks enough oxygen to allow for fires on its surface, so Martians would probably enjoy neither Boy Scout cooking merit badges nor high technology.

VENUS: LOVER OF LIFE?

Unlike Mars, Venus has a thick atmosphere. It has often been called Earth's sister planet. Now we know better: it is utterly hostile to life. Far from providing a loving home for a lush biosphere, as some astronomers believed only a few decades ago, Venus's surface temperature is a scorching 900°F. Needless to say, it completely lacks a hydrological cycle. The name the ancients gave this planet is not quite appropriate.

Like Mars, Venus lacks a planetary magnetic field, but for different reasons. Venus, though, may still have the liquid metallic core that Mars now lacks. But Venus's day is nearly the same length as its year, which means the planet is probably rotating too slowly to provide much of a magnetic field. The weak rotation of a circulating metallic outer core—assuming one exists—can't generate a planetary magnetic field by itself. Without it, the planet can't shield its atmosphere from direct interaction with the solar wind.

Venus, while nearly the same size as Earth, has a different system of internal dynamics. (Earth appears to be the only planetary body in the Solar System that still has plate tectonics.[10]) Its thick, immobile crust prevents the kind of tectonic movement we see on Earth; instead, Venus, like the other terrestrial planets in the Solar System, undergoes so-called stagnant lid convection.[11] Unlike the steady convective overturning of Earth's mantle and crust, which has slowly released internal heat, a large planet with stagnant lid convection can build up heat and, like a pot with a partially sealed lid, release it episodically—and catastrophically. Venus apparently underwent a planet-wide volcanic resurfacing event about five hundred to eight hundred million years ago, which we know from crater counts on its surface. The event completely erased any information previously recorded on its surface, and belched a great deal of greenhouse gas into its atmosphere.[12] Contrast this with the steady crustal recycling on Earth, which preserves information for long periods on its floating continents while regulating the amount of carbon dioxide in the atmosphere.

So why was Venus's history so different from Earth's? Perhaps stagnant lid convection is the full explanation, or perhaps its initial rotation state is involved as well. Astronomers believe the initial rotation rates and directions of the terrestrial planets resulted from the last large bodies that impacted them, imparting their angular momentum in the process. In this view, a retrograde rotation, which is backwards compared with its orbit, was just as likely as a prograde one, in which it orbits in the same direction, because the impactors should have impacted the early Earth from random directions. The direction of rotation is surprisingly crucial to the future state of a terrestrial planet, since it affects how a planet's liquid core, rotation, and orbit interact. Gravitational torques from the Sun—and for Earth, from the Moon—move the solid portion of a planet relative to its fluid core.

As a rough analogy, imagine a soft-boiled egg with a rigid shell, a firm white, and liquid yellow yolk that sloshes when the egg is spun. The fluid motions of a planet can be greatly intensified when the motions in the liquid part of the core "resonate" with one of the motions of the solid planet.[13] Friction converts this motion into heat at the core-mantle boundary at the expense of the planet's rotational energy. This additional heat is deposited at the base of the mantle, where the melting may eventually result in extensive volcanism at the surface and the accompanying release of greenhouse gas. Simulations indicate that Venus's interior would generate much more heat during a resonance passage than would Earth's.[14] This is because Venus's rotation is retrograde—its rotation is clockwise, while its orbit is counterclockwise, as viewed from the north—and the tidal forces on it are weaker. Earth's Moon more strongly decelerates its rotation, even though Venus is closer to the Sun. Because Earth's rotation is decelerating more quickly, it spends relatively little time at a resonance, and so suffers less destructive heat pulses from its interior. Even so, Earth apparently passed through a major resonance about 250 million years ago, when the length of its day was a few hours shorter. Some astronomers and paleontologists speculate that that event may have caused extensive volcanism and mass extinctions, though others believe they were caused by a large impact. It could have been even worse had Earth lacked a large moon.

WHAT ABOUT THE OTHERS?

Of course, our Solar System contains five other planets, or six if you count Pluto. We could spend the next several pages listing the differences between Earth and these other places. But Venus and Mars are as good as it gets. Mercury, scorching hot on its day side and with a rotation dragged almost to a halt by the Sun's gravity, is the only other terrestrial planet—the only other planet with a "stable observing platform." And as we move out into the Solar System, conditions for life only get worse. The more distant planets in the Solar System would also offer poorer vantage points than our Earthly platform. The outer planets are quite spread out, making it difficult to observe the other planets. And the inner planets would be largely lost against the bright glare of the Sun, making the first two tests of Einstein's General Theory of Relativity—the precession of the perihelion of Mercury and the bending of starlight—more difficult. The long orbital periods of the outer planets (decades to centuries) also require observations lasting more than a human lifetime.

In any case, most astronomers and astrobiologists don't deem these planets promising candidates for life. So since life is short and page print expensive, we won't beat a dead horse. Instead, let's consider the most popular alternative point of interest for life.

EUROPA: ANOTHER ABODE FOR LIFE?

Moving from the oven to the icebox, some astrobiologists are now focusing their attention on the second of the four Galilean Moons of Jupiter, Europa, which is about the size of Earth's Moon and may have an ocean.[15] The high-resolution images returned by the Galileo probe in the 1990s revealed a cracked, icy surface with evidence of recent melting and possible recent upwelling of fluids from beneath its surface. Based on the number of craters visible on the images, planetary geologists estimate the surface of Europa to be only about ten to twenty million years old. Thus, information on Europa's surface gets recycled on this timescale, which is about ten times faster than the recycling rate of Earth's sea floor. We can glean some information about the history of Europa's orbit from the pattern of cracks on its surface, but nothing like the data we can extract from Earth's ice and sea floor deposits. Nevertheless, it's interesting that in both cases (and for Mars) the key substance involved in information storage is water.

There has been considerable speculation about the possibility of life in Europa's ocean, much of it driven by the naïve assumption that the presence of liquid water virtually guarantees life, regardless of other environmental conditions.[16] We're skeptical. First, if there is an ocean under Europa's icy surface, it's probably about one hundred kilometers deep, twenty times deeper than the typical basin depth of Earth's oceans. The pressure at the bottom of Europa's ocean is about 2.5 times the typi-

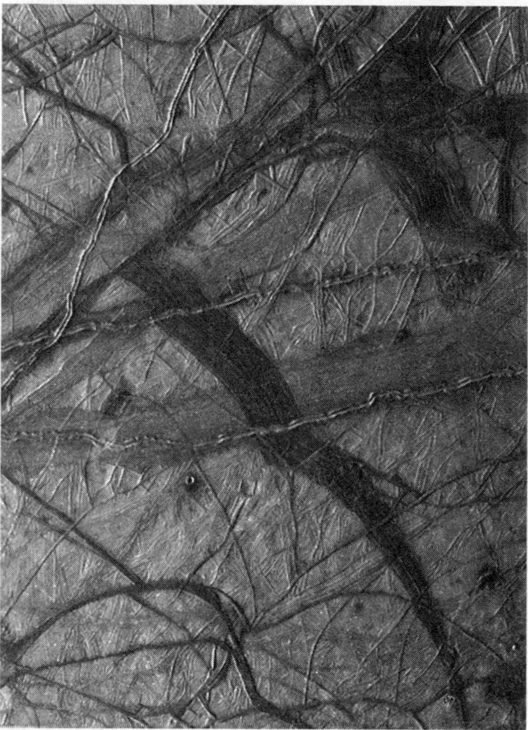

Figure 5.3: A 100 by 140 mile region of the icy surface of Jupiter's second Galilean moon, Europa. Many fractures and ridges are present throughout its surface. The relatively small number of impact craters visible on this moon implies a youthful surface, which recycles on a timescale of millions of years. The Galileo orbiter took this image in November 1996.

cal pressure in Earth's basins. Simple life can't tolerate an arbitrarily large pressure. Some microorganisms can grow at pressures up to one thousand times the pressure at sea level on Earth, but Europa's ocean bottom may exceed this limit by about 30 percent.[17]

Second, even if a hardy bug could survive this pressure, Europa has other problems. For instance, Europa's oceans have very little energy available for biological activity. Sunlight can't penetrate the thick ice. What liquid water exists results from tidal energy percolating up from the moon's interior. We don't know how the heat makes its way to the ocean, but it's probably very different from the vents on Earth. Earth's vents are near mid-ocean ridges, where ocean water circulates through freshly minted crustal rock, the result of plate tectonics. When mixed with the right minerals and gases, the heated water can provide chemical energy. On Europa, the vents are more likely to be tethered to old channels lacking such chemical richness. If a particularly robust organism managed to eke out an existence in Europa's ocean, it would probably be a lonely one, since the environment, with its paucity of land, air, and water interfaces, would prohibit a diverse collection of creatures. Even on Earth, vent systems are a minimal source of biological material; the surface waters produce about 100,000 times more biomass than all the vents combined. At most, Europa might support one thousand times less biomass than Earth, and probably far less than this.[18]

Third, Europa's oceans may be a planet-wide Dead Sea, too salty to sustain life.[19] On Earth, shallow seas evaporate near continents, keeping the oceans' salt content relatively low. After salt water enters a landlocked sea, its salt is deposited on the lake bottom as the water evaporates. Anyone who has visited a salt mine has seen firsthand how much salt is stored just beneath Earth's surface. A terrestrial planet without dry land can't store excess salts.

What's more, since its tidal heating is episodic, Europa slips in and out of orbital resonances with other nearby moons. So its ocean may sometimes freeze over nearly completely.[20] During such episodes its ocean grows much saltier (because the salt in liquid water is not incorporated in ice), killing off organisms that had been teetering on the edge of salinity tolerance.[21] So if you're hoping NASA will find the Europan equivalent of the lost city of Atlantis, or anything much more sophisticated than a few lonely and almost unimaginably robust microorganisms, don't hold your breath.

EARTH IN CONTEXT

This comparison of Earth with the other planets in our Solar System reveals its striking qualifications for most of the things we care about. It also illustrates the importance of trade-offs and thresholds for measurability. For instance, a planetary body without an atmosphere, such as the Moon or Mercury, obviously offers a clearer view of the universe from its surface than does Earth. A hypothetical Mercurian would have sharper, more constant views of the stars over a broader range of the electromagnetic spectrum. But our more restricted view of the distant universe has allowed surprisingly great advances in astrophysics. For example, some mountaintop observatories can correct for image motion caused by the turbulent atmosphere, producing images rivaling those of the Hubble Space Telescope. What's more, on clear, stable nights, astronomers can characterize the absorption properties of the atmosphere with enough accuracy to calculate the brightness of a star as if there were no atmosphere. This trick requires little more than measuring the changing brightness of a star as it makes its nightly trek across the sky; it would not be practical in a dusty atmosphere. Space-based observatories have helped mostly with exotic astronomical phenomena, such as those that do not emit much radiation in the optical range—X-ray and gamma ray bursts, for instance.

More important, without an atmosphere, we'd pay a heavy price for measuring other phenomena. We'd have no hydrological cycle, which is the basis for inorganic layered deposits and biological growth layers, and excessive impact tilling would degrade any fine sediments. Since Earth's atmosphere already meets the threshold of being fairly transparent to visible light, the benefits of lacking an atmosphere drastically outweigh its costs.

In contrast, for astronomers the difference between a relatively transparent atmosphere and a translucent or opaque one is like the difference between having one good eye and having no eyes. A gas giant planet like Jupiter or Saturn has a murky and dynamic atmosphere filled with opaque clouds. It also lacks a solid surface to provide a viewing platform and to impress and store information. The traces from the comet Shoemaker-Levy 9 fragment impacts on Jupiter in July 1994 are long gone; only our photographs remind us of that grand event.

Comparing Earth's geological activity with that of the Moon and other planets also helps put our planetary home in perspective. Little remains of Earth's Hadean period, between 4.5 and 3.8 billion years ago, given its active geology. (That might seem to count against our argument, but you'll see in the next chapter how it all works out.) At the same time, this geology allows geologists to map its interior effectively, while allowing them and everyone else to exist, as we discussed in Chapter Three. So it's hard to fault Earth for mucking up our Hadean period data. We can be grateful, moreover, that Earth has the atmosphere it has, and not just for the reasons already mentioned. Seismic waves are generated not only by subsurface quakes but also by surface impacts. When a body like the Moon lacks an atmosphere, excessively frequent surface impacts can become a source of noise for seismologists. Small impacts, which are far more numerous, generate weak seismic waves that are not very useful for probing the interior. They also partially mask the less frequent, larger impacts, which send seismic waves across the whole body.

Despite being spoiled by Earth's more accommodating seismic conditions, geologists have been able to overcome the Moon's frequent impact static and obtain enough seismic data to map its interior. It remains the only other internally mapped body, besides Earth, orbiting the Sun. (The Sun's internal structure has also been mapped, but not because we planted seismometers on it; see Chapter Seven.) By contrasting this data with Earth's, we get some idea of the likely seismic activity of the other planets and moons in our Solar System. Four of the six seismometers left on the Moon by Apollo astronauts recorded over twelve thousand seismic events between 1969 and 1977. Only eighty-one of these were good enough to reveal the Moon's internal structure.[22] Of these, thirty-four were deep quakes and fourteen were shallow ones; the remainder resulted from artificial and natural impacts. The biggest lunar quakes are very weak by our standards, less than about three on the Richter scale. Without water, the lunar rocks allow the seismic waves to propagate with relatively little attenuation, but its very heterogeneous and fractured crust scatters them, making it more difficult to interpret the data.

You may be surprised that there still are quakes in the Moon, since it's no longer volcanically active. These quakes probably result from the tidal deformation caused by Earth's gravity. In general, smaller planetary bodies will have fewer quakes and fewer impacts than larger ones. Thus, Mars should have fewer non-impact-related earthquakes than Earth but more than the Moon. Europa, with the severe tidal stressings of its crust caused by its proximity to the giant Jupiter, should generate lots of quakes, but they are less useful. Seismic shear waves can't travel through its submerged ocean, and the ice might generate too much seismic noise, though we can't say for sure until we land a craft there.

Finally, there is Earth's hospitable orbit and tilt. (See Plate 11.) As we described in Chapter Two, even slight changes in these two variables affect a planet's climate. If either of these were only slightly greater, every part of Earth's surface would vary more in temperature, and would very probably be less habitable.[23] In fact, Earth would be likely to have a qualitatively different biosphere, since the biodiversity of a region depends on the productivity of its biological base. With a region experiencing larger swings in temperature over the course of a year, fewer species would survive. And the weaker the biosphere, the smaller its role in stabilizing the climate. All told, Earth might support only an anemic microbial community, and nothing more.[24] This isn't merely academic. Since Earth's axial tilt and the eccentricity of its orbit vary nearly as little as they can, both factors are probably larger on most other terrestrial planets.

CHAOTIC WANDERERS

Earth's orbit also enhances scientific observation—for Earthlings, that is. Earth's circular orbit allowed the ancients to use the Sun's apparent motions on the sky to mark the passage of time. A less circular orbit would have hindered the ancients from modeling the movements of the planets—a crucial rung on the ladder for developing the theory of General Relativity (see Chapter Six). Today, "paleoastronomy" benefits from the regular orbits of Earth, the Moon, and the giant planets, which impress their periodic patterns on Earth's layered deposits. Because a stable system repeats with the same period for many cycles, researchers can apply what are called Fourier analysis techniques to separate the chaff from the wheat, or in this case, the background noise from the repetitive pattern.[25] The regularities of the orbits also allow astronomers calculate the orbits of the Earth-Moon system and other planets back in time several million years.

Not all bodies in the Solar System move so regularly. The orbit of Pluto, the orbits of main belt asteroids near Jupiter resonances, the axial tilt of Mars, and the spin axis orientation of Saturn's irregular moon, Hyperion, are all strongly chaotic.[26] In a chaotic system, future motions depend very sensitively on initial conditions, like the proverbial flap of a butterfly's wings in China that leads to a thunderstorm in Wyoming. The classic laboratory demonstration of this phenomenon is a double pendulum, which has a joint in the middle of an otherwise stiff rod. No matter how carefully you try to start the pendulum in the same way in each trial, it very quickly develops different trajectories. Even a simple pendulum without a joint grows chaotic when large amplitude swings send the ball over the top.

Gravitationally interacting systems show some degree of chaos if they are more complex than a simple two-body system. Thus, a single planet orbiting a star acts like a simple pendulum, while several closely spaced planets are more like a double pendulum—they will tend to experience large and lethally rapid changes.[27]

It was only in the late 1980s that astronomers discovered that our Solar System is chaotic.[28] The orbits of the giant planets are the least chaotic, followed by Earth and Venus, while Mars, Mercury, and Pluto are the most chaotic. Generally, the less massive planets tend to have more chaotic orbits.[29] The massive planets are like big bullies who push around the smaller kids in the neighborhood. Thus, a planet needs to be a certain mass to maintain a long-term stable orbit and, by extension, climate. As noted, the orbits of Mercury and Mars are less stable than those of the more massive Earth and Venus. Although Mercury is not expected to collide with Venus or escape the Solar System before the Sun becomes a red giant and swallows it, simulations indicate that its orbit can sometimes be quite elongated.[30] Its more chaotic orbit and tilt would severely limit the information one could glean from any layering processes. What's more, because its orbit fluctuates wildly, so would its climate, so it probably wouldn't preserve layered deposits very well either.

Encouraged by Earth's fairly stable climate and information-rich layered deposits, scientists who study Earth's ancient climate, called paleoclimatologists,[31] are beginning to search in deep marine cores for indications of subtle changes in the orbits of the other planets and to extrapolate Earth's orbit back beyond thirty million years.[32] We can study the chaotic history of the orbits of the other planets only because we live on a stable platform. As a result, Earth's records may one day give us a unique look into the precise dynamical history of the Solar System going back tens of millions of years.

DISTANT WANDERERS

While we have gained unprecedented knowledge of our planetary neighbors in the last few decades, arguably the most exciting recent development in astronomy has been the discovery of planets orbiting nearby Sun-like stars.[33] As of July 1, 2002, one hundred giant planets have been found with the Doppler detection method, which makes use of the "wobble" of a star as a planet orbits around it. Small changes in the Doppler shift of a star's optical spectrum can reveal the presence of an otherwise invisible planet. Present technology permits the detection of only planets more massive than Uranus or Neptune, though planned space missions, such as NASA's Kepler mission, promise to detect Earth-size planets.[34]

Because it contradicted expectations, the first extrasolar planet discovered around a Sun-like star caught many astronomers off guard.[35] Additional discoveries in the last few years have been equally unconventional. The systems differ from ours in two significant ways. First, some of the

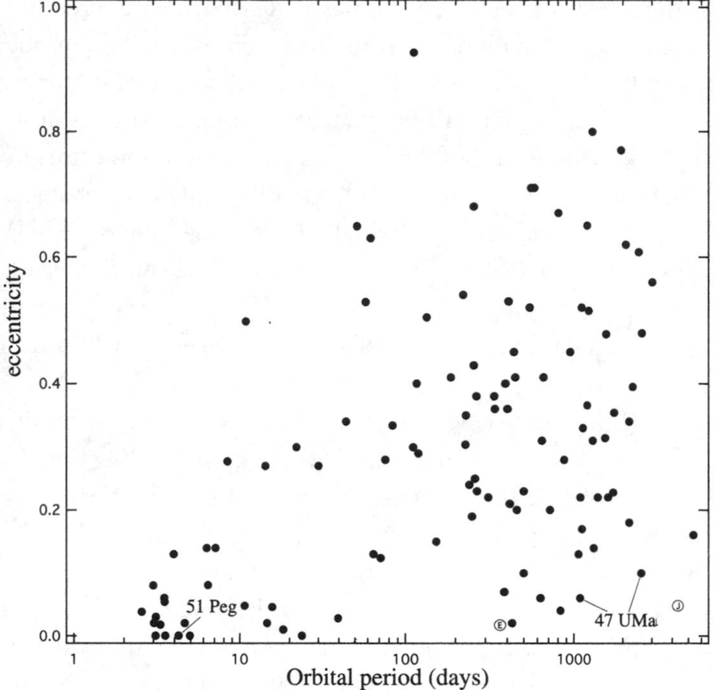

Figure 5.4: Orbital eccentricities and periods of all extrasolar planets discovered with the Doppler method as of June 15, 2003. Jupiter (J) and Earth (E) combine low eccentricities with relatively long orbital periods. (0.0 on the vertical axis is a perfect circle.) Only planets such as 51 Peg B with extremely short periods, so-called "hot Jupiters," have smaller eccentricities than Jupiter or Earth. (Orbital period is plotted on a logarithmic scale here.) Planetary systems with such hot Jupiters are probably hostile to life, as are systems with giant planets in highly eccentric orbits.

giant planets orbit very close to their host stars (so-called "hot Jupiters"), with periods as short as three days. Second, those with orbital periods greater than two to three weeks generally have highly elongated orbits, changing drastically in distance from their host stars; only seven giant planets with periods longer than one month have eccentricities less than 0.1. (A perfectly circular orbit has an eccentricity of 0, and a parabolic orbit, which is open at one end, has an eccentricity of 1.) In contrast, Jupiter's eccentricity is 0.05.[36]

To date, no true Jupiter twin—with a nearly circular orbit and an orbital period of about twelve years—has been found around a Sun-like star. The closest match is the third planet from the star 55 Cancri (55 Cnc), with an orbital period near fourteen years, a mass at least 4.3 times greater than Jupiter, and an eccentricity of 0.16. So apart from its orbital period, this planet bears little resemblance to Jupiter. The second planet in the system has an even more elongated orbit and takes a month to orbit its host star. Interestingly, 55 Cnc is one of the most heavy element–rich stars in the solar neighborhood, so it is hardly Sun-like.[37] Of the planetary systems discovered to date, the one around the single star 47 Ursa Majoris (47 UMa), about forty-five light-years from Earth, most resembles our Solar System. It contains two giant planets in fairly circular orbits with orbital periods near three and seven years; the outer planet was discovered about five years after the first.

Not surprisingly, the 47 UMa system has been touted as confirming that the Solar System is not "special" and, hence, as corroborating the Copernican Principle. While this discovery shows that the Solar System isn't unique in having giant planets in multiyear, nearly circular orbits, 47 UMa is the only system yet found that even slightly resembles our own. This suggests our planetary system may *not* be typical. Indeed, even this one may be lethally unlike ours. If Jupiter were as close to the Sun as the inner planet of the 47 UMa system is to its star, Earth would have our Solar System's big bully a little too close for comfort.[38]

Moreover, the fact that 47 UMa contains two planets with fairly circular orbits does not mean it is a habitable system. We do not yet know what else lurks there. We may later detect other planets in more eccentric orbits or too close to the Circumstellar Habitable Zone (we'll discuss this in Chapter Seven). Our Solar System has a striking regularity, with eight planets in low-eccentricity orbits. (Excluding Pluto, which is better classified as a Kuiper Belt Object, the average eccentricity of the planets in the Solar System is only 0.06.)

There's another interesting fact about the 47 UMa system—the host star is like the Sun. It joins 16 Cygni (Cyg) B and HD 222582 as the most Sun-like stars with planets, at least with regard to their mass and composition.[39] This doesn't mean that all solar analogs have giant planets in Jupiter-like orbits. Both 16 Cyg B and HD 222582 have planets in highly eccentric orbits, and other solar analogs have so far shown no evidence of planets. Thus, although it's still early in the game, it's beginning to look as if a star must be single and have very nearly the same composition and mass as the Sun merely to have a decent chance of being surrounded by Jupiter-like planets.

Astronomers have offered a number of theories to explain both the hot-Jupiter systems and the larger systems with elongated planetary orbits. The most popular explanation for the hot-Jupiter systems is that these planets migrated inward from a more outlying orbit. Several possible mechanisms could account for this.[40] But whichever is correct, they all hinder the formation of habitable terrestrial planets. A giant planet will scatter away any terrestrial planets in the Circumstellar Habitable Zone as it passes by,[41] and giant planets that perturb each other into more elongated orbits are better at perturbing smaller planets and generally wreaking havoc in their Solar System.[42]

Whether most other planetary systems are as friendly to life and discovery as our Solar System we can't yet say. If high eccentricities of giant-planet orbits continue to predominate as more systems are discovered, then it will become clear that systems like ours with many planets in stable circular orbits are rare, perhaps exceedingly so.

Moons as Alternative Habitats

Encouraged by all the new giant-planet discoveries, some astrobiologists have suggested large moons orbiting gas giant planets as alternative habitats to an Earth-like planet.[43] But such environments probably are much less life-friendly than our home world. First, comet collisions and breakups will be more frequent in the vicinity of a giant planet, because of its strong gravity.[44] (See Plate 12.) The capture and subsequent breakup of comet Shoemaker-Levy 9 in July 1994 illustrated this point, as did the subsequent discovery of impact crater chains on Jupiter's Callisto and Ganymede—116 craters in eleven crater chains, laid out like gigantic pearl necklaces.[45]

Second, compared with Earth, the impact velocities will be greater for a moon around a giant planet because of the stronger gravity of the giant planet and the orbital velocity of the moon around it.

Third, the particle radiation levels are higher near a gas giant with a strong magnetic field like Jupiter (although smaller gas giants, like Saturn, have weaker fields). As a result, a moon without a strong magnetic field[46] would more quickly lose its atmosphere, expose its surface to more radiation, and threaten and probably extinguish its surface life.[47]

Fourth, a moon's rotation around its axis will be synchronized with its orbit around its host planet in fairly short order causing one side to always face the host planet. This synchronized orbit, along with the slower rotation of a moon around a giant planet, would lead to greater temperature

Figure 5.5: Comet Shoemaker-Levy 9 in May 1993 after it had already started to break up following close passage to Jupiter in mid-1992. It eventually collided with Jupiter in July 1994.

swings between day and night. For environments similar to the Galilean Moon system, orbits in resonance can be beneficial, since they generate internal heat through tidal stressing.[48] But orbital changes on timescales of one hundred million to one billion years prevent tides from being continuously available as a source of heat.[49] If Europa's ocean periodically freezes over during lulls in tidal heating, for instance, it would probably devastate any complex life there.

These problems are somewhat mitigated if a moon orbits far from its host planet, but this situation would lead to greater fluctuation in the amount of sunlight it receives from the parent star. A moon's effective orbit about its host star would be much more elliptical. The present eccentricity of Earth's orbit is 0.017, only one-third that of Jupiter.[50] A Jupiter-like eccentricity would create further problems for life on the giant planet's moon. Moreover, it's not even clear that a giant planet can end up between 0.5 and five Astronomical Units from its host star in an orbit as nearly circular as Earth's. (Earth is one Astronomical Unit, or AU, from the Sun.)

Even if a giant planet can migrate inward and park itself in its star's habitable zone in a circular orbit, there's no guarantee its moons will come along for the ride. As a giant planet migrates inward, the gravity from its host star will compete ever more for its moons. Eventually, the planet may lose its moons to its greedy host star. Recent simulations indicate that the known hot-Jupiters wouldn't have retained any sizable moons as they

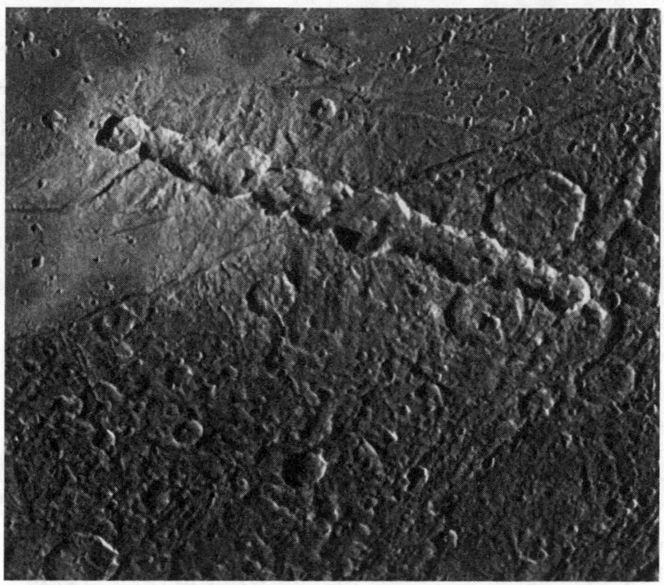

Figure 5.6: One of eleven known impact crater chains on the surfaces of the Galilean moons, Ganymede and Callisto. A comet probably created this chain on Ganymede after it was torn into pieces by Jupiter's gravity as it passed too close to the planet. Such crater chains are a stark illustration of the dangers to life on moons around giant planets.

migrated to their present locations.[51] How far a giant planet migrates before it loses a moon depends on the mass of the planet, the mass of the moon, and the initial direction of the moon's orbit; massive moons in large prograde orbits tend to be lost more easily. Tidal interaction might even draw some moons into their host planets. Regardless of the host planet's migration, moons will wander closer to or farther from their host planets depending on whether their orbital periods are shorter or longer than the planet's rotation period, respectively. Other moons in a system can complicate the migration if they are in resonance (such as Io and Europa around Jupiter).

Finally, such a moon, even if it survives, will probably have a quite different composition from that of Earth. Our Solar System, and presumably others, formed out of a protoplanetary nebula—a flattened disk of gas and dust. After hydrogen and helium, water vapor was probably the most abundant gas. The hotter interior of the nebula prevented some gases, like water vapor, from condensing and becoming part of forming planetary embryos. But planets outside this "water-condensation boundary" could incorporate large amounts of water. (In our early Solar System, that boundary was just inside the orbit of Jupiter.) That's why most of the outer moons in our Solar System are so rich in ice. So even if a system with a large moon orbiting an outer gas giant migrates inward toward its star's habitable zone, its waterlogged moon will probably have too large an inventory of volatiles, such as water and carbon dioxide, to have much hope for hosting life. If the giant planet can form *in situ* Sun-ward of the water-condensation boundary, then its moons will have the opposite problem. They will need a source of volatiles to be habitable. Capture by the giant planet is also possible, but all the examples of likely captured moons in our Solar System, with the exception of Triton around Neptune, are merely small, glorified asteroids.[52]

So although some giant planet-moon systems may avoid one of these deleterious factors, taken together, they show that an Earth-size moon around a giant planet offers a far poorer environment for even simple life than does an environment resembling the Earth-Moon system.

If life somehow managed to take hold and even thrive on such a forbidding moon orbiting a gas giant planet, the moon would offer a poorer overall setting for discovery. Observers on the hemisphere facing the planet would experience a dark sky only briefly each month as it passed into the planet's shadow. (Even this might be rare if the moon's orbit were highly inclined to the giant planet's orbit around the host star.) They would see their host planet going through a complete set of phases and several rotations each month while it seemingly floated motionless in their sky. Observers in the opposite hemisphere would never see their giant planet host, but they would get more dark nights. Observers on both hemispheres would also have to put up with the reflected light from other moons in the system. And as we discuss in the next chapter, the motions of the stars and other planets in the system would appear more complex. As hard as it was for Earth's inhabitants to figure out the true motions of their fairly simple platform, our moon dwellers would probably have an even more difficult task.[53]

Once they figured out the true geometry of their home world, however, moon dwellers would have two advantages to Earthlings. Their orbit around the host planet would serve as a baseline to measure parallaxes to other bodies in their planetary system (we'll explain this a little later). Second, their host planet-moon system would serve as a mini-analog of their star-planet system. We had to satisfy ourselves with measuring parallaxes to nearby asteroids and terrestrial planets using Earth's surface as the baseline. Nevertheless, the measurements from a moon wouldn't be straightforward, since the targets would be moving while their home world orbits about their host planet, and the host planet would be moving about its host star at the same time. To succeed, inhabitants would first have to understand the far more complex geometry of their system. They would have to overcome far greater geometric hurdles to exploit their advantages, and then one of those would be undercut by further calculation hurdles. So for scientific discovery, the costs of living on a moon would outweigh the benefits.

AT THE HEAD OF THE PACK

This brief tour of our planetary neighbors and their extrasolar cousins allows us to put the Earth-Moon system into proper perspective. Earth's long-lasting hydrological cycle, plate tectonics, oscillating magnetic field, continents, stable orbit, and transparent atmosphere together provide the best overall "laboratory bench" in the Solar System. Earth's surface strikes a balance between the permanence required to preserve patterns written on it and the dynamic yet gentle circulation that subtly sways these "recorder pens" without tearing up its paper-thin crust. The crust records and stores information while maintaining the most habitable environment for complex life in the Solar System. Continents amid oceans of water, enabled by plate tectonics—as Earth enjoys exclusively—seem to be the best overall habitat for observers. No other locations yet discovered hold a candle to this one blue dot, however pale it may appear to some. Of course, it doesn't follow that our Solar System has no role to play for life and discovery. On the contrary. But that's another story.

CHAPTER 6

OUR HELPFUL NEIGHBORS

We can . . . be thankful that the Solar System in which we live has been unreasonably kind throughout the long history of human efforts to understand its dynamics and to extend that knowledge to the rest of the universe. At each step along the way, it has served as a perspicacious teacher, posing questions just difficult enough to prompt new observations and calculations that have led to fresh insights, but not so difficult that any further study becomes mired in a morass of confusing detail.

—Ivars Peterson[1]

THE PLANETARY PLAYPEN

For the ancients, the planets were a fickle lot, seemingly drifting slowly and reliably across the sky, only to reverse course in an apparent attempt to defy expectations. It's easy to forget that before they referred to planets, the names "Mercury," "Venus," "Mars," and so on were the Latin names of the capricious Olympian gods of Greek and Roman mythology: Mercury, the messenger; Venus, the goddess of love and beauty; Mars, the god of war; and Jupiter, the thundering king of them all.

The planets have played an even more enduring role in astrology, in which they represent a person's various "energies," such as the soul, will, and mind: hooks on which hang our individual narratives, hopes, and destinies. While such flights of fancy are easy to dismiss, perhaps our race shares some real, if unarticulated and even misguided, intuition that these strange objects play a central role in our existence. Only recently, however, has that intuition found a scientific justification. Although the other wandering bodies in our Solar System aren't encouraging sites for complex life, they *have* contributed profoundly not only to Earth's own habitability but to the emergence of science, and therefore scientific discovery, as well.

The mere presence of the Moon and other planets in the Solar System fostered the development of celestial mechanics and modern cosmology.[2] Danish astronomer Tycho Brahe's (1546–1601) observations of the paths of the planets against the background stars allowed Johannes Kepler (1571–1630) to formulate his three famous laws of planetary motion. Kepler's discovery of his Third Law (that the square of the orbital period of a planet is proportional to the cube of its mean distance from the Sun) required several visible planets. It also helped that the planets span a large range of distance from the Sun. And Kepler's discovery of his First Law—that the orbits are not circles but ellipses—required that at least one such planet have a discernibly eccentric orbit. For Kepler, Mars served this purpose.[3] If, like many of the extra Solar Systems discussed in the previous chapter, Earth had lacked "wandering neighbors" (the word "planet," by the way, derives from the Greek word meaning "wanderer"), we might never have discovered these laws.[4]

Kepler's three empirical laws served as the foundation of Isaac Newton's more general physical laws of motion and gravity, which became the foundation for Einstein's General Theory of Relativity two centuries later. The planets may have inspired Kepler, but the Moon inspired Newton to apply his Earthly laws to the broader universe. Without the Earth-centered motion of the Moon, the conceptual leap from falling bodies on Earth's surface to the motions of the Sun-centered planets would have been much more difficult. By linking the motions of the Moon and planets to experiments on Earth's surface, Newton gave a physical basis to Kepler's Third Law. Otherwise, the Third Law would have remained a mathematical curiosity, more an indication of the cleverness of a mathematician with too much time on his hands than of a deep truth about the universe.[5] As it is, astronomy gave birth to physics.

The most habitable locale we know of happens to be near a star with several other planets whose orbital periods are substantially shorter than a human life span. Not only is a free-floating planet in interstellar space (or even in an open cluster) a poor home for complex life, it doesn't even provide the opportunity to discover these universal laws. Even geniuses like Kepler and Newton needed a planetary playpen to discover the laws of motion and gravity and to realize that they apply throughout the cosmos. And since General Relativity now forms the basis of cosmological models, once astronomers understood the motions of the planets, they were well on their way to understanding the structure and history of the universe.

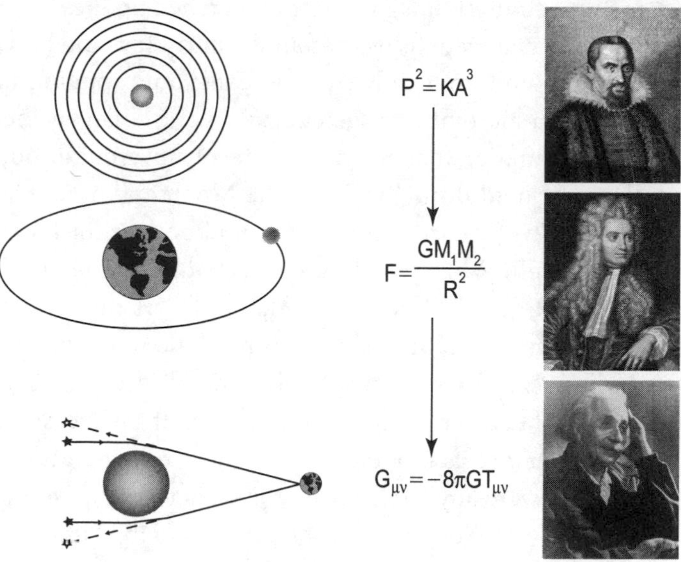

Figure 6.1: The particular configuration of the Solar System was important in each step of the development of gravitation theories. Kepler's Third Law of planetary motion was based on Brahe's naked-eye observations. Newton's Law of Gravitation was based, in part, on the orbit of the Moon about Earth and Kepler's Third Law. Einstein's General Theory of Relativity was built on Newton's Law and tested first with total solar eclipses.

Ivars Peterson, at the conclusion to his work *Newton's Clock: Chaos in the Solar System*, also notices this remarkable coincidence (while discussing dynamical chaos):

> A deep-seated puzzle lies at the heart of this newly discovered uncertainty in our knowledge of the Solar System. Was it an accident of celestial mechanics that the Solar System happens to be simple enough to have permitted the formulation of Kepler's laws and to ensure predictability on a human time scale? Or could we have evolved and pondered the skies only in a Solar System afflicted with a mild case of chaos? Are we special, or were we specially fortunate?[6]

For Kepler to formulate his laws, it helped that neither the orientation of Earth's axis nor the orbits of the planets exhibit observable chaos over a human lifetime. For contrast, imagine yourself perched on the surface of Hyperion as it tumbles along in its orbit around Saturn. It would be as disorienting as sitting on an erratically swirling chair on a rotating Ferris wheel on the edge of a spinning amusement park, while believing the reasonable but false assumption that everything is revolving around you.

Incidentally, you might think that since the Moon and other planets enhance measurability on Earth, the same would be true for the other planetary bodies as well. What's good for Earth should be good for Mercury, right? Well, not quite. Mercury, which completes three rotations every two orbits; Venus, which has a slight mismatch between its year and its day (never mind its nearly opaque clouds); and all the moons of the Solar System would offer more confusing vistas.[7] Even if a planet's dynamics aren't strongly chaotic, its motions might still be too complex for its inhabitants—if there were any—to discover the planetary laws with ease.[8] The length of Earth's year is quite different from the length of its day, making it easier to separate the effects of revolution and rotation.[9]

Moons would offer even more complicated views. As astronomers sometimes say, Earth is only one motion removed from the Sun. The Moon is two motions removed from the Sun, since it both revolves around Earth and around the Sun. Adding another nested layer of motion to an observer's platform would make the discovery and reconstruction of the true geometry of the orbits tortuously difficult. Despite the simplicity of our Earthly platform, it still took over a thousand years—from Aristotle and Ptolemy to Copernicus, Brahe, Kepler, and Galileo—for the human race to figure out the true geometry of the Solar System.

THE MOON: EARTH'S COMPANION

This insight would have taken even longer without a moon. Apart from the Sun, the Moon is the only body in the Solar System that is more than a nondescript point of light, one that an observer on Earth's surface can resolve without optical aid. (Apart from surface details, color, and size, Earth would look similar from the same distance; going through the same phases while apparently floating in space.) The Moon provides a conceptual bridge, allowing us to imagine Earth as a similar wandering planetary sphere. Since Earth's rotation is not yet locked with the Moon's orbit (as it is for the Pluto-Charon system), everyone gets to see the Moon revolving around Earth. Otherwise, the Moon would be visible from only one hemisphere, where it would appear motionless in the sky.

Still worse, consider observers on a planet without a moon. To associate their home planet with the other planets in their system, they would have to make a huge mental leap. The surface of a planet, up close, looks quite different from one tens of millions of miles away, which looks like a star to the naked eye. Certainly, the invention of the telescope helped wean astronomers from Aristotle and Ptolemy. But a telescope wasn't enough; we also needed an atmosphere that allows resolution substantially better than one minute of arc. (Venus reaches nearly this angular size on its closest approach to Earth.)

As we noted in Chapter One, the Moon also makes Earth friendlier to life. Studies of the Solar System have demonstrated that the Moon stabilizes Earth's axial tilt.[10] It now varies by a mere 2.5 degrees. Such a small variation produces, over thousands of years, the mild seasonal temperature changes we now enjoy. Temperature swings would be far greater without the Moon.[11] Climate fluctuations would be larger and thus easier to measure. But the larger climate swings would be more likely to destroy layers deposited in previous seasons. One pole could trade a large volume of ice with the other each half-year, without ever accumulating more than a year's worth of information. And the large variations in the tilt over a few thousand years would eliminate any long-lived ice deposits that might have accumulated near the equator. To be measurable, climate variations must be large enough in the climate record but not so large that they destroy it.

So the Moon's stabilizing influence preserves the polar ice, that great cache of historical information, while the extra-astronomical cycles it provides help climatologists calibrate their ice and marine sediment cores. Finally, the higher tides resulting from the Moon's gravity increase the land area subjected to periodic flooding. Without the Moon's effect on the tides, we would have fewer fossil tidalites today from which to reconstruct the history of Earth's rotation. On the other hand, if the Moon were much larger and induced stronger tides, it would have slowed Earth's rotation more quickly, perhaps synchronizing them by now. Earth's day would be the same length as its month, leading to large day-night temperature disparities and perhaps endangering the preservation of ice deposits. The Moon would loom larger than the Sun, and the overall tides would be weaker, since only the Sun would induce changing tides on Earth. Greater stability, then, doesn't simply and automatically make a place more habitable and measurable.

The relationship between Earth and its Moon is so intimate that it's probably best not to think of Earth as a lone planet, but as the habitable member of the Earth-Moon system. This partnership not only makes our existence possible, it also provides us with scientific knowledge we might otherwise lack.

THE HISTORY OF THE SOLAR SYSTEM

Earth's Moon is about the same size as Europa, but it's as bone-dry as Europa is wet.[12] Its crater-scarred surface preserves the traumas it has suffered over most of its history. Its face appears ancient, yet ageless. While far more craters should have formed on Earth over the same period, given Earth's larger surface area and stronger gravity, its active recycling has erased almost all the older ones.[13] This would seem to count against the correlation of habitability and measurability, except that the cratering record on the Moon is plainly visible from Earth's surface. As our nearest planetary neighbor, the Moon was the first to be mapped. We don't need to leave Earth to produce detailed maps of the Moon's near side, as anyone with a small telescope knows. We even have some lunar meteorites, though we might not have been able to identify their place of origin had we not already had lunar samples brought back by the Apollo astronauts. Nevertheless, since they come from all over the lunar surface, these meteorites actually sample the Moon better than the Apollo missions did.

The pattern of craters on the Moon allows us to reconstruct its cratering history—an important source of information on the early history of the inner Solar System.[14] (At the same time, the Moon has no atmosphere to burn up incoming meteoroids. Consequently, meteorites till[15] the upper surface so frequently and vigorously that they destroy older and more subtle structures, rendering the Moon generally less information-friendly than the gentle depositional processes we see on Earth's surface.[16]) Most of the Moon's visible maria, or large dark areas, were formed between 3.8 and 3.9 billion years ago, during the so-called "late heavy bombardment." The impact rate was high enough to erase most of the lunar surface prior to this period. On the positive side, since the Moon is Earth's close neighbor, the rich cratering record on its surface can be easily observed and translated to reconstruct Earth's cratering history, so Earth's active crustal recycling is not as problematic as it might seem.

Figure 6.2: Three-day-old Moon. The sunlit regions have been overexposed to reveal the detail in the regions illuminated by Earthshine. Astronomers observe Earthshine to monitor changes in Earth's albedo and to learn what Earth's spectrum would look like to a distant observer—and to learn what a distant Earth would look like to us.

Tilling notwithstanding, there's a unique treasure buried just beneath the Moon's surface. A large impact on one of the inner planets can blast a lot of stuff into space, and its neighboring planets and the Moon sweep up most of it. Therefore, fragments of Mercury, Venus, Earth, and Mars should be lying preserved on the Moon, most dating from the intense late heavy bombardment. Particularly intriguing is the likelihood that Earthly meteorites buried in the lunar regolith contain the remains of early life from 3.8 billion years ago. The Moon's lack of an atmosphere and its central location in the inner Solar System make it an ideal collector of planetary detritus.[17] Thus, the Moon serves as Earth's "attic," where relics from the early history of the inner Solar System are stored and preserved, waiting patiently for someone to climb up there and collect them.

Mirror, Mirror . . .

The Moon's present stable surface has also taught astronomers about real time changes in our global climate. As the Moon orbits Earth, it goes through its familiar phases. The more careful observers among us have noticed that the unlit part of the crescent Moon's face is not completely dark. This is called earthshine, and as the name implies, it is caused by sunlight reflecting off Earth and illuminating the Moon (and then returning back to your eyes). The Moon, in effect, acts like a huge mirror in space that allows us to see our reflection. A moon with an atmosphere, incidentally, would be far less useful for measuring our albedo (as it would be for study of the Sun's chromosphere during a solar eclipse).

We can't discern any details about Earth's surface with this method, but we can derive a surprisingly accurate measure of its albedo.[18] It is comparable to the best estimates derived from much more costly artificial satellite observations. Long-term measurements of Earth's reflectivity from space also require that instruments remain stable for several decades—no easy task. In contrast, the Moon's surface is a very stable mirror, which allows reliable measurements over several centuries. Because of our lunar mirror, we can reliably measure a global quantity, Earth's albedo, from a single observatory on the ground. Knowing the average albedo and its change over time is critical to understanding our climate. It also enables us to obtain a spectrum of Earth as it would look to a distant observer, and this contributes to the search for other Earths.[19]

A Lunar Telescope

The Moon also helps astronomers study more distant phenomena. As the Moon moves across the sky, astronomers can measure the angular sizes of stars and discover new binary stars by timing how long it takes to cover, or occult, them. Thanks to its large, angular size, the Moon occults many stars along its path. In this way, the Earth-Moon system acts like a giant telescope, allowing astronomers to resolve objects normally too small or close together to measure from the ground. A slow angular speed of a moon across its host planet's sky, like our own, allows for more detailed measurements. This method works best with a large moon without an atmosphere—which produces a crisp, knife-sharp edge on its limb—orbiting far from its host planet (but not too far, because the smaller a moon appears, the fewer stars it occults over a month).

Earth and its Moon together function as one system—a double planet, as it were. Geologically, they are near perfect opposites. The Moon lacks water, active geology, and an atmosphere, while Earth possesses all these in great abundance. This complementary relationship allows for far deeper and more comprehensive discoveries than would either set of conditions alone.

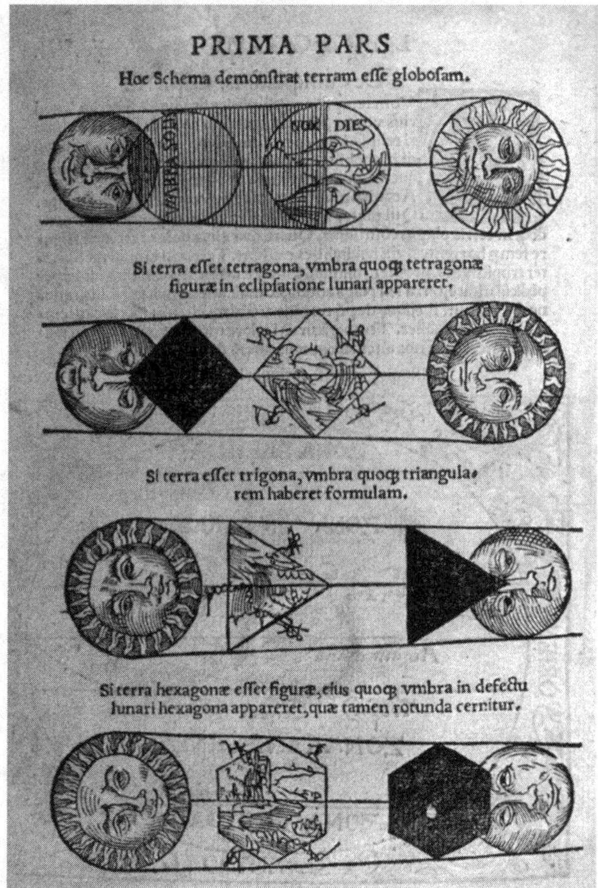

Figure 6.3: In this diagram from a popular sixteenth-century astronomy textbook (1540), Peter Apian illustrates an argument known since Aristotle: the shape of Earth's shadow on the Moon shows that Earth is a sphere. Apian compares the shadow produced by a spherical Earth with the shadows that would be cast if Earth were a square, a triangle, or a hexagon.

THE LOCAL DISTANCE LADDER

But merely discerning the laws of motion and gravity will not unlock the mysteries of the universe. Distances also figure very prominently. Historically, the Moon served as an important stepping-stone in establishing the size scale of the Solar System.[20] The story begins with Aristotle and the Athenians, who already knew of Earth's spherical shape from observing its curved shadow on the Moon during lunar eclipses.[21] (See Plate 13.)

Ascending the first rung in the so-called "cosmological distance ladder" requires understanding the shape and size of Earth.[22]

Around 200 B.C., Eratosthenes of Cyrene calculated Earth's size by measuring the Sun's position from the zenith—that is, directly overhead—at two widely separated cities along the Nile, which is helpfully aligned north-south. He noted that on a certain day of the year, the Sun shone directly into the bottom of a deep well in Syene, in southern Egypt, while it was seven degrees south of the zenith in Alexandria, in northern Egypt. Knowing the distance between the two cities, and assuming a spherical Earth and parallel light rays from the Sun, Eratosthenes estimated Earth's circumference at about 29,000 miles, only about 16 percent over the modern value of 25,000 miles.

The next step in the distance ladder is the distance to the Moon. The earliest attempt was by Hipparchus of Nicaea, who used observations of the total solar eclipse of March 14, 189 B.C.[23] He noted that the Sun was totally eclipsed in Hellespont, while it was only 80 percent obscured in Alexandria. This apparent difference is a parallax effect; the much closer Moon appears to shift much more than the Sun. You can see the same effect by holding your thumb about six inches in front of your face and alternately closing one eye at a time. Your thumb will seem to jump against the background. This is because the angle of view changes from one eye to another. The closer your thumb, the greater the apparent movement. To determine the distance to the Moon, one only needs to know how much the Hellespont/Moon/Sun angle differs from the Alexandria/Moon/Sun angle, and the linear distance between the two observers. Hipparchus estimated a lunar distance of about seventy-five Earth radii from us, not far from the modern value of sixty Earth radii.[24]

In the early third century B.C., Aristarchus of Samos devised a geometrical method to measure the distances to the Moon and the Sun.[25] He argued that Earth, the Moon, and the Sun formed a right triangle, with the Moon at the right-angle position when the phase of the Moon is in its first or third quarter. In principle, one can estimate the Sun's distance from Earth by measuring the angle between the Sun and the Moon during one of these two lunar phases. It's best to try this in the daytime, when both the Sun and Moon are high in the sky.[26] With this method, Aristarchus estimated the Sun to be about twenty times as far as the Moon; today, we know the value is 390. So Aristarchus's method can't yield the true distance to the Sun, only a lower limit. An observer would have to measure the angle between the Sun and Moon to within eight minutes of arc to get a useful distance.[27]

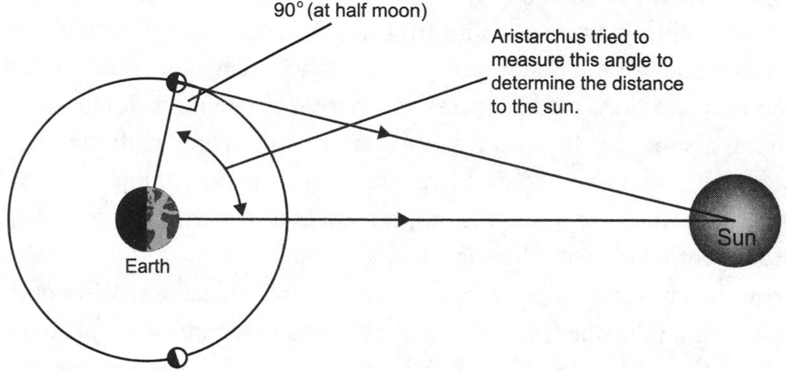

Figure 6.4: Aristarchus's method for estimating the distance to the Sun requires measuring the angle between the Sun and the first-quarter (or third-quarter) Moon. Using this method the Greeks determined that the Sun is much larger than Earth. Relative sizes and separations not drawn to scale.

Nevertheless, Aristarchus's result was good enough to convince him of the Sun's much larger size, and to support his heliocentric assumptions about its position and motion relative to Earth and the Moon. Combining the distances to the Sun and Moon with the size of Earth determined by Eratosthenes and Aristarchus's estimate of the relative sizes of Earth and Moon gleaned from lunar eclipses and the parallax of the Moon, the ancient Greeks were able to estimate the absolute sizes and separations of all three bodies. Some of their estimates were far off the mark by today's standards. Still, using only naked eye observations, they put Earth in the proper perspective compared with the two major lights in the sky. Aristarchus's method of measuring the relative distances to the Moon and Sun would be completely impractical from any other moon-bearing planet in the Solar System; the angles formed between a moon during first or third quarter and the Sun would differ from ninety degrees even less than they do for an Earth-bound observer.

Earth again provided a useful baseline in the nineteenth century, when astronomers attempted to establish the absolute size scale of the Solar System. Since Kepler's Third Law had given astronomers the relative distances of the planets, expressed in terms of the Earth-Sun distance—the Astronomical Unit, or AU—all they needed was a precise distance from Earth to one other body in orbit around the Sun. Aristarchus had given us only a crude distance to the Sun. The great hope among astronomers of the seventeenth to twentieth centuries was to measure accurately the parallax of a body in the Solar System other than the Moon. Parallax measurements are always made with respect to some much more distant object, which is assumed to have a negligible parallax compared with the nearer object. It is very difficult to measure directly the parallax of the Sun, given its brightness and size. A smaller, fainter object is much better, since it can be easily compared with the background stars.

Fortunately, the Solar System offers lots of potential targets. Astronomers looking to measure the scale of the Solar System first trained their telescopes on Mars, given its close approach to Earth every couple of years near opposition. The earliest serious attempts, with widely separated observers, began in the seventeenth century. Observers took advantage of those times when Mars passed especially close to bright stars. The most reliable pre-modern measurement of its parallax was achieved in 1877. But Mars's atmosphere and angular size limit the precision of its parallax. Even better is an asteroid that periodically comes close to Earth but has a significantly different mean distance from the Sun (which means it has to have an eccentric orbit). Of the few large asteroids that come close enough to Earth for high precision parallax measures, Eros has historically been the most important. In 1931, an international effort by astronomers to observe Eros during a close approach to Earth resulted in the most precise measure of the distance from Earth to the Sun until the advent of radar astronomy decades later.

Earth's surface is well suited to measure the size scale of the Solar System. Earth's size, the distribution of its continents, and the clarity of extraterrestrial light sources observed through its atmosphere, with the Moon and some asteroids serving as handy intermediate rulers in a well-appointed measuring kit, have allowed us to determine our place in the Solar System.

PLANETARY PROTECTORS

Once more, those features of our environment that are congenial to scientific discovery also promote Earth's habitability. We've briefly discussed the role of the Moon and the giant planets in all this. Jupiter and Saturn are probably the most significant planetary protectors, since they shield the inner Solar System from excessive comet bombardment.[28] And although this should be the subject of intense future research, we suspect the other terrestrial planets have played a role in maintaining Earth's habitability as well. First, the mere presence of other planets in the inner Solar System reduces the number of asteroids and comets hitting Earth, for the simple reason that an object that hits one of these other planets is no longer around to slam Earth. How much protection the other planets add depends on their combined surface areas and their proximity to Earth. Thus, Venus, the closest planet, and nearly the same size as Earth, offers the greatest protection in the inner Solar System. Mars, though a little farther than Venus, is closer to the main asteroid belt; it has almost certainly taken a few asteroids and comet hits on our behalf. The Moon, another Solar System vacuum cleaner, has only about 7 percent of Earth's surface area, so the protection it offers has been small but not insignificant.[29] To get a sense of the impacts that Earth might have endured, just look at the Moon's scarred face through a small telescope.[30] Yikes!

There is a second and admittedly speculative way the other planets might have helped Earthly life. Early in the Solar System's history, Earth probably experienced several collisions with enormous asteroids. Some may have vaporized its oceans and sterilized the entire planet.[31] These impacts would have hurled many fragments from Earth's surface into space. It's possible that some of these Earthly fragments could have seeded Mars with living organisms.[32] Thus, Mars, or even Venus, if it was hospitable to life early on, could have served as a temporary life storage surface while Earth recovered from a sterilizing impact; a subsequent impact on Mars would have reseeded Earth. An isolated planet would lack such temporary refuges for life (we'll return to this topic in the next chapter).

The same processes that form a planetary system also leave surplus asteroids and comets. A terrestrial planet with protective planetary neighbors is preferable to one in isolation around its host star. Too many planets, however, will make a system less stable. The most habitable and measurable system will be one with the most planets allowed by stability constraints. It appears that Earth belongs to such a system. Who knows what other peculiar features of our Solar System contribute to Earth's habitability? These await future research and discovery.

AT HOME IN THE SOLAR SYSTEM

Although few realized it until recently, it's good that Earth isn't an orphan. The Moon and other planets enhance both Earth's habitability and its inhabitants' ability to measure the universe and discover its laws. We're only now beginning to appreciate how much our Solar System's configuration is not only rare but also surprisingly crucial for life and scientific discovery. It has indeed been a perspicacious teacher. Given the recent trends in the planetary sciences, perhaps we should begin to view Earth and its immediate surroundings not as a carbon copy of systems bound to arise wherever stars and planets form, but as a finely tuned and interdependent system that together nurtures a strange little oasis. Like the baby bear's porridge, Earth is, once again, just right.

Section 2

The Broader Universe

CHAPTER 7

STAR PROBES

Within this unraveled starlight exists a strange cryptography. Some of the rays may be blotted out, others may be enhanced in brilliancy. Their differences, countless in variety, form a code of signals, in which is conveyed to us, when once we have made out the cipher in which it is written, information of the chemical nature of the celestial gases.... It was the discovery of this code of signals, and of its interpretation, which made possible the rise of the new astronomy.

—William Huggins[1]

STARLIGHT, STAR BRIGHT

When ancient peoples looked at the starry night sky, they saw vivid pictures, and wove those pictures into mythological tales. Many of these pictures, or "constellations," such as Leo, the lion; Taurus, the bull; and Scorpius, the scorpion, are recurring themes in otherwise diverse cultures. This is surprising, since constellations rarely resemble the objects associated with them. Besides their mythological significance, many people have also attached a cosmic significance to the movement of constellations across the sky, thus providing gainful employment to legions of astrologers. The Greeks and Romans went so far as to identify stars with various gods. In stark contrast, the austere biblical account of the creation of the stars describes them as mere created things:

> And God said, "Let there be lights in the expanse of the sky to separate the day from the night, and let them serve as signs to mark seasons and days and years, and let them be lights in the expanse of the sky to give light on the earth." And it was so. (Gen. 1: 14–15, NIV)

While human beings have speculated about the nature of stars for millennia, we have only recently discerned their true nature. During the early nineteenth century, about the time of Auguste Comte, the father of positivism, it was quite different. Spectroscopy was a young science, and the mathematical description of radiation emitted by heated bodies was still decades away. At that time, scientists measured the temperature of a substance with a thermometer, and they determined its composition by applying a series of laboratory tests involving most of their senses. But we can't put a star into a test tube and run chemical tests on it. We can't fly up to the Sun and stick a thermometer into it. Not surprisingly, then, Comte, with many of his contemporaries, assumed that the temperatures and compositions of stars would lie forever beyond the ken of science.[2] Why would the universe be constructed so that we could acquire such knowledge just as reliably as we do with objects we can hold in our hands?

But Comte was wrong. Stars are not merely hot, opaque balls of gas. William Huggins, the first stellar spectroscopist, saw stars as encoded books whose language we had to learn to translate. More literally, a star is a space probe that continuously broadcasts light in all directions, conveying information about itself and its local environment over vast distances—enough information, in fact, to keep thousands of astronomers quite busy. Stars are relatively simple, however. Most are nearly spherical, allowing astronomers to describe their structure with a few simple equations.

Two of the most fundamental properties of stars are luminosity—absolute brightness—and surface temperature. Around the time of World War I, American astronomer Henry Norris Russell and Danish astronomer Ejnar Hertzsprung independently realized that more luminous stars tend to be hotter. The diagram showing this, with luminosity on the vertical axis and temperature on the horizontal, has come to be called the Hertzsprung-Russell (or H-R) diagram. H-R diagrams have used a number of different kinds of temperature indices. The simplest is the photometric color index, determined by observing a star's brightness through two filters of significantly different colors. Another index is the spectral type, determined from optical spectra. When it was originally set up, the spectral type classification was based on the strength of the hydrogen absorption lines in the spectra. Astronomers put the stars with the strongest hydrogen lines at the beginning of the alphabet, thinking that strong hydrogen lines would translate into temperature. But this didn't work out, since they later learned that the order of hydrogen line strength doesn't correlate so simply with a star's

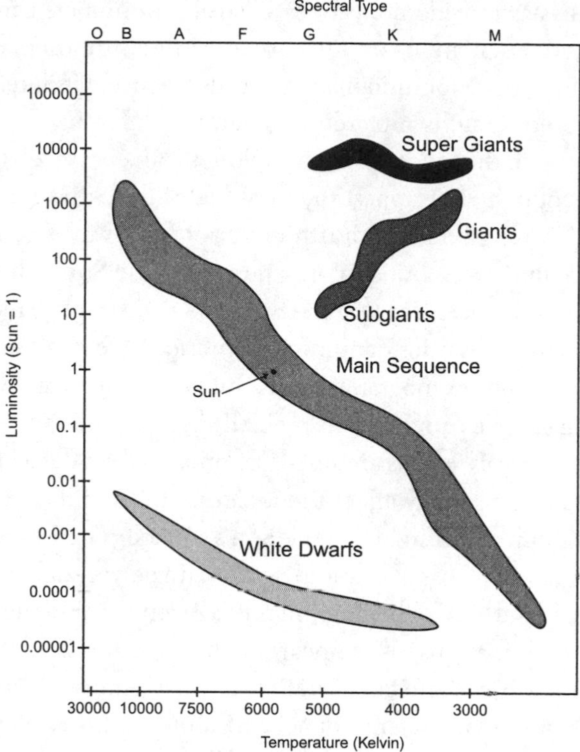

Figure 7.1: The Hertzsprung-Russell diagram for nearby stars. When stars are plotted according to their luminosity and temperature, they conform to a clear pattern. For stars on the main sequence, which includes about 90 percent of nearby stars, the brighter ones are hotter. Such orderly information allows astronomers to learn a great deal about stars, even over vast distances.

temperature. So now it's quite confusing. In order of decreasing temperature, the spectral type sequence for stars is O, B, A, F, G, K, M, for which Russell mercifully provided a mnemonic: *O*h *b*e *a f*ine *g*irl; *k*iss *m*e.[3]

INSIDE STARS

Deep Soundings

With a star's basic properties in hand, astronomers can infer its internal structure with the help of equations describing simple laws of physics discovered at Earth's surface. But how do they verify such calculations? Again, the laws of physics provide built-in tests. First, astronomers can measure small amplitude oscillations on the surface of the Sun, oscillations that result from sound waves of various wavelengths traversing its interior. Since waves of different lengths preferentially sample different regions of the Sun's interior (with the longer waves sampling the deeper layers), they give astronomers enough clues to test their models of the Sun's interior. Like the information we gain about Earth's interior from waves generated by earthquakes, sound waves bring information about the Sun's invisible interior to the visible surface. Because the Sun is a fairly simple ball of gas, it's easier to model than Earth's complicated interior.[4] We can't send probes to the Sun and plant seismographs on it. But because its visible surface is a hot, glowing gas, we can detect its oscillations from a distance.

Although the waves' effects are small, astronomers can detect them with spectroscopes because they affect the radial motions of the gas—that is, motions along our line of sight—at the Sun's surface. The waves also alter the compression of the gas as it undulates up and down, slightly changing its temperature, and hence the brightness of a given patch of the Sun's surface. So these oscillations also appear as small variations in the surface brightness of the Sun. The combination of solar astronomy and experimental physics makes the science of helioseismology a busy enterprise, and extends our knowledge to distant stars as well.

Some stars exhibit easily detectable pulsations. One type, the Classical Cepheid, is especially useful to cosmologists (we'll cover these in Chapter Nine). Small, highly dense white dwarfs, which are Earth-sized dying stars, pulsate and rotate with periods measured in minutes. Some display a rich pattern of pulsation periods, enabling astronomers to derive all their important structural properties.[5]

NEUTRINOS

Neutrinos, nearly massless neutral particles that hardly interact with matter, provide another test. According to the solar interior models, they should be produced at a certain rate in the core of the Sun to account for the energy leaving its surface. While the optical photons we observe from the Sun betray its surface temperature, the neutrinos, which we can also detect from Earth, tell us something about its core temperature, thus serving as a kind of long-range thermometer. Solar neutrinos also reveal something about the nature of the universe, but we'll leave that discussion for later.

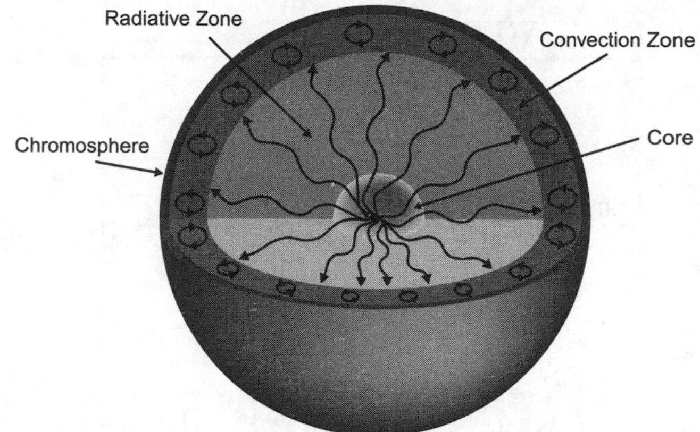

Figure 7.2: Cross section of a Sun-like star. Over most of its volume, the energy generated in the core through hydrogen fusion makes its way to the surface via radiative transport. Neutrinos are generated in the core. Over the outermost 20 percent of its radius, the star transports the energy by convection.

STAR SPECTRA

A star's spectrum abounds with all sorts of information about such matters as its surface temperature and gravity, its magnetic fields, and the elements and isotopes that make it up. As we mentioned in Chapter One, the optical spectrum of a Sun-like star is a smooth continuum interrupted by thousands of sharp absorption lines, regions of the spectrum where less light is emitted. Two forms of matter produce the absorption lines: neutral and ionized atoms, and simple molecules in the chromosphere—the thin, colorful layer of the Sun's atmosphere visible during total solar eclipses. The continuum is produced mostly by hotter gas in the underlying denser photosphere. An absorption line results when electrons absorb photons in a particular energy level in atoms of a particular element. Its strength is most sensitive to the number of atoms of the element in the gas (its "abundance") and the surface temperature of the star. In other words, the more atoms of an element that are present in a star's atmosphere, the darker its absorption line will be in the spectrum.

The high quality of the data astronomers derive from the Sun's spectrum is due not just to the wealth of photons available to their instruments; it turns out that the Sun's spectrum is nearly optimal for allowing them to extract its information. Their success depends on how precisely they can measure the individual absorption lines. This, in turn, requires that at least

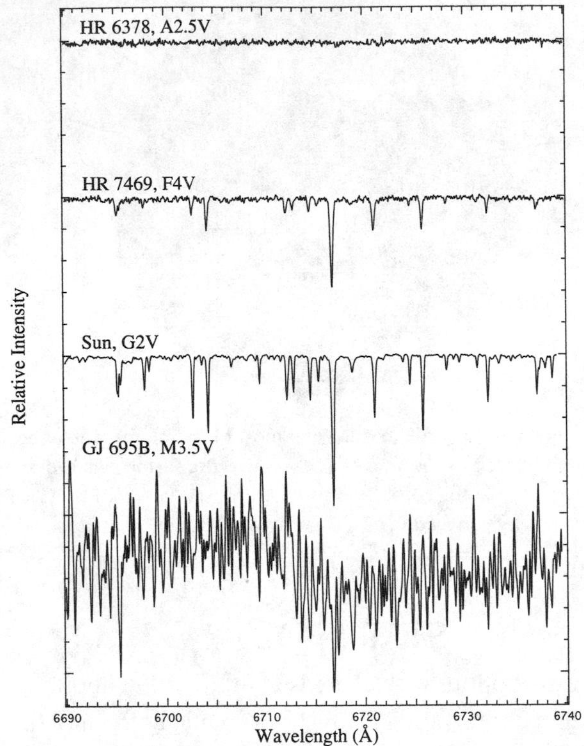

Figure 7.3: Comparison of high-resolution spectra of four nearby main sequence (dwarf) stars. The 50 Ångstrom slice corresponds to about 2 percent of the spectral range shown in Plate 14. The dips in the intensity tracings correspond to absorption lines (they are visible as the dark lines in Plate 14). Hot stars like HR 6378 have far fewer absorption lines than cooler stars like the Sun. But the strong molecular lines in very cool stars overwhelm most other spectral features and obliterate the smooth continuum visible in hotter stars. The Sun's spectrum is a "golden mean," displaying characteristics of both hot and cool stars, which allows astronomers to extract more information from the Sun than from either type alone.

some small "windows" of the spectrum's continuum be present among the "forest" of absorption lines. (See Figure 7.3.) Stellar spectra change their appearance drastically from O to M spectral types. Hotter stars have broader and fewer absorption lines; they have rather well-defined continua, but the small number of lines limits the quality of the analysis. Absorption lines become stronger and more numerous in the spectra of cooler stars, in part because molecular lines start to predominate.

Molecules in stellar atmospheres produce spectra that are rich in absorption lines.[6] Molecules, however, dominate the spectra of stars cooler than the Sun, even swamping the underlying atomic lines and obscuring the continuum. In other words, the extra information the absorption lines of a given molecule provide fails to compensate for the wealth of information they hide.[7] Although we can extract compositional information from the spectrum of a cool star, it is inferior to that obtained from Sun-like stars, which are the golden mean of measurability in this respect. Unlike those of hot stars, the optical spectrum of the Sun contains enough molecular lines for astronomers to derive useful data (such as isotope ratios) without dominating the spectrum. This makes the Sun's spectrum a nearly perfect compromise between the density of absorption lines and the integrity of the continuum. With the complementary information derived from meteorites, the information derived from the solar spectrum gives us a complete picture of the composition of the Solar System's birth cloud.[8]

The many sharp absorption lines in a star's spectrum also function as "velocity markers."[9] By monitoring the Doppler shifts in the wavelengths of the thousands of absorption lines in a star's spectrum, an astronomer can detect tiny variations in the star's velocity along his line of sight.[10] We've all heard a Doppler shift of sound waves from a siren, or a whistle from a passing train. The sound's pitch drops as it passes the observer.[11] The state of the art in Doppler measurement of starlight can detect bulk motions as small as one to two meters per second—a typical walking pace! It works best for stars at least as cool as the Sun; the absorption lines in the spectra of hotter stars tend to be too sparse and too broad. With this method, astronomers are discovering many giant planets orbiting other stars.

POINTS IN SPACE

If the usefulness of stars to science ended here, it would already be impressive. But stars don't only tell us about themselves, they are also especially useful probes of the broader universe. This is because their sizes are many orders of magnitude smaller than their separations, like tiny pinholes in a backlit blanket of dense, black felt. For example, the distance to the nearest star is about one hundred million times (or eight orders of magnitude) the Sun's diameter. This was particularly helpful to mariners of centuries past and astronomers of today, who have used stars as reference points.

Astronomers can measure their positions precisely only because starlight is very narrowly concentrated. Even today, the high-resolution Hubble Space Telescope in near-Earth orbit shows nearly all stars as unresolved points of light. Consider an interstellar nebula for contrast. A nebula is spread out over a large volume of space and is typically separated from other nebulae by distances only a few orders of magnitude greater than its size. The atoms in a nebula are relatively weakly bound by its self-gravity, so it doesn't take much to distort them—close encounters between clouds, close passages by stars, or the asymmetric pull from the large-scale gravitational field of the Milky Way galaxy will do the trick. This distributes light asymmetrically, making it difficult to define the nebula's center. Nebulae have low surface brightness and irregular shapes, making them very poor reference points compared with stars. Modern stellar astrometry has yet to reach a level of measurement precision limited by the angular sizes of stars.

Of course, there's one major disadvantage to seeing a star as a point: we can't study details on its surface. But there are several felicitous compensations. First, we can study the Sun's surface as a representative mid-range star. Second, most stars have uniform surfaces with uniform properties, so knowing their surface details would not add much important information. Third, we don't need to resolve stars to determine their basic properties, because they are spherically symmetric, simple, and thanks to their stellar spectra, can keep few of their secrets to themselves.

Today, astronomers use fast computers to simulate the gravitational interactions among thousands of stars, representing them in computer code as ideal mathematical points. Because of the great distances between them relative to their sizes, it turns out that stars are very close to this ideal, a fact that greatly simplifies simulations. The calculations would probably be intractable if stars were much larger, since they would distort one another on every close encounter, making it necessary to have detailed knowledge of the interior structure of all stars.[12] In effect, stars are very much like their representation in a computer. Like the virtual tag each virtual star carries with it, a real star contains and transmits information about its age, mass, position, and velocity. To put it in a topsy-turvy sort of way, the Milky Way galaxy is so weirdly accommodating to our efforts to measure its gravitational properties that it's like a gigantic simulation, sampling the local gravity field and transmitting encoded information to us, its interpreters.

TESTING PHYSICS

Stars are also remarkably useful testing grounds of the laws of physics. In Chapter One, we saw how the Sun has helped us to test some weak-gravity limits of General Relativity (via the bending of starlight). Astronomers also use a more exotic type of star, the pulsar. A pulsar has a mass comparable with that of the Sun compacted into a sphere about ten kilometers in diameter. It results from the explosion of a massive star, which blows its outer layers apart, leaving behind a very dense, furiously spinning neutron core. We know about pulsars because they emit highly directional radiation. Charged particles circulating around a pulsar's strong magnetic fields emit photons along narrow cones. This strong beaming allows us to measure a pulsar's rotation period precisely, using radio telescopes. In fact, some pulsars, especially those in binary systems, may be the most precise clocks in the universe—probably even more precise than atomic clocks. These stable pulses allow radio astronomers to study many types of phenomena that affect their arrival time from a given pulsar.[13] They can precisely determine the orbital characteristics of bodies in orbit about a pulsar,[14] test various aspects of General Relativity in the weak and strong gravity limits,[15] and learn about the properties of matter at nuclear densities. Even the way the radio waves from a pulsar interact with the free-floating atoms along our line of sight reveals a great deal about the intervening interstellar matter.

If the laws of physics had been such that stars were much larger or did not produce so many sharp absorption lines in their spectra, or if pulsars and white dwarfs were impossible, the universe would have been a far less measurable place. (How the properties of the universe change when the laws or constants of physics are slightly perturbed is a question we'll visit in Chapter Ten, where we'll discuss fine-tuning.) It turns out that the properties of stars are also delicately balanced for life. And as we hope to show, stars are key ingredients of habitable zones in the universe.

COZY LITTLE CIRCLES

In the late 1950s, astronomers introduced the concept of the Circumstellar Habitable Zone (CHZ).[16] While its definition has varied somewhat since then, they've generally defined it as that region around a star where liquid water can exist continually on the surface of a terrestrial planet for at least a few billion years. This definition is based on the assumption that life will flourish if this minimum requirement is met. Modelers have tended to focus quite narrowly, considering only a planet's distance from its host star, the composition of its atmosphere, and how these relate to the heating of its surface. They usually mark the inner boundary of the Circumstellar Habitable Zone as the point where a planet loses its oceans to space through a runaway greenhouse effect,[17] and define its outer boundary as the point where oceans freeze or carbon dioxide clouds form, both of which increase a planet's albedo and trigger a vicious cycle of increasing coldness until the oceans freeze over completely. (The location of the outer boundary is more difficult to determine, since it's hard to model the effects of carbon dioxide clouds.)

To understand the Circumstellar Habitable Zone we must know about all the processes that help maintain liquid water on a planet's surface. These include the greenhouse effect, the carbonate-silicate weathering cycle, myriad biological processes, ocean circulation, clouds, ice sheets, and plate tectonics and its associated phenomena.[18] To make the problem more tractable, researchers base their models on slightly perturbed versions of the present Earth. In other words, they assume an Earth-size (and Earth-like) planet in a circular orbit within a planetary environment similar to ours. Astronomers have slowly improved their models over time by adding factors. For example, by the 1970s astronomers recognized that the Sun's gradually increasing luminosity would cause the zone to move outward. And starting in the late 1990s modelers began treating plate tectonics as a changing rather than a steady process in equilibrium.[19] While these changes have made the models more realistic, other important factors, all dependent on the distance of the host star, are still missing.

THE DEVIL'S IN THE DETAILS

Consider the presence of an asteroid belt. You might think that the farther from our asteroid belt the better, assuming you remain within the CHZ. But the impact threat to a planet varies with distance from its host star in more than one way.[20] Planets closer to the main asteroid belt will suffer from more frequent asteroid collisions. (The peak of the asteroid distribution in our Solar System is at two Astronomical Units, AUs, from the Sun, about three AUs inside Jupiter's orbit.)[21] It also helps if a planet's orbit is nearly circular, lest it wander too close to the main asteroid belt at its farthest point from the Sun. Mars, in some sense, defines the inner edge of the main asteroid belt. It sweeps up lots of asteroids, making the inner edge sharper than it otherwise would be.[22]

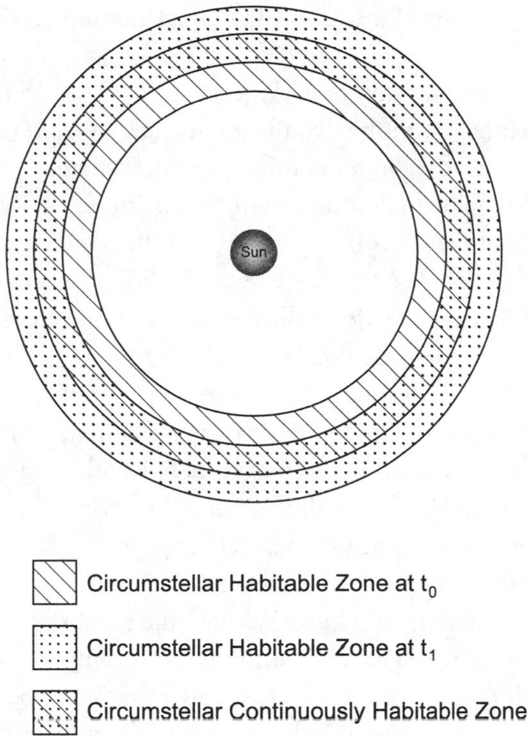

Figure 7.4: The Circumstellar Habitable Zone (CHZ) is that temperate region around a star wherein liquid water can exist on the surface of a terrestrial planet for extended periods. Since the luminosity of even a stable star like the Sun changes over billions of years, however, a star's CHZ will move outward over time. The Circumstellar Continuously Habitable Zone (CCHZ) is the overlapping region of various instantaneous CHZs.

Without Mars, Earth would be the closest planet to the main asteroid belt and bear the brunt of the impacts. As Kepler first showed us, the inner planets orbit the Sun at greater speeds than the outer planets. But since his laws apply to all bodies, not just planets, this means that asteroids and comets, too, orbit the Sun at greater speeds when they're close to it. You might assume that Mercury and Venus are safer from catastrophic asteroid impacts, since they're farther from the main asteroid belt. But since they're closer to the Sun, asteroids and comets in their vicinity are moving much faster. So when one of them does strike, it's likely to be a real whopper.[23] Among the terrestrial planets, then, there's an optimum range of distances, probably not far from Earth's location, that minimizes the threats from impactors.[24]

Next is the composition of a planet's atmosphere. To maintain liquid water on its surface with the extra heat provided by a stronger greenhouse effect, a planet near the outer boundary of the habitable zone needs much more carbon dioxide in its atmosphere. Thus, all else being equal, the farther a terrestrial planet is from its host star, the more carbon dioxide its atmosphere needs to keep it warm. We want liquid water at the surface, of course, since that's where life has the greatest access to energy-rich sunlight.

Cranking the carbon dioxide level way up, however, leads to other problems. A thick carbon dioxide atmosphere isn't a problem for some types of life forms, but large mobile creatures require an oxygen-rich atmosphere with a relatively low concentration of carbon dioxide.[25] Moreover, a planet near the inner edge of the habitable zone, like Earth,[26] will be more biologically fecund and diverse than one farther out.[27] The sunlight's energy maintains a large number of photosynthetic organisms, which, in turn, support a lush and diverse biosphere. While some autotrophs produce energy independently of sunlight, they're minor contributors to the parts of the biosphere relevant to complex life. A planet that produces biomass more slowly than Earth may not be able to ready itself for the arrival of complex life before its host star leaves the main sequence. It would also be more threatened by sudden external shocks to its ecosystem. So it may be no accident that Earth resides very close to the inner boundary of the Sun's habitable zone.[28] Habitability probably varies dramatically within the zone. Or to put it less charitably, it may very well be much narrower than currently assumed.

Moreover, just because someone can put an Earth twin in the right place in a model doesn't mean it can form there or that it will remain in a stable circular orbit. How a planet forms depends on its distance from the host star. While a protoplanetary disk is present in the early stages of a forming planetary system, gas and dust close to the star are heated more than material farther out in the disk. The more volatile (easily evaporating) elements remain in the gas phase in the hot regions, while the "refractory" elements, which tend to have higher melting points, can condense to form solid grains there.[29] We probably have asteroids and some comets to thank for the few volatile compounds, such as water and carbon dioxide, that abound in Earth's crust and outer mantle. Had Earth formed closer to the asteroid belt, however, its greater initial carbon and water endowment would probably have left a deep ocean, a thick carbon dioxide atmosphere, and a dead world.[30]

Besides having just the right amount of carbon dioxide in its atmosphere, a planet also needs just the right chemical ingredients in its core. Only certain kinds of cores can generate a life-protecting magnetic field. The makeup of a terrestrial planet's core probably depends on how it forms.[31] The inner terrestrial planets are more likely to have relatively larger iron-nickel cores and less sulfur mixed in, since sulfur is a volatile element. Adding sulfur to a planet's iron-nickel core reduces the core's melting point, much as salt reduces the freezing point of water.[32] Too little sulfur, and a liquid core will not only require a higher temperature to remain liquid but will be likely to freeze solid when the temperature drops. Too much sulfur, however, and a pure iron solid core may not form at all. To have a strong planetary magnetic field, a substantial fraction of a planet's core must circulate. A solid iron core probably helps too, just as an iron core in a solenoid strengthens the magnetic field.

Potassium is another key volatile element. Its long-lived radioactive isotope, potassium-40, is an important heat source in Earth that helps keep the mantle convecting and the crustal plates moving. As we noted in Chapter Three, plate tectonics, an essential part of the carbon cycle, keeps the continents above water. There may also be some potassium in the core, where it could help power the geodynamo. Its abundance in planetary cores probably increases with increasing distance from the host star. Just how much is in Earth's core, and how it's incorporated, are still controversial, but they both depend on a number of factors. Any sulfur in the core greatly increases the amount of potassium that can be sequestered there, as do higher temperatures coupled with oxygen.[33]

Each of these factors varies with distance from the host star in different ways. Some improve habitability with increasing distance, while others have the opposite effect. Asteroid impact rates must be fine-tuned to a narrow range of values and a specific time evolution, since they have both positive and negative effects. Others, like the resonant coupling between a planet's orbit and spin via the liquid portion of its core (discussed in Chapter Five) depend on distance in a more complex way. When multiplied out, these additional factors greatly reduce the best estimates for the width of the Circumstellar Habitable Zone for complex life and probably also for simple life. In short, a planet's true habitable zone depends on much more than just the intensity of light from its host star.

THE HOST STAR

Stars play two essential life-support roles: as sources of most chemical elements and as steady suppliers of energy. Under the pressure of gravity, stars fuse the nuclei of atoms in their hot interiors to build the chemical elements (we will return to this topic in the next chapter). Stars spend most of their lives fusing the abundant nuclei of hydrogen atoms (protons) in a phase of a star's life called the main sequence. Because hydrogen is so abundant, the main sequence is the longest-lasting phase of a star's life. While in it, a star's luminosity doesn't change much. The Sun, which has been a main sequence star for about 4.5 billion years, has brightened by about 30 percent since it first ignited its hydrogen. Once it leaves the main sequence, in about six billion years, the Sun will become several thousand times brighter in fairly short order.[34] The amount of time a star spends in the main sequence hinges on its mass. Stars with twice the Sun's mass only spend about a billion years in it, while those with half its mass last about 100 billion years.

Earth's geological record indicates that liquid water has been present somewhere on its surface continuously over most of its history. That means Earth must have stayed within the habitable zone even as the Sun brightened and the zone moved outward. The region over which all the instantaneous habitable zones overlap for some extended period of time is called the Circumstellar Continuously Habitable Zone (CCHZ). The CCHZ is narrower than the CHZ, especially if the time interval under consideration is a substantial fraction of the main sequence lifetime of the host star.[35] Of course, the position of this zone varies from star to star. Low-mass main-sequence stars, being less luminous than the Sun, have small, close-in CCHZs; the opposite is true for more massive stars.

But again, there's more to this game than just the light energy a planet receives from its host star. Stars more than about 1.5 times the Sun's mass are probably not viable habitats for complex life, because they spend relatively little time on the main sequence before they become red giants, and while in the main sequence, their luminosity changes relatively quickly. Such rapid changes are more likely to lead to drastic climate changes. But because of complex orbital dynamics, a rapidly increasing stellar luminosity leads to another threat—asteroids.[36] The scene around a star approaching the end of its main-sequence lifetime will be reminiscent of the early, violent stages of planet formation. So more massive stars will endanger their planets even before they leave the main sequence.

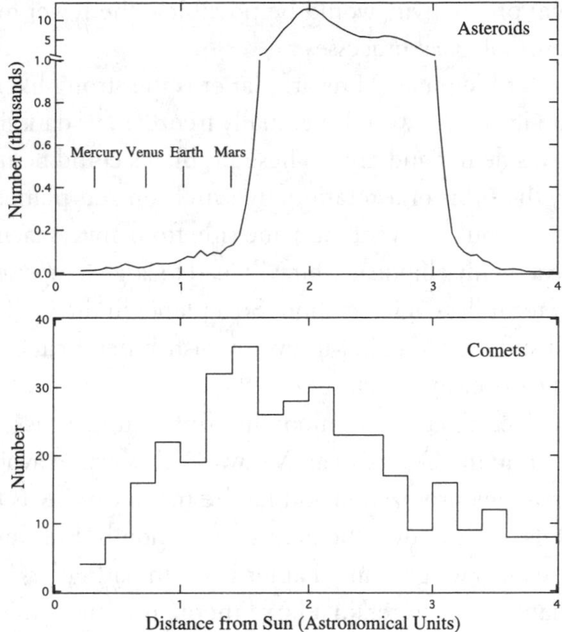

Figure 7.5: Distribution of the perihelia (closest approach to Sun) of nearly 200,000 asteroids and four hundred comets. The aphelia (farthest distance from Sun) of the four terrestrial planets are also shown. Asteroids are about 2.5 times more common near Mars compared with Earth. The comet perihelia distribution peaks near Mars. (The vertical axis of the top diagram is split for clarity.)

Near the opposite end of the scale, low-mass main-sequence stars (M dwarf stars) would also offer poorer habitats. An M dwarf star induces strong tides on a planet in its CCHZ, simply because of the planet's proximity to its host star, quickly braking its rotation (like Earth's Moon).[37] Why is this bad for life? If the planet's atmosphere is thin, it will freeze out on its dark side; a perpetually shadowed and cold region acts as a sort of "cold trap," not unlike cold traps used in vacuum pump systems to extract water from the air. High levels of atmospheric carbon dioxide could prevent this, but at the expense of animal-like life, which needs high oxygen and low carbon dioxide levels.[38] Even if we allow a thick carbon dioxide atmosphere, however, temperatures would be comfortable for life only in a narrow band along the planet's terminator—the line that separates the light from the dark side. And since the intensity of starlight is weak at the terminator, only weak biological productivity would be possible—the lower temperature would also slow biological processes.

Even worse for life on an M dwarf planet is the strong likelihood that all the water on its surface would eventually freeze on its dark side, leaving the illuminated side hot and dry.[39] These problems could be mitigated if, like Mercury, the orbit of a rotationally synchronized planet was quite eccentric, which would prevent the same side from always facing the host star. Of course a highly elliptical orbit will lead to large temperature swings on the planet, regardless of its rotation. So a planet in the CCHZ of an M dwarf star will suffer from either unevenly distributed heat or large temperature changes over the course of its year.

What if we place a planet-size moon in orbit around a gas giant planet, which, in turn, is in the CCHZ of an M dwarf? This would avoid the problem of rotational synchronization and let the moon expose its full surface to the light of its host star over the course of its month. But this still won't work, for the reasons we gave in Chapter Five. In particular, it's not clear that the gas giant can even retain a large moon as it migrates inward into the habitable zone of an M dwarf, where the gravity of the star struggles mightily for possession of the moon. And even if it does survive, the size of the moon's orbit around its giant host would be a significant fraction of the distance to the host star, creating large variations in temperature on the moon's surface.

M dwarf stars pose additional problems for life. Like the Sun, they exhibit flares. Some are stronger than solar flares, and because M dwarf stars are far less luminous, a flare's intensity compared with the star is that much greater. A strong flare on an M dwarf star can increase the relative X-ray radiation by a factor of one hundred to one thousand compared with strong flares on the Sun; the resulting increase in the ultraviolet radiation reaching the planet's surface would also be more intense.[40] Not only would such flares threaten surface life, they would probably strip away a planet's atmosphere more quickly as well. The large starspots associated with flares would cause the star's brightness to vary on longer timescales (by about 10 to 40 percent), mimicking an eccentric planetary orbit. Starspots and flares decline steadily as a star ages. So while the passage of time would mitigate these problems, at any age an M dwarf host star will be a less constant source of energy than a star like the Sun.[41]

Such bursts also would probably damage any existing ozone layer on a planet, because in its quiescent state an M dwarf star produces less ultraviolet compared with optical radiation than does the Sun. The steady flow of ultraviolet radiation from the Sun maintains the ozone shield in Earth's atmosphere, which offers some protection from modest increases in the extraterrestrial ultraviolet flux. So a planet orbiting in the habitable zone of an M dwarf star will be more susceptible to the damaging effects of short-lived ultraviolet and particle radiation events like stellar flares and nearby supernovae.[42]

Ultraviolet radiation is also crucial in oxidizing a planet's atmosphere. It was the steady dissociation of hydrogen-rich light molecules such as methane and water in Earth's atmosphere by the Sun's ultraviolet radiation, and subsequent loss of the hydrogen, that eventually allowed oxygen to become so abundant in its atmosphere. The process took about two billion years on Earth, about half its present age. Such a process will be slower where the intensity of ultraviolet radiation is weaker, as it is near a cool M dwarf star. One could counter that M dwarf stars last much longer in the main sequence than the Sun, but this does not much improve the odds for those M dwarf stars already formed, since the Milky Way galaxy is only about twelve billion years old—a tiny fraction of the theoretical lifetime of an M dwarf star.

The red spectra of M dwarf stars means that very little blue light will reach the surface of its orbiting planets. Although photosynthesis doesn't require blue light, it generally becomes less effective without abundant light blueward of 6,800 Å. Some bacteria can still use infrared light, but not to produce oxygen. Any marine photosynthetic organisms would have a hard time using red light as an energy source, since ocean water transmits blue-green light far better than blue or red light.[43]

Because the CCHZ around an M dwarf star is much closer in than the one around the Sun, any terrestrial planets within and near it will probably orbit closer together. This makes it more likely that such planets will perturb each other's orbits. The terrestrial planets in the Solar System are spaced far enough apart to have remained in fairly stable orbits for 4.5 billion years. Smaller orbits would have permitted only much shorter dynamical lifetimes. Ironically, then, the shortest dynamical lifetimes of terrestrial planets in habitable zones should be found around the longest-lived stars.

We can't assume that all stars will have terrestrial planets like Earth in their CCHZs. George Wetherill, of the Carnegie Institute of Washington, has produced computer simulations of the formation of terrestrial planets around stars of various masses.[44] He finds that during the late stages of planet formation, the distribution of terrestrial planets is relatively insensitive to a star's mass. If Wetherill is correct, then the most massive terrestrial planets will form at about Earth-Sun distance (give or take about half an AU) regardless of the star's mass. This suggests that fairly large terrestrial planets, like Earth, will reside in the CCHZ only when their host stars are similar in mass to our Sun. For a smaller star, they would tend to lie beyond this zone, and for a larger star, they would tend to lie inside it.[45] Although we're not yet sure, the high radiation levels present near a young M dwarf star might also prevent the coagulation of dust grains there, a necessary precursor to the formation of terrestrial planets.

GUIDANCE FROM THE WEAK ANTHROPIC PRINCIPLE

If we assume that our setting is a random sample of the observable universe, then we should expect that our local environment, the Solar System, and the Sun are not unusual in any important respects. Since many scientists today make that assumption, it's surprising to discover that in many important ways our local setting is quite unusual. One way of resolving such surprise is to consider whether some or all of these anomalous properties are in fact required for our presence as observers. We could interpret a large deviation from the average of a particular solar property as the result of an "observer selection effect." That is, perhaps that deviation is one of the necessary conditions for a habitable environment, and if our Sun didn't deviate from the mean, we wouldn't be there to observe it. Given this as a working assumption, we can determine habitability requirements by comparing the Sun's properties with those of other stars (we will consider the broader galaxy in the next chapter). This is a practical application of what is often called the Weak Anthropic Principle (WAP), or just the Anthropic Principle, which states that we should expect to observe conditions, however unusual, compatible with or even necessary for our existence as observers.[46]

Textbooks and science writers often assert that the Sun is an average or typical star. As Jay Pasachoff wrote in one of his popular introductory astronomy texts, "The Sun is an average star, because there are stars that are smaller than the Sun and stars that are larger."[47] But when we actually compare the Sun's properties with those of other stars in detail, we find that this is not the case.[48] Two of the most basic properties of a star are its mass and luminosity (though these are related for main-sequence stars through the mass-luminosity relation). The Sun is near the mid-range in mass and luminosity compared with the upper and lower extremes among nearby stars. But mid-range is not the same as average. Suppose we want to estimate the average adult male weight. If we consider the extreme ranges, and just take the mid-range value, we will get an answer just under 1,000 pounds; the average, however, will be somewhere between 150 and 200 pounds. Very rare extreme cases dominate the mid-range value. In the same way, the average star mass and luminosity are actually much smaller than the corresponding solar values. In fact, the Sun is among the 9 percent most massive stars in the Milky Way galaxy; most stars are M dwarfs. (See Plate 15.) Therefore, following the Anthropic Principle, we could interpret this as suggesting that a star at least as massive as the Sun is required for complex life to exist on an Earth-like planet, because we have "selected" it by our presence as observers.[49] This is consistent with our theoretical expectations that low-mass stars don't provide life-friendly environments. Of course we can narrow this range even further by applying our theoretical arguments against more massive stars (something the Anthropic Principle, as applied to the distribution of stellar masses, is silent about).

The Sun is a highly stable star. Its light output varies by only 0.1 percent over a full sunspot cycle (about eleven years), perhaps a bit more on century timescales. Most if not all of this variation is believed to be due to the formation and disappearance of sunspots and faculae (brighter areas) on the Sun's photosphere. At some level, of course, all stars are variable.[50] Among Sun-like stars of similar age and sunspot activity, however, the Sun's light varies much less than average,[51] preventing wild climate swings on Earth.[52] Taken together, then, these anomalies suggest that the Sun is atypical in ways that enhance Earth's habitability for technological life.

Another way to apply "Anthropic" reasoning is to consider what our home world is not like. For example, we're not living on a large moon orbiting a gas giant planet, a planet with a significantly eccentric orbit, a terrestrial planet lacking a large moon, or a planet orbiting an M dwarf star. These are all probably more common planetary configurations than ours. If our local environment were simply a random slice of space, we'd be much more likely to find ourselves in one of those other settings or, even more probably, in some vast and empty track of space between stars or even between galaxies. Quite reasonably the Weak Anthropic Principle says that we don't "because" none of those places is compatible with the existence of technological life. According to the Weak Anthropic Principle, and common sense for that matter, we can expect to find ourselves only in a place compatible with our existence. Apparently only improbable places qualify.

A LATENT RULER

As we argued earlier, the Sun's surface temperature and absolute brightness, or luminosity, are nearly optimum for extracting information from its spectrum. And as we argued in the previous section, the Sun also appears to be near optimum for supporting complex life.

As it turns out, the particular properties of the Sun, and our orbit around it, are important for scientific discovery as well. The method of stellar trigonometric parallax is one of the most important tools in the astronomer's tool chest. Earlier we showed how Earth and the Moon have been used as "rulers" to measure distances in the Solar System with the parallax method. Similarly, Earth's orbit serves as the baseline ruler for measuring distances to nearby stars. Because Earth's position changes during a year, nearby stars appear to wobble in reflex motion relative to distant background stars; this apparent motion is called annual stellar parallax.

Hypothetical Mercurians or Venusians would have a smaller baseline, since these planets have smaller orbits around the Sun, while Martians would have a larger one and Jovians—inhabitants of Jupiter—larger still. But a larger baseline comes at a cost: time. Jupiter orbits the Sun in about twelve years. To measure two or three back-and-forth wobbles of a star would take several decades. For measuring the distances to other stars, the ideal orbit of a planet should be large enough to provide adequate parallax but not so large that a single orbit takes too long to complete over an observer's lifetime.

Planets around the common M dwarfs would offer their inhabitants a much poorer measuring rod (assuming all the problems we noted above could be overcome). As we argued, since an M dwarf star is such a feeble light source, a planet would need a very tight orbit to maintain liquid water on its surface.[53] Astronomers on a planet around a star with 20 percent of the Sun's mass would have at their disposal a baseline only about 10 percent of ours. This means that they could measure only 0.1 percent as many stars as we can from our Solar System (assuming the same level of technology), and almost all of them will be other low-mass stars. Put another way, it would take stellar parallax observers of one thousand M dwarf star/planet systems to observe the volume of space we can survey from our single planet.

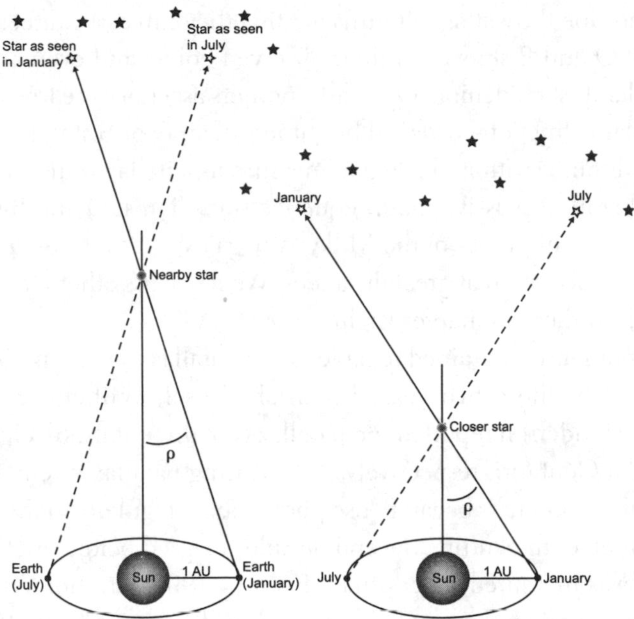

Figure 7.6: Earth's orbit serves as a ruler for measuring distances to nearby stars. Because Earth's position around the Sun changes during a year, nearby stars appear to move relative to distant background stars. The closer the star, the more it appears to shift against the distant background. The effectiveness of the method of trigonometric parallax depends on the luminosity of the host star. Hypothetical inhabitants of the much more common red dwarfs would be able to survey a far smaller volume of space. (Even the nearest stars are much farther away than illustrated here, and the angle ρ is much smaller than shown.)

Because we orbit one of the somewhat rare moderately massive and thus luminous stars, we can measure the parallaxes of the even rarer, and therefore more distant, very massive O and B stars. This is critically important if we are to understand the physics of stars, since the parallax of a star combined with its apparent brightness of a star gives us its luminosity. Luminosity tells us the total energy that a star generates. This, in turn, tells us about nuclear fusion and its dependence on temperature and pressure at the center of the star.

Once astronomers determined the luminosities of several O and B stars a few decades ago, they learned that they could use such stars as a kind of astrophysical "standard candle"—that is, as a source of light for which they can determine the wattage. It turns out that they can easily discern the true colors of O and B stars from their observed colors and even account for interstellar dust reddening. Once astronomers ascertain the true colors and interstellar extinction-corrected brightness of an O or B star, they can estimate its distance without having to measure its parallax. This method does not apply to less massive main-sequence stars. Thus, O and B stars have been very useful probes of the Milky Way galaxy, because their brilliance allows us to see them at great distances. We'll discuss other types of stellar standard candles in Chapter Eight.

Astronomers first attained reliable stellar parallax measurements in 1838 and 1839. In those two years Friedrich Bessel, Wilhelm Struve, and Thomas Henderson reported the parallaxes of three stars: 61 Cygni, Vega, and Alpha Centauri, respectively.[54] Measuring parallaxes remained difficult until astronomers began to use photography regularly in the latter half of the nineteenth century. Several decades elapsed before another stellar parallax was measured successfully. Even by 1900, astronomers had only determined the parallaxes of about one hundred stars; by 1950 they had measured several thousand.[55]

Several distinct characteristics of an environment must converge to make this possible: the observer's home planet must be far enough from the host star, the planet's atmosphere must not be too thick or murky, and there must be enough stars in the host star's vicinity. Earth qualifies on all counts. The first two characteristics are met within the Solar System (we'll discuss the third in Chapter Eight), and as we've seen, Earth's location in the Solar System is also vital to its habitability. Had Earth been closer to the Sun, it would probably have been a hothouse with a thick atmosphere like that of Venus. If it had been much farther from the Sun, Earth would have needed a much thicker—and more obstructive—atmosphere with a great deal more carbon dioxide, to keep water flowing on its surface. But such an atmosphere, even with surface water, would still be hostile to animal life. So in both cases, observations from the ground would be inferior to those we enjoy. Once again, the most habitable location also provides an excellent overall setting for scientific discovery.

A HANDY ASTROPHYSICS LAB

While Earth's orbit around the Sun offers a very good baseline for measuring the distances of nearby stars, the Sun itself serves as a kind of astrophysical laboratory. Yes, the Sun is vastly closer than other stars, so it's easier to study. But quite apart from its proximity to us, as we noted above, we can derive more high-quality information from the Sun's spectrum than from that of most other types of stars.[56]

Although the Sun is in several ways anomalous, it helps that some of its properties are near mid-range. In constructing theoretical models of stars, astronomers must start with the Sun and then extrapolate to lower and higher temperatures and luminosities in order to describe very different kinds of stars. If the Sun had been at either end of the temperature scale—either very hot or very cool—such extrapolations would be much more prone to systematic errors. As it is, the Sun's atmosphere has qualities of both hot stars (a well-defined continuum in its optical spectrum and absorption lines arising from transitions in ionized atoms) and cool stars (molecular absorption lines in its spectrum and a convection zone near its surface). If astronomers could only exist near a *very* rare class of star, such as a blue supergiant, then they would have had a hard time extending the knowledge of their home star to the more common main-sequence stars.

Astronomers have been able decipher clues about the Sun's internal structure for some time, but only very recently have they succeeded in detecting oscillations on the surfaces of a few other Sun-like stars.[57] The Sun's oscillations are excited by the convective currents of its outer layers, so other stars with similar currents should display similar oscillations. If theory holds, less luminous main-sequence stars should show smaller wave heights, called oscillation amplitudes, in their velocity and brightness. For example, a K5 dwarf star, about 60 percent the mass of our Sun, should exhibit velocity and brightness amplitudes only 0.3 and 0.5 times those in the Sun, respectively.[58] These values are small enough to create acute observational problems for astronomers who specialize in this area, called helioseismologists.

Low-mass stars also offer a poor prospect for the other major test of stellar models—neutrinos. In the early 1930s, physicists such as Wolfgang Pauli and Enrico Fermi first suggested that the universe might contain elusive particles now called neutrinos, produced prodigiously in the cores of stars. The particles would be so tiny, subtle, and noninteracting that most could pass through the entire Earth without hitting a single atom. Detecting them, even from an ideal star, would require massive, expensive underground detectors that provide many opportunities for neutrinos to collide with other atoms. But starting in1965, scientists began to detect neutrinos from the Sun, using just such detectors.[59]

Neutrino astrophysics would have been far more difficult had our Sun been only slightly less massive or less luminous,[60] or if we had lived much earlier in the Sun's history. Our location near the inner boundary of the Circumstellar Continuously Habitable Zone (as traditionally defined) helps, too. Even at Mars, the neutrino flux would be nearly 2.5 times less.[61]

Additional neutrino detectors that have come online since the late 1980s have resolved a long-standing solar neutrino mystery. Though some had theorized that neutrinos might have no mass, these more recent experiments suggest that they do have mass.[62] This is important for cosmology, since neutrinos have long been a candidate for solving at least part of the so-called dark matter problem. Neutrino telescopes built to study the Sun also led to the serendipitous detection of neutrinos from supernova 1987A in the Large Magellanic Cloud, one of the Milky Way's satellite galaxies. Such events are very rare in the Milky Way galaxy or its immediate vicinity. That event provided astronomers with a unique look into the workings of a supernova.

By producing neutrinos we can detect on Earth, the Sun's core serves as an indispensable laboratory to learn about the very small and the very large, while offering a direct test of our solar models.[63] The Sun is just barely luminous enough for neutrino astronomy and helioseismology to be practical scientific enterprises. Observers with our level of technology living on one of the far more common K or M dwarf stars would not have these important observational tests available to them.

So the Sun's local environment seems to offer the best type of habitat for complex life. At the same time, its particular properties disclose vital scientific information more abundantly than many more common types of stars, while also providing us with an excellent example of stars in general. Curiously, just when they have needed to test their theories, astronomers have discovered that the experiment is already set up. As we'll see, most places in the Milky Way galaxy offer a much less edifying vantage point, while being more hostile to life.

Chapter 8

Our Galactic Habitat

Ironically, our relatively peripheral position on the spiral arm of a rather ordinary galaxy is indeed rather fortunate. If we had been stationed in a more central position—say, near the galactic hub—it is likely that our knowledge of the universe of other galaxies, for example, might not have been as extensive. Perhaps in such a position the light from surrounding stars could well have blocked our view of intergalactic space. Perhaps astronomy and cosmology as we know these subjects would never have developed.

—Michael J. Denton[1]

OUR ISLAND UNIVERSE

Look at the sky on a clear night far from city lights and you'll see a fuzzy band of white light that looks like a wispy, luminous string of clouds. The ancient Chinese saw it as a river in the sky. The Greeks and Romans explained it with the myth of Heracles, the son of Zeus, who bit the breast of Zeus's wife Hera, spilling her white milk across the black firmament. The Romans called this part of the sky the Via Lactea, the Milky Way, and at least in the West, the name stuck. (The word *galaxy*, by the way, derives from the Greek words for Milky Way.)

There were some good guesses explaining this marvel in the night sky, but no one really knew what it was until Galileo pointed a telescope at the band in 1609, and resolved "a congeries of innumerable stars."[2] We now know that it is composed of stars in our home galaxy, densely concentrated because we are looking at the galaxy edge-on, into the disk, and so taking in a stretch of it many times thicker than when we look in directions away from the Milky Way band. Once believed to be the entire universe, our galaxy is actually just one of billions of galaxies, "island universes," which astronomers previously identified as mere fuzzy patches called nebulae.[3]

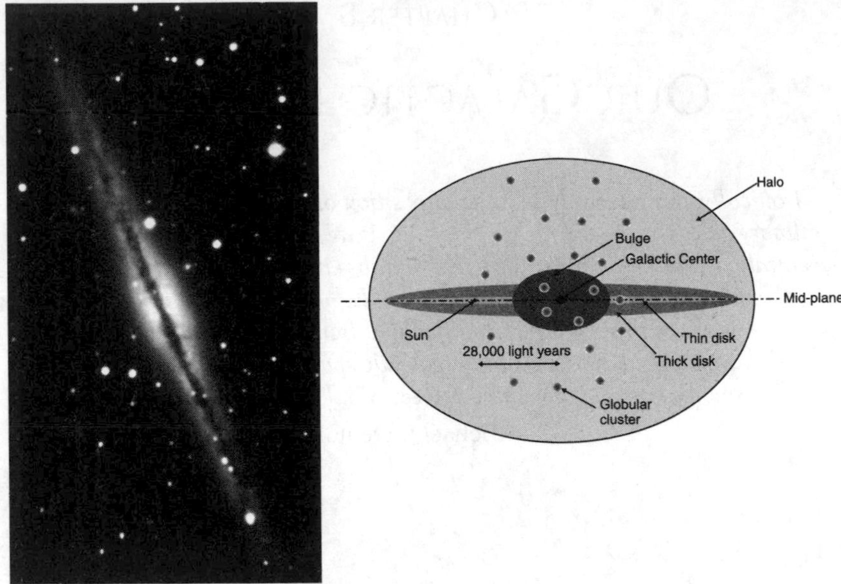

Figure 8.1: (left) Edge-on spiral galaxy NGC 891. Our Galaxy would probably look very much like this galaxy from the same vantage. Note the dust lane running across the full length of the disk. (right) The major components of the Milky Way Galaxy. The Sun is located at the mid-plane in the thin disk, about half way from the Galactic center to the edge of the disk. The globular clusters are spherically distributed around the Galactic center and are the most visible members of the halo.

Galaxies, like noses, come in many shapes and sizes. Despite such diversity, astronomers reduce them to three basic types: spirals, ellipticals, and irregulars—the latter admittedly a bit of a catchall. Ours is a spiral galaxy.[4] Most of its stars are located in its flattened disk, only about one percent as thick as its diameter. We live in the disk, very close to its mid-plane, about halfway between the galactic nucleus and its visible edge. Decorating the sky like heavenly pinwheels, spiral galaxies derive their popular name from the beautiful spiral pattern formed by their young stars and bright nebulae. Strikingly similar to the cross section of a Nautilus shell, the spiral pattern is thought to be a "density wave" phenomenon, somewhat like cars concentrating on crowded highways. The concentration itself progresses at a different speed from the individual cars that make it up, so snapshots at different times reveal different cars, but the overall pattern remains the same. We reside between the Sagittarius and Perseus spiral arms, a bit closer to the latter.

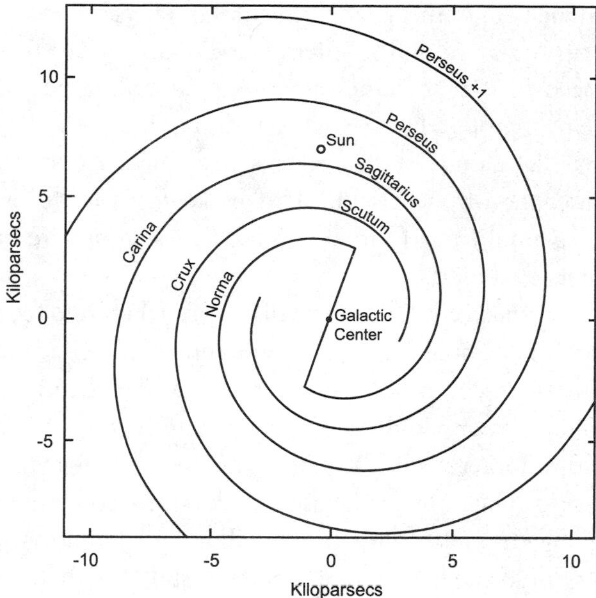

Figure 8.2: Face-on view of the Milky Way Galaxy. The Milky Way is a large, barred spiral galaxy. Shown are the outlines of the major spiral arms. We are located between two major spiral arms, Perseus and Sagittarius. This is one of the best places to learn simultaneously about stars, galactic structure, and cosmology. One kiloparsec is about 3,200 light-years.

Of the many denizens of our galactic neighborhood, the most obvious are the seemingly countless stars. Astronomers can see numerous star types in the immediate solar neighborhood, from the feeble brown dwarfs to the brilliant blue white O stars. They see stars in all their life stages, from the unborn pre-main-sequence stars to the long-deceased white dwarfs, neutron stars, and black holes. They see stars in isolation, pairs, triplets, and galactic, or "open," clusters.

Astronomers can also see a motley assortment of matter between the stars. These include ghostly giant molecular clouds, up to millions of times more massive than the Sun; diffuse interstellar clouds; supernova remnants; and the winds from dying red giant stars and their descendants, the planetary nebulae. They sometimes can see a glowing interstellar cloud when bright stars heat its inside. Such a fluorescing cloud is called an H II region, because its hydrogen gas is ionized. More indirectly, astronomers often see interstellar clouds called reflection nebulae illuminated from the outside, when the light from a nearby radiant star reflects off its dust. When astronomers view a star behind an interstellar cloud, they see the spectrum of the cloud's atoms and molecules superimposed against the spectrum of the star as sharp absorption lines. Hot O and B stars are the best type of star to study interstellar clouds in this way, because they have much broader absorption lines than these clouds. This makes it easier to distinguish between the spectral lines formed in the star's atmosphere and those formed by interstellar clouds.

Astronomers usually subdivide the Milky Way galaxy into four regions, or "populations": halo, bulge, thick disk, and thin disk. Each region is characterized by the ages, compositions, and movements of the objects within it. The halo contains only old metal-poor stars in highly elliptical orbits (recall that astronomers call all elements heavier than hydrogen and helium "metals"). The bulge, as fat as the disk is thin, contains stars spanning a large range in metal content, from about one-tenth to three times the Sun's. It is unlike the halo in that some stars still form there today. The orbits of its stars are also elliptical, but less so than those in the halo.

The flattened thick and thin disks overlap. The thin disk contains the greater diversity of objects, including most of the stars in the Milky Way, while the thick disk is more puffed up with older, more metal-poor stars. In the solar neighborhood, only a few percent of the stars are members of the thick disk; Arcturus, the brightest star in the constellation Boötes, is probably the most prominent thick-disk star in the solar neighborhood. Even within the thin disk, older objects are more spread out on both sides of the mid-plane. The most concentrated are very young "zero-age" objects such as giant molecular clouds, H II regions, and O and B stars, distributed in the thin disk with typical "scale heights" of about 150 to 300 light-years from the mid-plane. F and G dwarf stars reach heights of 600 and 1,100 light-years, respectively,[5] while the Sun reaches a maximum height of only about 250 light-years from the mid-plane.[6] If you think this seems like a dizzying array of galactic objects, you're right. We've been able to detect and study them all from our highly accommodating perch in the mid-plane of the Milky Way.

OUR GALACTIC PERSPECTIVE

Observers could not sample such diverse delights from every place in the Milky Way galaxy. The Sun, you see, is very near the mid-plane, having crossed it a mere three to five million years ago. The gas and dust in our neck of the woods is quite diffuse compared with other local regions in the local mid-plane.[7] This gives us a fairly clear view of objects in the nearby disk and halo, as well as distant galaxies. Since interstellar dust is concentrated in the galactic mid-plane, we don't get a clear view of objects behind the Milky Way band, looking edge-on into the disk. (For a time it was also called the "zone of avoidance," because optical sky surveys did not show galaxies in this region.) Fortunately, this represents only about 20 percent of the area of the optical sky.[8]

At optical wavelengths we can see only a few thousand light-years through the dust in the disk; at mid-plane the nearer dust is better at hiding everything behind it. This means that a good bit of the galaxy is hidden to an observer near the mid-plane. Now, imagine moving the Sun perpendicularly a little farther from the mid-plane. A little more of the previously hidden distant parts of the galaxy would now come into view, but the trade-off is that it would increase the area of the Milky Way band on the sky—increase, in other words, the zone of avoidance. And dust is not the only problem; more foreground stars would also interfere with our view of distant background objects. That's not a good trade. As we'll see later, this would profoundly hinder discoveries about the cosmos as a whole.

Of course, since we're very near the mid-plane, dust absorbs some light, and foreground stars interfere with our view of background objects. We can learn more about the distribution of dust in our galaxy by observing other, similar galaxies. Telescopic images of nearby spiral galaxies in the optical part of the spectrum clearly show bright spiral arms separated by darker regions. One might suppose that there are fewer stars between the arms. But in fact, stars are only about 5 percent more concentrated in the arms. The arms appear so much brighter because they contain star nurseries, where the brightest stars are born and die. Their intense radiation illuminates the surrounding gas and dust, both strongly concentrated in the arms. Therefore, while spiral arm dwellers would get a closer view of the rarer, very massive O stars, overall, this bright, thick dust would obscure their view of the local and distant universe.

The few examples of nearby superimposed galaxies give astronomers a direct measure of the transparency of spiral galaxies. A superimposed galaxy pair results when a foreground galaxy partially obstructs our view of the background one. The dust in the foreground galaxy absorbs and reddens some of the light coming from the background galaxy. Recent studies of such

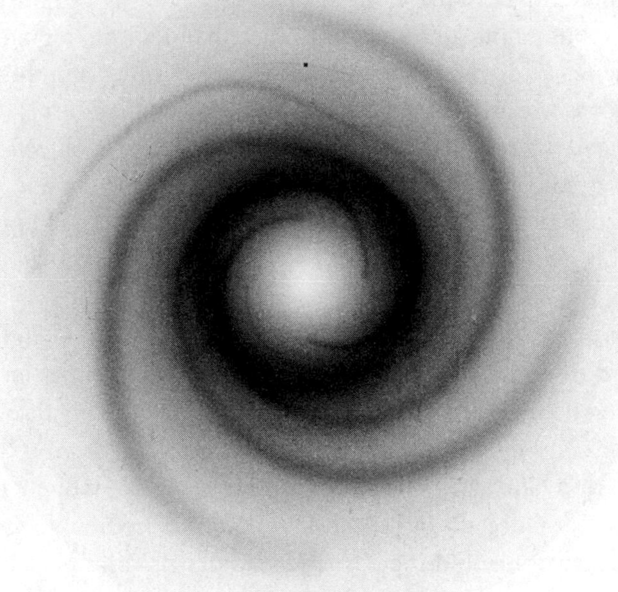

Figure 8.3: Face-on view of the Milky Way galaxy showing large-scale distribution of dust density. The bar and nucleus have been left out of the diagram. The Sun's location is shown as a black dot. Although a dust "hole" is shown in the center where the bulge dominates, there is some patchy dust there.

pairs demonstrate that the interarm regions are much more transparent than the dusty arms; they also show that the transparency of the interarm regions declines toward the center of a spiral galaxy.[9] Dust maps for our galaxy confirm this, though there is a "dust hole" at the galactic center.[10] So observers near the galactic center would have to deal with more foreground contamination from stars.[11] (See Plate 16.)

Observers closer to the densely packed galactic center would have a much brighter night sky, which might be pleasing to the eye but would block far more than it would reveal. We're quite far from the bulge. Moreover, large accumulated interstellar dust extinction crosses our line of sight to this galactic hub, where dust absorbs and scatters its light, greatly reducing its brightness. As a result, the otherwise bright galactic nucleus is invisible to the naked eye. Even so, when it's really dark we can see the bulge in the constellation Sagittarius as a brighter patch of the Milky Way band, since its light seeps through less dusty regions. Had we been located at a

Figure 8.4: The Great Cluster of Hercules, M 13. This is one of the better-known globular clusters to observers in the northern hemisphere. It contains several hundred thousand metal-poor stars about 25,000 light-years from Earth. Observers living in M 13 would get a poor view of the universe outside their cluster home. In 1974, its multitude of stars inspired astronomer Frank Drake to transmit a message toward it, hoping it would be intercepted by an extraterrestrial civilization (see Chapter 12).

quarter our present distance from the bulge, the nucleus would appear several million times *brighter*.[12] The bulge wouldn't be as bright as the full Moon, but it would cover nearly half of the sky, and brighten the rest.

Other problems would afflict astronomers in a globular cluster. They would get close up views of G, K, and M dwarfs and red giants, but they wouldn't see O, B, A, or F dwarfs, or interstellar clouds.[13] All the stars would have the same age and metal content. And of course the many bright foreground stars sprinkled all over their sky would profoundly compromise the views of the universe beyond the cluster.[14]

A view from the halo far from a globular cluster would offer a nice slice of the sky largely free of contaminating dust, gas, and stars. The view of the Milky Way's disk and bulge would be spectacular, with an easily discernible spiral structure. But a treasure to some is trash to others. The trade-offs for living in the halo would be a lack of close-up views of a great variety of stars, nebulae, and open clusters, and "light pollution" from the bright bulge and disk, which would hinder the view of faint objects. Most of what we know about stellar astrophysics depends on our ability to examine stars closely over a large range of masses, ages, and compositions. Above all, from the Solar System we can measure the parallaxes of many nearby stars and discern their true brightnesses (luminosities). Since stars are much more spread out in the halo, only a few M dwarf and perhaps some white dwarf stars would be close enough for our hypothetical halo dwellers to measure parallaxes using technology similar to ours. Important laws of stellar astrophysics, like the mass-luminosity relation, would be much more difficult to discover. Since young stars stay close to the mid-plane, observers in the halo, including those in globular clusters, would understand little about how stars form. The mid-plane offers the best vantage point to comprehend stars. It's only because we have a good grasp of the physics of stars that we can properly interpret the distant galaxies—after all, they're made of the same stuff we get to see up close in our immediate galactic neighborhood. In short, halo dwellers would have a hard time fathoming distant galaxies, and the cosmos as a whole.

Even though we don't have a bird's-eye view of the Milky Way, our perch is quite felicitous. First, we live among a small collection of galaxies called the Local Group. The Milky Way and Andromeda (Messier 31) Galaxies are its two largest members, followed by the galaxy in the constellation Triangulum (Messier 33). Both M31 and M33 are spiral galaxies, with the latter presenting us with a nearly face-on view. These teach us much about the overall structure of our home galaxy.

Second, we know we are living in a spiral galaxy because we can map its structure from the inside. This might seem impossible, like a tapeworm trying to describe the exterior of its host. But astronomers have learned to take advantage of the peculiar properties of the flattened rotating disk of gas and stars. So-called differential rotation allows radio astronomers to translate the measured radial velocities of interstellar gas clouds into distances from the Sun.[15]

For instance, imagine yourself as an ant riding on a spinning record and watching another ant on the record some distance from you. Although the record is spinning, the distance between you and the other ant does not change. This is because the spinning record is an example of solid body rotation: every part of the record is rigid relative to every other part, regardless of what the record is doing in space. The Milky Way disk does not rotate like a solid body. Instead, it rotates differentially, so that stars closer to the center complete an orbit more quickly than those farther out. Hence, observers living around a star in the disk would see stars or interstellar clouds elsewhere in the disk moving toward or away from them (yielding blueshifts or redshifts, respectively). The regularity of this rotation allows astronomers to convert the position on the sky and radial velocity of a given cloud into a distance.[16] Observers in our spinning-record galaxy would not have this extra information.

Comparisons with other types of galaxies also help us understand our own home in the Milky Way disk. Less flattened galaxies, like irregulars and ellipticals, lack the simple rotational dynamics of disk galaxies. Elliptical galaxies, which have less gas and dust, contain stars with a wide range of orbits—most highly inclined and eccentric. For observers inside this kind of system, no law relates the distance of a particular object to its observed radial velocity; they would need other means to determine its distance. Irregular galaxies, as the name implies, have very irregular dynamics and patchy distributions of stars and nebulae. Observers in such a galaxy would have an even more difficult time making sense of their neighborhood.[17]

Many galaxies are members of rich clusters containing thousands of gravitationally bound members. Life in a rich cluster of galaxies would be different from our home in the sparser Local Group. We'd get a close-up view of more nearby galaxies but at the expense of our views of the distant universe. It's not clear if our overall ability to study the universe would be worse, but given our limited knowledge of such a situation, the conjecture seems reasonable.

In short, settings in the halo, a globular cluster, the bulge, a spiral arm, an isolated galaxy, a dense cluster of galaxies, an irregular galaxy, or an elliptical galaxy would be less revealing than ours. We occupy the best overall place for observation in the Milky Way galaxy, which is itself the best type of galaxy to learn about stars, galactic structure, and the distant universe simultaneously; these are the three major branches of astrophysics. Does our location also offer the best overall type of habitat in the galaxy? Let's see.

HOME SWEET ZONE

Just as many other locations in the Milky Way galaxy are less amenable to scientific discovery, so too are they less hospitable to life. In sci-fi stories, interstellar travelers visit exotic places in the Milky Way and meet with interesting aliens. You name the place, and some intrepid writer has already put a civilization there: the galactic center, a globular cluster, a star-forming region, a binary star system, a red dwarf star, even the vicinity of a neutron star. *The Hitchhiker's Guide to the Galaxy* author Douglas Adams probably has a motley assortment of alien toughs smoking cigars and playing five-card stud inside a black hole somewhere. To their credit, probably only a small handful of these sci-fi authors are really earnest about the possibility of their far-flung and fanciful worlds. Most are just letting their imaginations run wild. But there was a time, not long ago, when prominent astronomers seriously speculated about intelligent beings on the Moon, Mars, Venus, Jupiter, and even the Sun.

Nowadays, thanks to our growing knowledge of these environments and the stringent requirements for life, canal-building Martians or Sun-dwellers sound quaint rather than futuristic. The Apollo, Mariner, Venera, Viking, and Voyager missions have put an end to those conjectures. We like a good far-flung space western as much as the next fellow, but when a self-declared expert starts talking about civilizations all over the galaxy not as whimsical fiction but as cosmic inevitability, we get skeptical. For just as most of the Solar System doesn't meet the strict requirements for complex or technological life, the same seems to be true for most of our galaxy.

Like the Circumstellar Habitable Zone in our Solar System, there is also a *Galactic* Habitable Zone (GHZ). (See Plate 17.) And its first requirement is to maintain liquid water on the surface of an Earth-like planet. But it's also about forming Earth-like planets and the long-term survival of animal-like aerobic life.[18] The boundaries of the Galactic Habitable Zone are set by the needed planetary building blocks and threats to complex life in the galactic setting. Let's discuss each in turn.

ASHES TO ASHES, DUST TO DUST

The story of how the elements came to be assembled at the place we call Earth is told by modern cosmology, galactic chemical evolution, stellar astrophysics, and planetary science. The Big Bang produced hydrogen and helium and little else. Over the next 13 billion years, this mix was cooked within many generations of stars and recycled. Beginning with the fusion of hydrogen atoms, massive stars make ever-heavier nuclei deep in their hot interiors, building on the ashes of the previous stage and forming an onion shell–like structure. Exploding as supernovae, the massive stars

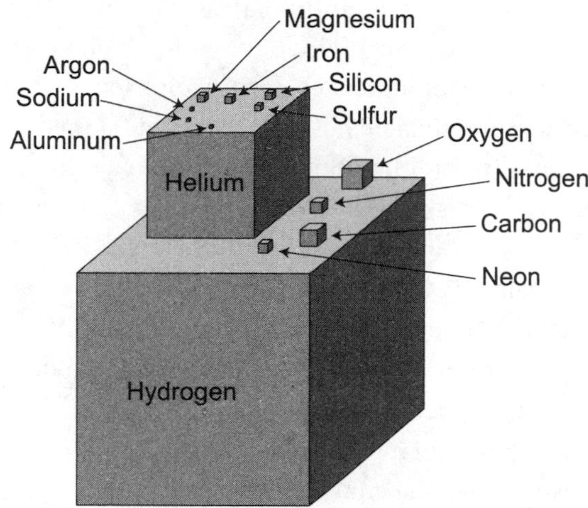

Figure 8.5: The relative proportion (by number) of the most abundant elements in the Sun. Hydrogen, helium, and the elements resting directly on top of the hydrogen cube dominate the composition of stars and Jovian planets. The terrestrial planets are composed mostly of oxygen and the other elements on top of the helium cube (but not including helium).

eventually return atoms to the galaxy. But they return them with interest, by producing heavy elements that didn't exist before. As a result, our galaxy's metal content—that is, its "metallicity," the abundance of heavy elements, or "metals," relative to hydrogen—has gradually increased to its present value, which is close to the Sun's. Today, metals make up nearly 2 percent of the mass of the Milky Way's gas and dust in the disk. Star dust literally courses through our veins.

Cloud collapse is one of the least-understood steps of star and planet formation. An interstellar cloud must contract by many orders of magnitude to form stars with planets. When contracting, a cloud will heat up as part of the gravitational energy is converted into the thermal energy of the motions of its atoms and molecules. The rising temperature increases the cloud's pressure support. It will stop contracting if it cannot cool by radiating heat in the far infrared parts of the spectrum, since infrared photons can escape the dense cloud. How well a contracting cloud can cool by radiation depends on its makeup. Hydrogen and helium lack very strong emissions in the infrared, so a cloud without metals must be fairly massive for its self-gravity to overcome the internal pressure. When metals are present, carbon and oxygen are the main coolants (via ionized carbon, carbon monoxide, and neutral oxygen). Thus, a metal-rich cloud is more likely to fragment into smaller cloudlets to form more low-mass stars.[19] Since massive stars make inhospitable hosts for life, to put it mildly, it follows that metal richness may be an important ingredient in a system's future habitability.[20] Specifically, it is our old friends carbon and oxygen that play a key role in forming stars, while serving as the best probes of the spiral-arm structure of the galaxy.[21]

Gas giant planets like Jupiter, for all their visual interest, are mostly hydrogen and helium. At present, the most popular model for the formation of a gas giant is called core instability accretion.[22] According to this model, the future gas giant must first form a rocky core of at least ten to fifteen Earth masses. Then the growing planet's gravity can attract and retain the plentiful hydrogen and helium in the protoplanetary disk. This can lead to a runaway growth of the planet. Thus, a minimum amount of metals is required so that the rocky core forms quickly before most of the gas is lost from the system (on a timescale near ten million years). Though we still have much to learn, the recent discoveries of planets outside our Solar System are helping us to get a handle on the value of this threshold metallicity.[23] Astronomers aren't finding giant planets around stars with less than about 40 percent the metal content of the Sun.[24]

Conversely, too much metal in the mixture can pose a different set of problems. The higher the initial allotment of metals in its birth cloud, the more planetesimals—early asteroid-sized planetary building blocks—and planets will form in a system, greatly complicating the orbits. In a planetary version of pinball, the giant planets might migrate inward by scattering the many tiny bodies. And if being metal-rich increases the chances that more giant planets will form, then they will also be more likely to perturb each other and destabilize the system, even dumping terrestrial planets into the host star or flinging them out of the system entirely.[25] Even if the giant planets don't migrate much, frequent bombardment from the many remaining asteroids and comets—which will be more abundant in a metal-rich system—will threaten planetary life.

Unlike gas giants, terrestrial planets contain very little hydrogen and helium, so they do not have to reach a critical mass to grow to their final state. Terrestrial planets in our Solar System apparently formed in the inner, hot regions where hydrogen, helium, and other abundant gases could not condense. Earth-size terrestrial planets have less chance of forming

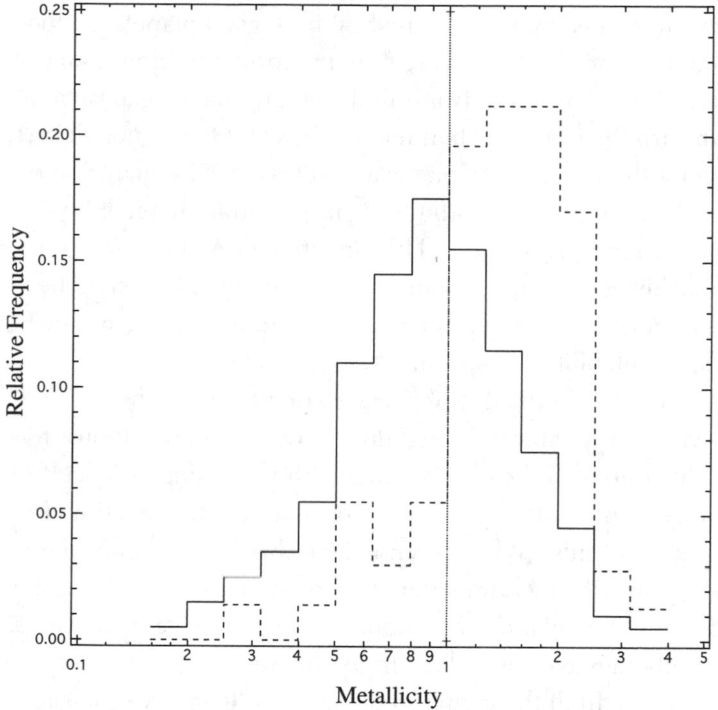

Figure 8.6: Metallicity distributions of seventy-one stars with giant planets detected with the Doppler method (dashed) and about one thousand randomly selected nearby Sun-like stars (solid). The relative incidence of stars with giant planets increases sharply for metallicity values greater than the Sun's (vertical dotted line). (The similar heights of the two distributions do not imply that stars with giant planets are as common as the randomly selected stars; they have been adjusted to similar heights for clarity.)

in a system condensing out of a more metal-poor cloud, such as a globular cluster.[26] Indeed, in the extreme case of a cloud without metals, where not even fist-sized rocks will form, making the right size terrestrial planets becomes impossible. On the other hand, with a high initial inventory of metals you're likely to end up with a busy, chaotic system with a bevy of large planets constantly irritating one another into a very inhospitable state.

Because different processes produced gas giants and terrestrial planets, differences in the initial metal content probably led to different ends.[27] And since it now looks as if you need both gas giants and terrestrial planets to form a habitable system, the optimum metal content for building a habitable planetary system might be quite narrow.[28] Recall that relatively few stars more metal-poor than the Sun have giant planets. Suppose that the amount of heavy elements required to form a minimum-mass habitable terrestrial planet is only a little less than that for giant planets that don't disturb their orbits. Then few systems will have enough metals to form an Earth-size terrestrial planet and at least one gas giant planet of the right mass and in the right orbit to do its important habitability-boosting work for the terrestrial planet.[29] The formation of asteroids and comets will also probably depend on the initial metal content differently. These need to be neither too common nor too uncommon, further narrowing the range for habitability in a system.

How does the abundance of heavy elements vary with time and place in the Milky Way galaxy? Several diverse types of observations show that the metal content of the disk declines with increasing distance from the galactic center.[30] Galaxies similar to ours appear to have similar "radial metallicity gradients." While it is not the only trend for heavy elements in the disk,[31] this is the most important one for our purposes.[32] The metal content of stars in the solar neighborhood is somewhat diverse, however, varying from about one-third to three times the Sun's.[33]

As we move from the center to the edge of the galaxy, the gas density drops, and with it, the rate of star formation. Since stars are the source of most heavy elements in the galaxy, this means that metallicity drops as we move from the center to the edge. Early in the history of the Milky Way galaxy, the metals built up quickly in the inner galaxy as star formation ramped up and peaked after only about two to three billion years. Since then, the star formation rate has been declining, interrupted by sporadic episodes of activity. Today, metallicity in the solar neighborhood is still increasing,[34] but it's expected to grow more slowly in the future as the supply of fresh gas dwindles. All this means that long ago, any region in the Milky Way galaxy was more metal-poor than that area is now. To find enough heavy elements billions of years before our system began to form, you should look at our galaxy's inner regions (see below). But our inner galaxy is one tough, crowded neighborhood, with little patience for some planet trying to pretty itself up with even a little primitive life.

The distribution of the chemical elements has a bearing on life, too. Earth is mostly iron by mass, with oxygen, silicon, and magnesium next in abundance. Surprisingly, the cosmically abundant and life-essential volatile elements—hydrogen, carbon, and nitrogen—are mere trace elements in the bulk Earth. Of course what really counts is their abundance in the crust, where they are much more common. Astronomers believe asteroids and comets delivered most of the life-essential volatile elements to Earth late in its formation, where they were eventually incorporated into its crust. So a terrestrial planet that forms like Earth should be supplied with the volatile elements, but the quantity is critical. The right elements are not always formed in the same proportion in all places and times in the Milky Way.[35]

To learn more about galactic chemical evolution, we must understand how the individual chemical elements are produced. The most abundant elements in Earth were produced primarily in supernovae. There are two basic types, Type Ia and Type II supernovae, which together provide the various elements needed to build a habitable planet.[36] A Type II supernova results when the core of a massive star collapses suddenly, releasing enormous quantities of energy that blow apart its outer layers. Many elements are synthesized in the ensuing maelstrom, and others produced earlier (in the so-called onion layers) are also ejected.[37]

The origin of Type Ia supernovae is less clear. The general consensus is that they result from the detonation of a white dwarf star in a binary star system. Unlike the progenitor of a massive star supernova, a white dwarf star does not have a prominent onion layer structure; it consists mostly of carbon and oxygen. The Type Ia supernova progenitor accretes matter from its companion until it reaches the so-called Chandrasekhar limit, near 1.4 solar masses.[38] At that point, runaway thermonuclear reactions in the interior convert much of the carbon and oxygen into heavier elements.[39]

Today supernovae pop off in our galaxy about once every fifty years. But judging from the present abundance of metals in the galaxy, the supernova rate must have been much higher early on. Their average rate over the entire history of the galaxy must be about one every three years.[40] Both the supernova rate and the relative numbers of Type II to Type Ia supernovae have been declining since shortly after the Milky Way galaxy began forming.[41]

Since supernovae return processed matter to the galaxy, stars forming today should differ from those formed in other times. In particular, there will be less oxygen, silicon, and magnesium relative to iron in interstellar matter as the galaxy ages. All else being equal, this implies that a terrestrial planet forming today will have a larger iron core relative to its mantle compared with one formed in the past. This may narrow the time frame for forming Earth-like planets if this ratio turns out to be critical for maintaining plate tectonics.

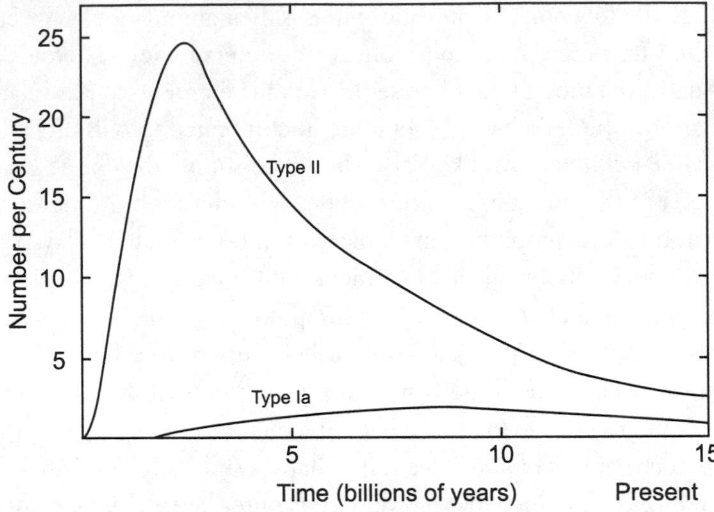

Figure 8.7: Calculated evolution of the supernova rate in the Milky Way galaxy since its formation. Type II (massive star) supernovae have always dominated over the Type Ia (white dwarf binary) supernovae. These reconstructions are based on Galactic chemical evolution models, which are calibrated using the observed chemical abundance patterns of stars of varying ages. In reality the evolution of the supernova rate was probably not as smooth as shown in the figure, but any additional variations were probably secondary in magnitude. These results of these calculations show that the Galaxy was a much more dangerous place early on.

Another important trend is the concentration of the long-lived radioisotopes in the interstellar medium. Compared with iron, radioactive isotopes of potassium, thorium, and uranium are becoming less abundant because the ratio of Type II to Type Ia supernovae is declining (only Type II events return these radioisotopes to the interstellar medium). So again, all else being equal, Earth-mass terrestrial planets forming today will, in 4.5 billion years, have about 60 percent of the internal heating available from radioactive decay in Earth today. Therefore, plate tectonics will probably not be as long-lived in future terrestrial planets, and a planet with less vigorous convection in its core may not develop a strong magnetic field.[42]

In summary, there weren't enough basic elements to build Earth-mass planets until the Milky Way was a few billion years old, and then there were enough only in its inner regions. Today regions of its disk within the Sun's orbit should contain plenty of building blocks to make Earth-mass planets. The formation of giant planets and comets, too, depends on the supply of ashes from supernovae, and also affects the habitability of Earth-

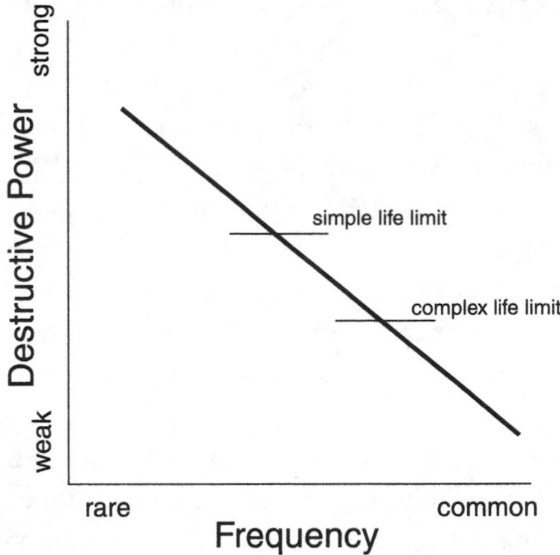

Figure 8.8: Most threats to life on Earth follow a simple rule: the more energetic, the rarer. Thus, solar flares, supernova radiation reaching Earth, asteroid and comet impacts, and volcanic eruptions pose major threats to life over long timescales. Since the threshold for extinction is higher for simple life than it is for complex life, a planet might be able to support simple life but not complex life. This figure also implies that a planet that experiences more frequent low intensity events of a given type, such as distant supernovae, will also experience more frequent high-intensity events of the same type (such as nearby supernovae).

mass planets. The relative abundance of magnesium, silicon, and iron also varies with place and time, leading to differences in planetary geology. Additionally, the typical Earth-mass planet forming in the future will probably have a smaller concentration of geophysically important radioisotopes and, as a result, weaker tectonic activity. By themselves, these factors limit the time and place in our galaxy wherein Earth-like planets can form.

Survival of Complex Life

Even if you manage to get all the right atoms in the right place at the right time to build a terrestrial planet of Earth's mass and orbit, you might not be justified in sticking a "habitable" label on it and shipping it out of your shop. A planet also must be largely free of threats to complex life over the long haul. There are two major long-range extraterrestrial dangers: impacts by large asteroids or comets, and transient radiation events.

Several large extinction events are evident in the geological record. After many years of debate, today most paleontologists are convinced that one of them, the Cretaceous/Tertiary (K/T) extinction sixty-five million years ago, was the result of an impact by a large extraterrestrial body. Very recent research on the 250-million-year-old Permian/Triassic boundary strata also points to an impact event, but its smoking gun is still missing (recall our discussion in Chapter Six on an alternative scenario).[43] (We don't yet know whether asteroids have impacted Earth more often than comets, so for now we'll assume roughly equal threats from each.)

Comets are quirky, visiting the inner Solar System like a transient uncle who drops by unannounced from who-knows-where. Astronomers think that comets spend most of their time in two long-term reservoirs, the Kuiper Belt and the Oort cloud. Until recently, the Kuiper Belt was only theoretical, a vast swarm of icy bodies in fairly low-inclination orbits just beyond Pluto's orbit. Today we know the orbits of about four hundred Kuiper Belt objects. Infrared observations of young nearby stars indicate that most are surrounded by excess dust, suggesting that they too possess Kuiper Belt objects.[44] Astronomers have interpreted changes in the shapes of certain spectral lines in Beta Pictoris, a young star with a dust disk, as infalling comets. Some have interpreted the discovery of water vapor around the aged star IRC+10216 as evidence of a large number of comets around it.[45] We infer the existence of the Sun's Oort cloud, its other major comet reservoir, from the long-period comets that visit our night skies. These comets have highly elliptical orbits—in fact, their orbits are almost parabolic—so they spend most of their time far from the Sun, beyond the Kuiper Belt, typically at about twenty thousand AUs. Since comets in the Oort cloud are only weakly bound to the Sun, it doesn't take much to perturb them.

While the creation of far-flung comet reservoirs around the Sun depends on gravitational deflections by the giant planets, once a comet is in one of these huge, nearly parabolic orbits, it's very sensitive to galactic-scale perturbations. These include the galactic radial and vertical tides, near encounters with giant molecular clouds (GMCs) and passing stars.[46] These galactic tides vary as the Sun oscillates up and down relative to the disk mid-plane on its trek around the center of the galaxy. (Though we can't yet verify it, some astronomers have argued that the period of this vertical oscillation matches Earth's cratering record.) A typical giant molecular cloud contains about half a million solar masses of gas and dust. Extrapolating the Sun's trajectory shows that we haven't passed near a GMC anytime during the last few million years, nor are we in danger of passing near one anytime soon. But stars can pass close to or even through the Sun's Oort cloud at any time. The expected combined effect of all these perturbers is an occasional spike in the comet influx into the inner Solar System superposed on more semi-regular variations.

Comets around a star are gravitationally sensitive to the star's position in the Milky Way galaxy. Stars in the disk are more densely packed the nearer they are to the center of the galaxy. Moreover, a planetary system forming out of a more metal-rich molecular cloud than did the Sun will be likely to form more comets. Therefore, planetary systems in the inner disk of our galaxy probably start with more populated Kuiper Belts and Oort clouds, and they should experience more frequent "comet showers" because of stronger galactic tides and more frequent nearby star encounters.[47] While the outer Oort cloud of an inner disk star will become depleted more quickly, it will also be replenished more quickly from its more tightly bound inner Oort cloud comets.

The threat posed by asteroids in our Solar System depends on the details of Jupiter's formation and orbit (as we discussed in Chapter Four). Galactic-scale disturbances probably have little effect on the dynamics of asteroids. The initial number of asteroids should be proportional to the initial endowment of metals, but their final number will depend on local factors. There is, however, one possible external factor relevant to asteroids: short-lived radioisotopes produced by massive stars in the Solar System's birth cloud. Chemical evidence from meteorites suggests that their parent bodies were heated early on. This additional heating hydrated the minerals in the asteroids, securing the water inside them. If it hadn't, the Sun's heat eventually would have sublimated the water. This is important, because many asteroids carried their water to Earth, where they helped fill its oceans. Without enough short-lived radioisotopes in the asteroids, Earth might never have received enough water.

High-energy radiation may work its destruction on unprotected life less dramatically than large impacts, but it is no less effective.[48] Earth's magnetic field and atmosphere shield surface life from most particle radiation and dangerous electromagnetic radiation. But certain extraterrestrial radiation bursts can damage the ozone layer in our upper atmosphere, resulting in more destructive radiation on Earth's surface.[49] Such "energetic transient radiation events," in order of decreasing duration, include active galactic nucleus (AGN) outbursts, supernovae, and gamma ray bursts. These and other radiation sources are more threatening in the inner regions of the Milky Way galaxy, simply because stars are more concentrated there.

Like a giant dragon sleeping in the heart of a mountain, a massive black hole probably dwells in the center of our galaxy, though at the moment it's fairly inactive. Like our Milky Way, most large galaxies in the nearby universe do not have active nuclei, but recent research with the Hubble Space Telescope is revealing that almost all probably have a giant black hole lurking there. Presumably, a dormant black hole "wakes up" when a star or cluster wanders too close and becomes disrupted. The radiation emanates not from the black hole itself but from the hot accretion disk around it. Fortunately for us, massive early-type galaxies—that is, giant ellipticals—seem to be the favored type of host of the most massive black holes.

Black holes are fearsome objects, distorting space, time, and common sense, so densely packed that not even light can escape their horizons. Massive black holes are thought to be the engines that drive distant quasars and the intensely luminous nuclei of nearby active galaxies, some with visible jets, like M87, a giant elliptical cD galaxy in the Virgo cluster. An AGN produces both high-energy electromagnetic and particle radiation. Most of it is emitted along the rotation axis—that is, "above" and "below" the galactic plane—but many of the charged particles will spiral along a galaxy's magnetic field lines and fill its volume. So, not surprisingly, an AGN can pose a substantial threat to life.[50] The safest place to be during an AGN outburst is in the outer disk, far from the nucleus, and probably close to the mid-plane, where Earth happens to be. The worst place to be is in the bulge, with scorching radiation and stars with highly inclined and elliptic orbits, which can come close to the energetic nucleus or pass through its jet.

Type II supernovae spew deadly radiation as well. Most are concentrated in the thin disk, inside the Sun's orbit, and especially along the spiral arms. The more energetic but less frequent Type Ia supernovae are more uniformly sprinkled in the disk and probably peak near the nucleus. Observations of supernova remnants indicate that the supernova rate peaks at about 60 percent of the Sun's distance from the galactic center, where they are 1.6 times more frequent than at the Sun's position. Supernovae of both types were also more frequent in the Milky Way galaxy during the first few billion years following its formation. Estimates of the rate of life-threatening supernovae in the Sun's neighborhood vary; they average one every few hundred million years.[51]

Finally, there are gamma ray bursts. Astronomers acquired evidence in the late 1990s that these enigmatic events are extragalactic, making them

Figure 8.9: M 87, a giant elliptical galaxy at the core of the Virgo cluster of galaxies, probably has a giant black hole in its nucleus. A "jet" is faintly visible coming out of the nucleus. Giant ellipticals like M 87 offer more dangerous environments and poorer platforms for scientific discovery than flattened galaxies like the Milky Way.

Figure 8.10: M 81, an elegant spiral galaxy in Ursa Major, is one of the nearest large galaxies. The arrow points to the location of supernova 1993J, which occurred inside the galaxy. The image was obtained in May 1993, about two months after the new star first appeared. All the other bright stars in the image are foreground stars in our Galaxy.

among the most energetic transient events since the Big Bang.[52] Wherever they come from, there is only one form of protection against them: location. Perhaps we're just lucky to have avoided a direct gamma ray burst jet, but it helps that we are in a region that's not densely packed with stars, as the galactic center is.[53]

We're also lucky we're not in a globular cluster, and not just for the reasons mentioned earlier. Because their orbits intersect the disk nearly perpendicularly, globulars pass through the disk at blistering speeds—about 250 kilometers per second or more. At such speeds, hydrogen atoms in the disk impacting the atmosphere of a planet in a globular would produce deadly X-rays. Impacting dust would deposit enormous quantities of energy as well—even worse than an Iraqi sandstorm.[54] Bulge and old disk stars would suffer from similar if somewhat less extreme threats. In contrast, the Solar System, with its very "cold" orbit in the thin disk (see below), is unlikely to suffer from such threats. Therefore, globular clusters are surely inhospitable environments for life, and bulge and old disk star systems, while not quite as bad, aren't exactly the Ritz Carlton of the Milky Way either.

We've now reviewed all the galactic-scale factors—at least all the ones we can think of—that set the boundaries of the Galactic Habitable Zone, largely from a theoretical viewpoint. But, we have not yet exhausted all sources of information on this topic.

APPLYING THE WEAK ANTHROPIC PRINCIPLE TO THE GALAXY

In the previous chapter we noted some ways in which the Sun is anomalous compared with other, nearby stars. We interpreted them within the framework of the Weak Anthropic Principle (WAP). We can play the same game with the Sun's composition, place, and orbit in the Milky Way galaxy.

Compared with other nearby main-sequence stars of similar age, the Sun is metal-rich. Ever since the Sun formed 4.6 billion years ago, our region has grown richer in metals, but the local interstellar gas is only now reaching the Sun's metallicity. Since Earth is made almost entirely of metals, building terrestrial planets requires a certain minimum metallicity of the interstellar gas. Moreover, the Sun's moderately high metallicity compared with that of other stars might mark the minimum required to build terrestrial planets as massive as Venus and Earth. At the same time, the Sun has a smaller metallicity than most stars observed to host giant planets. As we mentioned above, it appears that metal-rich planetary systems will have giant planets in either very short-period orbits or larger, eccentric orbits, both problematic for terrestrial planets hoping to host life. Thus, the Sun's metal content is anomalous compared with the general field star population and with the more select group of stars with giant planets. Perhaps its allotment of heavy elements, then, is near the golden mean for building Earth-like planets.

The Sun's orbit in the disk is also more nearly circular than most other stars of its age, and its motion perpendicular to the disk is less pronounced. Our location within the Milky Way galaxy might also be special. Of course, every place in our galaxy has its own peculiar features. Some places are less ordinary than others, however, such as the nucleus and what astronomers call the corotation circle. This circle is that place in the disk of a spiral galaxy wherein a star revolves around the galactic nucleus with the same period as the spiral arm pattern. If a star's orbit is near but not too near the corotation circle, then it will cross spiral arms less often than it would otherwise. But stars actually at the circle will resonate with the spiral arm pattern, eventually getting sent on large excursions[55] and invariably visiting spiral arms. So for maximizing the time away from spiral arms, a star should be close to, but not actually at, the corotation circle.

This matters because the giant molecular clouds lurking in spiral arms pose several threats to life. Complex life couldn't recover from extinction events if spiral-arm crossings occurred too often. The Sun's nearly circular orbit and proximity to the corotation circle make it less likely that it has recently crossed or will soon cross a spiral arm.[56]

Still, the fact that we're nestled between the major spiral arms doesn't guarantee that we will avoid all supernova threats. Spiral arms are ragged, with spurs that protrude into the areas between the arms. For example, the nearest cluster of massive stars is the Scorpius-Centaurus OB association (Sco-Cen), at about four hundred light-years; it's not inside a main arm. Fortunately, this is too far away to pose a serious threat to us. (Interestingly, it was closer to the Sun's present position five to seven million years ago, but the Sun was then somewhere else.[57]) Antares, a bright-red star visible in the constellation Scorpius, should be the next Sco-Cen member to go supernova, safely distant from us. Perhaps the association's most massive stars blew away excess gas and dust from the region we're presently passing through, giving us a clearer view as a result. We might not be here today if the timing of our passage through the disk had been a little different.[58]

The Weak Anthropic Principle also suggests that the Sun's gentle vertical motion and present position relative to the mid-plane may contribute to Earth's habitability. As we noted above, the maximum height the Sun reaches above the mid-plane is less than that of the typical star of its age.[59] Right now we're about fifty light-years from the mid-plane, on our way out. During a full vertical oscillation cycle, a star spends most of its time near the extremes, where it is moving slowest. Hence, a random sampling of a star's position should find it far from the mid-plane. One would think that staying close to the mid-plane, with all its massive stars and spiral arms, would be dangerous. On the other hand, interstellar dust and gas stay very close to the mid-plane, offering significant protection from the UV radiation emitted by nearby Type II supernovae. But large excursions from the mid-plane lead to another threat. Like a ball on a spring, a star with larger vertical amplitude than the Sun will pass through the disk at greater speed. This may pose a problem if the star plunges quickly through a dusty region, since the dust will heat up a planet's atmosphere.

AN EXCLUSIVE COUNTRY CLUB

In light of all these factors, we see that the Galactic Habitable Zone is a rather exclusive country club for observers. Compared with our present location, the inner ghetto of the Milky Way suffers from greater radiation threats and comet collisions, and an Earth-size planet is less likely to form there in a stable circular orbit. The outer regions are safer, but stars there will be accompanied by only fairly small terrestrial planets, planets too small to retain an atmosphere or sustain plate tectonics. While we can't yet say how wide it is, our best guess is that the GHZ is a fuzzy annulus (or ring) in the thin disk at roughly the Sun's location, a ring whose habitability is compromised where the spiral arms cross it. If proximity to the corotation circle is important for habitability, then this thin and often broken ring could be narrower still. (See Plate 17.)

At the same time, the Galactic Habitable Zone—and, more specifically, our location in it—offers one of the best overall locations to be a successful (galactic and stellar) astronomer and cosmologist. Even though we're near the mid-plane, there's very little interstellar extinction in the solar neighborhood. The disk is highly flattened and so less obstructive, and we're far enough from the galactic center to keep it from excessively obscuring our view of the distant universe.

Our model of this habitable zone is still incomplete, as is our understanding of the threats to life and of its basic requirements. But we're hopeful that advances in astronomy will answer many of the questions we've posed.[60] If the trend that has held since galactic astronomy came of age a few decades ago is any indication, the estimated size of the GHZ will continue to shrink.

OTHER GALAXIES

As if our galaxy's habitable zone weren't exclusive enough, the broader universe looks even less inviting. About 98 percent of galaxies in the local universe are less luminous—and thus, in general, more metal-poor—than the Milky Way.[61] So entire galaxies could be devoid of Earth-size terrestrial planets.[62] In addition, stars in elliptical galaxies have less-ordered orbits, like bees flying around a hive minus a bee's capacity to react to impending collisions. Therefore, they are more likely to visit their galaxy's dangerous central regions.[63] They're also more likely to pass through interstellar clouds at disastrously high speeds (though such clouds are less common in elliptical galaxies). In many ways, ours is the optimal galaxy for life: a late-type, metal-rich, spiral galaxy with orderly orbits, and comparatively little danger between spiral arms.

Interactions between galaxies can also affect habitability. For example, the Andromeda Galaxy is predicted to have a close encounter with our galaxy in about three billion years. Such an event will dislodge most stars in the disks from their regular orbits. It may also feed fresh fuel into our galaxy's central black hole and bring it back to life, making the inner galaxy even less desirable. So our Galactic Habitable Zone may only last another three billion years. Of course, lots of stars will be flung into intergalactic space. Any surviving inhabitants in such systems could remain safe, unless they get too close to the active nucleus of a large galaxy.

The local density of galaxies also affects habitability. The Milky Way galaxy is part of the Local Group, a collection of about thirty-five galaxies; the largest three are spirals, and the remainder are small dwarf spheroidal and irregular galaxies. Compared with other concentrations of galaxies in the local universe, ours is somewhat sparse. Rich clusters, like Virgo and Coma, each contain thousands of galaxies. Close encounters, called "galaxy harassment," are common among these galaxies, and galaxies aren't restrained by good manners. The smaller galaxies can lose much of their gas during the close encounters, like the two Magellanic Clouds, both irregular galaxies orbiting our galaxy. The motions of galaxies in clusters also heat up the gas between them, which, in turn, strips gas from the outer disks of the galaxies. Massive cD elliptical galaxies are found in the centers of many rich clusters, presumably having grown at the expense of many hapless smaller galaxies. With their intense nuclear activity, such super galaxies would be anything but super places to live. Overall, then, rich clusters are probably less habitable than sparse groups.

Clearly, other regions of the Milky Way galaxy and other regions of the nearby universe are, by and large, quite different from our present location. Precious few plots of galactic real estate are as amenable to complex life as ours, to say nothing of its value for observation. Our home is a fairly comfortable porch from which we can gaze out to the ends of space and the beginning of cosmic time. And as we'll see, not all times in the history of the universe are like our present.

Chapter 9

Our Place in Cosmic Time

The progress made in our understanding of the universe during the twentieth century is nothing short of stunning.
—Michael S. Turner[1]

COSMIC DISCOVERY

Finding Our Place and Time

In the 1920s, American astronomer Edwin Hubble began a careful study that led to a rediscovery of the reality of cosmic time. It began as a fairly mundane research project. Using the one hundred–inch Hooker telescope at Mt. Wilson in California—then the world's largest—he was studying the Andromeda nebula, and later extended his work to other so-called spiral nebulae. At least since the time of Immanuel Kant scientists had wondered whether these football- and cigar-shaped objects were nearby and smallish, or distant and enormous. Kant had conjectured that they might be "island universes" in their own right.[2] With cutting-edge telescopic power, Hubble was able to resolve individual Cepheid variable stars—a type of astronomical standard candle—in the spiral nebulae M31 (Andromeda), M33, and NGC 6822.

With further study, Hubble noticed that the spectra of many nebulae tended to be "shifted" toward the longer-wavelength, red end of the electromagnetic spectrum compared with our Sun and nearby stars. By combining these redshift data with distance measurements, he eventually

Figure 9.1: Edwin Hubble, seated in a bentwood chair, looking through the Newtonian focus of the Mt. Wilson one hundred–inch telescope (c. 1922).

discovered that the more distant a nebula, the greater its redshift.[3] His discoveries suggested shocking and revolutionary answers to three contested issues in astronomy and cosmology. His measurements of Cepheids showed that the Milky Way, contrary to the common assumption at the time, was not to be equated with the universe. They further demonstrated that the spiral nebulae are really distant galaxies, perhaps as our Milky Way galaxy would appear from a distance. (See Plate 18.)

Finally and most significantly, the combined discoveries implied that the universe itself was expanding. In an historical instance of the left hand not knowing what the right hand is doing, Albert Einstein's General Theory of Relativity had already predicted that the universe was either expanding or contracting. Unfortunately, Einstein found the notion so distasteful that he had introduced a "fudge factor," a variable called a cosmological constant, theoretically retrofitted to keep the universe in steady, eternal equilibrium. But upon learning of Hubble's discovery, Einstein made a widely publicized trip to California to see Hubble's data for himself. As a result of Hubble's discoveries, and the works of Georges Édouard Lemaître, a Belgian Roman Catholic priest and physicist who had studied under Arthur Eddington, and Soviet Aleksandr Friedmann—whose solutions to Einstein's theory implied an expanding universe—he repented of his cosmological constant, famously calling it the "greatest blunder" of his career.

Like Einstein, most astronomers of the early twentieth century, including the young Hubble, believed in a static and eternal universe. Even after Einstein conceded his error in the late 1920s, many scientists would not accept the implications of an expanding universe—namely, that it came into existence sometime in the finite past. (One critic, Fred Hoyle, dubbed such an event the "Big Bang," and the name stuck.) For example, consider the account C. F. von Weizsäcker gives of a discussion he had with the physical chemist Walther Nernst in 1938:

> He said, the view that there might be an age of the universe was not science. At first I did not understand him. He explained that the infinite duration of time was a basic element of all scientific thought, and to deny this would mean to betray the very foundations of science. I was quite surprised by this idea and I ventured the objection that it was scientific to form hypotheses according to the hints given by experience, and that the idea of an age of the universe was such a hypothesis. He retorted that we could not form a scientific hypothesis which contradicted the very foundations of science. He was just angry, and thus the discussion, which was continued in his private library, could not lead to any result.[4]

Most scientists trusted observations more than Nernst's definition of science, and eventually accepted the expanding-universe model. It helped that Hubble's discoveries came at just the right time, when theoretical cosmologists were just beginning to ponder the overall properties of the universe within their new mathematical framework.

It also marked a striking convergence of theory and discovery. In the 1920s, Lemaître and Friedmann had first proposed expanding models of the universe derived from Einstein's equations. Friedmann saw that General Relativity implied that "at some time in the past (between ten and twenty thousand million years ago) the distance between neighboring galaxies must have been zero."[5] Lemaître was the first to describe an early version of the Hot Big Bang model (although he didn't give it that name). As he put it, "The evolution of the world can be compared to a display of fireworks that has just ended: some few red wisps, ashes and smoke. Standing on a well-chilled cinder, we see the slow fading of the suns, and we try to recall the vanished brilliance of the origin of the worlds."[6]

Stellar Standard Candles

But Hubble's insight rested on a prior discovery. In 1908, Henrietta Leavitt was diligently studying the periodic light variations of Cepheids in the Magellanic Clouds for Harvard College Observatory. (The Large and Small Magellanic Clouds are the Milky Way's largest satellite galaxies and are only visible from the Southern Hemisphere.) She eventually discovered that those Cepheids with longer periods were brighter. Because all the Cepheids in a cloud are nearly the same distance from Earth, the correlation between period and apparent brightness translates into a relation between period and luminosity (*absolute* brightness). Like cosmic lighthouses, Cepheids communicate through the simple ebb and flow of light: *Slower is brighter*. Satellite galaxies where stars were still being born provided the critical clues for discovering this period-luminosity (P-L) relationship. Classical Cepheids are massive stars that last only a few million years, and so dwell only where stars are forming. After astronomers had calibrated the P-L relation by observing a few nearby Cepheids,[7] Hubble and other astronomers were able to determine the distances to the spiral nebulae. Those Cepheids with the longest period, nearly fifty days, were among the most luminous stars known. For these reasons Cepheids were the first practical extragalactic standard candles. And as with much of the evidence on which modern cosmology is based, if our setting had been different, we might never have seen them.

We mentioned in Chapter Seven that astronomers have used massive main-sequence stars as standard candles. And early in the twentieth century, astronomers such as Harlow Shapley had some success in estimating distances with the low-mass variable stars known as the RR Lyraes.[8] But

Figure 9.2: Albert Einstein (1879–1955).

none could compete with the Cepheids. The brightest are over four hundred times brighter than the RR Lyrae variables.[9] As a result, Hubble was able to measure the brightest Classical Cepheids in the Andromeda Galaxy. Since then, astronomers armed with better telescopes and detectors have observed them in ever more distant galaxies, the greater reach allowing them to refine the relation between distance and redshift.

Over the past five or six years, astronomers have concluded that Type Ia supernovae are also exceptionally good standard candles. As we explained in the previous chapter, one of the end states of a moderate-mass star in a binary system is a Type Ia supernova, which can pop off a billion years or so after the birth of its progenitor. This means that Type Ia supernovae can be seen in galaxies where stars are no longer forming. They are extremely luminous, outshining all the other stars in their host galaxies for a few weeks. And since they occur in both nearby and distant galaxies, and in spirals and ellipticals, they probe a substantial fraction of the visible universe.

Once astronomers discover a Classical Cepheid or Type Ia supernova in a galaxy, they can establish its distance. Just how reliably they can do this depends largely on how "standard" the standard candle is and how well they have calibrated it. To calibrate a cosmological standard candle, an astronomer must first establish a "distance ladder." Earth's surface is the lowest rung on that ladder. Next come the Moon's and then Earth's orbit, which serves as the baseline for stellar parallax measurements. Once astronomers determine the distances to a few luminous "secondary candles" from their parallaxes and other, less direct means, they can measure much greater distances[10] and calibrate other types of standard candles.[11]

DISCOVERING THE ECHOES OF CREATION

While most astronomers were convinced by Hubble's observations that the universe was expanding and that this implied a beginning, a few still held out for an eternal past. Eddington, in speaking for many, said that "philosophically, the notion of a beginning to the present order of Nature is repugnant."[12] In the early 1930s, Eddington still resisted Lemaître's hypothesis of a dense, hot beginning. Instead, he opted for a universe with a cosmological constant, one that had fallen out of balance from an eternal static state in favor of expansion. Eddington's model suffered from obvious logical problems, so most astronomers had grudgingly accepted the Lemaître model.

The first serious challenge to a finite past came in 1948, when Hermann Bondi, Thomas Gold, and Fred Hoyle independently proposed Steady State models, postulating matter that spontaneously appears in the space between the galaxies as the universe expands. This would maintain a constant matter density on large scales and accommodate the observations of an expanding universe, but avoid the troublesome idea of a beginning. Steady State, as the name implies, described a universe without significant change when averaged over long timescales. For several decades, no direct evidence was available to allow astronomers to decide between Big Bang and Steady State models.

In 1965, however, two engineers at Bell Telephone Lab, Arno Penzias and Robert Wilson, noticed excess noise in a radio antenna at seven centimeters wavelength. They found it coming from all directions of the sky with equal strength, and couldn't attribute it to any known sources of radiation. Cosmologists quickly interpreted this finding as the long-sought relic radiation from the Big Bang, which, unbeknownst to Penzias and Wilson, some physicists had already predicted. (Because it was detected in the microwave part of the electromagnetic spectrum, it's called the cosmic microwave background radiation, or CMBR. Although we tend to associate "microwave" with boiling water and burnt popcorn, the background-radiation spectrum corresponds to a blackbody emitter 2.7 degrees above absolute zero.) Because it was a prediction of the Hot Big Bang model and not of the Steady State model, Penzias and Wilson's discovery effectively sounded the death knell for the latter.

Ralph Alpher, George Gamow, and Robert Herman presented the first modern picture of the Big Bang's residual radiation in the late 1940s. They understood that the expanding-universe models implied a much denser and hotter time in the distant past and that if we went back far enough, we would find a time when the density of matter prevented the free flow of photons, when matter and radiation were in "thermal equilibrium." The temperature and pressure were high enough to ionize most of the hydrogen, the most abundant element. Prior to that, photons did not travel far before being scattered by free electrons. In a sense, the radiation "decoupled" from the matter after the protons and electrons recombined. Most photons produced during the decoupling era (CMB photons)—when the gas in the universe was about four thousand degrees Kelvin (6,700 degrees Fahrenheit)—traveled unimpeded as the universe continued to expand. Since then, the expansion of space and time has stretched the photons, redshifting them about a thousand times. These cosmic microwave background photons—the oldest in existence—have been traversing space and time for most of the history of the universe. Thus, the CMB sky is sometimes called the surface of last scatter, since it represents the last time these photons strongly interacted with matter before setting off on their long journey.

Current models indicate that the CMB photons were liberated about 380,000 years after the Big Bang event. It took that long for the universe to expand and cool enough for protons to combine with electrons to form neutral hydrogen. This cosmic expansion has preserved the shape of the spectrum of the cosmic microwave background radiation—a critical clue to its origin. This is because the universe today is far too transparent to photons for the background radiation to have been produced in its present state; in other words, the radiation is too far out of equilibrium with the matter in the present universe to account for its origin. The present background radiation points back to a time when the universe was much denser and hotter.

The Steady State models failed to predict not only the background radiation but also the more recent observations showing reverse galaxy evolution with distance. When we look farther and farther into space, we are really looking back in time. And as we peer deeper and deeper into the past, we see galaxies in much earlier stages—just what we would expect if the Big Bang theory is correct. In the famous images taken by the Hubble Space Telescope called Hubble Deep Fields, we see galaxies whose redshifts place them some nine billion light-years away. This means we are seeing distant galaxies not in their "present" state but as they existed some nine billion years ago. Similarly, the background radiation delivers information about the newborn universe. John Barrow and Frank Tipler describe the significance of this:

> The background radiation has turned out to be a sort of cosmic "Rosetta stone" on which is inscribed the record of the Universe's past history in space and time. By interpreting the spectral structure of the radiation we can learn of violent events in the Universe's distant past.[13]

At the same time, we learn something of the expansion of the universe, since changing space and time modified the CMB photons on their way to us. Cosmologists especially value the subtle fluctuations in the intensity, and hence temperature, of the background radiation across the sky. Caused by sound waves traveling through the ionized gas just prior to decoupling, these fluctuations are an unexpectedly rich source of encoded information about the early universe. There were regions of higher and lower gas densities and temperatures, which are visible to us today as variations in the surface brightness of the background radiation on the sky. (See Plate 19.) This is like the information we gain about the Sun's internal structure by studying oscillations on its surface, except that the background radiation is a snapshot in time. Cosmologists can extract at least ten cosmological parameters—that is, overall properties of the universe—from these variations.[14]

But since they don't reveal every cosmological property equally well, these data don't make other observations obsolete. For example, Type Ia supernovae are still especially useful, because they probe so much of the observable universe. Among the most important cosmological properties are the Hubble constant, the matter-energy density, and the cosmological constant, also called the vacuum-energy density. The Hubble constant is the present rate of expansion of the universe, which astronomers determine by measuring the distances and redshifts of galaxies. The matter-energy density is basically the total amount of matter and energy in the universe—not including dark or vacuum energy—which we discern from its light emissions and gravitational effects. The cosmological constant, or vacuum energy, is more mysterious. Although most of us think of a vacuum as pure nothingness, vacuums in the universe contain energy, which, at very large scales, can actually counteract gravitational attraction. Astronomers have

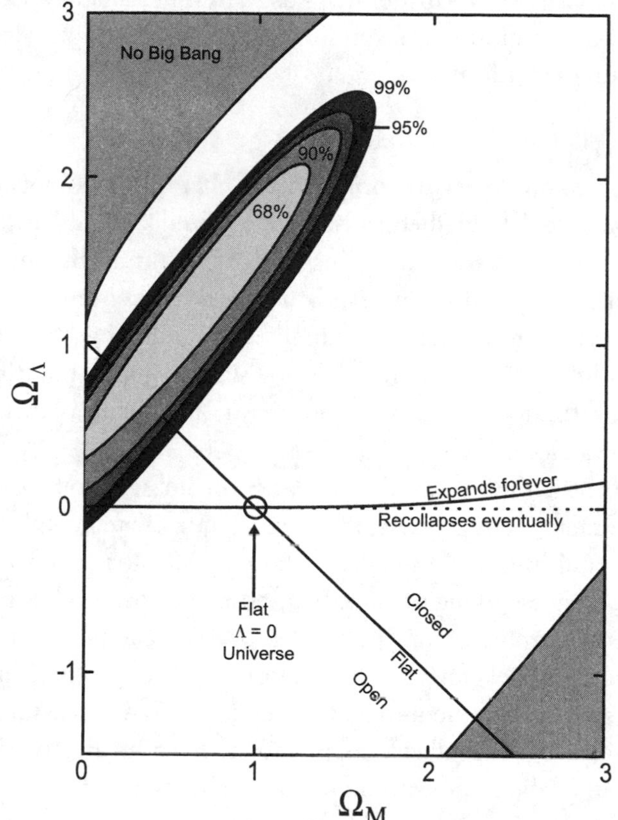

Figure 9.3: Two of the most important cosmological parameters for understanding the geometry of the universe are the matter-energy density (Ω_M) and the cosmological constant-energy density (Ω_Λ). The region of the diagram constrained by Type Ia supernovae data (contours) is at right angles to the region constrained by the microwave background data (near "Flat" line). Because of this, the two sets of data complement each other in determining the values of these two parameters.

recently detected its effects from observations of distant standard candles. We'll discuss this below.

For discerning the latter two, the data gleaned from Type Ia supernovae are nearly perfect complements of certain data derived from the background radiation.[15] Neither is redundant, since together the two sets of data substantially narrow the possible values of these two important parameters. This narrowing permits the most efficient use of our measurements, and demonstrates that all visible matter helps us learn about the universe.[16] And the observed distribution of galaxies fills in gaps in information left from these two sources of information.

COSMIC STEW

In the 1940s, no one knew the origin of the chemical elements. As Big Bang cosmology took hold, theorists soon recognized that the universe had cooled too quickly to form anything beyond lithium. Hydrogen and helium predominate in the universe at large. But whence came all the other elements? Astrophysicists eventually concluded that the elements heavier than lithium were produced inside stars well after the Big Bang. By the late 1950s they had worked out many of the details of this theory, called stellar nucleosynthesis.

More recently, the observed abundance of helium compared with hydrogen in metal-poor galaxies, the deuterium in Lyman-alpha clouds (see below), and lithium in nearby metal-poor stars has confirmed the expectations from Big Bang models. In a sense, the cosmic abundance of the light-element isotopes is a type of telescope that can peer just beyond the time when the background radiation formed. Particle physics and cosmology merge in the moments after the Big Bang. So we have three independent observations that all give support to the Big Bang theory: Hubble's galaxy redshift relation, the cosmic microwave background radiation, and the relative abundances of the light-element isotopes.

FURTHER TESTS OF COSMOLOGICAL EXPANSION

Discovering the background radiation and measuring the light-element abundances added additional support to the Big Bang theory, but they did not demonstrate unambiguously that the observed redshifts are due to cosmological expansion. Fortunately, our universe provides additional tests for the theory, and Earthly astronomers have access to them.

Allan Sandage, an observational cosmologist working at the Observatories of the Carnegie Institution of Washington in Pasadena, is perhaps best known for his decades-long pursuit of an accurate estimate of the Hubble constant. Another of his pursuits is an independent test of cosmic expansion. Richard Tolman proposed the first such test in 1930 with a crucial prediction,[17] which Sandage recently confirmed with observations of thirty-four early-type galaxies.[18] At the same time Sandage was able to eliminate the "tired light" hypothesis, which posits that redshifts result from photons losing energy over vast distances, by showing that galaxies do not change with redshift as the hypothesis predicted.

Three other tests have confirmed the reality of cosmological expansion. One is provided by the apparent broadening of the light curves of the distant Type Ia supernovae compared with nearby ones. This means that the light from a distant supernova seems to wax and wane more slowly than that from a nearby one. According to the Big Bang standard model, this phenomenon is due to time dilation. As the cosmos expands, it leaves behind "stretch marks" in the fabric of space-time. This time dilation was first confirmed in 1995 from supernova light curves.[19]

The last two tests involve the cosmic microwave background radiation: one confirmed the change in its temperature with redshift,[20] and the other the shape of its spectrum.[21] All but one of these tests is possible only because we can see other galaxies. Access to these distant galaxies has allowed cosmologists to eliminate many theoretical models of the universe. What a loss to our inquisitive spirit if we couldn't choose among static, Steady State, and Big Bang models. That our time and place in the universe are apparently fine-tuned not only for life but also for observation has, in one of the most important questions about the universe, made all the difference.

Quasars and Intervening Matter: A Record of Cosmic History

Quasars are another useful probe of the distant universe, since they report about an epoch shortly after matter and radiation decoupled. These powerful beacons are the most luminous and distant objects in the visible universe. Quasars are believed to be galaxies in their early stages, when their central black holes were growing rapidly by accreting gas. Before disappearing into the black hole, the gas forms a very hot, bright accretion disk around it. It is this accretion disk that makes quasars so bright even from vast distances.

Quasars tell us about themselves as well as about intervening matter. As the light from a distant quasar travels through space and time, gas clouds along the way absorb some of it, impressing absorption lines in its spectrum. The specific wavelength of the absorption depends on the amount of cosmic expansion between the background quasar and the absorbing gas, and to a lesser degree, on the relative velocities between them. Quasars emit light most strongly in the ultraviolet part of the spectrum, and the

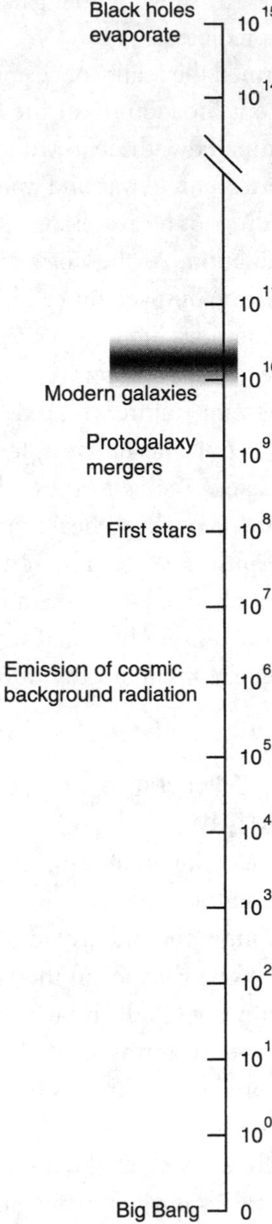

Figure 9.4: Important events in the history of the universe on the way to habitability. The Cosmic Habitable Age (shaded) is a tiny segment of the total span of time that cosmologists often consider (up to 10^{150} years). It coincides with the best time to discover the geometry of the universe.

intervening gas absorbs most strongly at a distinctive spot in the electromagnetic spectrum called Lyman-alpha, due to an atomic transition in neutral hydrogen. In laboratories, its wavelength is 1,216 Ångstroms—the far ultraviolet. But as we look into the heavens from our vantage point, each intervening gas cloud or galaxy has a different redshift and thus produces a Lyman-alpha absorption line at a *different* wavelength. Together, the intervening gas clouds and galaxies produce a Lyman-alpha "forest" of absorption lines.[22] This cosmic forest, like ancient petrified forests on Earth, presents a time-ordered fossil record of cosmic history. Conveniently, cosmic expansion redshifts the ultraviolet forests into the region of the spectrum where Earth's atmosphere becomes transparent (starting at 3,000 Ångstroms). Only this happy circumstance allows astronomers to study them from ground-based observatories. A contracting universe would shift the absorption lines into the far ultraviolet end of the spectrum, making them invisible from the ground.[23]

THE COSMIC HABITABLE AGE

We've discussed the Circumstellar Habitable Zone in our Solar System and the larger-scale Galactic Habitable Zone. But there's a still larger-scale framework, which we can call the Cosmic Habitable Age (CHA). When we consider the universal properties of the observable universe, age is more basic than location. Not all places and times around a star or within a spiral galaxy are equally habitable. Similarly, not all ages of the universe are equally habitable.[24] This is obvious in the very early universe prior to decoupling. At that epoch the universe was a dense, hot plasma of elementary particles and light nuclei. Stars had not yet synthesized the heavy elements that make up our bodies. It was a dreadfully hostile environment for life of any sort. But the beginning is not the only no-man's-land. If we think of everything an environment needs to support life, especially complex life, then, cosmically speaking, probably only a fairly short period in the history of the universe is habitable.

The life-essential elements heavier than helium weren't present in the universe until they were made in the first stars and then ejected from their interiors. The first generation of stars began seeding their environment perhaps a few hundred million years after the beginning of cosmic time. The life-essential elements concentrated more quickly in the larger galaxies, especially in their inner regions. So even if some stars had Earth-size planets only a few billion years after the beginning, they would have been stranded in the most dangerous neighborhoods. Unlike our present, the early universe was poor in heavy elements and rich in high-energy quasars, star births, and supernovae. Early-forming planets in the inner regions of galaxies would have been bathed in lethal levels of gamma ray, X-ray, and particle radiation.

Much of our discussion in Chapter Eight about chemical evolution in the Milky Way galaxy also applies to the broader universe, with some important differences. Since the Milky Way is more massive than most other galaxies, it has accumulated heavy elements more quickly. As a result, planets probably formed around stars in the Milky Way earlier than they did in most other galaxies.

In short, the universe has been getting more habitable. But this trend won't continue indefinitely. We also mentioned in Chapter Eight that in the interstellar medium the long-lived radioisotopes that are so important to geology are declining (compared with stable heavy elements). This decline limits the maximum future birth time of a habitable planet. Some smaller galaxies may grow as metal-rich as the Sun in the distant future, but by then these radioisotopes will be far less abundant. Their planets will produce less internal heat. The typical Earth-size planet forming today will not have as much geological activity in 4.5 billion years as the present Earth does. Of course, a larger planet could compensate for this. The average terrestrial planet should be more massive in the future as the metals continue to build up in the galaxy. But as we've argued, many processes on a terrestrial planet are sensitive to its size—asteroid and comet impact rate, surface relief, ocean depth, internal heat retention, atmosphere loss rate, time to oxygenate atmosphere, and orbital stability. It may be possible to buy some time with a slightly larger terrestrial planet, but much larger planets will create big problems for life.

We can't be precise about the maximum future birth time of a habitable planet, but it's notable that the three most important radioisotopes for heating Earth—potassium-40, uranium-238, and thorium-232—have half-lives of 1.3, 4.5, and 14 billion years. Perhaps it's a coincidence that their average half-life is similar to the lifetime of Sun-like stars in the main sequence. One of these corresponds to the age of the universe, and another to the age of Earth. In any event, the long-lived radioisotopes, like the long-lived stars, effectively delay the release of energy originally produced by the Big Bang. On this evidence alone, we can estimate that the future birth time of habitable planets is probably less than ten billion years. This is a tiny slice of time compared with the timescales cosmologists are accustomed to—up to 10^{150} years into the future.

Although the buildup of metals has made the universe a more habitable place until now, this trend may not continue indefinitely. We noted in previous chapters that those stars that are more metal-rich are more likely to host giant planets. Nevertheless, most of the giant planets around these stars have orbits that are hostile to habitable planets. It's beginning to look as if habitability is optimized for a relatively narrow range of metallicity.[25] This translates into a narrow range of ages of the universe that are acceptable for building habitable planets.

If we combine all the relevant properties of the universe that vary over time—declining star-formation rate, declining high-energy radiation levels, increasing metals, declining radioisotopes—we begin to get a picture of the Cosmic Habitable Age. We can easily rule out the extremes as hostile to life. The very early universe lacked galaxies and planets altogether. And even after they formed, there were still not enough metals to build habitable planets and organisms in safe places. In the distant future, the universe will be dominated by black holes, neutron stars, white dwarfs, and red dwarfs. Any terrestrial planets remaining in close orbits about these stars will have tidally locked rotations and lack tectonic activity.

But the Cosmic Habitable Age may be narrower still, since its defining factors depend on time in different ways and vary from galaxy to galaxy. As a result, many galaxies may have no habitable planets for much of their history. For example, low-mass galaxies with modest star formation may not amass enough heavy elements to build Earth-size planets for another five or ten billion years. By then, long-lived radioisotopes may be too dilute to sustain plate tectonics in their planets. Some small galaxies might already have stopped forming stars. If so, then perhaps only galaxies at least as massive as the Milky Way with the right star-formation histories enjoy the Cosmic Habitable Age. Perhaps we *must* live at just this time in the history of the universe, in the sense that it is the only time compatible with our existence.[26]

A PRIVILEGED TIME AND PLACE

Astronomers entered the twentieth century not knowing the true nature of the Milky Way, spiral nebulae, stellar energy sources, or the origin of the elements or the cosmos. Is the Milky Way the whole show, with the spiral

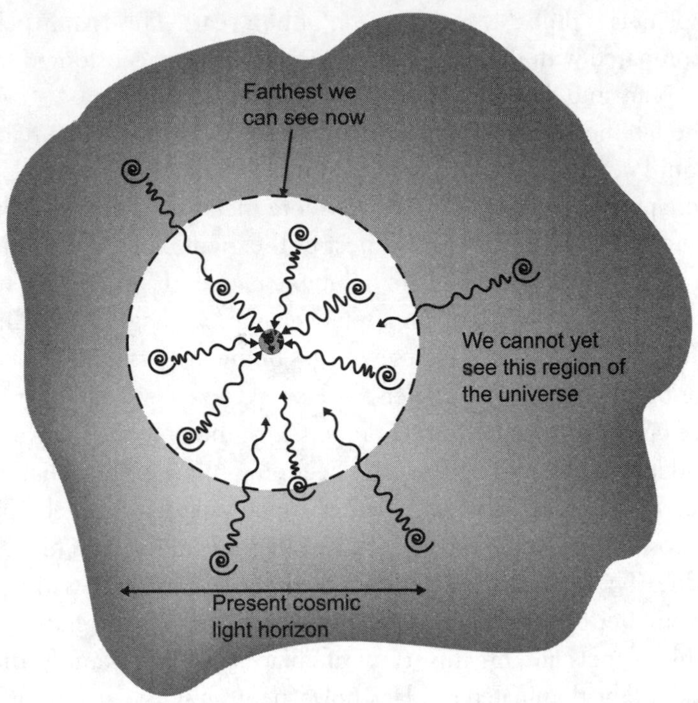

Figure 9.5: The distance we can see at any one time, called the particle horizon, is set by the speed of light and the age of the universe. As the universe ages, the particle horizon will continue to expand, revealing a larger volume. But eventually the effects of the event horizon will begin to reduce the number of objects we can see.

nebulae small objects within it? Does gravity power the stars, or does some new source of energy need to be invoked? Were all the chemical elements formed inside stars? Is the universe static and eternal, or is it dynamically changing and finite? Why is the night sky dark?

Today, we have very good, if sometimes general, answers to all these questions. It is a fascinating time to be an astronomer. Hardly a month passes without the announcement of an important new experiment in astrophysics. As we write this, the new Wilkinson Microwave Anisotropy Probe is measuring the background radiation with unprecedented precision, promising to give us an even more detailed idea of the nature of the universe.[27] (See Plate 19.) We have seen in the preceding chapters how the universe has revealed itself to us in a few brief centuries on size scales ranging from Earth's surface to the nearby galaxies. Our vision now extends to

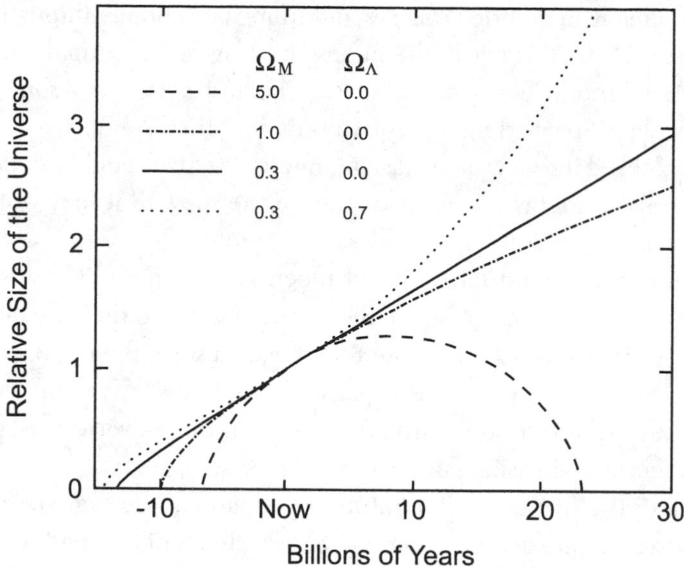

Figure 9.6: Four possible histories of the universe shown in terms of the matter energy density (Ω_M) and the cosmological constant energy density (Ω_Λ). A closed universe (dashed) has enough matter energy density to cause it to collapse. A flat, or critical, universe (dashed-dotted) is just barely able to avoid collapse and expands forever, but at an ever decreasing rate. An open universe (solid) expands forever and experiences less deceleration than a flat universe. An open universe with a cosmological constant (dotted) accelerates indefinitely. This last case best matches what we know about the universe.

the universe at large. As we gaze out into the distant universe, we now know we are peering back in time to an epoch close to the Big Bang event.

We're inclined to marvel at the scientific ingenuity that has allowed us to decode such information. But we shouldn't forget the remarkable conditions necessary for such ingenuity. Our location in the Milky Way allows us to view the distant universe and also the many different kinds of nearby stars, a prerequisite for understanding other galaxies. But for scientific discovery, time may be as important as location at the largest scales.

Hypothetical and bizarrely hearty residents of the early universe—say, a billion years or two billion after the beginning—would have had a front row seat to a spectacular fireworks display of nearby supernovae and quasars, their central black hole engines fed by abundant gas falling in toward their deep gravity wells. The Hubble Deep Fields reveal a young universe filled with distorted galaxies, disturbing one another through close encounters. Partly as a result, the intense heating from the many massive stars and supernovae bequeaths to the galactic dust a bright and sometimes beautiful glow. So when most of the stars in the Milky Way galaxy formed, hot dust blocked the view of the distant universe. If they could have existed, early cosmic residents might have enjoyed the show. But they wouldn't have seen far beyond it.

The universe would have looked much smaller then. The distance observers can see at a given age of the universe is called the particle horizon. It's as large as the distance light has traveled since the Big Bang, taking into account the intervening expansion, and limits the information we can directly glean from the universe. It swells as the universe ages, giving astronomers an increasing sample of the universe.

Whether the universe will continue to expand at the same rate, slow down, or accelerate depends on the contrary effects of the matter/energy density and the dark or vacuum energy density. (Vacuum energy is often called the "cosmological constant," which, unlike Einstein's concept with the same name, does not maintain the universe in static equilibrium.) Imagine the attractive force of gravity competing with the repulsive force of dark or vacuum energy. The central question is, Which one now predominates at the cosmic scale? According to the best current measurements of Type Ia supernovae light curves, dark energy is now overwhelming gravity's tug at cosmic scales, and *accelerating* the universe's expansion. Until about six billion years ago, gravity still ruled the roost, and the cosmic expansion was decelerating.[28] Although this is new evidence, if the prevailing interpretation of the Type Ia supernovae observations is correct, the universe should continue to expand and even accelerate indefinitely.

Today, the universe presents astronomers with about 40 percent of the sources that are in principle observable—quite a large sample.[29] But as the expansion continues to accelerate, objects in the universe will appear with ever greater redshifts and gradually fade from view. The most distant objects will be receding the fastest and will fade first. It is like what a stationary observer would see as an object passes through the event horizon of a black hole. In fact, cosmologists call this cosmic space-time limit of vision an event horizon. The effects of the event horizon are not yet visible, since it is beyond the particle horizon, but we are surprisingly close to seeing them, at least on a cosmological timescale. If the best estimates are correct, then the event horizon will begin affecting our view of the universe in twenty to thirty billion years. After that, the amount of accessible information in the universe will start to taper off. The first to fade from view will be the most distant parts of the universe, such as the background radiation.[30]

Even before the background radiation becomes undetectable, however, it will become ever harder to measure, since it gets fainter as the universe expands and redshifts it. This began at the decoupling era and will continue indefinitely. Moreover, right now we have an advantage: The various independent radiation emissions—cosmic rays, the microwave background, the galaxy's starlight, synchrotron emission from its magnetic fields, and warm dust—*happen* to be enough alike to allow us to do both galactic and cosmic astronomy. About this striking fact, mathematician and cosmologist Michael Rowan-Robinson says, "Possibly we just have to accept this as a coincidence, as we have to accept the similar apparent sizes of sun and moon."[31] But this "coincidence" is an outcome of the particular age of the cosmos, the age that is also the most habitable. At this most habitable of cosmic moments, distinct emissions in a large galaxy like ours are enough like the background radiation in brightness to make them all accessible to us, while differing enough to allow us to distinguish them. If the background radiation were much stronger than the local emissions, galactic astronomy would suffer; if the local emissions were much stronger, cosmology would suffer. As it is we can disentangle the cosmic background and the galactic sources. But this happy situation will not persist forever. Rowan-Robinson is right to compare it with the match between the Sun and the Moon in our sky, but not, perhaps, to shrug it off as a coincidence.

Early in the modern debates in cosmology, the similarities between the cosmic and local sources of radiation suggested to some astronomers, like Fred Hoyle, that the background radiation could have a local origin. Then why are astronomers so confident today that they really have detected the *cosmic* microwave background radiation, rather than a galactic imposter? Thankfully, each of the contaminants has a unique spectral signature. Foreground contaminants include interplanetary dust, warm interstellar dust, radiation emitted by free electrons spiraling along the magnetic field lines, stars, and distant galaxies and quasars.[32] These variously hinder our ability to reconstruct the true brightness patterns of the background radiation on the sky.[33] But our setting greatly helps astronomers separate the background radiation from its foreground contaminants. By themselves, spectra couldn't provide this, but with additional information on the distribution of the radiation sources on the sky, astronomers can subtract the foreground contaminants and produce a "cleaned" CMBR map. Since the background radiation is cosmic, it is expected to be isotropic on the sky, which means that it looks pretty much the same in every direction. In contrast, most local sources of radiation are highly anisotropic. Especially because we live in a highly flattened galaxy far from its hub, we can tease out the anisotropic galactic foreground contaminants from the isotropic background radiation. (See Plate 19.)

Had we been much closer to the galactic center, trying to detect the background radiation would have been, well, hellish. All the local radiation contaminants not only get stronger but also appear more isotropic closer to the galactic center, since the ellipsoidal bulge dominates the disk there. The very worst place for observers to discover and measure this evidence of the Big Bang (or any other cosmological phenomenon) is at the center of the galaxy, since they wouldn't know whether they were detecting local or cosmic radiation. They would get a close view of the giant black hole at the galactic center—like the benefit of a fly getting a brief close-up of a windshield on a racing car. That would be a bad trade, because it's not a unique object; there are many smaller and less hazardous black holes distributed throughout the Milky Way galaxy. The background radiation is unique.

We also benefit from the fact that the Milky Way's nucleus is currently inactive. An active galactic nucleus would emit lots of obtrusive charged-particle radiation. Similarly, studying the background radiation is much easier between spiral arms, where there is less interstellar dust, than inside a spiral arm. In general, then, those very places in the galaxy most threatening to complex life are also the poorest places to measure this echo of the Big Bang (and any other cosmological source).

Most other types of galaxies would also offer poorer views of the background radiation than we get. The worst would be the highly spherical ellipticals and their smaller cousins, the globular clusters. Observers in such systems would be surrounded by nearly uniformly distributed sources of radiation, benefiting from an anisotropy only near their peripheries. Irregular and active galaxies would also be fairly low-rent districts. With intense star formation, active galaxies produce high infrared fluxes from supernova-heated dust. And like adding fog to rain, supernovae themselves and galaxies with active nuclei spew high-energy particles, producing intense radio emissions. You might suppose that this radiation would be easy to detect and correct for, since it comes from a single direction. But charged particles permeate a galaxy along its magnetic field lines, creating the illusion that they originated from multiple directions.

Like everything else, foreground contamination changes with time. The very early universe, before the first stars were born, contained essentially no metals, and hence no dust. Shortly after the first stars seeded the interstellar gas with metals, dust began to form. The frequent supernova explosions warmed it, making it glow brightly. There were fewer stars to contaminate the foreground. But particle emissions would have been more intense, and the background emission would peak closer to the optical region of the spectrum, where more of it would be absorbed by interstellar dust. With all these competing factors, it's hard to say without detailed calculations whether the background radiation was easier to measure in the early universe than it is now. Today dust and particle emissions probably interfere less, but as stars and metals continue to accumulate in our galaxy, and the universe continues to expand, it will become ever more difficult to measure. Our educated guess is that apart from the epoch just before stars began forming, the background radiation is more accessible now than it was in the past or will be in the future (this would be an interesting question for an aspiring young researcher to investigate).

After the background radiation, the distant quasars will be the next objects to fade from view. Future observers will not have quasars to probe matter over vast distances. The continued cosmic expansion will spread the galaxies ever farther apart.[34] In about 150 billion years, galaxies presently beyond about thirty million light-years will fade as they approach the event horizon.[35] Before that happens, these nearby galaxies will be spread out past what are now the farthest known quasars. Again, this is a surprisingly "short" time, compared with the timescales cosmologists often ponder.

We live in a time when there are still plenty of stars forming, but as fewer and fewer stars are produced, there will be fewer massive stars, and thus, fewer Classical Cepheids and supernovae, all valuable measuring tools for astronomers. Millisecond pulsars await a similar fate. Typically, pulsars spin down in a few hundred million years.[36] And since their progenitors were massive stars, millisecond pulsars—the best cosmic clock in the business—will closely track the declining rate of star formation. Type Ia supernovae will keep popping off for a few billion years after stars cease to form, but by then these valuable measuring rods also will be less common.

Since, over time, large galaxies like the Milky Way tend to cannibalize their satellites, the future will have fewer of these galactic groupies. (The Milky Way galaxy is presently consuming the Sagittarius Galaxy.) Future observers may have nothing like our Magellanic Clouds, which serve as middle rungs in the cosmological distance ladder. And not just any satellite galaxy will do. To contain Classical Cepheids, some stars must still be forming. How long would Henrietta Leavitt's discovery have been postponed without something like the Magellanic Clouds?

The distant future will contain fewer objects relatively uncontaminated by stellar nucleosynthesis, making it hard to ascertain the ratios of the light-element isotopes produced in the very early universe. Today we still have a few such samples, such as nearby old stars in our galaxy's halo and ionized nebulae in low-mass nearby galaxies. But as the universe ages, stars will continue to process gas and alter the original light-element abundances away from their original values, and the old, metal-poor stars will continue to die away.[37] And as quasars fade, studying the intervening Lyman-alpha clouds (a useful source of data on the original deuterium abundance) will prove ever more difficult. Much of the information now available will disappear into the recesses of cosmic history.

What if the first observational cosmologists had not been born until, say, A.D. 10 billion? Well, they probably wouldn't have the period-luminosity relation for Classical Cepheids at their disposal, because there probably wouldn't be anything like the Magellanic Clouds orbiting the Milky Way galaxy. With few stars forming, Classical Cepheids would probably be rare in the Milky Way galaxy, too. These observers might know about the less luminous Type II Cepheids and maybe even about their period-luminosity relation, but these standard candles would be faint and thus more difficult to detect in other galaxies. Most galaxies (and galaxy clusters) would be farther away. Nearby galaxies would be more diffuse, and their redshifts greater. Such scientists might eventually discover the Hubble Law and set out to search for the elusive background radiation, by then much weaker than the foreground contaminants. Cosmology would be a much more ponderous and difficult endeavor. Some profound discoveries might simply never be made.

Closer to home, there are several direct connections between Earth's surface and the broader universe. For example, we mentioned in Chapter

Three that radio astronomers measure the motions of the continental plates by observing distant quasars. Cosmic ray particles impacting Earth's atmosphere produce carbon-14 (as well as beryllium-10 and chlorine-36), which is very useful in dating the remains of once living things. Of course Earth's atmosphere does blur the images of astronomical bodies, but an interesting cosmic coincidence reduces its negative impact on our observations of distant galaxies. The angular sizes of distant galaxies do not just decrease linearly with increasing distance, as nearby objects do. Paradoxically, angular size declines until a redshift of about one and then increases for larger redshifts. The minimum angular size of a Milky Way–like galaxy is just about the same size as the blurring effect of our atmosphere.[38] Our atmosphere is not nearly the impediment for viewing the most distant galaxies that one might expect.

In sum, over the next few billion years most galaxies will be farther apart, the particle horizon farther away, the cosmic microwave background radiation dimmer, cosmological standard candles and pulsars rarer, and quasars fainter. Fewer samples of the original light-element abundances will be available, and most profoundly, once the effects of the event horizon start to become apparent in about a couple of Hubble times (a Hubble time is roughly the present age of the universe), some important diagnostics about the universe will gradually disappear. Observers living in the "near" future will enjoy a more distant particle horizon, but at the cost of most of the astrophysical tools astronomers use today. We already have access to a sizable fraction of the maximum theoretical number of radiant sources—about 40 percent. All told, we are living within the best overall age of the universe to do cosmology. What took us only a century to learn would have taken observers in other ages much longer. The Big Bang hypothesis might never have occurred to those living during the ages when the background radiation is not easily observable; unlike us, they might never advance beyond the Steady State hypothesis. Of course, if we're correct about the Cosmic Habitable Age, then it's unlikely that observers could exist at ages distant from ours, so they would be spared the frustration of their ignorance.[39]

PARADOX SOLVED

In 1826, the German astronomer Heinrich Wilhelm Mathäus Olbers gazed at the heavens and asked a deceptively modest question, "Why is the

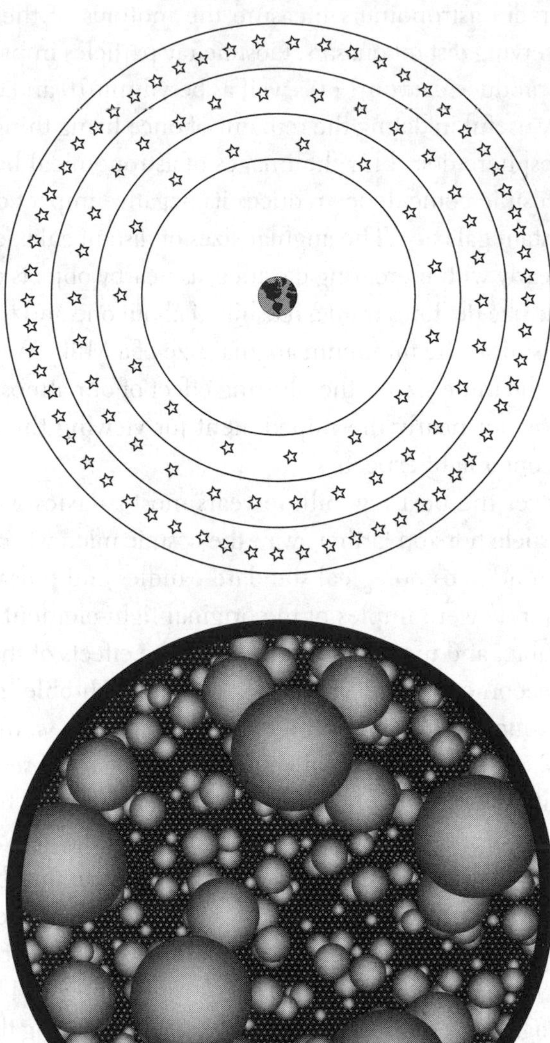

Figure 9.7: Top: Although Kepler had first described it, Olbers wrote more clearly on the paradox of the darkness of the night sky. Consider a spherical shell of stars around Earth. Compared with a closer, equally thick shell, each star in a more distant shell will be fainter, but it will contain more stars; so each successive shell will be as bright as the interior shell. Therefore, if the universe is infinite in age and extent, why is the night sky so dark? **Bottom:** Hypothetical high-powered telescopic view in an infinite and eternal universe. In such a universe every line of sight will encounter the surface of a star. Every patch of the sky will be as bright as the surface of the Sun. Such a universe would be hostile to life and scientific discovery.

night sky dark?"[40] Generations of astronomers learned to call his question Olbers' Paradox. The dark sky seemed to be a paradox because a static and eternal universe uniformly filled with an infinite number of stars—just the universe assumed by many scientists—should produce a uniform and intensely bright sky, day and night. Many proposed solutions to the paradox failed. For instance, some claimed that intervening gas and dust would block much of the starlight. But it became clear that given enough time, even such inert matter would begin to glow hot and bright with the energy absorbed from an infinite swarm of stars.

Figure 9.8: Heinrich Wilhelm Mathäus Olbers (1758–1840).

The modern understanding of a universe with a finite past, initiated by Hubble's discoveries early in the last century, has allowed astronomers and cosmologists to solve this long-standing conundrum. Simply put, there's no paradox if the universe is neither eternal nor infinite in the requisite sense. In fact, a dark night sky is itself evidence for a beginning.

Besides answering a popular scientific riddle, however, this fact is obviously important for both life and discovery. Just imagine the obstacles to observation posed by a sky as bright as the surface of the Sun. It is precisely because the universe is *not* infinite and eternal that we can discover so much about it, despite its enormous size. We can distinguish and separate the variety of information that is transmitted to us from the heavens. And lest we forget the obvious, life in a static and eternal universe bathed in intense radiation would be unlikely to prosper.

As astonishing as the discovery of a changing universe with a finite past was to many scientists at the beginning of the twentieth century, it has conspired with another set of discoveries to transform the way many see the universe and our existence itself.

CHAPTER 10

A UNIVERSE FINE-TUNED FOR LIFE AND DISCOVERY

There is for me powerful evidence that there is something going on behind it all.... It seems as though somebody has fine-tuned nature's numbers to make the Universe.... The impression of design is overwhelming.

—Paul Davies[1]

A UNIVERSE-CREATING MACHINE

Imagine you're taken captive by some powerful aliens, like Q on *Star Trek*, a group of highly intelligent if utterly obnoxious beings who exist as a sort of unified community called the Q continuum. Among their many qualifications, the Q can travel back in time. In the story we're concocting, imagine that the Q transport you back to the moment of the Big Bang. After arriving, one Q takes you into a spacious room, with a large, complicated device on one side, adorned with scores of enormous dials not unlike the dials on a Master padlock. On closer inspection, you notice that every knob is inscribed with numbered lines. And above each knob are titles like "Gravitational Force Constant," "Electromagnetic Force Constant," "Strong Nuclear Force Constant," and "Weak Nuclear Force Constant."

You ask Q what the machine is, and after some snide and dismissive comments about the feebleness of the human mind, he tells you that it's a Universe-Creating Machine. According to Q, the great collective Q continuum used it to create our universe. The machine has a viewing screen that allows the Q to preview what different settings will produce before they press Start. Without going into detail about how it works, Q explains that

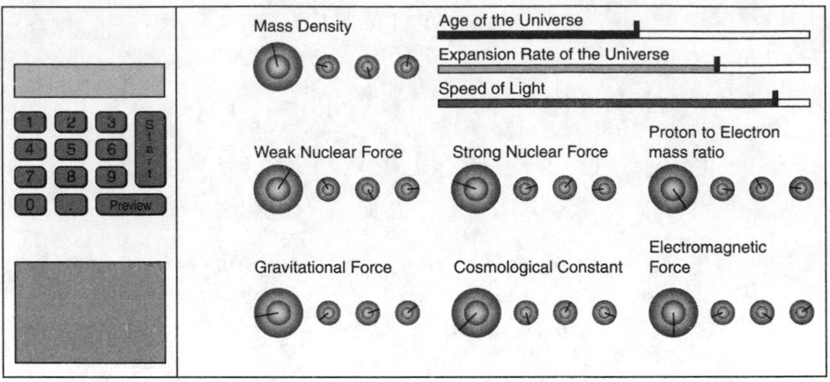

Figure 10.1: One way of understanding fine-tuning is to imagine a Universe Creating Machine with numerous dials, each of which sets the value of a fundamental law, constant, or initial condition. Only if each of the dials is set to the right combination will the machine produce a habitable universe. Most of the settings produce uninhabitable universes.

the dials must all be set very precisely, or the Universe-Creating Machine will spit out a worthless piece of junk (as shown on its preview screen), like a universe that collapses within a few seconds into a single black hole or drifts along indefinitely as a lifeless hydrogenated soup.

"Well, how precisely do the knobs have to be set?" you ask. With some embarrassment, Q tells you that, so far, they've found only one combination that actually produces a universe even mildly habitable—namely, our own. "So," you ask, "do you mean that there are only two habitable universes, the one that the Q exist in, and ours that you created?" In a volatile mixture of anger and chagrin, he admits, "Um, no, there's just this one." This arouses your suspicions: "Now, what sort of bootstrapping magic allowed you to create the universe you yourself exist in?" Crushed by your keen command of logic and highly sensitive baloney detector, Q finally admits, "Well, we didn't actually find the right combination ourselves. In fact, the machine doesn't exactly belong to us. We merely found it, with the dials already set. The machine had done its work before we arrived. Ever since then, we've been looking for another set of dial combinations to create another habitable universe, but alas, so far we haven't found one. We're certain that other habitable universes are possible, though, so we're still looking."

This fanciful story illustrates one of the most startling discoveries of the last century: the universe, as described by its physical laws and constants, seems to be fine-tuned for the existence of life. In the nineteenth century, many scientists thought the universe and our existence had little or no fundamental connection, and that our existence was simply one of the events that happen in an infinite and eternal universe. The apparent fine-tuning of the universe obviously contradicts this.

Physicists sometimes refer to this fine-tuning of the laws and constants as "large number coincidences," or somewhat imprecisely, "anthropic coincidences." But what are they referring to? When physicists say, for example, that gravity is "fine-tuned" for life, what they usually mean is that if the gravitational force had even a slightly different value, life would not have been possible.[2] If gravity were slightly weaker, the expansion after the Big Bang would have dispersed matter too rapidly, preventing the formation of galaxies, planets, and astronomers. If it were slightly stronger, the universe would have collapsed in on itself, retreating into oblivion like the groundhog returning to his hole on a wintry day. In either case, the universe would not be compatible with the sort of stable, ordered complexity required by living organisms.

Specifically, physicists normally refer to the value of, say, gravity relative to other forces, like electromagnetism or the strong nuclear force. In this case, the ratio of gravity to electromagnetism must be just so if complex life as we know it is to exist. If we were just to pick these values at random, we would almost never find a combination compatible with life or anything like it. Given the prevailing assumptions of nineteenth- and twentieth-century science, discovering that the universe is fine-tuned was a *surprise*. Underlying the astonishment is the implication that the range of uninhabitable (theoretical) universes vastly exceeds the range of universes, like our own, that are hospitable to life. Thrown to the winds of chance, an uninhabitable universe is an astronomically more likely state of affairs.

In the previous pages we've detailed many examples of what we might call "local" fine-tuning: those features of our particular setting *within* the universe that make it highly conducive to both life and scientific discovery. Most discussions of fine-tuning in physics, astrophysics, and cosmology, however, have focused only on those features of the overall cosmos that must be just right in order for it to be habitable. Scientists have devoted long books to cases of such fine-tuning.[3] Let's consider just a few.

FINE-TUNING IN CHEMISTRY

Although some scientists had suspected that the universe is fine-tuned for life, it was not until the nineteenth century that they began to gather specific examples.[4] In his 1913 work *The Fitness of the Environment*, Lawrence

Henderson reviewed the known properties of the environment particularly relevant to life, focusing on carbon and water. By drawing on a wealth of data from chemistry and comparing properties of carbon and water with those of other substances, he demonstrated how remarkably well these two chemicals suit living organisms, actual or theoretical. Slight changes in their chemical properties would have a profound effect on the fitness of the environment for life, as we noted in Chapter Two. Certain other elements also seem to be uniquely suited for their biological roles.[5]

Besides being essential for life, carbon and water have other important properties. For example, we have easily minable ores because liquid water is a remarkable solvent (as we discussed in Chapter Three)—a fact that has made high technology possible. And although we have avoided discussing the "laboratory sciences" until now, we can't avoid mentioning water's role in the development of chemistry. Throughout the nineteenth and much of the twentieth centuries, chemists spent most of their time working on reactions in aqueous solutions. If water could not dissolve such a broad range of substances, chemistry would have developed at a snail's pace,[6] perhaps never reaching the status of a science.[7] Thus, the conditions that make our planet congenial to life have also contributed to the rapid development of chemistry as a science.

CARBON AND OXYGEN, ACT II

But such fine-tuning isn't limited to chemistry. In 1952–1953 Fred Hoyle discovered one of the most celebrated examples of fine-tuning in physics.[8] In contemplating the required pathway for the production of carbon and oxygen in nuclear reactions in the hot interiors of red giant stars, Hoyle correctly predicted that carbon-12 must have a very specific nuclear energy resonance not known at the time.[9] A nuclear resonance is a range of energies that greatly increases the chances of interaction between a nucleus and another particle—for example, the capture of a proton or a neutron. An energy resonance in a nucleus will accelerate reactions if the colliding particles have just the right kinetic energy. Resonances tend to be very narrow, so even very slight changes in their location would lead to enormous changes in the reaction rates. This may seem obscure, but think of a wineglass shattering when just the right acoustic note is played. That's a resonance.

The relevant nuclear reactions occur in the stage of a star's life following the hydrogen-burning main sequence, during so-called helium-shell burning. Recall that our Sun will not enter this latter stage for another four to five billion years. Fortunately for us, many moderate- to high-mass stars have already reached this stage and have seeded our galaxy with a healthy dose of carbon and oxygen. During this advanced, helium-shell burning stage, alpha particles (helium nuclei) abound in a star's deep interior, creating frequent high-energy collisions. When two helium nuclei collide, they form an unstable beryllium-8 nucleus; even this is only possible because the mass of two helium nuclei is very close to that of the beryllium-8 nucleus. It remains bound long enough (just 10^{-16} seconds) to collide with another alpha particle to form carbon-12.[10] But this result is not quite sufficient. Because it is effectively a three-body reaction, carbon-12 won't be produced without a resonance.

It was the lack of a known resonance at the energy level required to produce carbon that led Hoyle to make his famous prediction. Since the universe contains plenty of carbon, Hoyle deduced that such a resonance must exist. Had the resonance been slightly lower, the universe would have far less carbon. In fact, the observed abundance of carbon *and* oxygen depends on a few other coincidences. It turns out that the lack of a resonance in oxygen at the typical alpha particle energy in a star prevents all the carbon from being used up to make oxygen (thankfully, the closest resonance is just a little bit too low). But if the fine-tuning stopped there, the universe would have squandered most of its oxygen well before any star system had time even to think about hosting life. You see, certain conservation laws prevent easy capture of alpha particles by oxygen-16 to form neon-20, even though a resonance exists in neon-20 at just the right place. Otherwise, little oxygen would remain.

As a result of these four astounding "coincidences," stars produce carbon and oxygen in comparable amounts. Astrophysicists have recently confirmed the sensitivity of carbon and oxygen production to the carbon-energy resonance; a change in the (strong) nuclear force strength (the force that binds particles in an atomic nucleus) by more than about half a percent, or by 4 percent in the electromagnetic force (the force between charged particles), would yield a universe with either too much carbon compared with oxygen or vice versa, and thus little if any chance for life.[11] Including the other three required fine-tunings further narrows this range.

Notice that the two substances Henderson singled out in 1913 as having the most extraordinary life-friendly chemical properties—carbon and

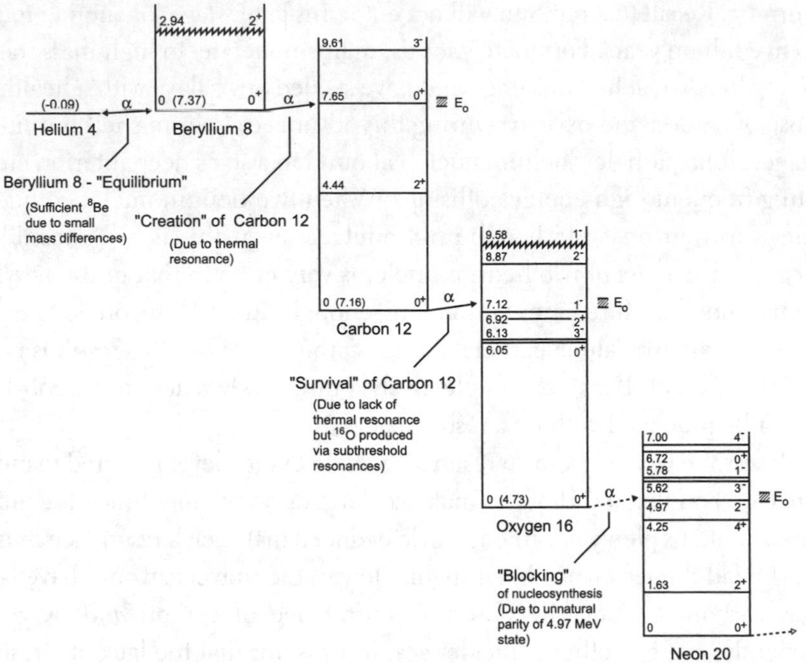

Figure 10.2: The primary nuclear reactions involved in producing nuclei heavier than helium in red giant stars. Four remarkable coincidences give the universe comparable amounts of carbon and oxygen. Energy levels (in MeV) and spin-parity (J^π) are shown for each nucleus. During "helium burning," light elements are built up by acquiring α particles (helium nuclei), starting with two α particles combining to form the very short-lived beryllium-8 nucleus. A "coincidence" is present at each of the four α capture steps from beryllium to neon. The most famous is the presence of a nuclear energy resonance in the carbon nucleus at 7.65 MeV above the ground state. This resonance greatly improves the probability that the very short-lived beryllium nucleus will capture an energetic α particle.

water—would later surprise scientists in a different area of study, as carbon and oxygen nuclei in stars. The first surprise falls under the jurisdiction of chemistry, since it involves only electromagnetic forces. The second brings in nuclear physics. In fact, the most celebrated examples of fine-tuning in nuclear physics relate to the so-called fundamental forces of nature.

LET'S START AT THE VERY BEGINNING . . .

The strengths of the "fundamental forces"—the gravitational, strong nuclear, weak nuclear, and electromagnetic (these last two now often combined into one force called electroweak)—are perhaps the most popular examples of fine-tuning. These forces affect virtually everything in the cosmos. And like those individual dials on the Universe-Creating Machine, each one must take a narrow value to render a life-friendly universe.

The Strong Nuclear Force

The strong nuclear force (often called just the nuclear force) is responsible for holding protons and neutrons together in the nuclei of atoms. In such close quarters, it is strong enough to overcome the electromagnetic force and bind the otherwise repulsive, positively charged protons together. It is as short-range as it is strong, extending no farther than atomic nuclei. But despite its short range, changing the strong nuclear force would have many wide-ranging consequences, most of them detrimental to life. A good example is its role in forming the periodical table of the elements, that friend of seventh-graders everywhere.

Like carbon and oxygen synthesis in stars, the abundance of the heavier elements turns on the details of the nuclear and electromagnetic forces. Because there are no stable elements with atomic weights of five and eight, carbon-12 can only be synthesized through the pathway described above. Stars build the heavier elements with carbon and oxygen ashes. Irrespective of the carbon bottleneck, the periodic table of the elements would look different with a changed strong nuclear force.[12] If it were weaker, there would be fewer stable chemical elements.[13] The more complex organisms require about twenty-seven chemical elements, iodine being the heaviest (with an atomic number of 53). Instead of ninety-two naturally occurring elements, a universe with a strong force weaker by 50 percent would have contained only about twenty to thirty.[14] This would eliminate the life-essential elements iron and molybdenum.

If this were the only consequence of a weaker strong nuclear force, then we might conclude that our universe has two to three times more heavy stable chemical elements than complex life requires. But in a universe with a shorter periodic table, one or more of the isotopes of the light elements would probably be radioactive. The most abundant elements in Earthly life are hydrogen, carbon, nitrogen, and oxygen. If any of their main isotopes were even slightly unstable (with half-lives measured in billions or tens of billions of years), the radiation produced from their decay would pose a serious threat to organisms.[15] In our universe, potassium-40 is probably the most dangerous light radioactive isotope, yet the one most essential to life. Its abundance must be balanced on a razor's edge. It must be high enough to help drive plate tectonics but low enough not to irradiate life.[16]

Further, in a universe with a weaker strong nuclear force, each element would have fewer stable isotopes. As we argued previously, the rich variety of chemical elements and stable isotopes significantly helps researchers measure Earthly and cosmic phenomena. So a large periodic table isn't simply a dirty trick for seventh-grade science students. It makes life possible while greatly enhancing the measurability of the universe.[17]

THE WEAK FORCE

Several key processes relating indirectly to life are particularly sensitive to the weak-force strength.[18] For instance, the weak force governs the conversion of protons to neutrons and vice versa, and the interaction of neutrinos with other particles. The weak force comes into play when a massive star explodes as a supernova—via the energy deposited by neutrinos on the expanding shock front—and when protons and electrons combine in a star's core. This process precipitates the initial collapse, which allows such stars to return their metal-enriched outer layers to the galaxy. Without it, there would not be enough essential elements available for life.

The weak force is also critical in producing primordial helium-4 soon after the Big Bang, in a cosmic cauldron hot and dense enough for brief nuclear reactions. Slight tweaks in the cosmological expansion or in nuclear physics would lead to a quite different end. In our universe the early Big Bang produced about 25 percent helium-4 by mass. Changes in the weak force would produce a universe with a different fraction of helium. Although stars have been cycling the interstellar gas for about 13 billion years, the fraction of helium in the universe has only increased by a few percent. So all stars that have ever formed in our universe have started with similar amounts of helium.

This variable determines a star's luminosity, its lifetime on the main sequence, and the so-called stellar mass-luminosity relation—all important for a planet's habitability.[19] Boiled down to basics, the only property of a star that affects a planet's orbit is its mass, while the only property of a star that affects a planet's surface heating is its luminosity. These two properties are linked through the mass-luminosity relation, which, in turn, depends on the composition of a star's core. Helium stars, like flashbulbs, burn brightly and quickly. Hydrogen stars, in contrast, are more like wax candles. Our hydrogen-burning Sun consumes its nuclear fuel more than one hundred times more slowly than a pure helium star of comparable mass. A helium star of an appropriate mass wouldn't last nearly long enough for life to develop. Not that life would ever develop around such a star anyway: it would contain no water or organic compounds, making the formation of life on any timescale impossible.

It's not even clear that stars could form from contracting clouds of gas in a universe of pure helium. Unlike hydrogen, helium does not form molecules, which are the primary means by which dense interstellar clouds cool, and thereby contract to form stars.

Gravity

Gravity is the least important force at small scales but the most important at large scales. It is only because the minuscule gravitational forces of individual particles add up in large bodies that gravity can overwhelm the other forces. Gravity, like the other forces, must also be fine-tuned for life. Think of its role in stars. A star is in a state of temporary balance between gravity and pressure provided by hot gas (which, in turn, depends on the electromagnetic force). A star forms from a parcel of gas when gravity overcomes the pressure forces and turbulence and causes the gas to coalesce and contract. As the gas becomes more concentrated, it eventually becomes so hot that its nuclei begin to fuse, releasing radiation, which itself heats the gas.

What would happen to stars if the force of gravity were a million times stronger? Martin Rees, Britain's Astronomer Royale, surmises, "The number of atoms needed to make a star (a gravitationally bound fusion reactor) would be a billion times less . . . in this hypothetical strong-gravity world, stellar lifetimes would be a million times shorter. Instead of living for ten billion years, a typical star would live for about ten thousand years. A mini-Sun would burn faster, and would have exhausted its energy before even the first steps in organic evolution had got under way."[20] Such a star would be about one-thousandth the luminosity, three times the surface temperature, and one-twentieth the density of the Sun. For life, such a mini-Sun is a mere "shooting star," burning too hot and too quickly. A universe in which gravity was weaker would have the opposite problem.

Most stars transport the energy generated deep in their interiors to their surfaces by two processes: radiation and convection. In the Sun's case, radiation transports energy most of the way, but convection mostly takes over for its outer 20 percent. Cosmologist Brandon Carter first noticed the interesting coincidence that mid-range mass stars are near the dividing line between convective and radiative energy transport. This dividing line is another razor's edge, a teetering balance between gravity and electromagnetism.[21] If it were shifted one way or the other, main-sequence stars would be either all blue or all red (convection resulting in red stars). Either way, stars in the main sequence with the Sun's surface temperature and luminosity would be rare or nonexistent.

This would surely be a loss for Martha Stewart and other lovers of yellow. But would a universe so well adorned for the Fourth of July be less habitable than ours? Red stars, certainly, would make for less habitable conditions, for some of the reasons we gave in Chapter Seven (such as the slowdown of oxygen buildup in a planet's atmosphere). Very blue stars would be hostile to life, since they would produce too much harmful ultraviolet radiation, though moderately bluer stars might still support life.[22] And we have already shown how much more useful to science the Sun's spectrum is than are bluer or redder stars.

What about planets? A stronger gravity would result in a stronger surface gravity for a planet the mass of Earth, and would also boost the planet's self-compression, increasing the surface gravity even more. Martin Rees notes that a strong-gravity terrestrial planet would prevent organisms from growing very large.[23] Such a planet would also suffer more frequent and higher-velocity impacts from comets and asteroids. Perhaps such a planet also would retain more heat, possibly leading to too much volcanic activity. Of course, these problems could be avoided by having a smaller planet with a surface gravity comparable to Earth's. But a smaller planet would lose its internal heat much faster,[24] preventing long-lived plate tectonics.

Not only would tinkering with gravity change the stars and planets, it would also alter the cosmos as a whole. For example, the expansion of the universe must be carefully balanced with the deceleration caused by gravity. Too much expansion energy and the atoms would fly apart before stars and galaxies could form; too little, and the universe would collapse before stars and galaxies could form.[25] The density fluctuations of the universe when the cosmic microwave background was formed also must be a certain magnitude for gravity to coalesce them into galaxies later and for us to be able to detect them.[26] Our ability to measure the cosmic microwave background radiation is bound to the habitability of the universe; had these fluctuations been significantly smaller, we wouldn't be here.

The Cosmic Tug-of-War

As we discussed earlier, on the largest scales, there's also a cosmic tug-of-war between the attractive force of gravity and the repulsive dark or vacuum energy. Often called the cosmological constant, which is theorized to be the result of a nonzero vacuum energy detectable at cosmological scales, it's one of the few cosmological parameters that determine the dynamics of the universe as a whole. By observing Type Ia supernovae, astronomers have determined that today it contributes about as much to the dynamics of the universe as the gravitational attractive force from visible and dark matter combined. This coincidence remains unexplained, but some cosmologists suspect it's amenable to "anthropic explanation."[27] There's only one "special" time in the history of the universe when the vacuum and matter energy densities are the same, and we're living very near it. If the vacuum energy had become prominent a few billion years earlier than it did in our universe, there would have been no galaxies. If it had overtaken gravity a little earlier still, there would have been no individual stars.

A few billion years might seem like a lot of room to maneuver, but there's an even more striking level of fine-tuning here. The second "cosmological constant problem" is that the observed value of the vacuum energy is between 10^{53} and 10^{123} times smaller than that expected from theory. The vacuum energy density is, basically, the energy density of spacetime in the absence of fields resulting from matter.[28] Until the Type Ia supernovae results demonstrated a few years ago that the cosmological constant is something other than zero, most cosmologists hoped that some undiscovered law of physics required it to be exactly zero. They already knew that its observational upper limit was much smaller than the "natural" values expected from various particle fields and other theoretical fields. These particle fields require an extraordinary degree of fine-tuning—at least to one part in 10^{53}—to get such a small, positive, nonzero value for the vacuum energy. At the same time, its value must be large enough in the early universe to cause the newborn universe to expand exponentially, as inflation theory postulates. How the present value of the vacuum energy relates to the early expansion is yet another issue of debate.

MULTI-TUNING

In most analyses of the fine-tuning of the force strengths and constants of nature, only one parameter is adjusted at a time (to make the problems more tractable). This would correspond to changing one dial at a time on our Universe-Creating Machine while leaving the other dials unchanged. Even taken individually, each of these examples of fine-tuning is impressive. But in the real universe the values of all the constants and force strengths must be satisfied simultaneously to have a universe hospitable to life.

So for instance the strong nuclear force must be set to certain narrow limits for stars to produce carbon and oxygen in comparable amounts, for beryllium-8 to remain bound at least 10^{-16} seconds, to keep the deuteron bound, to allow a minimum periodic table for life, to keep the light abundant isotopes stable, and to keep the di-proton unbound. The range for each of these parameters is narrow. The range within which all of them are satisfied simultaneously is much smaller, like the bull's-eye in the middle of an already tiny target. Add the required range for the weak force strength and the bull's-eye becomes smaller still, and so on for the other forces.[29] Add the specific requirements of simple life (water and carbon chemistries) and it becomes even smaller, and more so for advanced and then technological life.[30]

Eventually, we will have a set of equations, each describing a different constraint on the laws of nature that allows them to permit life.[31] (Determining this complete set of equations may just be the single most important goal of science. We'll leave that as an exercise for the reader.)[32] While physicists still lack the theoretical know-how to do the complete calculation, it's unlikely that simultaneously changing several physical constants or fundamental forces will lead to a universe as habitable as ours. Astronomer Virginia Trimble observed early in the debate on fine-tuning:

> The changes in these properties required to produce the dire consequences are often several orders of magnitude, but the constraints are still nontrivial, given the very wide range of numbers involved. Efforts to avoid one problem by changing several of the constraints at once generally produce some other problem. Thus we apparently live in a rather delicately balanced universe, from the point of view of hospitality to chemical life.[33]

John Gribbon and Martin Rees reach a similar conclusion:

> If we modify the value of one of the fundamental constants, something invariably goes wrong, leading to a universe that is

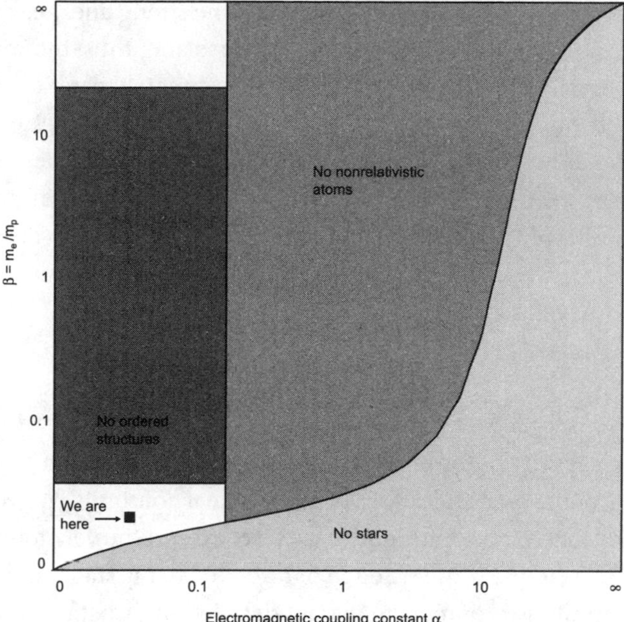

Figure 10.3: Multiple-tuning plot for the electron-to-proton mass ratio (β) and the electromagnetic coupling constant (α, also called the fine structure constant). Universes with ordered structures occupy a tiny region. They require that β be much less than one, otherwise the nuclei in atoms would have unstable locations. Large values of β appear consistent with ordered structures, because the electron is assumed to take the place of the nucleus, but this is probably impossible for elements more complex than hydrogen. α must also be much smaller than one to keep electrons in atoms from being relativistic. The third major exclusion region in the diagram shows where stars cannot exist. (The axes are scaled by the arc tangent of the logarithm of β and α.)

> inhospitable to life as we know it. When we adjust a second constant in an attempt to fix the problem(s), the result, generally, is to create three new problems for every one that we "solve." The conditions in our universe really do seem to be uniquely suitable for life forms like ourselves, and perhaps even for any form of organic chemistry.[34]

We tend to think of laws and parameters as governing the cosmos in general, but as we've seen, changes in these universal variables have profound consequences on particular objects within the universe. Changes in the relative strengths of gravity and electromagnetism affect not only cosmological processes but also galaxies, stars, and planets. The strong and weak nuclear forces determine the composition of the universe and, thus, the properties of galaxies, stars, and planets. As a result, we ultimately can't divorce the chemistry of life from planetary geophysics or stellar astrophysics.

Although we have only scratched the surface, it should be clear that there are many examples of "cosmic-scale" fine-tuning in chemistry, particle physics, astrophysics, and cosmology. Most published discussions of such fine-tuning are limited to the requirements for life, but cosmic fine-tuning extends well beyond mere habitability.

FINE-TUNING FOR LIFE *AND* DISCOVERY

The striking capacity of the cosmos for discovery also depends on the particular forms of the laws and values of the physical constants. We've hinted at this already, but it is astonishingly pervasive, even at the very foundations of matter. For example, the existence of discrete energy states at the quantum level permits astronomers to extract detailed information from light-emitting bodies with spectroscopic analysis. Atomic regularity results in a distinctive spectroscopic signature for each element and molecule. As we showed in Chapter Seven, stellar spectra tell us a great deal about the universe. The astronomer E. A. Milne once noted[35] that if the laws of nature had produced a nonquantum microscopic realm characterized by continuous energy distributions—had atoms behaved more like planets in orbit about stars, lacking discrete energy states—spectroscopy would have been a far less useful tool.

The fact that each type of fundamental particle has a universal mass also greatly enhances scientific measurement. This permits astronomers to apply the results of laboratory experiments on Earth's surface to even the most distant and early parts of the universe. Since the existence of the elements, and hence life, depends on distinct quantum states and the mass constancy of fundamental particles, habitability and measurability are yoked, it seems, all the way down.

CARBON AND OXYGEN, ACT III

We have discussed several examples of the life-facilitating qualities of carbon and water, the existence of which is highly sensitive to the laws of physics. And as noted above, water, the universal solvent—which contains two hydrogen atoms for every one oxygen atom—has nourished the seedbed of both chemistry and high technology. But carbon and oxygen individually contribute to the measurability of the cosmos as well. The isotope carbon-14, for instance, helps us date once-living things, because it conveniently has a half-life similar to the time it takes for buried organic matter to decompose. Carbon-14 also provides information about the recent history of our extraterrestrial environment, since it is sensitive to radiation from the galaxy and the Sun and its interaction with Earth's magnetic field. Silicon carbide grains in meteorites are the primary sources of samples from previous generations of stars, and the encoded carbon-12 to carbon-13 isotope ratios are the primary means to tie the presolar grain record to that obtained from stellar spectra. Without these isotopes, meteorites would be much less useful historical records. Finally, carbon atoms seen in distant intergalactic gas clouds against background quasars are the best probes of the temperature of the microwave background at various redshifts.

Oxygen also has some endearing qualities. An oxygen-rich planetary atmosphere, unlike many alternatives, is transparent to optical light, allowing a clear view of the broader universe. And the carbon monoxide molecule (CO), with one carbon and one oxygen atom, is the best tracer of dense molecular gas in the interstellar medium and around young stars.

DIMENSIONS

A bit more abstractly, it appears that a habitable universe must have exactly three space dimensions.[36] This insight was actually one of the earliest examples of anthropic reasoning applied to physics. In 1955 G. J. Whitrow asked whether the dimensionality of our universe is related to our existence, though P. Ehrenfest had already asked in 1917 how the laws of physics depend on the dimensionality of space.[37] A surprisingly diverse array of phenomena hinges on this fact: the inverse square law of gravity, the stability of atoms, and wave equations, among others.

Alternative, "dimensionally challenged universes" are so debilitating that they are not just hostile to life; they even thwart the mere transmission of information. John Barrow writes, "Only three-dimensional worlds appear to possess the 'nice' properties necessary for the transmission of high fidelity signals because of the simultaneous realization of reverberationless and distortionless propagation."[38] For audiophiles these terms refer to sound quality in stereos, but they apply equally to all wave phenomena. Reverberation occurs when signals emitted at different times arrive simultaneously; signal distortion is an alteration of the form of the wave as it propagates. Astronomers take for granted the remarkable fidelity of information carried by light across the universe. Life, too, almost certainly requires high fidelity in neurological signal transmission, as Whitrow suggested. A three-dimensional universe, unlike the alternatives, allows information to flow with a minimum of fuss and bother.

Theorists have pondered alternatives to our one time dimension as well.[39] As we saw in Chapter Nine, the finitude and direction of cosmic time contributes to the habitability of our universe. It seems reasonable to conjecture that altering the number of time dimensions (assuming this is even possible) would also enormously complicate cause-and-effect relationships and consequently make prediction much more difficult, if not impossible. Indeed, the only safe prediction in such a place might be that accurate predictions weren't possible.

DISCOVERABILITY

In Chapter One we briefly distinguished among observability, measurability, and discoverability. Paul Davies, perhaps more clearly than anyone, has pointed to the features of our universe that have facilitated the discovery of the laws of nature.[40] Discovering a law has much to do with its simplicity. The inverse-square laws of gravity and electric fields helped lead to the early discoveries of the universal Law of Gravitation and Maxwell's Equations. The trek from Kepler to Newton to Einstein was facilitated, then, not only by the particular characteristics of our local environment but also by the mathematical simplicity of gravity. At the same time, the laws of physics are not so simple that they prohibit the variety and complexity necessary for life.

Kepler's formulation of his laws required that macroscopic bodies in our universe be well described by "classical" laws—that is, laws with distinct and measurable positions and motions. Davies argues that one should not just assume that any universe appearing from a quantum initial state would later exhibit classical properties.[41]

Harlow Shapley once marveled, "It is amazing what grand thoughts and great speculations we can logically develop on this planet—thoughts about the chemistry of the *whole* universe—when we have such a tiny sample here at hand."[42] Similarly, Paul Davies contends that the success of science depends on the "locality" and "linearity" of the physical laws. Concerning locality, Davies writes:

> It is often said that nature is a unity, that the world is an interconnected whole. In one sense this is true. But it is also the case that we can frame a very detailed understanding of individual parts of the whole without needing to know everything. Indeed, science would not be possible at all if we couldn't proceed in bite-sized stages.[43]

In other words, we can discover truths that hold throughout the universe simply by investigating a representative local sample. Most of the laws of physics take simple linear forms. If they did not, chaos would reign, and we could not extrapolate such "local" observations in time or space. We could not reliably predict the orbits of the planets very far into the future or reconstruct their past motions. We would be imprisoned in the cage of our direct and parochial observations. But we are not.

Linearity and locality are closely related to nature's long-term stability—another prerequisite for life and discovery. Our very ability to establish the laws of nature depends on their stability.[44] (In fact, the idea of a law of nature implies stability.) Likewise, the laws of nature must remain constant long enough to provide the kind of stability life requires through the building of nested layers of complexity. The properties of the most fundamental units of complexity we know of, quarks, must remain constant in order for them to form larger units, protons and neutrons, which then go into building even larger units, atoms, and so on, all the way to stars, planets, and in some sense, people. The lower levels of complexity provide the structure and carry the information of life. There is still a great deal of mystery about how the various levels relate, but clearly, at each level, structures must remain stable over vast stretches of space and time.[45]

And our universe does not merely contain complex structures; it also contains elaborately nested layers of higher and higher complexity. Consider complex carbon atoms, within still more complex sugars and nucleotides, within more complex DNA molecules, within complex nuclei, within complex neurons, within the complex human brain, all of which are integrated in a human body. Such "complexification" would be impossible in both a totally chaotic, unstable universe and an utterly simple, homogeneous universe of, say, hydrogen atoms or quarks. Moreover, surprisingly, our universe allows such higher-order complexity alongside quantum indeterminacy and nonlinear interactions (such as chaotic dynamics), which tend to destabilize ordered complexity.[46]

Of course, although nature's laws are generally stable, simple, and linear—while allowing the complexity necessary for life—they do take more complicated forms. But they usually do so only in those regions of the universe far removed from our everyday experiences: general relativistic effects in high-gravity environments, the strong nuclear force inside the atomic nucleus, quantum mechanical interactions among electrons in atoms.

And even in these far-flung regions, nature still guides us toward discovery. Even within the more complicated realm of quantum mechanics, for instance, we can describe many interactions with the relatively simple Schrödinger Equation. Eugene Wigner famously spoke of the "unreasonable effectiveness of mathematics in natural science"[47]—unreasonable only if one assumes, we might add, that the universe is not underwritten by reason. Wigner was impressed by the simplicity of the mathematics that describes the workings of the universe and our relative ease in discovering them. Philosopher Mark Steiner, in *The Applicability of Mathematics as a Philosophical Problem,* has updated Wigner's musings with detailed examples of the deep connections and uncanny predictive power of pure mathematics as applied to the laws of nature.[48]

HIERARCHICAL CLUSTERING

The delicate balance of our universe's forces and constants determines the way in which matter is distributed. That distribution strikes a felicitous balance between uniformity and diversity, homogeneity and "clumpiness." In his book *Just Six Numbers,* Martin Rees asks, "Why does our universe have the overall uniformity that makes cosmology tractable, while nonetheless allowing the formation of galaxies, clusters and superclusters?"[49] His point is significant. If matter were evenly distributed in the universe, as might result if the expansion rate of the universe were initially faster, then life-essential structures like galaxies, stars, planets, and rocks could not have formed and measurability would suffer, well, immeasurably. At the other extreme, if an imbalance of the forces caused all matter to clump together into a homogeneous neutronium mass or a giant black hole, life and measurability would be impossible. One might think that a broad range of tunings would avoid either of these two extremes, but the distribution as well as the properties of matter in the universe actually must be fine-tuned for both.

In fact, the cosmological distance ladder described earlier works not only because matter tends to clump in our universe neither too little nor too much. It also works because the clumps are themselves clumped, and so on over many orders of magnitude. This hierarchical clustering of matter produces peaks in the distribution of matter on size scales of moons, planets, stars, star clusters, satellite galaxies, galaxies, and clusters of galaxies. These bodies are the rungs on which astronomers have climbed the distance ladder from Earth's surface to the edge of the visible universe. If matter did not clump on size scales between small planets and galaxies, there would be no stars. The universe would be a far less measurable place, not to mention being considerably less habitable.

Clustering helps discovery in other ways. A universe with clustered matter is more transparent than a less clustered one. Consider a comet. Why does a comet like Hale-Bopp appear so bright in the night sky? The solid nucleus of a typical comet may be only a few tens of kilometers across. But a comet brightens as it approaches the Sun because parts of it disintegrate into myriad tiny dust particles. These particles reflect far more sunlight because of the eternal truth of geometry that surface area to volume ratio increases as the size of a body decreases. Thus, when a comet begins to disintegrate, its total surface area grows dramatically. If the universe were a similarly diffuse sea of solid particles, radiation would have trouble making it even one light-year before it got hassled to death by the cosmic dust bowl.

In fact, in such a dusty, nebular universe, measuring the distance of anything would be tough. As we noted in Chapter Seven, astronomers can effectively measure the positions of stars in part because their sizes are much smaller than their separations, providing a striking contrast in the night sky. On the other hand, the diffuse nebulae in galaxies provide much less of this kind of contrast. If we had found ourselves in such a nebula, or in a comparably nebular universe, the sky would be a confused mess.

Clumping helped Kepler derive the simple laws of planetary motion because the typical mass of a planet is only a tiny fraction of the Sun's mass. When Newton derived his Law of Gravity a few decades after Kepler formulated his laws, he discovered that Kepler's mathematical models of our Solar System were imperfect. Contrary to Kepler's assumption, the Sun is not exactly located at one focus of an ellipse but wobbles a bit in reaction to the bodies around it. And the square of the period of a planet is not exactly proportional to the cube of its semi-major axis. Fortunately, the planet masses are low enough and they are far enough apart that their mutual gravity is very small. Consequently, Kepler's approximations were more than adequate to explain Tycho Brahe's observations until later, when more precise observations could be made. Thanks to the degree of hierarchical clustering in our Solar System, Kepler's equations were good enough to allow astronomy to continue its progress until these minor complications could be dealt with, and then, happily, resolved within the framework of Newton's Laws (through perturbation theory). If our universe didn't clump most of its matter in a series of generously separated points along its size scale, then the size difference between stars and planets probably would have been too small to allow Kepler to succeed.

Philosopher Robin Collins has observed that hierarchical clustering also allowed physicists to distinguish the fundamental forces. He argues (following physicist A. Zee) that it's only because these four forces—gravity, electromagnetism, the weak force, and the strong nuclear force—have widely varying strengths and scales of effect that we can disentangle the effects of each and, hence, study one force at a time. For instance, at the range of scales from planetary masses to galaxies, the only significant force is gravity, whereas at the level of atomic structure, the only significant force is electromagnetism. Echoing Davies, Zee says, because of this "we can learn about Nature in increments. We can understand the atom without understanding the atomic nucleus. . . . Physical reality does not have to be understood all at once."[50]

HIERARCHY AND SIMPLICITY

One final example combines discoverability, hierarchy, and simplicity. As we noted, the simplicity of laws has greatly aided in their discoverability. Robin Collins argues, however, that such simplicity comes in multiple varieties, each of which has proved essential to discovering the laws of nature. Besides the simplicity discussed above, such as the simple form taken by Newton's law of gravity, the laws of nature display what Collins calls "hierarchical simplicity," in which simplicity of the laws arises at each conceptual level within the history of physics. Consider the transition from the Newtonian theory of gravity to that of General Relativity.[51] These two laws inhabit radically different conceptual and mathematical frameworks: one describes forces between particles; the other, curved-space time. And yet we can translate each theory into the other. Such translation, however, obliterates the simplicity of the theories. Nevertheless, both gravitational laws—Einstein's and Newton's—are surprisingly simple laws within their respective conceptual frameworks.

What this suggests is that for some reason, the ultimate laws of physics give rise to mathematically simple theoretical laws at each conceptual level, even for those later judged inadequate, such as Newtonian space-time. This odd truth allows each conceptual level to serve as a ladder to the next level. If the theoretical laws could not be simple and yet relatively precise at each conceptual level, we could probably not discover them for that level, and hence could not progress from level to level toward the fundamental laws of physics.

Collins argues that General Relativity would have been almost inconceivable without the Newtonian theory of gravity already in place; even as it was, developing General Relativity took a true act of genius. Developing Newton's law of gravity also demanded an act of genius and required not only that the laws of gravity be simple,[52] but also that Newton's law reduce to simple rules of planetary motion—namely, Kepler's three laws. Even with simple laws of planetary motion, it took Kepler fifteen years of trial and error to discover them. Like an excellent tutor, the universe has not been so demanding as to ensure failure but rather has allowed us to succeed while still presenting us with worthy challenges.

OUR PLACE IN THE COSMOS

So the laws of physics allow a universe with a hierarchy of levels and diverse sizes of clumps within it. Of course, we're the tiny clumps that we're most concerned about. And it seems no discussion of the cosmos is complete without a typically reproachful sermon about our size compared with that of the universe. Most such discussions imply that our size is somehow a mark of disrepute. Astronomer Stuart Clark probably spoke for many when he said, "Astronomy leads us to believe that the Universe is so vast that we, on planet Earth, are nothing more than an insignificant mote."[53] Actually, over the extreme range of size scales from quarks to the observable universe, the range from humans to Earth is smack in the middle on a logarithmic scale. But more important, our middling size actually maximizes the total range of structures we can observe, both large and small. We're really a very nice fit in the cosmos.

If we were only a few orders of magnitude smaller, the realm beyond Earth's surface might be largely unavailable. Imagine yourself the size of an ant. The maximum optical resolution of an eye is determined by the size of the pupil; an ant-sized being with an eye 0.1 millimeters in diameter would

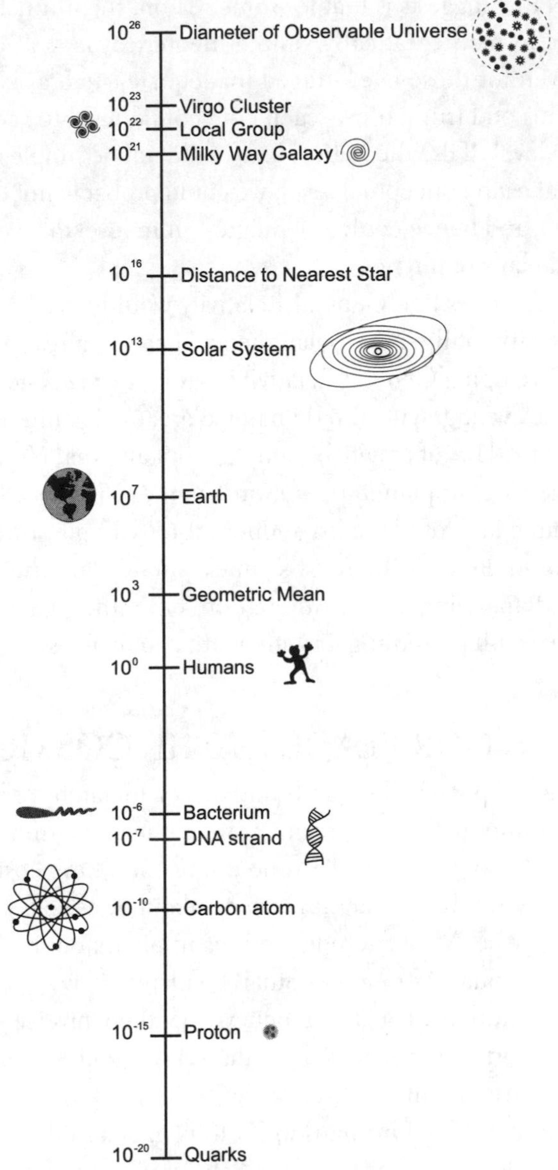

Figure 10.4: Size scales in the universe. Although Earth and human beings are often considered "insignificant" compared to the vastness of the cosmos, the Earth-human size range actually includes the geometric mean of extremes of sizes in the universe. We are near the middle of the size scale ranging from quarks to the observable universe, when figured on a logarithmic scale (in powers of ten). For scientific discovery of both large- and small-scale structures, this middling size is near optimum. (Approximate sizes and distances are indicated in meters.)

have a maximum resolution near one degree,[54] not quite enough to resolve the full Moon. Humans can discern details of about one minute of arc in angular size. Astronomers can greatly improve this limit by building telescopes. The diameter of a telescope's objective (its primary light-collecting lens or mirror) determines its resolution. Our size allows us to make telescope objective mirrors up to about twelve inches in diameter fairly easy to grind—amateur astronomers often make them in their garages. The optical resolution of a twelve-inch objective is a little better than one second of arc, comparable to the blurring effect of Earth's atmosphere on distant celestial sources. A technologically adroit ant-sized being might easily be able to construct a telescope objective half a millimeter in diameter, giving a resolution of about eight minutes of arc, only good enough for observing the Moon, the Sun, and bright stars.

Ironically, a tiny being would also have a harder time detecting the smallest building blocks, the fundamental particles. Physics laboratories built to measure fundamental particles are typically a couple of orders of magnitude larger than humans. The smaller the constituent of matter, the more energy is needed to detect it. To muster high energy, physicists must use large machines to accelerate particles to near the speed of light (usually using large electromagnets arranged in sequence).

We're near the upper end of the animal size scale. Larger animals are generally less dexterous than we are, without having significantly better vision. The largest animals, not surprisingly, live in the oceans, where buoyancy provides additional support. Although some probably have more sensitive vision, that vision is not necessarily optimal for viewing the heavens, to say nothing of the lack of dexterity among large marine life. Of course, we can always imagine some larger, more ethereal life form, such as the ghostly electromagnetic beings in Fred Hoyle's *The Black Cloud* or the giant floaters in Jupiter's atmosphere, as Carl Sagan imagined in *Cosmos*. But even if such creatures could exist, they would probably not be as manipulative of their environment as we are.

In his book *Nature's Destiny*, biologist Michael Denton has emphasized the ways our size seems to be well adapted for technology. In particular, he makes much of how our size allows us to control fire, a necessary step toward high technology. Sustainable fires, he argues, are impractical for organisms significantly smaller than we are.[55] This point may be debatable, though artificial and natural fires never seem to be smaller than a few centimeters. Fire is unusual in that it is quite hot yet non-explosive, because of the relative inertness of carbon and oxygen. This makes it useful as a simple energy source. Biology and controllable fire both need an environment with plenty of carbon and oxygen. Once again, qualities necessary for life also benefit technological progress and discovery.

After a long discussion linking the human dimensions and the environment, Denton concludes:

> It would appear that man, defined by Aristotle in the first line of his *Metaphysics* as a creature that "desires understanding," can only accomplish an understanding and exploration of the world, which Aristotle saw as his destiny, in a body of approximately the dimensions of a modern human.[56]

Theoretical physics and cosmology were surely unrelated to humanity's needs for survival for most of our history. Our ancestors got along quite well without Bell's theorem of quantum mechanics, Big Bang cosmology, and the theory of stellar nucleosynthesis. Nevertheless, in many ways we seem to be curiously overprepared, enough so that when the opportunities availed themselves, we could discover the laws of the universe, even in their most distant and obscure manifestations. This curiosity fits hand in glove with the other surprising fact that we've spent a good bit of ink and paper developing. We've moved from the details of Earth's geophysics and atmosphere, to the beginning of cosmic time and the forces and constants that apply throughout the universe. Over and over we've seen a pattern: the rare conditions required for habitability also provide excellent overall conditions for discovering the universe around us. At some point, this pattern should lead us to not only re-evaluate certain entrenched assumptions about the universe but even to reconsider our very purpose on this tiny speck orbiting a seemingly inconsequential star between the spiral arms of one ordinary galaxy among billions.

SECTION 3

IMPLICATIONS

Chapter 11

The Revisionist History of the Copernican Revolution

If Copernicus taught us the lesson that we are not at the center of things, our present picture of the universe rubs it in.

—Robert Kirshner[1]

HISTORY DISTORTED

By now, the result of the preceding chapters should be clear: Our local environment, centering on the near-present time and Earth's surface, is exceptional and probably extremely rare, with respect to both its habitability and its measurability. Further, the evidence suggests that in our universe these two properties are yoked, that those highly improbable places best suited for the existence of complex and intelligent observers also provide the best overall conditions for making diverse and wide-ranging scientific discoveries. We think this evidence has profound implications, which we'll develop later.

Standing audaciously in the way of our argument, however, is a tradition, especially popular in astronomy and cosmology, that we *don't* occupy a privileged place in the cosmos. The advocates of this tradition ground their heritage in the founding of modern science, with Galileo, Kepler, and especially Copernicus. They also claim the four hundred years of scientific discovery that followed that founding. Given such lineage, if we left this claim unchallenged, our argument would look pretty shaky. Fortunately, the historical pedigree of the so-called Copernican Principle is quite suspect. Indeed, in its acquisitive current form it would have found few detractors more adamant than Copernicus himself.

THE OFFICIAL STORY

> Ptolemy (second century) was the first and boldest in a long succession of spin doctors for the primacy of human beings. The whole universe, he postulated, rotated around us, with Earth sitting at the center of Heaven itself. Any marketing consultant will tell you that positioning is everything, and center-of-the-universe is hard to beat. A Polish astronomer named Copernicus (1473–1543) rudely pointed out: Sorry Earthlings, we spin around the Sun, not vice versa.... Giordano Bruno, a sort of sixteenth-century Carl Sagan, popularized these concepts... saying, among other things, that "innumerable suns exist. Innumerable earths revolve around those suns. Living beings inhabit these worlds.... "Bruno's crime, like Galileo's, was to undermine the uniqueness of our planet, and by doing so, to threaten the intellectual security of the religious dictatorships of his time.... Over time, advances in astronomy have relentlessly reinforced the utter insignificance of Earth on a celestial scale.

So wrote Nathan Myrhvold in a *Slate* article, aptly titled "Mars to Humanity: Get Over Yourself."[2] Myrhvold has swallowed this stereotype whole. The story is so common that it's tedious to repeat it in great detail. As we all think we know, ancient superstition put Earth and its inhabitants at the physical and metaphysical center of a small, anthropocentric—that is, "human-centered"—universe. The benighted masses thought Earth was flat,[3] while the educated elites, following Ptolemy and Aristotle, imagined it as a sphere, with the Moon, planets, Sun, and stars revolving around it.

Copernicus, according to the popular story, demoted us by showing that ours was a sun-centered universe, with Earth both rotating around its axis and revolving around the Sun like the other planets. This claim is sometimes accompanied by still more egregious factual errors. For instance, Bruce Jakosky explains in *The Search for Life on Other Planets*, "Because of this tremendous change in world view, Copernicus' views were not embraced by the Church: the history of his persecution is well known."[4] Never mind that Copernicus wasn't persecuted and died the same year (1543) that his ideas were published, not at the oil-soaked stake but peacefully and of natural causes. Since these historical facts muddy the popcorn-

Figure 11.1: This famous image is often mistaken for a seventeenth-century woodcut. In fact, its earliest known appearance is in Camille Flammarion's *L'Atmosphere: Météorologie Populaire* (Paris, 1888), 163. Flammarion and many others have used it as evidence for the incorrect claim that the medievals believed Earth was flat.

movie simplicity of the Official Story, with its cast of intrepid, steely eyed scientific heroes on the one hand and its one-dimensional villain priests on the other, the historical facts are garbled. (Understand, we don't believe this is part of a willful conspiracy. Jakosky is a well-known and respected scientist, and the publisher, Cambridge University Press, is a respected publisher of astronomy books. Such a mistake could only survive the editorial process because a great many intelligent people simply assume the stereotype.)

The popcorn movie continues on from Copernicus's persecution with a bravura medley of fact and fiction: The messiah Copernicus leaves his even less fortunate followers, like Bruno, the first martyr, and Galileo, the first saint, to suffer even more hideous consequences. In time, however, the brave and unflagging march of scientific evidence overwhelms the darkness and idiocy of religious superstition—swelling and triumphant musical score followed by cheers and the film's credits. The test audience loves it; everyone goes home fat and sassy in the knowledge of modern man's incalculable superiority to the superstitious fools of a dead and defeated past.

Thus is the story purged of its cumbersome subtleties. The Copernican Revolution, we're led to believe, was the opening battle in the ongoing war between Science and Religion. Textbooks and science writers on the subject display varying degrees of reductiveness and aversion to detail, but with few exceptions, the central message is the same: Religious superstition maintained the myth that Earth and human beings are the center of the universe, both physically and metaphysically, but modern science has taught us otherwise. Copernicus is the enduring symbol of science's unflinching commitment to the facts, even when it means displacing humanity from our false sense of uniqueness and importance. As astronomer Stuart Clark puts it: "Astronomy leads us to believe that the Universe is so vast that we, on planet Earth, are nothing more than an insignificant mote."[5] Strangely, some even see Copernicus's work as playing the role of moral teacher. Philosopher Bertrand Russell once said, "The Copernican Revolution will not have done its work until it has taught men more modesty than is to be found among those who think Man sufficient evidence of Cosmic Purpose."[6]

The intended subtext, of course, is that one will be scientific only to the extent that one is nonreligious. To be "religious," in the narrow sense intended here, is to believe that there is something unique, special, or intentional about our existence and the existence of the cosmos. "Science" here has a special definition as well. Rather than a search for the truth (*scientia* means knowledge) about nature—based on evidence, systematic study, and the like—science becomes applied naturalism: the conviction that the material world is all there is, and that chance and impersonal natural law alone explain, indeed must explain, its existence.

Toward this end, the official story line comes close to reversing the most important historical points. Predominant in that story line is the link between our "central" location and our importance in the overall scheme of things. As planetary scientist Stuart Ross Taylor puts it:

> Copernicus was right after all. The idea that the Sun, rather than the Earth, was at the centre of the universe caused a profound change in the view of our place in the world. It created the philosophical climate in which we live. It is not clear that everyone has

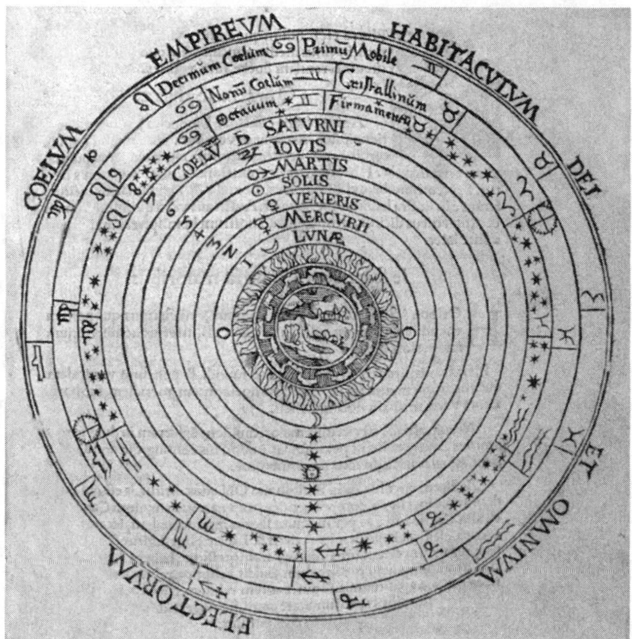

Figure 11.2: This cosmic section from Peter Apian's popular astronomy textbook (1540), shows the Aristotelian/Ptolemaic world picture that most medievals took for granted: a spherical Earth in the undignified middle, surrounded by elemental spheres and spheres for each of the planets, the Sun, the Moon, the fixed stars, and additional astronomical periods, and finally, as Apian puts it, the "empyrean heaven or habitation of God and of all the elect."

> come to grips with the idea, for we still cherish the idea that we are special and that the entire universe was designed for us.[7]

Historians of science have protested this description of the development of science for decades, but so far their protests have not trickled down to the masses or the textbook writers.

In a nutshell, what these protesting historians understand is that the ancient cosmology was a combination of the physical and the metaphysical vision of the Greek philosopher Aristotle (384–322 B.C.), and the careful observations and mathematical models of Ptolemy (circa A.D. 100–175) and other astronomers. The universe they envisioned was a set of nested, concentric spheres that encircled our spherical, terrestrial globe, a model that nicely explained a whole range of astronomical phenomena in the pre-telescope era. The crystalline spheres were thought to connect so that the movement of the outer, stellar sphere of the stars moved the inner spheres that housed the planets, the Sun, and the Moon. Such a model gave order to the east-to-west movement of the Sun across its daily path and the Moon at night, the celestial sphere encircling the poles, and the perplexing and somewhat irregular paths of the known planets.

This view had in its favor the collaboration of commonsense observation of the apparent movement of the skies, the apparent stability of Earth itself, and a number of plausible arguments. For example, if Earth were in motion, one would expect a stiff east wind, and that an arrow shot straight in the air would come down west of the archer. Although it appears naïve to modern minds, this cosmology is distinct from many other cosmologies, because it took account of observations of the heavens in trying to discern the structure of the cosmos.[8] In this sense, it reflected what we now see as a virtue of science—namely, openness to observation of the natural world.

Contrary to popular impression, neither Aristotle nor Ptolemy thought that Earth was a large part of the universe. Aristotle considered it of "no great size" compared with the heavenly spheres,[9] and in Ptolemy's masterwork, the *Almagest*, he says, "The Earth has a ratio of a point to the heavens."[10] Both of them reached this conclusion because of observations of Earth's relation to the stars, from which they surmised that the stellar sphere was an enormous distance from Earth. Copernicus based one of his arguments for the rotation of Earth on this shared assumption, observing, "How astonishing, if within the space of twenty-four hours the vast universe should rotate rather than its least point!"[11]

More important, the "center" of the universe was considered no place of honor, any more than we think of the center of Earth as being such. And Earth was certainly *not* thought to be "sitting at the center of Heaven itself," as Myrhvold puts it. Quite the opposite. The sublunar domain was the mutable, corruptible, base, and heavy portion of the cosmos. Things were thought to fall to Earth because of their heaviness, and Earth itself was considered the "center" of the cosmos because of its heaviness. The modernist interpretation of geocentrism has it essentially backwards. In our contemporary sense of the words, Earth in pre-Copernican cosmology was the "bottom" of the universe rather than its "center."

In contrast, Aristotelians considered the heavens immutable in their regularity and composition. Whereas the sublunary regions were composed of the four mutable elements of earth, water, air, and fire, the heavens were composed of a "fifth element," called "quintessence," or ether. Heavenly

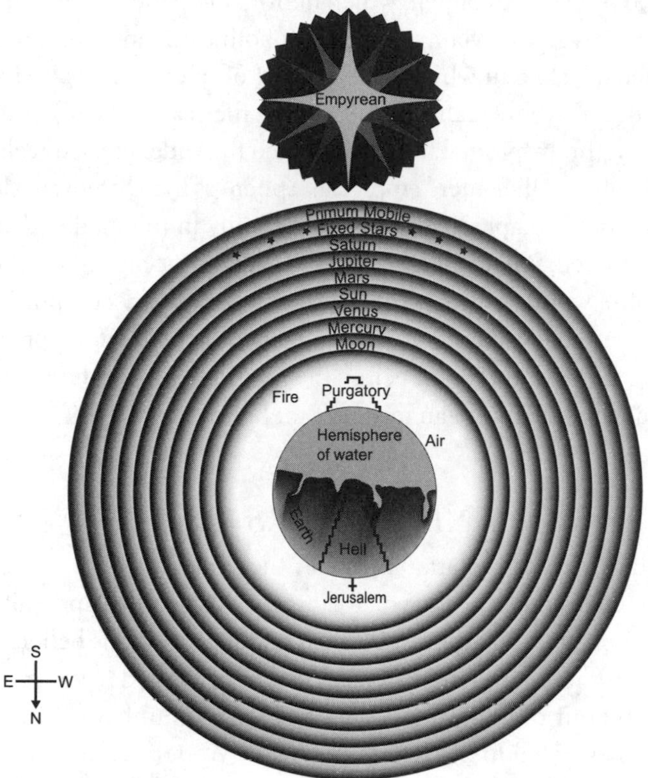

Figure 11.3: The Aristotelian/Ptolemaic universe according to Dante. Contrary to the common stereotype, the pre-Copernican cosmology did not view the "center" of the universe as the most important place. For Aristotle, Earth was a sort of cosmic sump, where earth, air, fire, and water mix to cause decay and death. The spheres of the Moon, the planets, and the stars, were the realm of the eternal and immutable. In the *Divine Comedy*, Dante describes the levels of hell mirroring the celestial spheres, with Satan's throne at the very center of Earth. Not surprisingly, against this backdrop, Copernicus, Galileo, and others could argue that the heliocentric view elevated the status of Earth, while also reinterpreting the meaning of the center.

bodies were perfectly spherical, and moved in a circular way befitting their perfection. From this it followed that the laws governing the heavenly realms were quite different from and superior to the laws governing the sublunary regions.

When Christian theology was added to the mix in the Middle Ages, the center or bottom of the universe became, quite literally, *hell*. Dante's *Divine Comedy* immortalized this vision, taking the reader from Earth's surface through the nine circles of hell, which mirror, and hence reverse, the nine celestial spheres above. Man, composed of both earth and spirit, occupied an intermediate state in which he was a sort of microcosmos.[12] He could ascend to the heavenly realm, or descend to the realm of evil, death, and decay. Other purely spiritual beings populated the wider created reality, and God dwelt "above" the outer "empyrean" sphere as the Unmoved Mover of everything else. Metaphysically speaking, reality in the medieval scheme was God-centered, not man-centered. Thus, Augustine argued that God did not create the world "for man" or out of some necessary compulsion, but simply "because he wanted to."[13] It's false, then, to say that the pre-Copernicans gave Earth and human beings the position of highest esteem while Copernicus relegated us to an insignificant backwater.[14]

WAR AND RUMORS OF WAR

Added to this misinterpretation is usually the simplistic warfare metaphor between science and religion, which distorts a complicated and quite interesting relationship.[15] While we are often told that religious beliefs tend to hinder scientific inquiry, many recent historians instead have argued that science grew out of a theistic milieu and that a number of theological beliefs were essential to the rise of science in Western Europe in the sixteenth and seventeenth centuries. Human beings in every culture could observe the natural world around them. Only one culture had the philosophical and theological pre-conditions that gave birth to modern science. The Judeo-Christian tradition, quite contrary to modern stereotype, helped correct and transform the Greek tradition when the two worldviews began to interact in the Middle Ages. As science historian Reijer Hooykaas puts it, "Metaphorically speaking, whereas the bodily ingredients of science may have been Greek, its vitamins and hormones were biblical."[16]

One of modern science's most important biblical inheritances is the notion that linear time is fundamental to the physical universe rather than an illusion. In other words, that cosmic history actually goes somewhere rather than merely in circles.[17] In this way it distinguished itself from the cyclical view of time held by the Greeks. Also important was the biblical distinction between Creator and creation, which had at least two profound results. First, in contrast to the Greek philosophical tradition, the biblical authors considered matter to be good and manual labor to be ennobling.[18] Second, human beings were to respect nature as God's good creation, but not revere it as a god or a divine offspring.[19] This break with the Greek tradition helped pave the way for scientists to submit nature to experiment and investigation, and to develop technology.[20]

Another significant biblical contribution is that since God was free in creating the world,[21] nature is *contingent*. It might not have existed, or it might have had different properties from the ones it has. As a result, nature's properties must be discovered rather than merely deduced from the principles of logic or mathematics.[22]

Balancing the view that God is free is the belief that he is good and rational rather than fickle, irrational, or even malicious, as the pagan gods so often were. Because of this, Jews and Christians could expect nature to be orderly and even lawful in its structure and behavior. Nature as a whole is contingent, but in its regular operations it is not capricious or perversely deceptive. The origin of modern science drew from this careful balance of contingency and order. Science emerged, not surprisingly, in a culture that held these ideas in careful balance.[23]

Finally, since they believed that God is one and that human beings are created in God's image, medieval Christians and Jews could expect nature to have a sort of unity (to be a *uni*verse)[24] and to be accessible to the human mind. These ideas, brought to fruition by interaction with the Greeks, were the seedbed from which natural science slowly grew. It's hardly a coincidence that science emerged in the time and place where these many factors converged. Although they are now forgotten, modern science draws on the interest of specific theological convictions.[25]

CHRISTIAN THEOLOGY AND ANCIENT COSMOLOGY

To discuss the largely pagan Greek culture does not, of course, give us a full and fair view of the Greeks' contribution to modern science. Plato wanted to kick the poets out of his ideal republic precisely because they so often depicted the gods as amoral, fickle, and irrational. Both Plato and Aristotle groped their way to the notion of a unifying cosmic principle or being in some sense superior to the rest of the cosmos, a progress that certainly encouraged them to see the universe as a unified cosmos rather than a confused chaos. But science didn't fully develop until Western civilization had been leavened with these specifically theological convictions. We've grown accustomed to attributing the rapid series of discoveries and inventions from the seventeenth century to the present to the discovery of the so-called "scientific method," a simplistic phrase we often use without understanding the soil from which modern science sprang. Certainly Aristotle's writings were part of that soil, but the relationship between his cosmology and Christianity is far more complicated than the modernist trope reveals.

For centuries, the "Christian" West had no single, official cosmology, in part because the biblical texts and imagery lack the sort of explicit detail to provide a model of the cosmos without other sources. Over time, however, the exposure of Western scholars to Aristotle's thought—mostly transmitted from the Muslim world—led to an integration of the Aristotelian/Ptolemaic cosmology with Christian theology. This was not initially a happy marriage. In 1277, the bishop of Paris, encouraged by Pope John XXI, issued a series of decrees denouncing orthodox (or "radical") Aristotelianism.[26] The central problem was that the orthodox disciples of Aristotle restricted God's freedom in creating the world. Aristotle viewed the cosmos as eternal, while the Christian saw it as the free creation of God, who spoke it into being. Nevertheless, Thomas Aquinas's moderation and domestication of Aristotle, for good or for ill, eventually won the day, and the synthesis became the foundation of university teaching across Europe.[27]

By the time of the Renaissance, some three centuries later, Christians understood the main themes of their faith in an Aristotelian/Ptolemaic framework. An obvious point of contact was, to use the philosophical term, teleology, intrinsic to both the Judeo-Christian tradition and Aristotle. Everything in the Aristotelian universe was "purposeful" in the sense that it was directed toward some end or goal. Everything moves, finally, because of the movement of the Unmoved First Cause, which is also the Final Cause toward which everything tends. Once "baptized," such a concept provided a plausible argument for the existence and activity of God based on the features of the natural world.

At the same time, there was nothing essential to Christian theology in the specifics of Aristotelian cosmology. Christianity had existed without Aristotle's cosmological model; it would do so again.[28] Christian doctrine did not require teleology in Aristotle's somewhat organic sense. In fact, many later non-Aristotelians—like Galileo, Kepler, Newton, and Boyle—continued to apply teleological concepts to the natural world by pointing to the particularly intricate and orderly configuration of natural features, which suggested the activity of an intelligent designer.[29] But by the time of

Boyle, nature was seen more as an intricate and interlocking machine, an artifact of intelligence rather than an organism whose individual parts themselves tended toward some internally motivated end.

And last but certainly not least, the key figures who proposed the new cosmology were religious. Nevertheless, once grown up together, wheat and tares are difficult to separate, and the stage was set for discord when the old, essentially Aristotelian/Ptolemaic cosmology began to crumble.

THE (REAL) COPERNICAN REVOLUTION

Copernicus lived during the Renaissance and the early Reformation (1473–1543), periods of enormous change and upheaval, which helped foment Copernicus's revolutionary ideas. Besides his intellectual context, he was aware of an empirical problem: the motion of the planets never fit comfortably in the Aristotelian framework. Explaining planetary motion required postulating complicated and inelegant "epicycles" and "equants" to accommodate their apparent variation and reversal as they crossed the sky.[30] Published while he was on his deathbed, Copernicus's mathematical model, *On the Revolution of the Heavenly Spheres* (*De Revolutionibus Orbium Caelestium*), proposed that Earth rotates around its axis and revolves around the Sun along with the other planets (although, contrary to popular impression, he didn't put the Sun in the precise center of Earth's orbit). While this scheme promised to resolve some of the complexities in the planetary orbits, it didn't dissolve the need for epicycles and other ad hoc explanations. Observations accurate enough to verify Copernicus's scheme weren't available when he proposed it in 1543;[31] nevertheless, Copernicus made an important imaginative leap.[32] Unlike most of his predecessors, he gave priority to the anomalous evidence of the planets, and suggested a theoretical solution that to many seemed to contradict the straightforward evidence of the senses. History would show that on this point, he had chosen correctly.

He also initiated a vision of an even larger cosmos, one that culminated in the infinite universe of Newton. But Copernicus wasn't the cool, detached empiricist of modernist lore, and he had no intention of "demoting" man's status.[33] On the contrary, as a man of the Renaissance, he sincerely appealed to anthropocentric premises in making his case.[34] Anthropocentrism was always much more at home in Renaissance humanism influenced by certain

Figure 11.4: Nicolaus Copernicus (1473–1543).

classical and specifically Stoic traditions than in the medieval Aristotelian synthesis that preceded it.

The neo-Platonic strain of Renaissance humanism motivated his search for mathematical simplicity and "harmony" in nature. It instilled in him an aesthetic sense that the apparently inelegant motion of the planets was a problem to be solved. "A felt necessity was the mother of Copernicus' invention," says Thomas Kuhn. "But the feeling of necessity was a new one."[35] Neo-Platonism also inspired his reverence for the Sun, and allowed him to transform the status of the physical center from that of the earlier

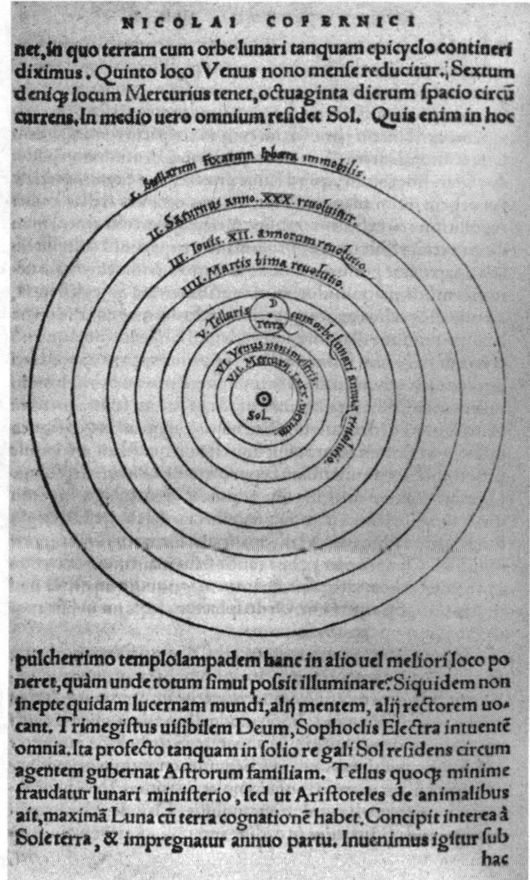

NICOLAI COPERNICI

net, in quo terram cum orbe lunari tanquam epicyclo contineri diximus. Quinto loco Venus nono menſe reducitur. Sextum deniq; locum Mercurius tenet, octuaginta dierum ſpacio circũ currens. In medio uero omnium reſidet Sol. Quis enim in hoc

pulcherrimo templo lampadem hanc in alio uel meliori loco poneret, quàm unde totum ſimul poſſit illuminare? Siquidem non inepte quidam lucernam mundi, alij mentem, alij rectorem uocant. Trimegiſtus uiſibilem Deum, Sophoclis Electra intuentẽ omnia. Ita profecto tanquam in ſolio regali Sol reſidens circum agentem gubernat Aſtrorum familiam. Tellus quoq; minime fraudatur lunari miniſterio, ſed ut Ariſtoteles de animalibus ait, maximã Luna cũ terra cognationẽ habet. Concipit interea à Sole terra, & impregnatur annuo partu. Inuenimus igitur ſub hac

Figure 11.5: This heliocentric cosmic cross section, from Copernicus's *On the Revolution of the Heavenly Spheres* (1543), shows Earth with its Moon circling around the Sun, along with the other then known planets.

Aristotelian cosmology. Otherwise, heliocentrism would have *demoted* the Sun. As science historian Dennis Danielson puts it, Copernicus "needed to renovate the [cosmic] basement, because the basement needed renovating."[36] So Copernicus enthuses thus:

> In the middle of all sits Sun enthroned. In this most beautiful temple could we place this luminary in any better position from which he can illuminate the whole at once? He is rightly called the Lamp, the Mind, the Ruler of the Universe: Hermes Trismegistus

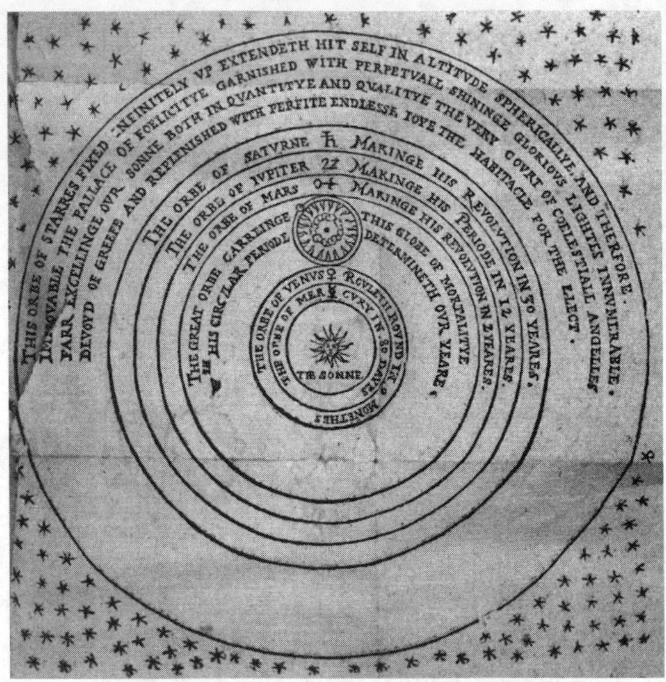

Figure 11.6: An early English Copernican cosmic section by Thomas Digges (1576). Although Digges's Copernican model was otherwise conservative, he was apparently the first to break through the finite outer wall of the cosmos and posit an infinite expanse of stars.

> names him the Visible God; Sophocles' Electra calls him All-seeing. So the Sun sits as upon a royal throne ruling his children the planets which circle round him.[37]

At the same time, he grounded his insights in specific observations, thus departing from the fundamental neo-Platonic distinction between the worlds of sensible, illusory appearances on the one hand, and intelligible reality on the other.[38] His was a Christian neo-Platonism in which the material world itself, as the creation of an omnipotent God, could reflect the precision of mathematics, something unthinkable to the strict neo-Platonist. Without this modification of pure Platonism, the seemingly inelegant and mathematically messy movement of the planets would not have struck Copernicus as a problem to be solved. This neo-Platonism, tempered by Christianity, was an important guiding influence, then, for Copernicus and for other Renaissance scientists who followed him, including Galileo and Kepler.[39]

Figure 11.7: Tycho Brahe (1546–1601).

Despite his innovation, Copernicus retained the notion of perfectly circular, celestial spheres, along with other elements of the old Aristotelian cosmology. As Thomas Kuhn says, "The significance of *De Revolutionibus* lies . . . less in what it says itself than in what it caused others to say."[40] Ironically, one of Copernicus's most important successors was the anti-Copernican Tycho Brahe (1546–1601).[41] Brahe, a Danish naked-eye astronomer, proposed a cosmological model that made Earth the stationary centerpoint of a rotating stellar sphere, Sun, and Moon, with the other planets rotating around the Sun.[42] His enduring legacy, however, was his remarkable observational record, which allowed his Copernican understudy, Johannes Kepler (1571–1630), to resolve many of the problems that beset Copernicus's own proposal. The appearance of a "new star" (a nova) in 1572, and Brahe's observations of comets, previously mistaken for atmospheric phenomena, helped precipitate the demise of two other Aristotelian convictions: the immutability of the heavens and the existence of concentric heavenly spheres.

Kepler's access to Brahe's detailed observations allowed him to correct the often inaccurate ancient records that Copernicus had inherited. A

Figure 11.8: Frontispiece of 1660 book by Jesuit Athanasius Kircher, depicting the Tychonic system. In Tycho's system, the Sun continued to circle around Earth, as in the Ptolemaic scheme, but the other planets circle around the Sun, as in Copernicus's scheme. While Tycho's system is often considered a compromise between Ptolemy and Copernicus, it signaled a radical break with the traditional idea of solid celestial spheres.

Figure 11.9: Frontispiece of *Almagestum novum,* by Giovanni Battista Riccioli, 1651. In this richly detailed image, Urania, the Muse of astronomy, is holding a scale weighing the Tychonic and Copernican systems. Tycho's system outweighs the Copernican one. For many years, the Tychonic system was the chief alternative to Copernicanism. Notice that the Ptolemaic system in the bottom right corner has clearly been rejected, and is not even weighed in the scales.

deeply religious Lutheran,[43] Kepler internalized the logic that Copernicus had only partially embraced. He had a mystical attraction to harmonic regularities that inspired him to search for mathematically simple laws of celestial motion. At the same time, his commitment to realism and his access to reliable data eventually led him to propose elliptical planetary orbits that vary in speed, both ideas repugnant to the strict Aristotelian.[44] Moreover, Kepler anticipated our argument, by realizing the value for astronomical observations of being somewhere other than the center of the universe:

> Thus it is apparent that it was not proper for man, the inhabitant of this universe and its destined observer, to live in its inwards as though he were in a sealed room. Under those conditions he would never have succeeded in contemplating the heavenly bodies, which are so remote. On the contrary, by the annual revolution of the earth, his homestead, he is whirled about and transported in this most ample edifice, so that he can examine and with utmost accuracy measure the individual members of the house. Something of the same sort is imitated by the art of geometry in measuring inaccessible objects. For unless the surveyor moves from one location to another, and takes his bearing at both places, he cannot achieve the desired measurement.[45]

Figure 11.10: Johannes Kepler (1571-1630).

Plate 1: The famous *Earthrise* photo taken by Bill Anders on December 24, 1968. Anders took the picture with the Moon situated to the right, since he understood them to be orbiting the Moon at its equator. Moments later Frank Borman took a black and white photograph with the lunar horizon at the bottom. Popular reprints of the color Anders photo usually turn it 90 degrees so that it is oriented like the Borman photo. The image has become a moving if ambiguous symbol of Earth's place in the universe.

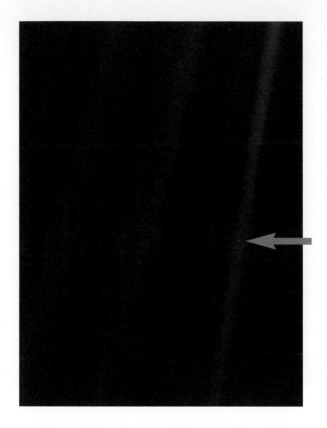

Plate 2: Earth as a "Pale Blue Dot." Voyager 1 took this image in 1990 while looking back at the Solar System from a distance of some four billion miles. Earth is immersed in one of the rays of sunlight scattered in the camera's optics, resulting from the small angle between Earth and the Sun.

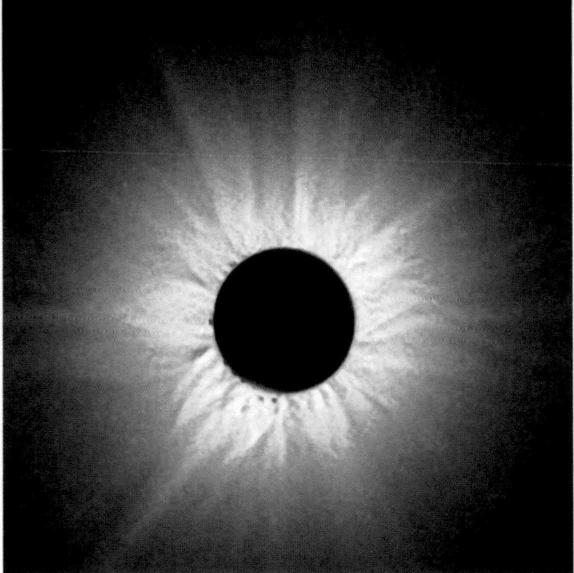

Plate 3: Two total solar eclipses. The first is a photo of the October 24, 1995, total solar eclipse from Neem Ka Thana, India. Six photos taken over a range of shutter speeds were combined to reveal detail in the bright and faint regions of the Sun's atmosphere. Notice the pink chromosphere hugging the dark lunar disk. The long coronal streamers were plainly visible to the unaided eye. The second is a ten-picture, composite image of the June 21, 2001, eclipse, taken from Chisamba, Zambia (Africa). The less flattened appearance of the corona compared to the earlier eclipse is caused by the changing solar magnetic field over the sunspot cycle.

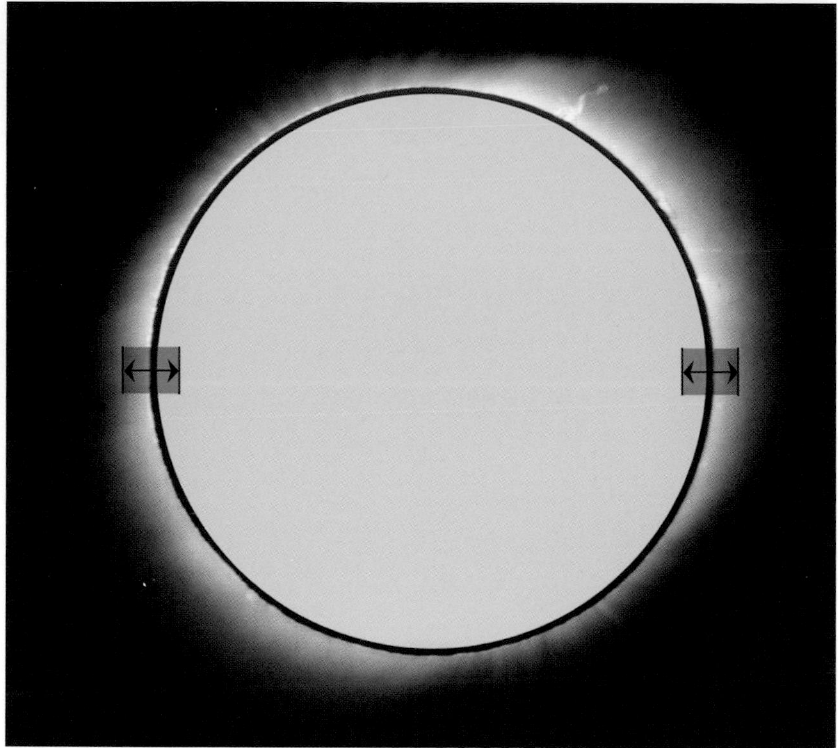

Plate 4: Colorized image of the October 24, 1995, total solar eclipse. The yellow disk, representing the occulted solar photosphere, is superimposed over the image. The red markers indicate the range of apparent sizes of the Moon. Notice the black outer edge of the Moon's disk, which just covers the Sun's bright photosphere, revealing portions of the thin chromosphere and prominences.

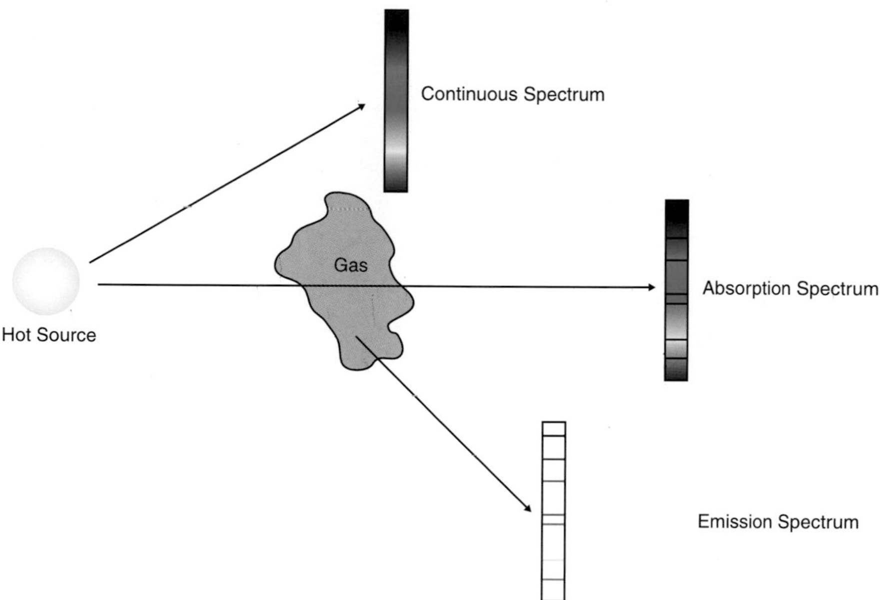

Plate 5: Gustav Kirchhoff first recognized the basic types of spectra. A hot, dense object emits a continuous spectrum. If a cooler, low-density gas is placed between a continuous source and the observer, atoms in the gas remove light at specific wavelengths, producing an absorption spectrum. The same gas observed from a different direction (without the hot source behind it) produces an emission spectrum.

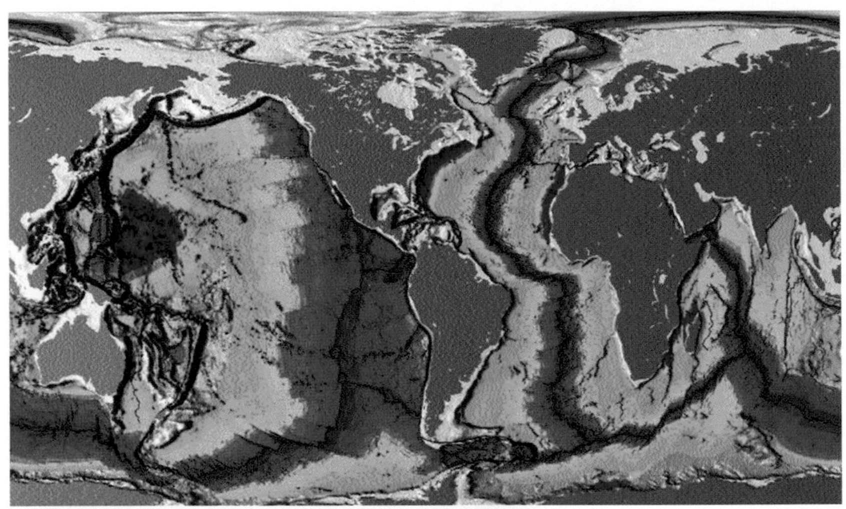

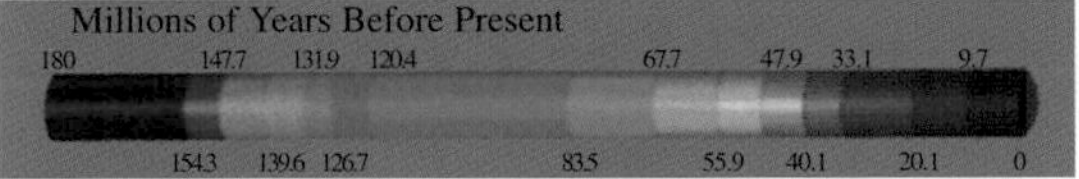

Plate 6: Crustal age of the sea floor as determined from oceanographic magnetic measurements. Note the symmetry of the ages on opposite sides of the mid-ocean ridges. This map was constructed mostly from oceanographic magnetic measurements.

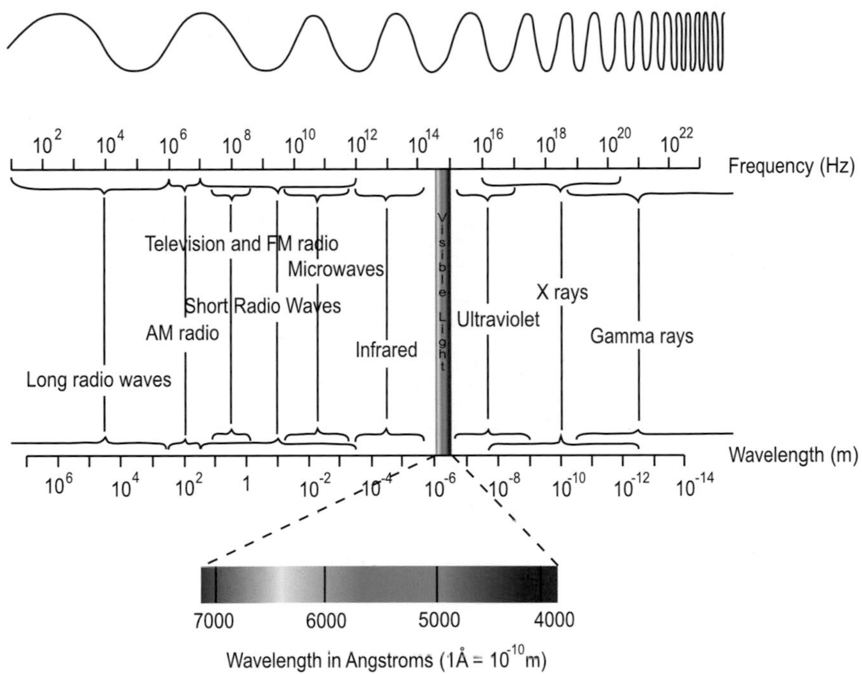

Plate 7: Earth's atmosphere is transparent to the narrow sliver of optical light and to radio waves. Shorter wavelengths are visible as blue, and longer ones as red. (This image is called a spectrum.)

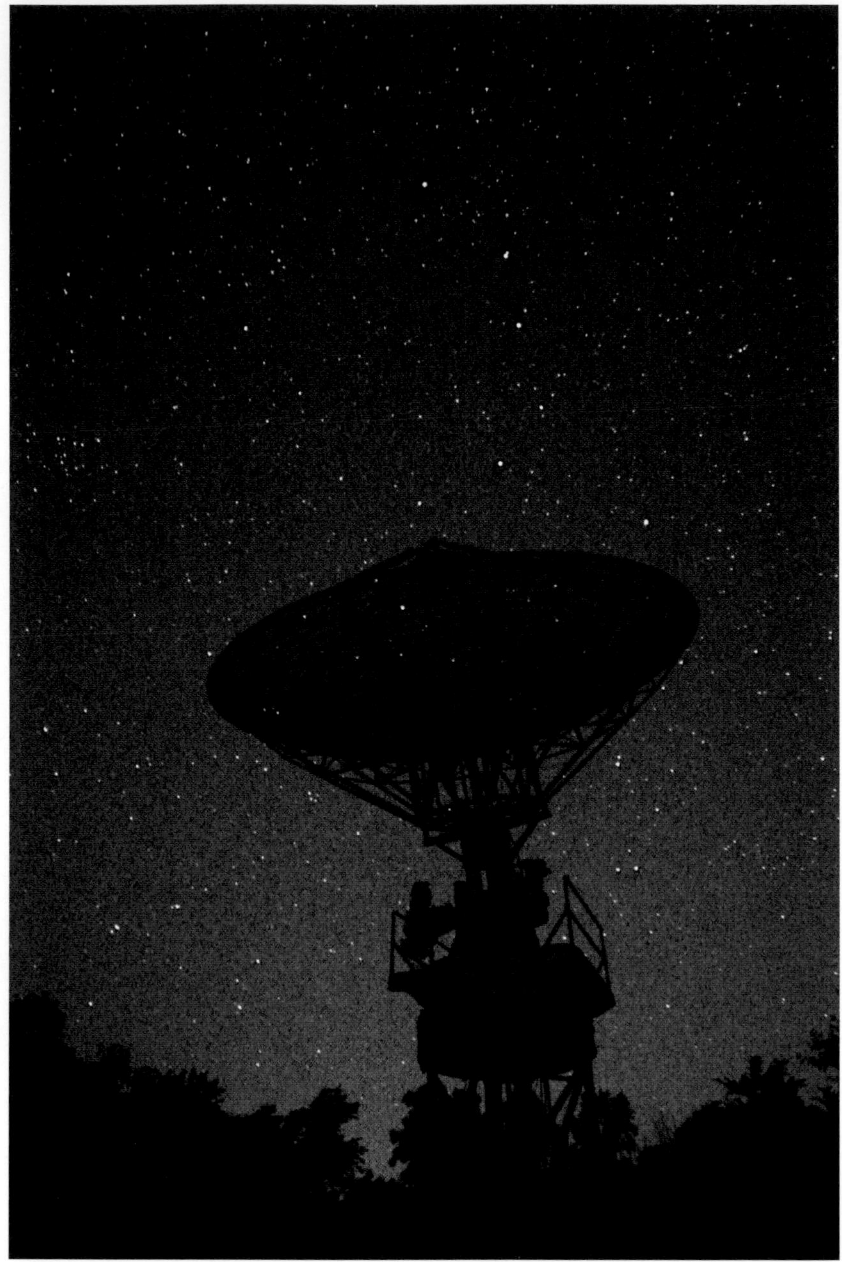

Plate 8: Earth's transparent atmosphere allows us to see objects beyond Earth with remarkable clarity. The atmosphere is also transparent to the radio region of the electromagnetic spectrum. A radio telescope antenna at Fick Observatory in central Iowa is shown against the backdrop of the northern constellations.

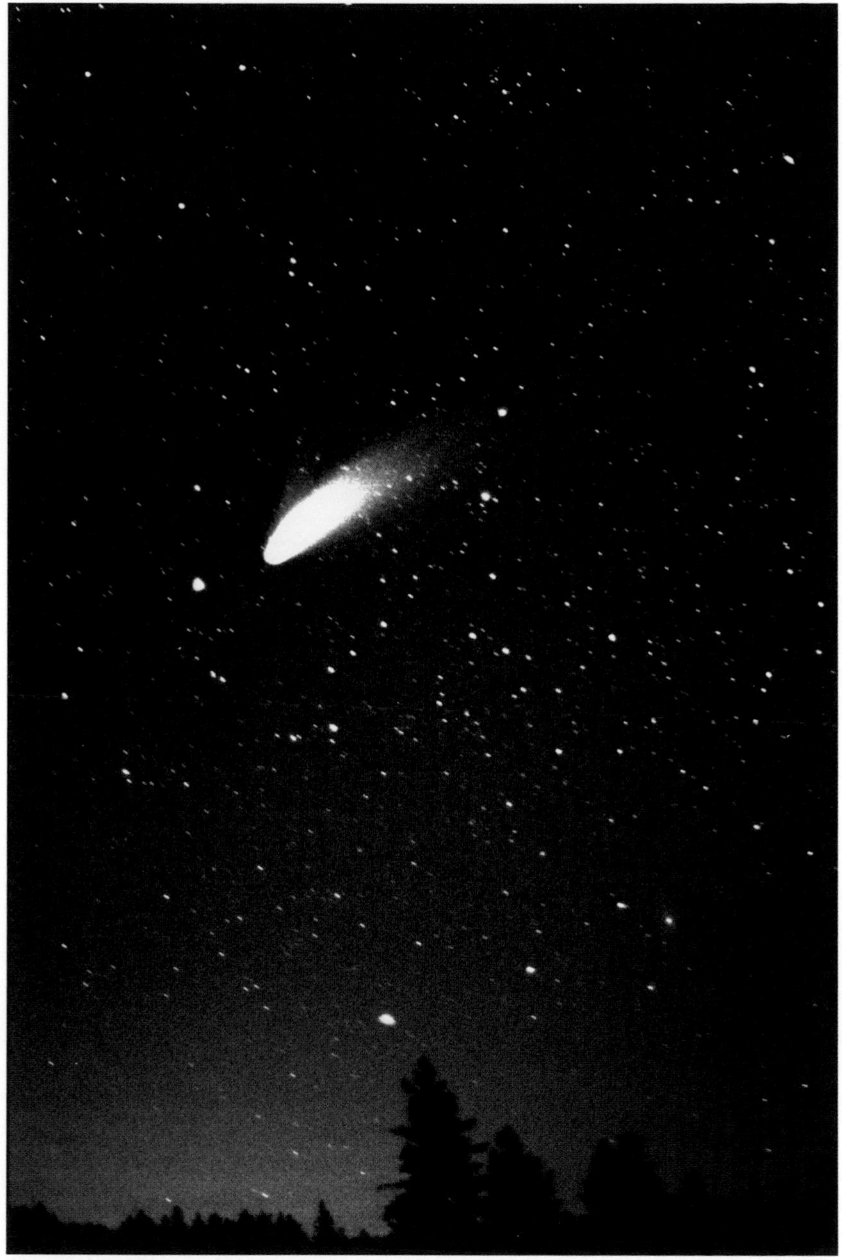

Plate 9: Hale-Bopp as viewed from the Washington Cascade mountains on April 1, 1997. Hale-Bopp was a large comet, which would have caused mass extinctions if it had hit Earth.

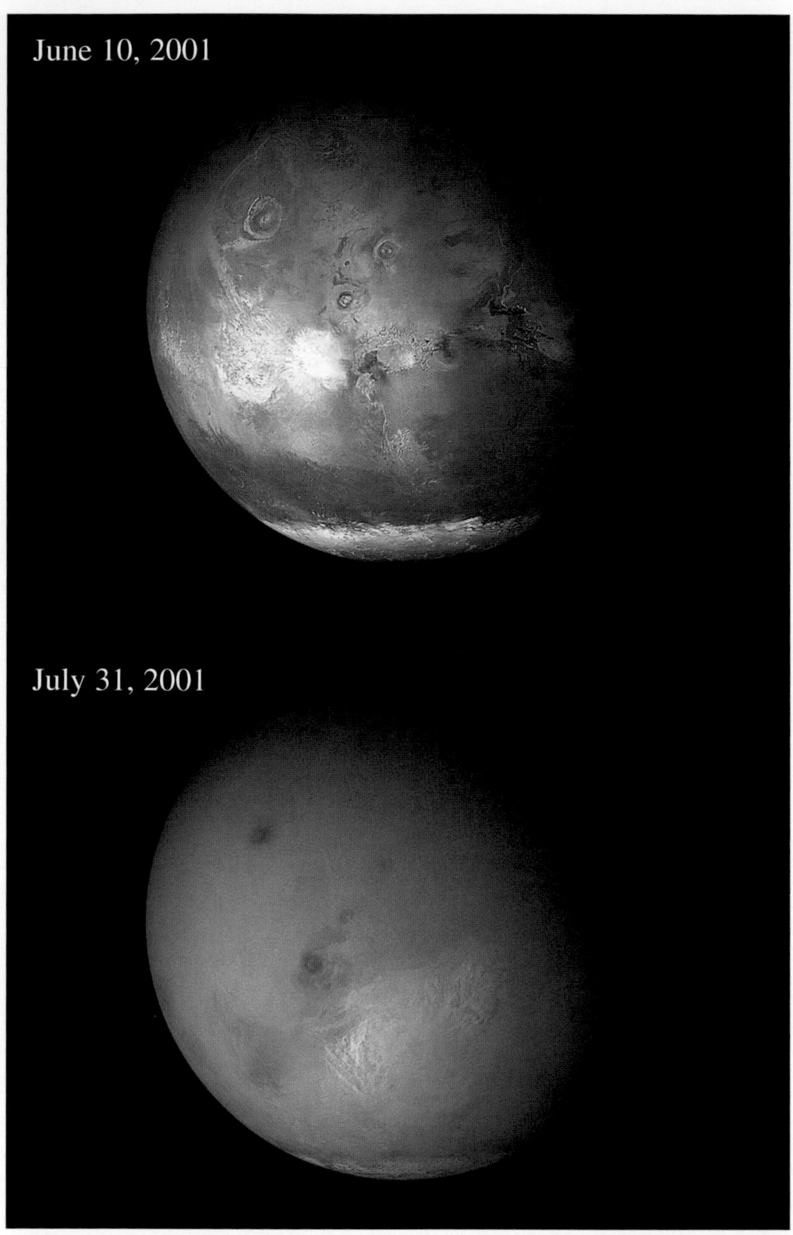

Plate 10: Mars Global Surveyor images of Mars just before a global dust storm **(top)** and shortly after it completely engulfed the planet **(bottom)**. The storm continued to intensify through September. A single dust storm can hide Mars's face from our view. A planet-wide Martian dust storm prevented the Mariner 9 probe from mapping its surface for a few days in 1973. An even stronger dust storm greeted the Mars Odyssey probe upon its arrival in October 2001.

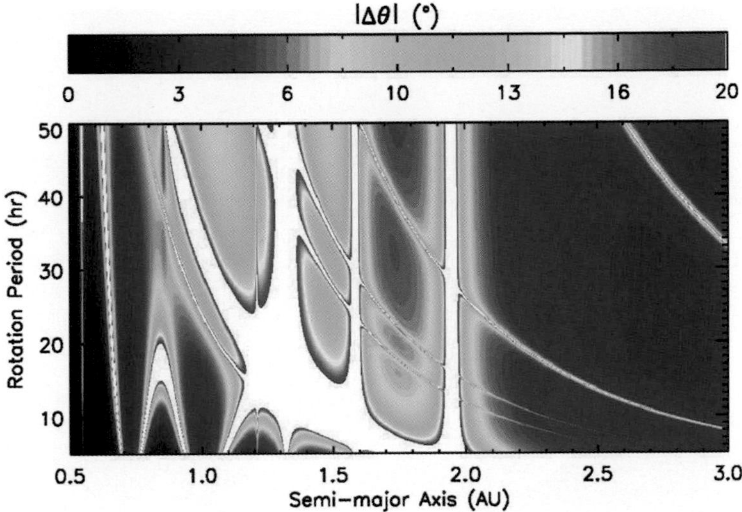

Plate 11: The range over which a planet's obliquity (axis tilt) varies depends on several factors, including the planet's internal structure, rotation period, and distance from its host star, as well as the locations of other planets, and the presence of a large moon. Shown in the figure are the results of numerical experiments on the expected obliquity variations of Mars with various combinations of rotation period and distance from the Sun; Mars's actual period Is 24.6 hours and it is 1.52 AUs from the Sun. White areas correspond to variations greater than 20 degrees. Stability comparable to Earth's (about 2 degrees) is rare for a Mars twin inside about 2.5 AUs.

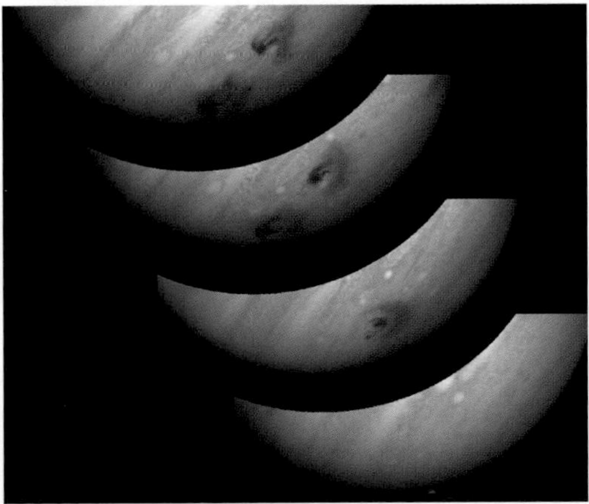

Plate 12: Jupiter takes some hits for the Solar System. This composite shows several impacts of comet Shoemaker-Levy 9 on Jupiter. The images from lower right to upper left show: the impact plume on July 18, 1994 (about five minutes after the impact); the fresh impact site 1.5 hours after impact; the impact site three days after the first G impact and 1.3 days after the L impact; and further change of the G and L sites from winds and an additional impact (S) in the G vicinity, five days after the G impact. Within a few years, all traces of the impacts had disappeared.

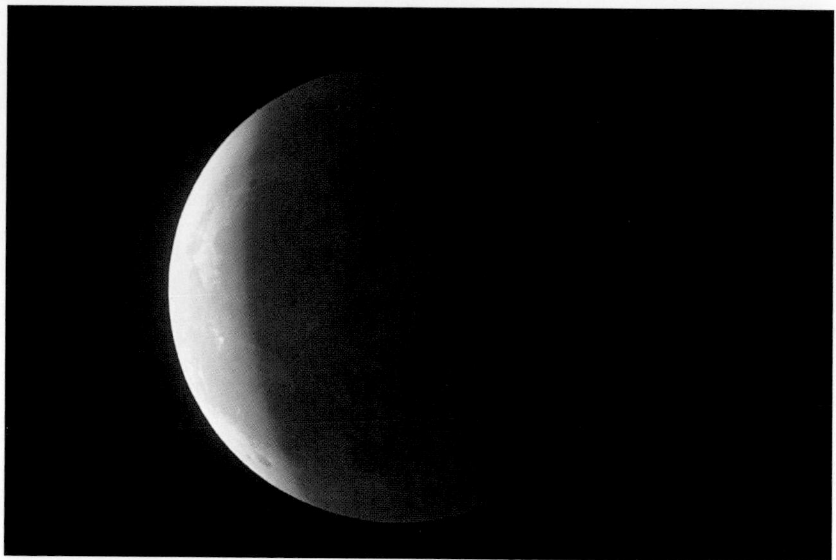

Plate 13: Partial phase of the total lunar eclipse of May 15, 2003. Notice the curved shape of Earth's shadow. Aristotle cited observations of lunar eclipses as evidence that Earth is round.

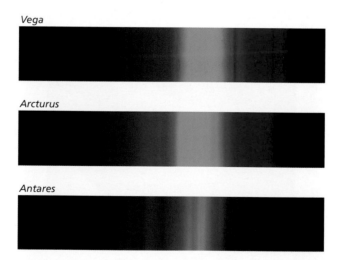

Plate 14: Low-resolution spectra of three bright stars. Antares is a very cool red supergiant. Molecules in its atmosphere produce the broad dark bands in its spectrum. The warmer red giant star, Arcturus, is slightly metal-poor. Absorption lines are faintly visible in its spectrum. Vega is a hot main sequence star (a dwarf). Hydrogen atoms in its atmosphere are responsible for the prominent absorption lines in its spectrum.

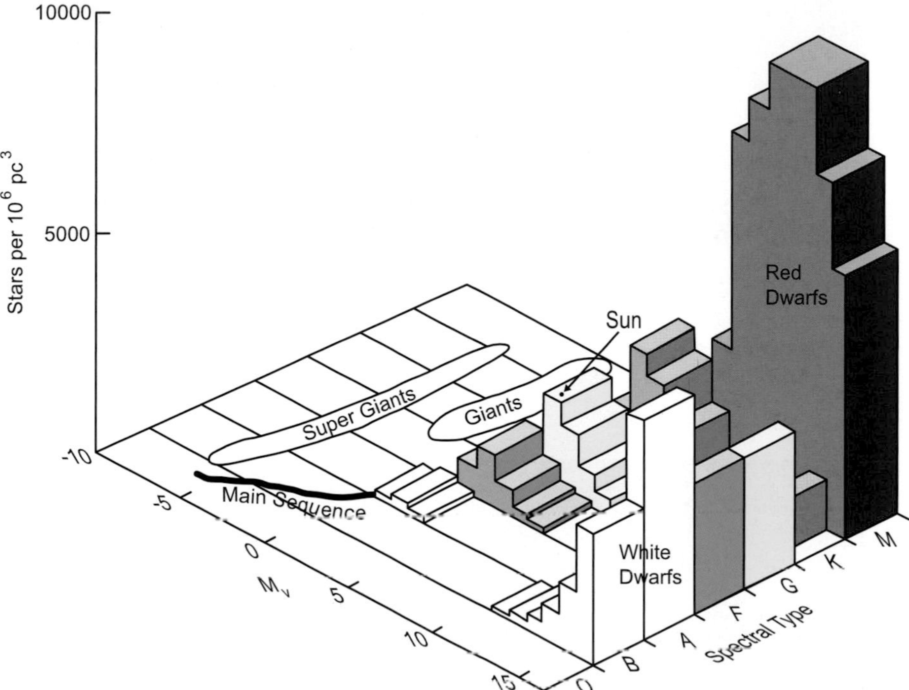

Plate 15: Space densities of nearby stars over a range of spectral types. The majority of stars are dim red dwarfs, which are unlikely to support complex life. The Sun is one of the relatively rare luminous main sequence stars.

Plate 16: Hubble Space Telescope image of the aligned galaxies NGC 2207 and IC 2163. This and other similar pairs of spiral galaxies have allowed astronomers to verify that spiral arms are very dusty, while interarm regions are transparent in the outer disk. But even the interarm regions become very dusty toward the center of the galaxy.

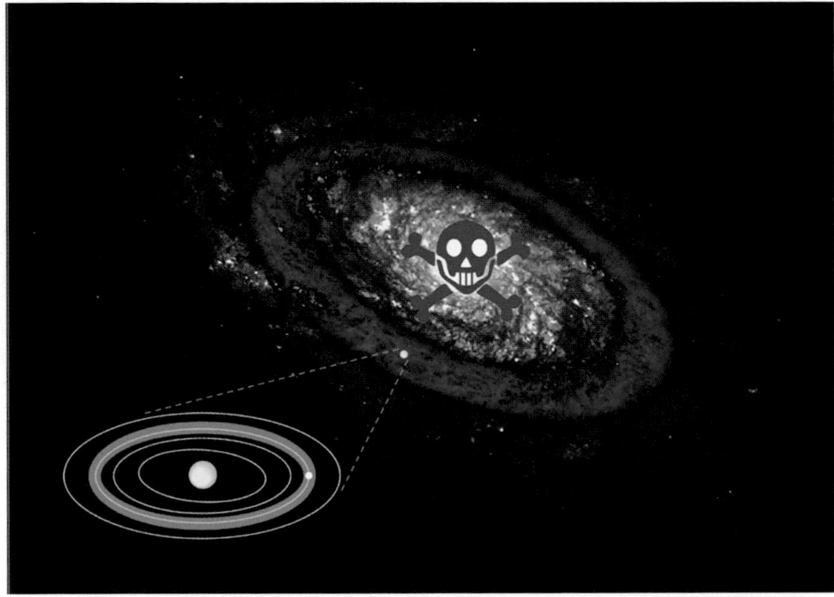

Plate 17: The Galactic Habitable Zone. Illustration of the approximate location of the most habitable places in the galaxy. The inner disk is more dangerous, and the outer disk lacks enough heavy elements to build Earth-size planets. The Circumstellar Habitable Zone is shown in the lower left.

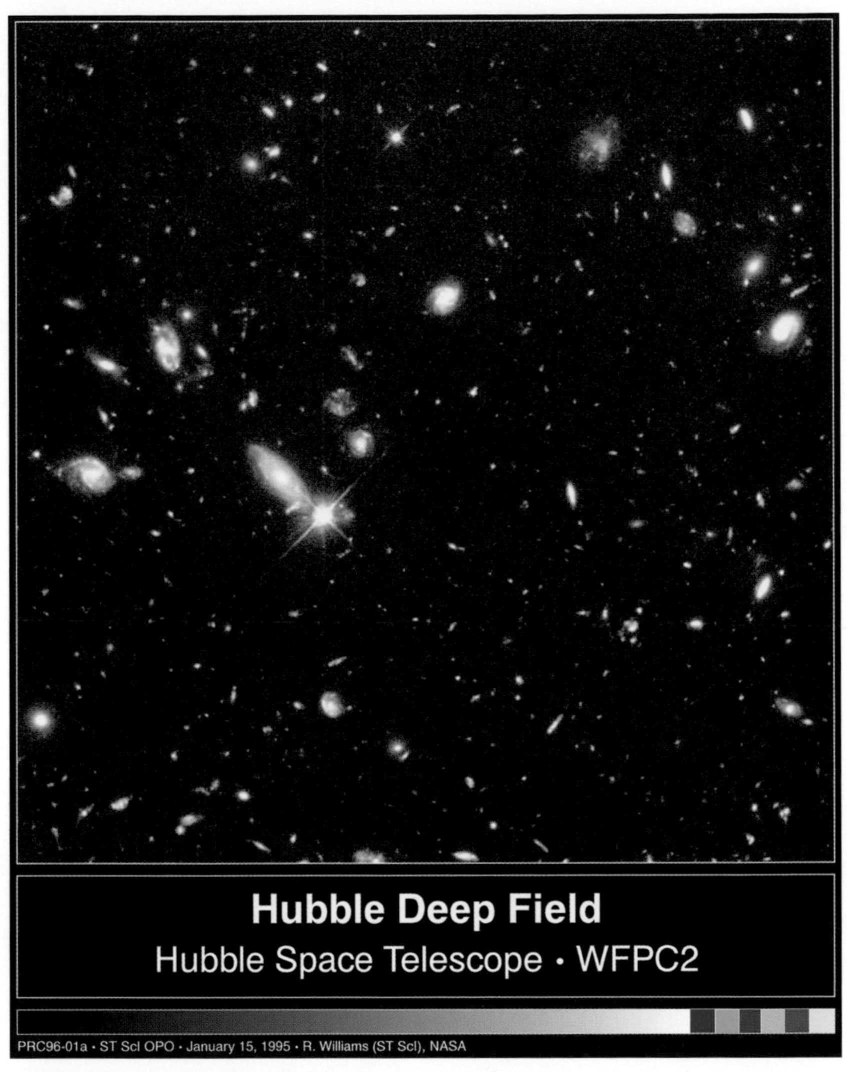

Hubble Deep Field
Hubble Space Telescope • WFPC2
PRC96-01a • ST ScI OPO • January 15, 1995 • R. Williams (ST ScI), NASA

Plate 18: The first and most memorable "Hubble Deep Field," imaged by the Hubble Space Telescope in 1995. This image reveals myriad distant galaxies as they existed billions of years in the past.

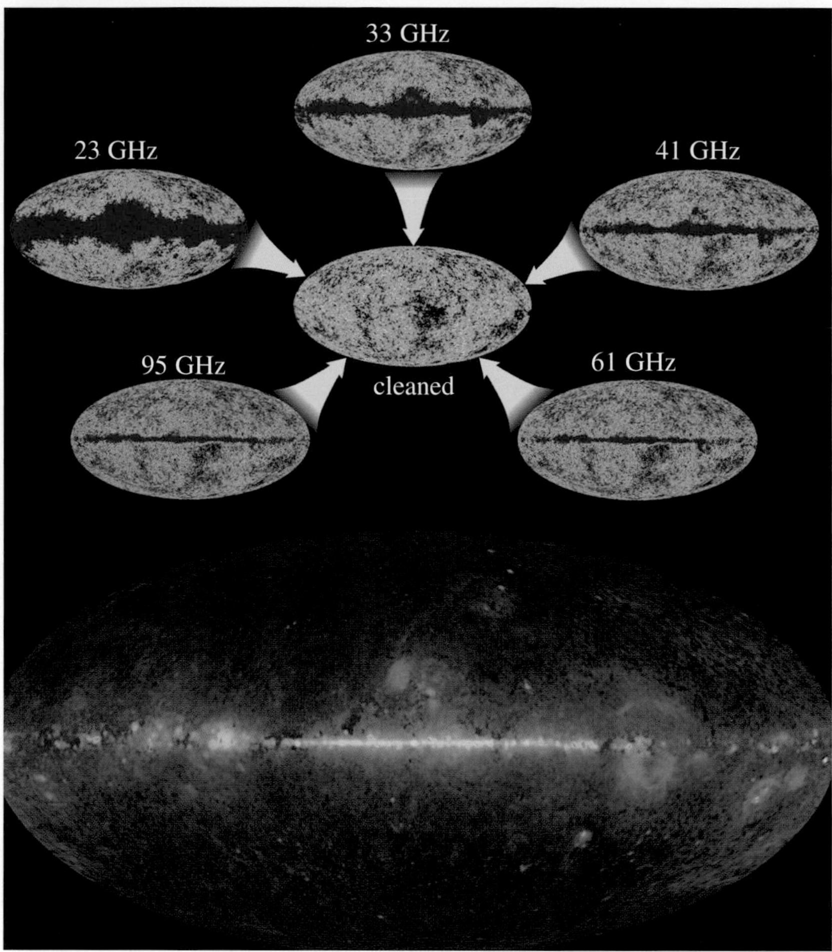

Plate 19: (Top) Full-sky maps at five frequencies in the microwave region of the spectrum (red is bright, blue is faint). The horizontal band running across each of the five maps is the Galactic plane. The final "cleaned" map with the foreground contaminants removed is shown in the center; the remaining variations in brightness are due to the cosmic background radiation. **(Bottom)** The 41 GHz map with the Galactic foreground contaminants color-coded: synchrotron radiation is red, free-free radiation (interacting charged particles) is green, and thermal dust is blue.

Galileo added observations enhanced by the telescope. Peering through his portal to the heavens, Galileo resolved a dizzying panoply of stars invisible to the naked eye. He discovered the four "Galilean" moons of Jupiter and saw the phases of Venus. Both of these discoveries showed that not all bodies have their motions centered on Earth. He also observed craters, valleys, and mountains on the Moon and found spots on the Sun. Such "blemishes" contradicted the belief that all heavenly bodies are pristine spheres. In the Aristotelian cosmology, these were anomalies to be ignored or somehow explained away. Galileo attempted to do neither. On the contrary, his revelations enhanced the plausibility of the new, emerging cosmology.

So, far from demoting the status of Earth, Copernicus, Galileo, and Kepler saw the new scheme as exalting it. Galileo in particular defended the notion of "earthshine," in which Earth reflected the light and glory of the Sun more perfectly than the Moon did. He thought that Earth's new position removed it from the place of dishonor it occupied in the Aristotelian universe, and located it in the heavens:

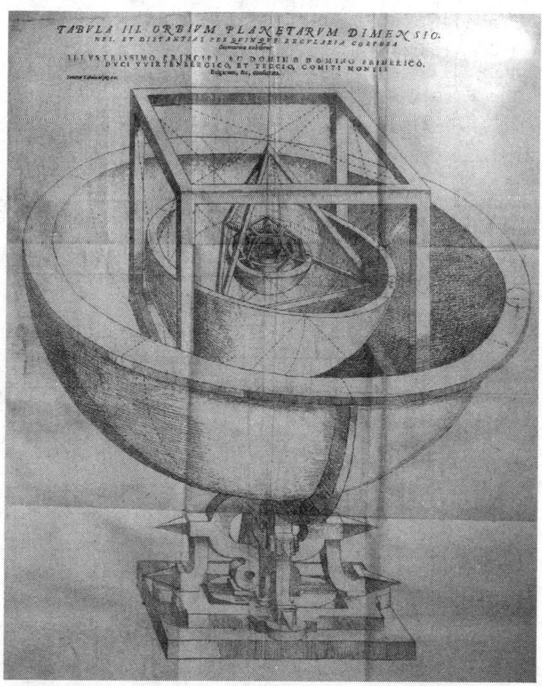

Figure 11.11: This "geometric planetarium" by Johannes Kepler (1596) illustrates Kepler's mystical attraction to mathematical harmonies in nature.

Figure 11.12: Galileo Galilei (1564–1642), from the 1613 frontispiece of his discourse on sunspots.

> [M]any arguments will be provided to demonstrate a very strong reflection of the sun's light from the earth—this for the benefit of those who assert, principally on the grounds that it has neither motion nor light, that the earth must be excluded from the dance of the stars. For I will prove that the earth does have motion, that it surpasses the moon in brightness, and that it is not the sump where the universe's filth and ephemera collect.[46]

Galileo's recantation of his views before the Inquisition in 1633 is a profoundly misunderstood episode in the history of science. It can't be reduced to a simple conflict between the scientific pursuit of truth and religious superstition. The event involved a number of complicating factors and historical quirks, with one-dimensional heroes and villains in short supply. Galileo not only insisted that the Church immediately endorse his views rather than simply allowing his insights gradually and inevitably to gain widespread acceptance, he also mocked Pope Urban VIII in his 1632

Figure 11.13: The figures on this frontispiece of Galileo's *Dialogue Concerning the Two Chief World Systems, Ptolemaic and Copernican* (1632), which led to Galileo's infamous and widely misunderstood trial before the Inquisition, are often misidentified as the characters in the dialogue. In fact, it shows Copernicus (right) holding a sun-centered model, in dialogue with Aristotle and Ptolemy (left) holding an Earth-centered armillary sphere. Galileo choose an easy target by making the Ptolemaic system the foil against Copernicanism. He ignored the Tychonic system, which was actually the leading alternative to Copernicanism at the time.

Dialogue Concerning the Two Chief World Systems.[47] At the same time, he angered the secular Aristotelians who ruled the universities and whose careers were dedicated to the older cosmology. He further irritated them by publishing his ideas in the popular Italian vernacular, rather than in the usual scholarly Latin. In any case, Galileo was primarily reproved not for teaching Copernicanism but for going back on his previous promise not to teach that it was literally true.[48] Even after his censure he continued to receive a Church pension for the rest of his life, hardly the sort of treatment one would expect from a monomaniacally oppressive Church authority. These are just a few of the complexities that make the incident poor material as a symbol of an eternal warfare between science and religion.[49]

Dominican monk Giordano Bruno's execution in Rome in 1600 is another misunderstood event, actually having little to do with our story. It should go without saying that his execution is a dark spot on the history of the Church. His Copernican views were incidental, however, since he also defended pantheism and an eternal universe with an infinite number of occupied worlds, neither of which Copernicanism entailed. Bruno wasn't even a scientist. Nevertheless, writers frequently call him into service to frame a discussion on the possibility of extraterrestrial life, or to prove that religion sought to destroy every brave and clearheaded empiricist it could find.[50] In fact, Bruno was almost certainly not executed for his "Copernican" opinions but for his heretical views on the Trinity, the Incarnation, and other doctrines.[51]

It's also important to note that much of the resistance to what we call the Copernican Revolution derived from the fact that for some time it left many important questions unanswered—in particular, how the planets and stars moved and cohered without the celestial spheres. One central insight was the switch from Aristotle's belief in projectile motion, in which a moving object must be acted upon directly to keep moving, to the modern concept of inertia,[52] in which a moving object keeps moving unless stopped by wind drag or something else. A related insight also contrary to Aristotle was Newton's mathematical understanding of gravity, which allowed bodies to act on one another from a distance without direct contact. Without the earlier dichotomy between the terrestrial and celestial realms, Newton could describe, even if he could not explain, both the fall of objects to Earth and the revolution of the planets around the Sun.

For Newton, natural laws were God's ordinary or regular ways of acting in the world. Miracles were God's extraordinary or irregular ways of doing

Figure 11.14: Isaac Newton (1642–1727).

so.[53] According to Newton's way of thinking, besides the ordinary actions of physical laws, God acted by sustaining the motion of celestial spheres, and by setting up the initial orbits of the planets[54] and later preventing them from disintegrating. Newton didn't hesitate to appeal to extraordinary acts of God to explain features of the natural world. Nevertheless, many of Newton's successors thought Newton was suggesting that God had to "correct" his own regular actions. They preferred instead the notion of God manifesting his powers not with irregular actions in nature but strictly by establishing regular "laws" that governed the entire cosmos.

With the skeptical arguments of the later Enlightenment and the Epicurean materialist strain of thought revived during the Renaissance (present already in Galileo and others[55]), the concept of nature's Law*giver* slowly disappeared, and with it Newton's concept of natural law.[56] In its place stood an atomistic and impersonal concept of Law as mechanistic natural necessity, and for a time this view threatened to crowd out all other explanations (such as chance and design) in natural science. Newton's deistic successors left a picture of nature as an eternal, infinite, and self-sustaining clockwork machine. Others, having abandoned Aristotle's teleological views, and

Newton's specific design arguments, threw the baby out with the bathwater and banished any notion of design from the natural world altogether.

The concept of contingency in nature, so hard-won in the Middle Ages, and defended by many of the founders of modern science, was again in danger of succumbing to a deterministic system. Darwin's theory of evolution in the nineteenth century, and the mysterious disclosures of the quantum world in the twentieth century, led some to reintroduce chance as a scientific explanation[57]—and increasingly, despite intense opposition, there are whispers that design, too, may have a role to play in the science of the twenty-first century.

WHERE ARE WE AND WHY DOES IT MATTER?

This is merely a rough sketch of the origins of modern science. But it's enough to expose the received version as so much mythology. The centrality of Earth in pre-Copernican cosmology meant something entirely different to the pre-Copernicans from what it means in the textbook orthodoxy we've all learned.[58] There is no simple inference from central location to high status any more than a modern person would privilege the center of our Earth as the ideal terrestrial place to be. Geocentrism did not imply anthropocentrism. As Dennis Danielson puts it, "The great Copernican cliché is premised upon an uncritical equation of *geo*centrism with *anthro*pocentrism."[59] Denying either or both did not automatically disprove the existence of purpose or design in nature. The official story gives the false impression that Copernicus started a trend, so that removing Earth from the "center" of the universe led finally, logically, and inevitably to the scientific establishment of our insignificance. By sleight of hand, it transformed a series of empirical discoveries with ambiguous metaphysical implications into the Grand Narrative of Naturalism.

That said, there are two important issues lurking in the neighborhood—namely, the question of design and purpose in nature, and the uniqueness and significance of Earth and its intelligent inhabitants. While distinct one from the other, these issues are usually part of a package. Their main Western expression is the biblical depiction of man, who, while created along with the heavens, Earth, and animals, and formed from the dust of the ground, is nevertheless specially endowed to perceive the starry heavens above and the moral law within (to paraphrase Immanuel Kant). Man is made in the divine image, free to choose the good, or deny it. Christianity adds to this the doctrine of the Incarnation, whereby God himself became human in order to reconcile man and all creation to himself.

Of course, the conviction that humanity plays a part in some cosmic purpose is not exclusively Christian or uniquely biblical. It's not even limited to the main "religions of the book"—Christianity, Judaism, and Islam. For instance, the Roman philosopher and statesman Cicero argued, far more strongly than most Christian and Jewish theologians, that since the gods and man alone share the gift of reason, and are aware of the passing of time and seasons, "all things in this universe of ours have been created and prepared for us humans to enjoy."[60] Cancel the theological differences, and what remains is the common conviction that the cosmos is designed, our place in the cosmos is suffused with purpose and, whether we're the central characters or not, we play a part in some grand cosmic drama.[61]

Nonetheless, one might still argue that the series of insights Copernicus initiated are suggestive. Forget the myths and muddle about the "center of the universe." The discoveries of other planets, one of which is Earth, the sight of moons encircling many of those planets, the realization that the Sun is one of hundreds of billions of stars in the Milky Way, which is one of hundreds of billions of galaxies in a very large, very old universe, are hardly trivial. Something in this overall picture, even without false stereotypes, can leave us with a gnawing sense of insignificance and isolation. It doesn't refute the idea that we hold a special place in the cosmos, but it does seem to weigh against it.

In the middle of the last century, astronomer Harlow Shapley transformed this metaphysical angst into a scientific rule, which Carl Sagan subsequently popularized. Shapley called this rule, for obvious but historically inaccurate reasons, the Copernican Principle. It is now often called, more descriptively, the Principle of Mediocrity.

Chapter 12

The Copernican Principle

Because of the reflection of sunlight... the Earth seems to be sitting in a beam of light, as if there were some special significance to this small world. But it's just an accident of geometry and optics.... Our posturings, our imagined self-importance, the delusion that we have some privileged position in the Universe, are challenged by this point of pale light. Our planet is a lonely speck in the great enveloping cosmic dark. In our obscurity, in all this vastness, there is no hint that help will come from elsewhere to save us from ourselves.

—Carl Sagan[1]

METAPHYSICAL PRETENSIONS

Every occupation has its risks. Coal miners, if they aren't careful, can contract lung disease and develop emotional disorders related to lack of sunlight. Transcribers, if they don't take frequent breaks, can develop carpal tunnel syndrome. Crop dusters, if they let their minds drift, can have short careers. Other professions have less obvious risks. For instance, some modern physicists, astronomers, and cosmologists, despite the fact that they're *natural* scientists, risk drifting into far-flung speculations, and confusing observation with assumption. The Russian physicist Lev Landau once observed, "Cosmologists are often wrong, but never in doubt."[2] Their talk about such esoteric matters as multiple universes and time travel may bring to mind Mark Twain's quip in *Life on the Mississippi*: "There is something fascinating about science. One gets such wholesale returns of conjecture out of such a trifling investment of fact."

But the risk comes with the territory. Many physicists, and all astronomers and cosmologists, deal with very large-scale phenomena, at times even considering the history of the entire universe over billions of years. As a result, they must use certain guiding rules that have wide-ranging scope and generality. Because these rules exceed what we can observe, they are difficult if not impossible to establish and often rely on explanatory power, aesthetic sensibility, intuition, and speculation. This doesn't mean scientists should disavow the rules. It just means they have to beware of the dangers of their profession.

In the early twentieth century, Albert Einstein used one such idea, called the "cosmological principle," to expand the reach of his General Theory of Relativity (GR). The principle was simply this: We should assume that at very large scales, the universe is homogeneous and isotropic—that is, that matter is evenly distributed and that the universe looks the same in every direction. Nothing in GR implied that the universe conforms to this principle, but the principle allowed Einstein to apply GR to the universe as a whole.[3] And while it has needed ongoing qualifications, it has allowed cosmologists to produce mathematical models of the universe that fit our best observations.

Sometime in the twentieth century, however, Einstein's cosmological principle came to be identified with a subtly different idea, the Copernican Principle, also known as the Principle of Mediocrity or Principle of Indifference.[4] In its modest form, the Copernican Principle states that we should assume that there's nothing special or exceptional about the time or place of Earth in the cosmos.[5] This assertion has a certain plausibility, since without any other information, it's reasonable to suppose that our location is a random sample of the universe as a whole. And there will obviously be more ordinary than extraordinary places to be.[6] Besides, it need not be merely an assumption, since one can formulate it as a scientific hypothesis, make predictions, and compare those predictions with the evidence.

It has a closely related but more expansive philosophical or metaphysical expression, however, which says, "We're not here for a purpose, and the cosmos isn't arranged with us in mind. Our metaphysical status is as insignificant as our astronomical location." Metaphysically, this denial of purpose is usually accompanied by naturalism, the view that the (impersonal) material world is all there is and that it exists for no purpose. Although a minority opinion throughout most of Western history, this view has had adherents from the very beginning. In its early pre-Socratic form among Epicureans and others, it amounted to a conviction that the apparent order of the universe emerged from an infinite and eternal chaos, without purpose or design. Given enough time, space, and matter, these thinkers supposed, anything that can happen, will. Even some theists, like Descartes, seemed to prefer this view.[7] Still, only in the modern age has such a denial of design and purpose in nature enjoyed official majority status among the cultural elite.[8] To question it publicly is virtually to guarantee an end to cocktail party conversations—and invitations, for that matter.

The metaphysical Copernican Principle often guides scientific research. By itself, of course, it's not irrational or "unscientific" to allow one's philosophical convictions or preferences to guide one's research. If it were, no one would be scientific, since it's unavoidable. All scientists are thinking human beings, and all thinking human beings have points of view, which shape how they see the world. Any definition of science that rules out all scientists isn't worth the trouble. Besides, the history of science is filled with examples, some of which we saw in the previous chapter, in which philosophical predilections led scientists to new discoveries. The problem is not when scientists express their points of view in their scientific work, but when their points of view blind or distort their perception of the evidence. There's an important difference between allowing one's philosophical beliefs to guide one's research or suggest lines of inquiry, and imposing those beliefs on the evidence or ignoring contradictory evidence altogether.

What makes natural science admirable is that, at its best, it provides us with a way to publicly test what we believe against the natural world, while allowing us to overlook our individual motives and opinions. One way to do this is to consider the empirical consequences of our assumptions. What, for instance, would count against the Copernican Principle? To insist that nothing could because it's simply a fundamental, commonsense understanding of modern science is to reduce the principle from scientific hypothesis to untestable dogma. Fortunately, such a response is unnecessary. It's fairly easy to imagine what observations would count against it: If human beings, Earth, or our immediate environment were highly unusual or unique in some important ways, then we would have reason to doubt it. If the cosmos seemed specially fitted for our existence, or the existence of life, then that would also count against it. Conversely, evidence that confirmed the mediocrity of our surroundings, or the cosmos itself, would count in its favor.

This means that if it is to be anything more than an arbitrary assumption, we will seek out empirical tests for the Copernican Principle, much as Einstein sought and suggested effective means for testing his General Theory of Relativity. Einstein didn't ask us to simply presuppose it. Nor should we simply presuppose the Copernican Principle, short of rigorous testing. Such an approach to the principle may seem commonsensical, but some among its advocates tend to identify the principle with science itself. For instance, astronomer Mario Livio recently has defended—quite correctly, in our opinion—the concept of beauty as a central ingredient of good scientific theories. He even goes so far as to propose the "cosmological aesthetic principle." In a startling bait-and-switch, however, he identifies "beauty" as literal conformity to the Copernican Principle. To be beautiful, he says, a theory "should be based on symmetry, simplicity (reductionism), and the generalized Copernican principle."[9] A theory is beautiful to the degree that it avoids taking us into account or implying that we are in any way exceptional.

However one judges Livio's quirky concept of beauty—and we'll avoid discussing the fact that in the mainstream history of aesthetics, only the first two elements have even provisional support—clearly he's trying to acquire by definition what must be won by evidence and argument. Scientists don't have to presuppose the Copernican Principle, and there's certainly no argument that establishes it from the basic rules of reasoning. Furthermore, just assuming it is true violates a central scientific virtue, which is to be open to evidence from the natural world. We can't determine the nature of reality by imposing a definition of science that restricts the questions we can ask. And we shouldn't arbitrarily protect our assumptions with definition games. In any case, the important issue is not whether the Copernican Principle is beautiful but whether it is *true*. And one way to determine this is to consider its predictions.

PREDICTIONS OF THE COPERNICAN PRINCIPLE

Once considered, it's fairly easy to produce some general predictions of the Copernican Principle. In practice, these are usually unstated expectations rather than actual predictions. This has the effect of protecting them from critical scrutiny—all the more reason, then, to make them explicit.

We all take some of its implicit predictions for granted. For instance, we think that the same laws of physics and chemistry govern both the heavens and the Earth. We're reasonable in concluding that nature's laws are uniform, so that the law of gravity doesn't differ on Earth and the Moon, or on Mondays. Moreover, there are lots of stars and galaxies, and we can expect that many of those stars will have planets circling them. In at least these ways, then, Earth is not unique. This is the firm legacy of the Copernican Revolution. If we stopped here, the Copernican Principle might appear to be well founded. But on closer examination, many of the important predictions turn out to be false or at least questionable.

Here let's consider the Copernican Principle in its natural jurisdiction: astronomy. It manifests itself in cosmology, physics, and biology as well. But we'll hold those issues for the following chapters.

Because astronomy considers objects as small as meteorites and individual planets, and as large as clusters of galaxies, the Copernican Principle has generated the most predictions in this field. Let's scrutinize the major ones in turn. Appropriately, we'll begin with one of the earliest:

Prediction 1:
Earth, while it has a number of life-permitting properties, isn't exceptionally suited for life in our Solar System. Other planets in the Solar System probably harbor life as well.

This was one of the earliest expectations of modern astronomers. When only scant evidence was available, many respected scientists expected to find intelligent life on other planets in our Solar System. Kepler famously conjectured that the structures on the Moon were built by intelligent beings. More recently, Giovanni Schiaparelli (1835–1910) described Martian "channels," which to Percival Lowell (1855–1916) suggested the existence of a Martian civilization. Translating, or mistranslating, Schiaparelli's "channels" as "canals," Lowell founded his own observatory in Flagstaff, Arizona, and dedicated his time to gathering evidence to support his belief.

Lowell is important because of his influence and because he explicitly linked the idea of Martian life to his opposition to anthropocentrism, thus embodying the spirit of the Copernican Principle: "That we are the sum and substance of the capabilities of the cosmos is something so preposterous as to be exquisitely comic. . . . [Man] merely typifies in an imperfect way what is going on elsewhere, and what, to a mathematical certainty, is in some corners of the cosmos indefinitely excelled."[10] According to Carl Sagan, Lowell's enthusiasm "turned on all the eight-year-olds who came after him, and who eventually turned into the present generation of astronomers."[11]

But the Mariner, Viking, and Sojourner missions to Mars revealed a barren and inhospitable environment, and dampened enthusiasm for Martian civilizations. Yet the belief that Mars once harbored life lives on, most recently in the excited announcement of the discovery of microscopic magnetite crystals in the Martian meteorite ALH84001 and the discovery of vast water-ice fields under the Martian surface.

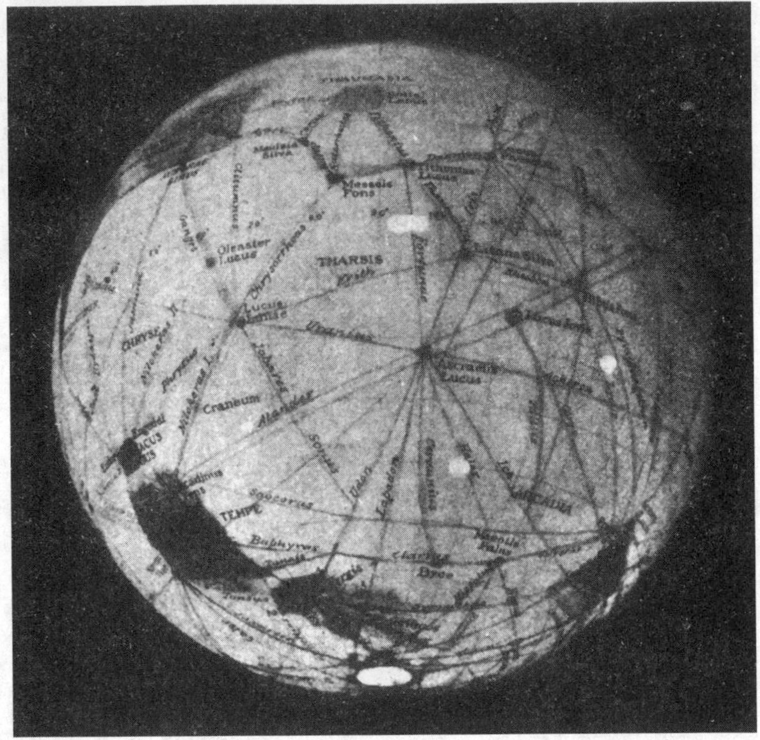

Figure 12.1: Percival Lowell's drawings of Martian "canals" and "oases" from *Mars as the Abode of Life* (1908). With the arrival of the space age, it became clear that Lowell's canals were a combination of optical effects, wishful thinking, and naturally occurring structures.

Most now recognize that the other planets in the Solar System are not good candidates for life. As we discussed in Chapter Five, however, the expectation that extraterrestrial life exists in our Solar System has not disappeared; it has shifted to a few outlying moons orbiting Jupiter and Saturn, such as Europa, where liquid water may exist below the surface. Although we have no evidence for life of any sort in the outer reaches of the Solar System and virtually no one expects to find intelligent life there, speculations abound for the type of exotic creatures that may dwell in the deep, icy crevices of Europa.

Much of this optimism ignores the myriad ways in which Earth is exceptionally well suited for the existence, and persistence, of life, as we detailed in Chapters Two through Six. No other place in our Solar System comes close to providing the astronomical and geophysical properties that make Earth habitable. If anything, the other planets show how narrow the conditions for habitability are, even for planets in an inhabited Solar System.

The basic pattern is worth repeating, because it's so often forgotten or ignored. From the seventeenth to the twentieth century, many expected to find intelligent, even superior life on the Moon, Mars, and other planets in the Solar System. This expectation required direct contrary evidence to suppress it. Now, at the beginning of the twenty-first century, despite PR blitzes from Martian-life enthusiasts, the search has moved from the planets to a few obscure outlying moons. At the same time, the aspirations have been substantially downgraded. No one today expects to find advanced or intelligent life elsewhere in the Solar System. ET advocates now argue that finding the Europan equivalent of slime mold would be just as significant as finding intelligent Martians.

Add to this pattern the evidence of Chapter Six: some of the planets once said to diminish Earth's status now seem to be the guardians of her habitability. Finally, recall that these rare properties discussed in Chapters Two through Six have been crucial in a diverse array of scientific discoveries here on Earth, from the nature of gravity to the internal structure of our planet revealed by seismic activity. Surely these facts about Earth's superiority both for living and observing should count as a sobering contradiction of the Copernican Principle.

Let's try another prediction on for size, though, in this case, as it applies to our host star:

Prediction 2:
Our Sun is a fairly ordinary and typical star.

Textbooks and science writers have repeated this claim so often that it has become an entrenched dogma. This isn't the result of conscious deception, of course, but of the power of the Copernican Principle to shape expectations. We often see what we expect to see. As we discussed in Chapter Seven, however, we now know that our Sun has a number of important, and anomalous, life-permitting properties that simultaneously contribute to its measurability. These include its luminosity, variability, metallicity, and galactic orbit. So our Sun doesn't fit this prediction well at all. Even among the minority of stars in the same class as our Sun, it seems to be exceptional in several important ways, again in contrast to the expectations nurtured by the Copernican Principle.

Maybe the Copernican Principle's prediction about our Solar System will fare better. Or maybe not.

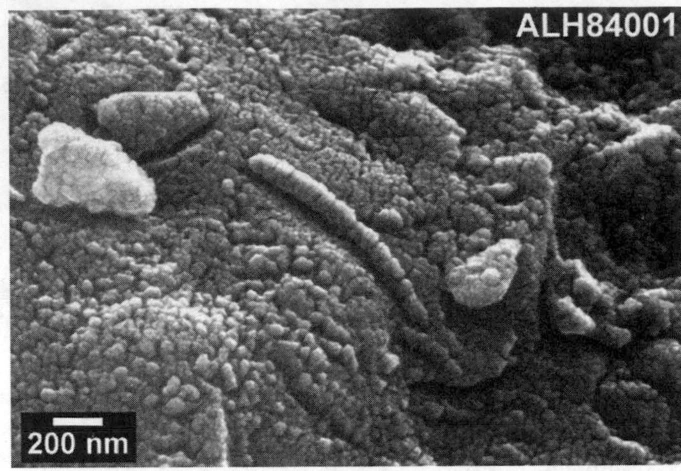

Figure 12.2: Electron micrograph of sausage-shaped structure in the carbonate minerals in the 4.5 pound Martian meteorite, Allen Hills 84001 (ALH84001). This image and other data pertaining to ALH84001 were cited in 1996 as evidence for ancient life on Mars. Today this evidence remains highly controversial.

Prediction 3:
Our Solar System is typical; we should expect other Solar Systems to mirror our own.

This belief has been so common in the past few decades that, until recently, it inspired entire research programs in astronomy and planetary science. Before 1995, when Swiss astronomers Michel Mayor and Didier Queloz discovered the first planet orbiting a Sun-like star, 51 Pegasi, in the autumn constellation Pegasus, many theorists had constructed computer models that "predicted" that other Solar Systems would be more or less like our own. In particular, they predicted that small terrestrial planets would have fairly circular orbits close to their host star, with large gas giants farther out, also in fairly circular orbits.

We now know this is incorrect. The planet 51 Pegasi B is a gas giant about half the mass of Jupiter. It whips around its host star every 4.2 days, only one-eighth the distance from its star that Mercury is from our Sun. It's not what the theorists expected. After the discovery, Mayor noted, "It was very strange to consider the attitude of people facing something completely in disagreement with theory." According to Mayor, some astronomers even said things like "Oh, this is not a planet, because you cannot form Jupiter-like planets close to their stars."[12] The discoveries of other extrasolar planets since then

Figure 12.3: Illustration of Martians for a nonfiction article by H. G. Wells, "The Things That Live on Mars," in *Cosmopolitan* magazine, March, 1908. The article appeared during the controversy over Lowell's alleged canals on Mars, when it was widely believed that Mars was home to intelligent life. In the following century, expectations have progressively diminished as we have learned more about the Red Planet.

have continued to contradict conventional wisdom. In particular, most have highly eccentric orbits, quite in contrast to the planets in our Solar System.

Of course, we still don't have enough hard data to say just how unusual our Solar System is. Because of the time required and the nature of the radial velocity search method, detection of extrasolar planets is biased toward systems containing massive planets with relatively short orbital periods. To detect a true Jupiter analog around a Sun-like star, astronomers have to monitor it for about twelve years (the time period of Jupiter's orbit). The required data are just now reaching this time base. In the next few years, astronomers will begin to have a rough estimate of the percentage of planetary systems similar to our own. SETI enthusiasts are holding out hope that these "silent" data will yet vindicate the Copernican Principle. Nevertheless, what appears most likely, at the moment, is that our Solar System will turn out to be highly atypical with respect to its habitability.

If this turns out to be the case, advocates of the Copernican Principle have a failsafe backup prediction:

> **Prediction 4:**
> **Even if our Solar System is not typical, there are lots of planetary configurations that are consistent with the presence of biological organisms. Variables like the number and types of planets and moons are mainly contingencies that have little to do with the existence of life in a planetary system.**

This proposition has the endearing quality of being impossible to falsify, since it's always possible that an unknown organism exists in some bizarre, far-flung system. More important, it provides an escape hatch if the previous prediction turns out to be false, thus protecting the Copernican Principle from an otherwise embarrassing contradiction.

In any event, the evidence from Chapters One through Six weighs heavily against it. The existence of a large, well-placed moon, of circular planetary orbits, of a properly placed asteroid belt with felicitous properties, of the early bombardment of these asteroids on Earth, of the outlying gas giants to sweep the Solar System of sterilizing comets later on—all these and more are profoundly important for the existence of complex life on our planet.

Ironically, despite the positive claims made for the Copernican Principle, this prediction actually may have slowed the progress of science, by leading astronomers to underestimate the importance for life of seemingly trivial details like comets, asteroids, moons, and outlying planets. Similarly, it may have discouraged astronomers from giving the concept of our Solar System's habitable zone due credit. If earlier astronomers had taken these matters more seriously, they might have put much less energy into quixotic searches for extraterrestrial life in our Solar System that turned out to be wild-goose chases, and focused instead on the role of our Solar System's specific configuration for the habitability of Earth itself. This is a topic scientists are just now beginning to explore; it shouldn't be stymied by the assumption that our Solar System's peculiar features are unimportant for Earthly life.

Nor should a closely related and similarly doubtful prediction of the Copernican Principle:

Prediction 5:
Our Solar System's location in the Milky Way is relatively unimportant.

This inspires the belief that all or most stars, regardless of their location, are receptive hosts for habitable planets, including, for instance, binaries and stars in globular clusters or near the galactic center. In 1974, using the Arecibo radio telescope in Puerto Rico, Frank Drake and other astronomers transmitted a radio message to the globular cluster M13, Great Cluster of Hercules.[13] With perhaps 300,000 old, densely packed stars, this was a very unlikely place for planets in general, let alone habitable ones. While they may not have stated it explicitly, Drake and his colleagues were clearly assuming this prediction of the Copernican Principle. For that reason, they aimed their telescope at a small target with lots of stars, to maximize their chances of contacting an intelligent race. But the number of stars is irrelevant if the stars aren't compatible with life.

When discussing the history of Milky Way astronomy, writers often tell a story parallel to the Copernican stereotype we criticized in Chapter Eleven. Because of miscalculations due in part to cosmic dust that obscured the light from distant stars, earlier Milky Way astronomers William Herschel and Jacobus Kapteyn had placed our Solar System near the galactic center. Using evidence from globular clusters, however, Harlow Shapley later determined that our location is actually thousands of light-years from the center and near a spiral arm.[14] As we might expect, advocates of the Copernican Principle see this as another confirmation of their theory. As one science writer says, "Just as Copernicus had removed the Earth from the center of the Solar System, so Shapley would yank the Sun from the center of the Milky Way and put it in the celestial equivalent of a suburb."[15] The literary parallel may be too good to resist, but once again, the imagery is deceptive. As we now know, the galactic center, like Dante's hell, is the very last place we'd want to be. Our Solar System is located within what may be a quite narrow Galactic Habitable Zone, and far from dusty, light-polluted regions, permitting an excellent overall view of both nearby stars and the distant universe.

The Copernican Principle also has something to say about our galaxy as a whole:

Prediction 6:
Our galaxy is not particularly exceptional or important. Life could just as easily exist in old, small, elliptical, and irregular galaxies.

In a recent study of the size of the Milky Way, astronomers S. P. Goodwin, J. Gribbin, and M. A. Hendry argued that our galaxy is slightly smaller in diameter than its closest large neighbor, Andromeda (M31), and is, "at most, an averagely sized spiral galaxy." They concluded that their finding provides further support for the "principle of terrestrial mediocrity" (aka the Copernican Principle), by which they mean "that there is nothing special about where and when we live and observe from. We seem to live on an ordinary planet orbiting an ordinary star, and it is natural to infer that the Solar System resides in an ordinary galaxy."[16] But one might question the significance of mere galactic size when seeking support for that principle. As we've seen, the fact that some stars are larger than the Sun hardly counts against the Sun's life-friendly properties. Similarly, when we consider the properties of a spiral galaxy most relevant to its habitability, such as mass and luminosity—which correlate with metal content—we find that the Milky Way is exceptional. In fact, astronomers have known since the early 1980s that the Milky Way and Andromeda galaxies have comparable luminosities,[17] and recent research indicates that the Milky Way's mass probably exceeds Andromeda's, even though Andromeda is itself quite massive and luminous.[18]

As we've argued, age and galactic Hubble type are also important. All galaxies in the early history of the universe, and low-mass galaxies forming now, are metal-poor, making them unlikely habitats for life. Similar problems attach to globular clusters and irregular galaxies. We now have reason to suppose that large spiral galaxies like the Milky Way that formed at about the same time are substantially more habitable than galaxies of different ages and types. The metal content of a galaxy is highly dependent not just on its age but also on its mass. Without enough metals, there aren't enough materials for building terrestrial planets. And without terrestrial planets, there are no environments suitable for life. Our very massive, spiral galaxy, then, is an especially suitable home for a habitable planet and Solar System, while providing an optimal position for viewing and discovering both our own galaxy and the wider universe.

One might think the Copernican Principle had nowhere else to go, no bigger field in which to be wrong, but one does remain: the universe as a whole.

CHAPTER 13

THE ANTHROPIC DISCLAIMER

[T]he next battle of the Copernican revolution is thrust upon us. Just as our planet has no special status within our Solar System, and our Solar System has no special location within the universe, our universe has no special status within the vast cosmic mélange of universes that comprise our multiverse.
—Fred Adams and Greg Laughlin[1]

HEDGING THE BETS

The Copernican Principle isn't unique to astronomy. It manifests itself in physics and cosmology as well. But in recent decades, the Copernican Principle has fallen on hard times in these areas, so much so that an interesting modification of it, which uses the Anthropic Principle, has been called in to shore it up. This response is a result of two of the most significant scientific advances of the twentieth century (both made possible by the exceptional measurability of the cosmos): the dual discoveries that the universe is both fine-tuned for life and finite in age. Its finitude may seem like a relatively innocuous discovery. In fact, it's almost as damaging to the Copernican Principle as the fact that the universe is fine-tuned for life and discovery.

THE COPERNICAN PRINCIPLE IN COSMOLOGY

Although there had been a protracted debate about the age of the universe in Europe since the mid-seventeenth century, which later included the United States, by the nineteenth century, orthodox scientific opinion held the universe to be both infinite and eternal. The assumption provided an easy way to avoid questions about the origin of matter, space, time, and natural laws. Newton had initiated this development, but not to avoid philosophically worrisome questions. For him, an infinite universe answered the question of why gravity did not cause the constituents of the universe to collapse in on one another. Moreover, he viewed the universe as the "divine sensorium"—the medium through which God acted in the world. To be adequate to this task, the cosmos needed to be infinite. Needless to say, later advocates, who added eternity to infinity, did not share this motivation; so what Newton suggested became still another prediction of the Copernican Principle, but this time applied to the entire universe:

Prediction 7:
The universe is infinite in space and matter and eternal in time.

This assumption continued into the twentieth century, until Edwin Hubble discovered the relation between the redshifts and distances of galaxies and inferred from this the expansion of the universe. His findings provided a way to reconcile "classical" Newtonian physics with a finite universe. And perhaps more important, they confirmed Einstein's General Theory of Relativity (although, as mentioned in Chapter Nine, Einstein himself had initially resisted the notion of an expanding universe). Hubble's insight led to the development of Big Bang cosmology, which implied that the universe, including cosmic time itself, had a beginning. There could be no greater contrast between assumption and observation. The difference between a temporally finite universe implied by Big Bang cosmology, and the eternal universe assumed by Aristotelian cosmology and two centuries of modern science, trivializes all other shortcomings of the Copernican Principle. This is because, among other things, it's difficult to resist the conclusion that anything that begins to exist must have some "outside" cause to bring it into existence.[2] This is not the sort of evidence advocates of the Copernican Principle anticipated or particularly liked.[3]

So as we discussed in Chapter Nine, before contrary evidence made it untenable, scientists such as Fred Hoyle, Hermann Bondi, and Thomas Gold developed the Steady State models to preserve an eternal universe without a beginning.[4] According to the theory, the universe was expanding, but new matter was continually coming into existence. In this way, it preserved the observation of an expanding universe but avoided the need for a beginning to the universe as a whole. This virtue alone won it some adherents. In 1961, for instance, Denis Sciama used the Copernican Principle to choose the Steady State model over the Big Bang model, arguing that an important goal of science "should be to show that no feature of the universe is accidental."[5] In this case, at least, the principle misguided him. As we noted previously, the discovery of the cosmic background radiation and the explanatory power of the nucleosynthesis of the light elements led to the demise of the Steady State model.

Still in pursuit of an eternal past, some scientists have also suggested an "oscillating-universe" model. The idea is that our universe is just one episode in an eternal cycle of Big Bangs, expansions, and collapses. But this proposal also had problems. Philosopher of science Stephen Meyer summarizes the two important ones:

> [A]s physicist Alan Guth showed, our knowledge of entropy suggests that the energy available to do the work would decrease with each successive cycle. . . . Thus, presumably the universe would have reached a nullifying equilibrium long ago if it had indeed existed for an infinite amount of time. Further, recent measurements suggest that the universe has only a fraction . . . of the mass required to create a gravitational contraction in the first place.[6]

We can add to these problems a third: the evidence of a re-acceleration of the cosmic expansion discussed in Chapter Nine. Apparently, not only is the universe's expansion not slowing down enough to imply an eventual contraction, it's actually speeding up. If one were hoping to smuggle infinite time with an exotic dash of Far Eastern reincarnation into modern cosmology, this recent observation would spell trouble. If this view of an accelerating expansion prevails, it will render the idea of a cosmic recollapse unthinkable, thus leaving the notion of a beginning intact.[7]

THE COPERNICAN PRINCIPLE IN THE PHYSICS LAB

With no place further to go, our metaphysically bloated Copernican Principle can lumber back to the lab and try its hand at physics. Here we get one final prediction:

Prediction 8:
The laws of physics are not specially arranged for the existence of complex or intelligent life.

But the Copernican Principle gets no help here either. The universe not only appears to have a finite past, from which it is still expanding and ever more rapidly; its laws also appear intricately fine-tuned for life, in contrast to the prediction above. As we discussed in Chapter Ten, it has become apparent that if the numerous physical laws, constants, and derived parameters of our universe did not take on the very precise values they have, nothing even roughly resembling our habitable universe would have existed.

Although it's easy to get lost in a swamp of inconceivable numbers and obscure philosophical nitpicking, the idea of fine-tuning is fairly easy to grasp intuitively. Recall the story of the Universe-Creating Machine from Chapter Ten.

If that illustration doesn't work for you, however, here's another one. Imagine an enormous white wall extending in every direction, up to the clouds and in either direction to the horizon, which represents some range of possible universes, with their respective laws and constants, or lack thereof. Then imagine that wall covered with countless black, red, and green dots. Around the outer boundary receding from the visible horizon are black dots, which represent chaotic and disorderly "universes" that don't follow regular laws. Nearer in are the red dots, which represent universes that have stable laws and constants but are not compatible with the existence of complex, living observers anything like us. Within this sea of red dots, you see a green one labeled "The Actual Universe." Although we take it for granted, order is just not what we would expect if we were thrown to the winds of contingency. So the vast majority of this enormous wall will be black. A much smaller proportion of the wall will be red, and an almost inconceivably smaller proportion will be green. The question is, Why is the green dot the only one, so far as we know, that exists?

Now, saying that our universe's laws, constants, and parameters take very narrow values does not mean that no other universe of any sort could be habitable by any sort of organism.[8] What it means is that if we were to fiddle with the values that actually hold in our universe—if, for instance, we were to adjust the value of gravity just slightly in either direction—the universe becomes uninhabitable for any sort of complex organisms. In other words, for as many dots as we care to check in the general neighborhood of our home dot, many, many more will be red than green. And that's just in the area of universes that conform to regular laws. Whether a green dot appears in some far-off section of the wall doesn't affect the fact that our universe falls within a sea of red dots.

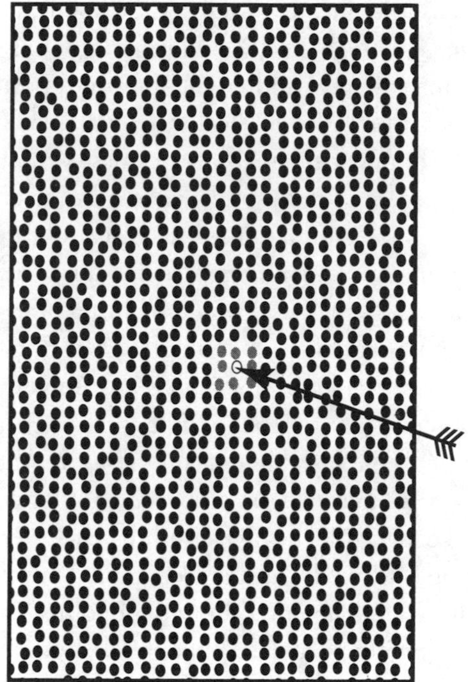

Figure 13.1: Of the many possible universes within our "universe neighborhood," only those with laws and constants very much like our own would be compatible with the existence of complex life.

That this one green dot was selected out of a vast area otherwise populated with red dots requires some adequate explanation. And a surprising number of physicists and astronomers have concluded from this evidence that the universe is in fact designed. Physicist Paul Davies, for instance, reversed his earlier views, saying, "The impression of design is overwhelming."[9] And the late astrophysicist Fred Hoyle, one of the founders of the Steady State model and an intransigent atheist, admitted, "A commonsense interpretation of the facts suggests that a superintellect has monkeyed with physics, as well as chemistry and biology, and that there are no blind forces worth speaking about in nature."[10]

Some scientists, such as Cambridge cosmologist Stephen Hawking,[11] hope the problem will be resolved by collapsing the fundamental forces, and everything they entail, into a single Grand Unified Theory. Given such a theory, the various forces that now seem fine-tuned relative to one another will be, he supposes, the inevitable outcome of some single overarching law. While this might resolve the appearance of fine-tuning between *independent* variables, it would inherit the same problem it was supposed to fix, for any such unified theory would posit some particular value, or number, or formula. While all the actual laws of physics might follow necessarily from it, the higher-level formula would not be necessary. The fine-tuning, it seems, would simply get moved up one level. In fact, it seems that the situation would get worse. Instead of multiple variables, there would be a single, *grand* one, from which the array of sublaws would produce our habitable universe. It would be like a billiard play on a table with a countless number of balls that sinks every ball in one shot.

Others argue that no matter how much our universe may seem fine-tuned, we can't get an actual probability out of it because we can never completely check all the possible dots on the wall. We can't because no matter how closely we look, we can always fit another dot on, in the same way that we can keep hypothetically perturbing laws ever so slightly on our Universe-Creating Machine. There are a potential infinity of possible universes, just as there are a potential infinity of points on a number line.[12] How can we check them all to determine the proportion of habitable and uninhabitable ones? There may be some truth in this complaint, and it is what distinguishes cosmological fine-tuning arguments from comparisons of local environments within the cosmos. But even without the comforting precision that an actual probability might provide, everyone recognizes that the dizzying number of so-called coincidences is fishy and needs explaining, or at least explaining away. Clearly, the many scientists who will go to unsurpassed speculative lengths to salvage the Copernican Principle recognize this need for an explanation (as we will see below).

As before, a story might help clarify. Return again to the Universe-Creating Machine. Perhaps Q, your host, had withheld one very important detail. Although initially satisfied with his truthfulness, you decide to take a closer look at the mysterious machine. In addition to the many dials, you now notice a camouflaged cover. Under it, you find a hidden keypad and display screen with one giant number on it, say, 91215225.79141425. Again embarrassed, Q confesses that that one number, the Grand Unified Combination, in fact controls the whole show. With the keypad, one can change the number, *without limit*, in either direction on either side of the decimal. Unlike the individual dials, neither the number itself nor its number of digits is fixed. But any change to that Grand Combination affects all those combination dials designating the different forces. Q, it turns out, has

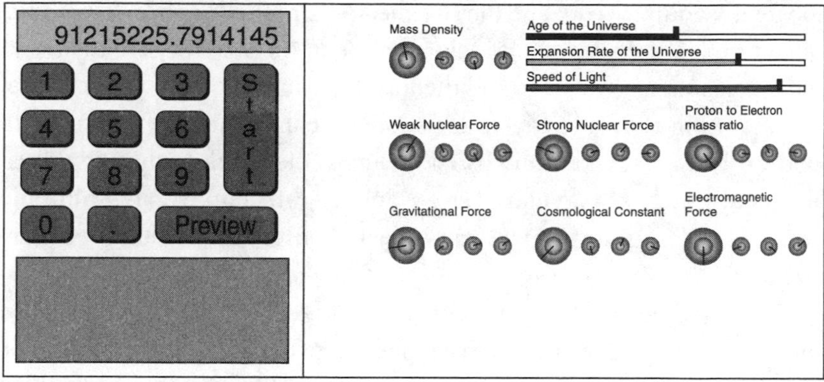

Figure 13.2: Even if all the various cosmological parameters were determined by a single unifying law, and even if the range of possible universes were infinite, we could still recognize that the universe is fine-tuned. Such a universe would be like a universe produced with a digital Universe-Creating Machine, which requires a very specific numerical combination to produce a habitable universe.

spent his time trying to change this number, rather than bothering with the individual dials. He is still convinced that there are other numbers that will produce habitable universes; he just hasn't found any yet.[13]

Now, the fact that there is only one combination, designating one unified force, means that the various forces aren't really independent. And the fact that the possibilities for it are potentially infinite will prevent you from running a probability calculation on the total. You could only consider the (many) numbers near the original combination that have been tried so far, and run a probability on those. Would this prevent you from concluding that the original Grand Combination had been intentionally set? Surely not.

THE ANTHROPIC PRINCIPLE AND THE MANY-WORLDS HYPOTHESES TO THE RESCUE

So on the surface, this series of false predictions looks like a stunning defeat for the Copernican Principle, especially since many of these large number coincidences are so large that they trivialize even the possibilities for habitable settings available within our large, ancient universe. But the metaphysically ambitious Copernican Principle, once degenerated into dogma, is resilient and highly adaptable. Thus adapted, it assimilates an apparently contrary concept, the so-called Anthropic Principle. The somewhat misnamed Anthropic Principle has been the subject of many books and articles since the 1970s, and theorists have formulated it in a variety of ways, ranging from the trivial and tautological to the obscure and outlandish. The most widely discussed version is the Weak Anthropic Principle mentioned in Chapters Seven and Eight. The Weak Anthropic Principle (WAP) states that we can expect to observe conditions necessary for our existence as observers.[14]

In a sense, the WAP is a check on an overly zealous and unsophisticated application of the Copernican Principle, since it restricts the possible laws, constants, and parameters to that range compatible with the existence of complex, carbon-based observers. In that sense, it reminds us that we shouldn't expect to see those laws, constraints, and parameters having just any random value.[15] Because of a selection effect, we will see only those local conditions that allow us to exist. This is handy, since it helps explain why, if the Copernican Principle is true, we don't find ourselves in a much more probable slice of the universe, such as intergalactic space sometime in the distant future.[16] *The Copernican Principle, then, explains all those ways in which our setting is commonplace; the Anthropic Principle accounts for the exceptions.* Using both these principles together, however, is a bit like the story of a railway stationmaster who claimed that all the trains were running on time. When passengers complained that their trains were hours late, he qualified his assertion: "Actually, what I meant to say was that the trains all run on time—except when they don't."[17] The Anthropic Principle is a sort of epicycle that prevents the Copernican Principle from dying the death of a thousand qualifications. The Weak Anthropic Principle may go some way toward explaining our strangely habitable circumstances without recourse to a cosmic intelligent designer, but it does little to salvage the Copernican Principle.

The Strong Anthropic Principle (as some define it) applies this same reasoning to the laws, constants, and initial conditions of the universe as a whole: We can expect to find ourselves in a universe compatible with our existence.[18] True enough. One problem, however, is that some think the Anthropic Principle in its universal application is a *sufficient explanation* for such fine-tuning. But by itself the Anthropic Principle is not an explanation. It simply states a *necessary condition* for our observing the universe. It's no explanation for why that universe exists, or is fine-tuned. What is surprising is *not* that we observe a habitable universe, but that a habitable universe is, so far as we know, the only one that exists.

Think, for instance, of a present-day detective investigating a murder, in which the victim has died by being decapitated in her apartment by a light saber. One of the conditions necessary for the decapitation was the presence of a light saber in her apartment. But no one would be satisfied if, after finding a fully functioning light saber in her kitchen, the detective summarily dismissed the investigation, concluding, "There's really nothing surprising about this murder, since a light saber was present in the victim's apartment. In a crime scene with a light saber and light-saber decapitation wounds, we can expect to find a light saber and light-saber decapitation wounds." If he persisted, making no effort and seeing no reason to figure out how a weapon far beyond the technological capabilities of our culture ended up in the woman's kitchen, or to determine who wielded the weapon, he would be looking for a new job in short order.

Another analogy is the popular story of the firing squad.[19] Imagine an American intelligence officer, captured by the Nazi SS during World War II, who is sentenced to death by a firing squad. Because of this officer's importance, the SS assign fifty of Germany's finest sharp shooters to his execution. After lining him up against a wall, the sharpshooters take their positions three meters away. Upon firing, however, the officer discovers that every sharpshooter has missed, and that instead, their fifty bullets have made a perfect outline of his body on the wall behind him.

What would we think if the officer reflected on his situation, and then responded, "I suppose I shouldn't be surprised to see this. If the sharpshooters hadn't missed, I wouldn't be here to observe it"? We would rightly wonder what he was doing in intelligence, since the more sensible explanation would be that, for some reason, the execution had been rigged. Perhaps the sharpshooters had been ordered to miss, or they had colluded with one another for some unknown reason. In short, the best explanation would be that the event was the product of intelligent design. Shrugging one's shoulders and concluding that it's a chance occurrence is just dense.

Sensitive advocates of the Anthropic Principle have realized that more is needed. What is needed, they argue, is recourse to other universes. But not just any set of universes will do. For instance, following the so-called Everett Interpretation of quantum physics, some physicists suggest that other "universes" exist in the sense that different lines of universes split off as a result of the collapse of "wave functions" at the quantum level. At every collapse, a different branch splits off and goes its own way. Such a process, needless to say, would leave a bunch of universes in its wake. Similarly, some "inflationary" cosmological models postulate a vast array of distinct "universes," or discrete domains, which were produced very soon after the Big Bang. After expanding exponentially, each such domain slowed down to the sort of modest expansion we now detect in our universe.[20] But whatever the virtues of these theories, they won't do the job, and not merely because they lack observational support. The problem is more fundamental than that: both theories presuppose the very fine-tuned laws that need explaining.

What is needed, many suppose, is a full-blown World Ensemble, as postulated by a many worlds hypothesis (MWH), with a more or less exhaustive (that is, infinite), random distribution of different laws in separate, causally disconnected "universes."[21] In such a scenario, innumerable universes (a "multiverse," or "World Ensemble") exist. The only ones observed will be those capable of having observers. In just those universes, observers will look around, and the more benighted ones will marvel that their universe is so well suited for them. In this way, and only in this way, does the Anthropic Principle have any hope of saving the Copernican Principle from a stinging defeat.[22]

In other words, if all this were true, we would observe a habitable universe merely because of a "selection effect." Its improbability would be apparent, not real. We would be like a tiny dandelion growing in a thin crack in the middle of an otherwise arid parking lot around one of those sprawling Texas malls. The dandelion might look around and find it to be a remarkable coincidence that it was planted in the very narrow area that allows it to grow and prosper. What are the chances? If it knows nothing about dandelion reproduction, it might think that it was intentionally planted in the crack. But it would probably be mistaken, because it doesn't recognize that it is captive to a selection effect. Dandelions release lots of seeds into the wind. Some are bound to land in narrow patches of dirt in the middle of big parking lots. Without more information, it should assume that its location is the result of chance, wind, and some other impersonal processes, not of an underpaid landscape engineer.

But there's a whopping difference between this story and our situation. First, it's not at all obvious that an *actual* infinite set of anything, including universes, is even possible.[23] And second, we know that there are gobs of dandelion seeds floating around in the wind every spring, but we have no independent evidence to think other universes exist, except that fine-tuning contradicts the Copernican Principle and lots of people don't like what that might suggest.[24] Astronomers Fred Adams and Greg Laughlin are quite explicit on this point:

Figure 13.3: This image by Thomas Wright (1750) illustrated Wright's "Original Theory of the Universe," and in particular, his ideas on the structure of the Milky Way. To modern readers, however, it might suggest the more modern and more speculative idea that our universe is only one member of a multiverse, which is made up of many different "universes."

The seeming coincidence that the universe has the requisite special properties that allow for life suddenly seems much less miraculous if we adopt the point of view that our universe, the region of space-time that we are connected to, is but one of countless other universes. In other words, our universe is but one small part of a *multiverse*, a large ensemble of universes, each with its own variations of physical law. In this case, the entire collection of universes would fully sample the many different possible variations of the laws of physics.... With the concept of the multiverse in place, the next battle of the Copernican revolution is thrust upon us. Just as our planet has no special status within our Solar System, and our Solar System has no special location within the universe, *our*

> *universe has no special status within the vast cosmic mélange of universes that comprise our multiverse.*[25]

Notice that they don't say that the next *task* of the Copernican Revolution is to find evidence for this so-called multiverse. They say "battle," which is more appropriate for an ideological crusade than a scientific hypothesis. We have come full circle to the infinite chaos with its chance pockets of order posited by certain pre-Socratic philosophers. This multiverse argument, infinitely more fanciful than most fanciful science-fiction stories, simply presupposes the Copernican Principle, along with the grand mythology that has built up around the Copernican Revolution. But why do that? Not only do we have good reason to doubt the Copernican Principle, but using it to summon other universes into existence is clearly bad faith, since we would never accept such reasoning in any other area.

Recall the officer who has survived the Nazi firing squad. How reasonable would it be for him to conclude, "Well, apparently there must be millions and millions of similar executions going on simultaneously all around the globe. Given so many tries, it's inevitable that some group of sharpshooters will all miss their target. I'm just the lucky one." No one would accept this because, one, such an ad hoc inflation of the available "tries" is without evidence, and, two, such a response to events would destroy our ability to make practical judgments. If the firing squad survivor persisted, he could always attribute every experience in his life to chance,[26] no matter how strong the evidence of intelligent design. In fact, *no amount of evidence for apparent design could ever count as evidence of actual design*. But if science is a search for the best explanation, based on the actual evidence from the physical world, rather than merely a search for the best naturalistic or impersonal explanations of the physical world, how responsible is it to adopt a principle that makes one incapable of seeing an entire class of evidence?

Many worlds hypotheses are diverse, and any adequate treatment would require an entire book. But most suffer from the same dilemma: If the alternative universes are causally disconnected from our own, then merely postulating them isn't a causal explanation, and doesn't even compete with, say, intelligent design, since an intelligent designer would be offered as a causal explanation for fine-tuning *this* universe (or a habitable universe like this universe). One still needs a causal explanation for the whole ensemble. If, on the other hand, the many worlds theorist does offer some causal mechanism, some equivalent to a Universe-Creating Machine, then he has merely shifted the problem back one level, since the "machine" itself would need fine-tuning.[27]

One can of course simply posit that the entire multiverse has existed necessarily and eternally, without a cause. After all, we must all stop the regress of explanation somewhere. But this strategy exposes the many worlds hypothesis as essentially the final deductive consequence of the Copernican Principle, with little more basis than the strength of its motivating premise. The fact that those committed to that principle will continue to look for theoretical justifications for a multiverse doesn't obligate those who aren't so committed. And given the persistent failure of the Copernican Principle's testable predictions, we may be wise not to place much confidence in its most speculative consequence.[28]

THE FAILURE OF THE COPERNICAN PRINCIPLE

The preceding list of Copernican Principle predictions, by no means exhaustive, has two implications: When we can test the Copernican Principle against the evidence, it tends to fail.[29] And when it does not fail, it's often because it retreats to a position that makes it virtually unfalsifiable. Although we should expect speculative rearguard defenses to continue, the actual evidence still points in a problematic direction for the Copernican Principle: toward a single, expanding, fine-tuned universe with a finite past, which has changed profoundly over time. We not only occupy an exceptional location within that universe, we also occupy a special moment in cosmic history. While we and our environs are not literally the physical center of the universe, we are special in other, much more significant ways. In a sense, we are nestled snugly in the "center" of the universe not in a trivial spatial sense but with respect to habitability and measurability. This fact stands in stark contrast to expectations nurtured by the Copernican Principle.

In addition, these findings cast a different light on the general narrative of discovery from Copernicus to the present. The existence of other planets, stars, galaxies, and the like is unambiguous (though limited) support for the Copernican Principle only if these are not at all relevant to our own existence. But as we've seen, there's no reason to assume this. With respect to habitability, our existence depends on such local variables as a large stabilizing moon, plate tectonics, intricate biological and nonbiological feedback, greenhouse effects, a carefully placed circular orbit around the right kind of star, early volatile elements–providing asteroids and comets, and outlying giant planets to protect us from frequent ongoing bombardment by comets. It depends on a Solar System placed carefully in the Galactic Habitable Zone in a large spiral galaxy formed at the right time. It presupposes the earlier explosions of supernovae to provide us with the iron that courses through our veins and the carbon that is the foundation of life. It also depends on a present rarity of such nearby supernovae. Finally, it depends on an exquisitely fine-tuned set of physical laws, parameters, and initial conditions.

But why such a large universe? Why all those other galaxies? Surely our argument breaks down here, one might insist. Don't the sheer number and size of galaxies imply, despite the narrowness of conditions required for our existence, that we're simply the lucky recipients of a giant cosmic lottery? In a sense, one could argue that, so far as we know, even our universe is just barely habitable. Some have even concluded that if life is rare in the universe, then the universe is generally "hostile" to life, since it contains so many threats and so few apparently hospitable places. Ignoring for a moment the fine-tuning of laws and constants that constrain the entire universe, and ignoring that such a claim tacitly assumes that the universe exists for no purpose, this conclusion might be plausible. There is certainly a selection effect at the local level, since, unlike undetectable alternate universes, there are lots of other places we know to be within the actual universe. The Weak Anthropic Principle attempts to capture this: If we're living in and observing a just barely habitable universe, we shouldn't be surprised that we will only find ourselves on a habitable planet in a narrow, habitable orbital zone around an unusually habitable star in just the right place in a just barely habitable galaxy at just the right time in a just barely habitable universe just the right age to produce terrestrial planets from an initially dense primordial state. There are other interesting issues here, but that alone isn't surprising.

But notice how the correlation between habitability and measurability reverses the implication of a large universe. The immensity of the universe gives us a wide field for discovery. The very existence of the planets, stars, and galaxies allows us to compare and quantify the conditions for measurability and habitability. This would be impossible if Earth were a lone planet encircling a lone star in an otherwise empty universe. It would have been intellectually impossible without the (real) Copernican Revolution, which led to the realization that the heavens and Earth follow the same laws and are composed of the same matter. Compared with the starry and sprawling firmament that beckons us and stirs our intellectual curiosity, and that finally points us beyond the universe itself, a single isolated planet would be a cosmic cage. There would be far less to discover and inspire us, to draw us out of ourselves, to cause us to appreciate that our existence balances on a razor's edge.

If there were no other galaxy clusters, Edwin Hubble could never have inferred the expansion of the universe from the ruddy hue of distant galaxies. Gravitational attraction among the galaxies within our immediate galactic neighborhood, the Local Group, overwhelms the otherwise detectable effects of the cosmic expansion. An observer living in a universe containing only the rich Virgo cluster would also fail to discover Hubble's Law. At the level of the local supercluster, which contains the Local Group and the Virgo cluster, the cosmic expansion is detectable, although the effects of gravity remain. It is only above the level of galactic clusters that the cosmic expansion is significant enough to substantially overcome the force of gravity and the "peculiar velocity" noise of nearby galaxies. In general, the farther we can sample galaxies, the more reliably we can determine the Hubble constant.

Similarly, if we had existed in the distant future, we would never have detected the cosmic background radiation, the linchpin in deciding between the Steady State and Big Bang models of the universe. These two pieces of evidence have forced modern people, contrary to their philosophical preferences, to contemplate the shocking notion that the universe actually had a beginning.

Scientists have often formulated the Copernican Principle in terms of measurability. Stuart Clark, for instance, defines it in this way: "A cornerstone in every cosmological model of the Universe: it states that there are no preferred vantage points in the Universe. In other words, from the Earth we see everything that could be seen from any other place within the Universe. . . ."[30] Well, cornerstone or not, this is clearly wrong for both our location and our time. If we sat atop a neutron star ten billion years ago, the universe would look quite different. And if we were presently orbiting a red giant in the center of an elliptical galaxy, there would be many things we would not be able to see (we'd also be dead, but let that pass for now).

Still, lurking in the background is the impression that since Earth and its inhabitants are tiny in comparison with the entire universe, they are insignificant. This is a deep-rooted intuition. Even an Old Testament

Psalm expresses it in the theological idiom:

> When I consider your heavens, the work of your fingers,
> the moon and the stars which you have set in place,
> what is man that you are mindful of him,
> the son of man that you care for him? (Psalm 8:3–4, NIV)

Of course once this impression becomes an argument, it doesn't amount to much. Physical size is hardly a reliable indicator of significance. Miners sift through tons of rock to discover a single, small diamond. They discard the tons of rock and keep the diamond. Besides, as we discussed in Chapter Ten, one could just as well argue that we (or Earth) must be really important, since on the size scale leading from quarks to the universe, we're strangely close to the middle.[31] This sort of double-edged reasoning gets us nowhere. Nevertheless, it's good to remind ourselves that we can't answer such questions with size comparisons. The answers will hinge on much more subtle factors.

WHAT DOES IT MEAN?

Since many have used the Copernican Principle as an argument for naturalism—the idea that the (unintended) physical universe is all there is—one might suspect that arguments against the Copernican Principle would automatically count against it. What's sauce for the goose is sauce for the gander, right? Well, the issue is more complicated. It's not a simple two-horned dilemma, since those who reject the Copernican Principle have another option. Perhaps, they will conclude, the origin and evolution of life are just astronomically rare accidents. Maybe life is a thin veneer on an isolated speck of dust in an otherwise meaningless and impersonal expanse of space, time, matter, and energy.

But before we consider that possibility, and what might count against it, let's consider a popular and initially more attractive manifestation of the Copernican Principle—namely, the search for extraterrestrial intelligence.

Chapter 14

SETI and the Unraveling of the Copernican Principle

[N]othing in the universe is the only one of its kind . . . there must be countless worlds and inhabitants thereof.
—Lucretius (98–55 B.C.), *De Rerum Natura*

ENTER E.T., STAGE LEFT

It was bound to happen. As soon as Copernicus suggested that Earth was a planet, and that the other planets, too, were actual places and not mere points, someone was sure to ask whether those planets, like ours, were inhabited. Of course, debate about other worlds had been a topic of philosophical speculation since the dawn of recorded history.[1] But now the speculations had a place to land.

Given the limited knowledge of the other planets in the Solar System, it was a natural inference. Kepler imagined occupants on the Moon, as did William Herschel. There were those Martian "channels" that Schiaparelli spoke of which Lowell took to be Martian construction sites—"canals." And as late as the 1950s, some believed that Mars had vegetation.[2] A series of discoveries during the late twentieth century ended the idea of intelligent life in our Solar System anywhere but Earth. Nevertheless, the hope of extraterrestrial life continues to live on in the popular and scientific imagination. Once established, the idea made an easy leap from our Solar System to other suns with other inhabited planets. Our planetary neighborhood is apparently barren, but to quote the opening line of *The Hitchhiker's Guide to the Galaxy*, "Space . . . is big. Really big."[3] That bigness encourages speculation.

By the late twentieth century, interest and belief in extraterrestrial life had permeated Western society. The fascination first emerged in the literary world at the beginning of the century through writers such as H. G. Wells, who helped herald the new literary genre of science fiction. It was only with the advent of popular media like radio, however, and especially motion pictures and television, that such fascination became a true mass cultural force. A frighteningly realistic 1938 radio adaptation of H. G. Wells's *War of the Worlds* by Orson Welles, which described an invasion by hostile Martian forces, inadvertently sparked a nationwide panic. Since the 1960s, alien life has become one of the most common themes in entertainment. From Arthur C. Clarke and Stanley Kubrick's 1968 classic *2001: A Space Odyssey*, to *Star Trek*, *Star Wars*, *Close Encounters of the Third Kind*, *Alien* and its three sequels, *E.T.*, *Men in Black*, *Independence Day*, and *X-Files*, aliens inhabit the modern consciousness as much as angels and demons inhabited the medieval consciousness. Technological advances, especially the Apollo Moon landings from 1969 to 1973, gave the topic the aura of reality. As a result, various recent polls indicate that over 60 percent of Americans now believe in extraterrestrial life of some sort.[4] For years, we too were part of that majority.

The increased fascination with alien life has been accompanied by increased reports of sightings of UFOs (with a presumed extraterrestrial origin), alien artifacts like crop circles, alien encounters, and even "alien abductions." In many cases, it's easy to see quasi-religious overtones in this fixation.[5] Publications such as *UFO* magazine refer to advocates of alien visitation as "believers." In more mainstream scientific circles, belief in extraterrestrial life has become so prevalent that astronomer and science historian Steven Dick has described it as the dominant myth of our age, symbolizing what he calls the new "biophysical cosmology."[6]

PARADOX LOST

None of this, of course, discredits the idea. While it's easy to dismiss late-night alien abductions as fantasies or delusions, the question whether extraterrestrial life, and in particular extraterrestrial intelligence, exists is surely a legitimate one. Its main problem, at the moment, is that we have no evidence for it. This is more than just an observation. When beefed up, it becomes a formidable argument against the existence of extraterrestrial intelligence (ETI), usually called Fermi's Paradox. In 1950, Nobel laure-

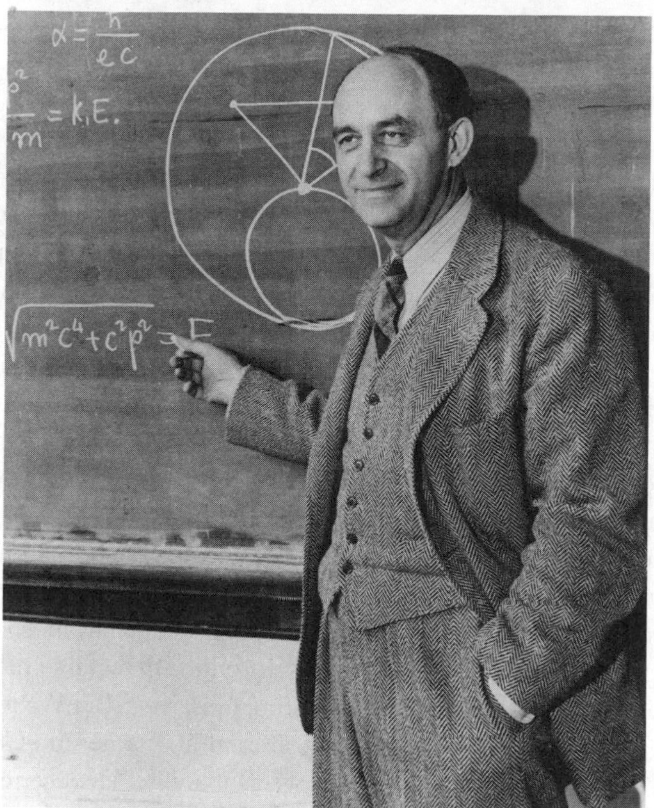

Figure 14.1: Enrico Fermi (1901–1954).

ate and physicist Enrico Fermi (1901–1954) turned to his colleagues at Los Alamos in New Mexico and asked, "If there are extraterrestrials, where are they?"[7] His argument—not really a paradox unless you assume that ETIs must exist—is simple. If there were numerous other intelligent civilizations like our own in the Milky Way, some of them would surely have had a head start on us. They would eventually run out of room or encounter local hazards or simply grow curious, which would encourage migration. Within a few million years—a mere blink of the eye on galactic timescales—they would have colonized the rest of the galaxy or, as some later argued, sent self-replicating robots.[8] They would target habitable systems, and either colonize other habitable planets, terraform nearly habitable planets, or mine asteroids. While our galaxy may be twelve billion

years old, there's no trace of such colonization, either now or in the past. The reasonable conclusion: They're not here because they don't exist.

Notice that Fermi used a touch of "Copernican" reasoning. He assumed that if a multitude of advanced civilizations exist, we probably would *not* be the first one on the scene. This is an intriguing twist, since most scientists who believe extraterrestrial civilizations exist have also gained inspiration from the Copernican Principle. (As we'll see, the Copernican Principle gets rather complicated when it enters the biological realm. In fact, it further unravels.)

Fermi's question reminds us that we don't have any hard, publicly available evidence of any sort for life anywhere other than Earth. Granted, it's not a knockout argument,[9] since absence of evidence isn't necessarily evidence of absence (although it might be). Believers in ETI have offered abundant responses to Fermi's question. Perhaps interstellar travel is more of a barrier than we suppose, or intelligent aliens are not the inveterate colonialists that Fermi assumed. Perhaps the colonization wave passed us by.[10] Or perhaps they're very good at hiding, or are so advanced that they no longer use radio communications or produce detectable radio signatures, or have quarantined us like a nature preserve.[11] Bill Watterson perhaps inadvertently parodied such responses in his comic strip *Calvin and Hobbes*, where Calvin says, "The surest sign that intelligent life exists elsewhere in the universe is that it has never tried to contact us."

The responses are so diverse, and at times so far-fetched, that one might wonder whether belief in ETIs is not, for some, unfalsifiable. Still, there's no reason to suppose that the methods of science cannot be applied to the question. Surely some event could *verify* it. (The opening scene of the movie *Independence Day* comes to mind.) Clear verification is precisely the hope of the Search for Extraterrestrial Intelligence (SETI). Although the Copernican Principle is at home in cosmology and astronomy, its most popular current expression is at the intersection of astronomy and biology, now often called astrobiology, and particularly in SETI.[12]

SEARCHING THE SKIES

Early SETI researchers such as radio astronomer Frank Drake sometimes transmitted radio signals into space in the hope that an alien civilization might happen to intercept them. But most researchers now spend their time trying to detect intentional or unintentional radio (and more recently, visible light[13]) transmissions from extraterrestrials. Using sophisticated arrays of radio telescopes and pattern-recognition software, they hope to separate intelligent signals (if any ever appear) from the background radio noise that permeates the universe.

As with any research, limited resources[14] have imposed some discipline on SETI. Since they require expensive telescope time, SETI researchers want to aim their radio telescopes at a star system with some chance of harboring intelligent life. To get some purchase on this question, Frank Drake devised an equation, now called the Drake Equation. He first proposed it at a 1961 meeting as a simple way to calculate the number of advanced civilizations in the Milky Way galaxy able to communicate with radio signals.[15] He devised the equation not to determine the chances of ETI generally but to determine the chances of us detecting radio-communicating ETIs. Nevertheless, with some adjustments, it can address the more general question as well.

To the uninitiated, the Drake Equation can look intimidating, but it requires nothing more complicated than third-grade-level multiplication and fractions. Although a few symbols have changed over the years,[16] the clearest version is as follows:

$$N = N_* \times f_p \times n_e \times f_l \times f_i \times f_c \times f_L$$

N, the product of the equation, is the total number of radio-communicating, technological civilizations in the Milky Way at any one time. The answer derives from the total number of stars in the galaxy (N_*) times the fraction of stars with planetary systems (f_p) times the number of habitable planets in each system (n_e—*e* stands for "Earth-like") times the fraction of habitable planets on which life emerges from inorganic matter or organic precursors (f_l) times the fraction of those planets on which intelligent beings also evolve (f_i) times the fraction of those planets on which sufficient communications technology arises (f_c) times the fraction of the average planetary lifetime during which there is an advanced civilization (f_L). Simple.

Since every *f* is a number between 0 and 1, the product of the equation will be vastly lower than the total number of suitable stars in the galaxy (N_*). In addition, many of the variables are unknowns. So the numbers we plug in depend profoundly on the assumptions we bring to the problem. In many cases, we have little more than an educated guess. Nevertheless, walking through the equation helps illuminate the array of factors that a

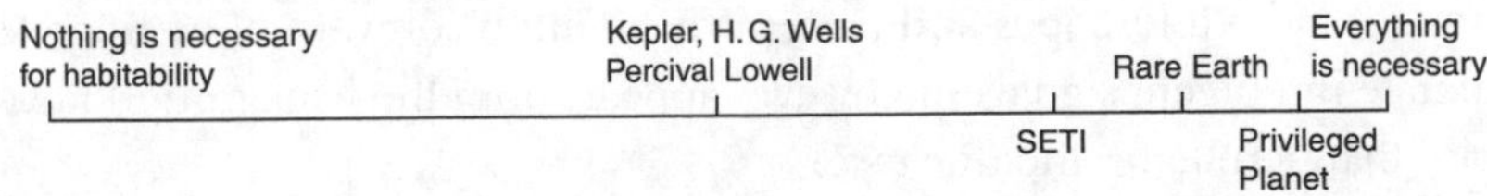

Figure 14.2: On a line representing all possible opinions on habitability, the contemporary debate about the frequency of habitable planets in the universe occupies only a tiny segment. It's widely recognized that life requires specific conditions, and that, compared with the total volume of the universe, or even the total number of stars, habitable planets (and civilizations) are uncommon. The debate concerns how many factors are needed to get a habitable planet, and thus, how uncommon such planets are.

single environment needs to host life. (For the curious, in Appendix A we offer our own revised version of the Drake Equation.)

Think of the conditions for habitability spread out on a number line, with different views represented by points on the line. The point farthest to the left represents the view that there is nothing special about Earth, its local environment, the Milky Way, the number and types of galaxies, the size and age of the universe, and the laws of physics as a whole. While these are sufficient for the present habitability of Earth, none are necessary. There are innumerable other, equally sufficient pathways that have the same outcome—complex, intelligent life. The point farthest to the right represents the view that everything about our cosmos, locally and universally, is necessary for our existence. We depend not just on the precise value of gravity and a terrestrial planet but also on the exact number of atoms that make up the planet Mercury and the temperature on a gas giant around a star in the Andromeda Galaxy. Now, so far as we know, no one seriously defends either of these extreme positions. The entire debate takes place somewhere to the right of center.

Compared with the ET enthusiasm of a century ago, in fact, we could cast the current debate as a question of how *uncommon* habitable planets, and complex life, are in the universe. Carl Sagan sanguinely estimated in the 1960s that there might be one million civilizations in the Milky Way. But even that would mean that only one in one hundred thousand star systems contained ETIs—still a small percentage. In any case, what we want to know is just *how* much, locally and universally, is necessary for our existence, and for the existence of complex life forms similar to us. On the line of habitability, most SETI researchers are well to the left of us, but even they are right of center.

The Drake Equation has two modest qualities: It restricts the search to carbon-based life and requires that a habitable planet maintain liquid water on its surface, which, as we've argued, are eminently reasonable propositions rather than mere failures of the imagination. Moreover, it considers only the number of civilizations in the Milky Way, since even the Andromeda Galaxy, the closest large galaxy to ours, is two million light-years away. It would be extremely difficult if not impossible to detect radio signals from isolated planets at such a distance, not to mention that any signals we did detect would be two million years old. If we want to estimate the probability of the existence of other civilizations outside the Milky Way, then, we'll have to add another variable: the number of (habitable) galaxies in the universe.

Although the equation adds a bit of realism to the debate over extraterrestrial life, it's easy to detect the bias of most SETI researchers.[17] The Drake Equation is now forty years old, but until recently only the first term was based on actual observations not limited to our own Solar System. But this limitation has not prevented SETI researchers from assuming wildly sanguine numbers for the variables.[18]

Moreover, it glosses over the complexity of the problem, since every factor is really the product of another, hidden equation. For example, n_e, the average number of habitable planets per system, depends on many factors, ranging from the metallicity of a host star to planetary size and Solar System configuration. And as we discover more about habitability requirements, the values of the equation draw closer and closer to zero. As SETI supporter Bernard Oliver once said, the Drake Equation, despite its mathematical façade, is "a way of compressing a large amount of ignorance into small space."[19]

SETI has drawn much of its inspiration from the Copernican Principle, which to SETI advocates implies that intelligent extraterrestrial life is relatively common. This is obvious in Carl Sagan's famous conjecture, using the Drake Equation, that there are perhaps one million advanced civilizations in the Milky Way.[20] Little more than a guess, it remains entrenched in both the popular and the scientific imagination. As we have seen, however, the recent trend of discoveries, and its pace, has increased our appreciation of the many requirements for complex, nonmicrobial life.

Drake couldn't take account of these discoveries, of course, so he vastly underestimated the barriers for success. But since his equation has become a standard part of the debate, it's a good place to start.

Stars

This version of the Drake Equation requires the total number of stars in the Milky Way galaxy, estimated at around 100 billion. But as we discussed in Chapter Seven, that's not a very useful number, since the vast majority of stars in our galaxy are not likely candidates for life. To get in the ballpark, we need to include additional factors. Over 80 percent of the stars in the Milky Way are low-mass red dwarfs, which, as we argued, probably lack habitable zones. One to 2 percent are short-lived, massive blue giants that aren't around long enough for complex and technological life. About 4 percent of stars are early G-type, main-sequence stars like the Sun.

Since we know that our Sun is capable of supporting life, we might suppose that other stars the size and type of the Sun will be equally habitable. But as we now realize, our star seems to contain just the right amount of life-requiring metals, even compared with otherwise similar nearby stars. It contains enough metals for building a habitable planet, but it doesn't have too much, which might have produced an unstable planetary system with too many massive planets. As we discussed in Chapter Eight, overly massive planets would lumber obtrusively through a planetary system, making the presence of terrestrial planets in stable circular orbits less likely. Additionally, at least 50 percent of main-sequence stars are born in binary systems, most of which are inhospitable for similar reasons.

That might still seem to leave a lot of candidate stars, until we remember that most of those stars don't occupy the Galactic Habitable Zone (GHZ), where it is safer from comet impacts and sterilizing radiation while containing proper amounts of metals for building the right-size terrestrial planets in stable orbits and for building organic beings. Now, these are only the factors related to stars that we've discovered so far. Given the trend, there are probably others. In any case, compared with the rest of the factors in the Drake Equation, this pruning is trivial.

Earth-like Planets

The discovery of extrasolar planets has always been "one of the holy grails of the extraterrestrial life debate."[21] Until recently, however, determining the fraction of stars with planetary systems (f_p) was pure speculation. In an intellectual culture steeped in the Copernican Principle, estimates have usually erred wildly on the side of life's commonality. Presuming that our planetary system was ordinary, many astronomers expected most, if not all stars, or at least Sun-like stars, to have planets. Many astrophysicists, like

Virginia Trimble, used "common sense" and "computer models" to show that "the Milky Way probably still contains at least 10^{10} stars that could have harbored habitable, terrestrial planets for more than five billion years."[22] After decades of failure, careful searches in the last few years have finally begun to pay off. We now know that at least some other Sun-like stars do have planets.

These discoveries, however, have been mixed blessings for SETI enthusiasts for two reasons. First, we now know that our Solar System probably is not typical, even compared to other stars known to have planetary systems. In fact, the features that make it habitable may be extremely rare. Second, the same methods that have detected extrasolar planets around some Sun-like stars have not detected planets around most others. Of course, some of this failure is due to the limitations of the present detection methods. Still, it suggests that many stars do not have giant planets. We do not yet know enough to say just what the percentages are, but we can already place some limits. Microlensing searches toward the galactic bulge show that giant planets accompany no more than about 30 percent of low-mass stars in the inner Milky Way.[23] As of early 2002, astronomers using the Doppler method estimated that about 4 percent of Sun-like stars have giant planets at least as massive as Jupiter.[24] Research in the next few decades will give us more confidence in this factor for both terrestrial and gas giant planets.

Chapters One through Eight make clear that n_e, the average number of habitable planets per system, is likely to be extremely low. Since theorists had assumed that our Solar System is typical, they believed that most planetary systems would have about nine planets, with one in the habitable zone. It's starting to look as if this confidence was premature. Like the others, this one factor is really the product of another, larger equation, hidden within it. With the discovery of every new extrasolar planetary system, it becomes clearer that they can assume a variety of configurations. No law of physics or celestial mechanics requires that all smallish terrestrial planets orbit relatively near their star, with all gas giants perched farther out, all in fairly circular orbits. In fact, gas giants can orbit in much smaller and much larger elliptical orbits. Such planets would almost surely disturb the orbits of any hapless terrestrial planets unlucky enough to form. No law of physics requires that a well-placed asteroid belt always remain after the other planets have formed, or that the one terrestrial planet in the Circumstellar Habitable Zone must have a large moon to stabilize the tilt of its axis. The law of gravity, which controls the orbits of bodies around a star, allows an enormous degree of freedom in the natures, shapes, and sizes of orbits.

Until recently, most astronomers restricted this factor primarily to the average number of planets in a star's Circumstellar Habitable Zone. We now appreciate that many other properties of our Solar System contribute significantly to Earth's habitability, as we covered in previous chapters. Just since the early 1990s, astronomers have discovered the importance of the Moon for stabilizing Earth's tilt, that gamma ray bursts are extragalactic, highly luminous events, which would sterilize any nearby life if one were to occur in the Milky Way, and that there are hazardous giant black holes in the nuclei of nearly all nearby massive galaxies. Just in the last few years astronomers have detected the destruction of dust disks around young stars in the Orion Nebula, which implies that many stars will not be accompanied by outer planets. How many other such factors await discovery? Given present trends, we're willing to bet that there are more to come.

THE ORIGIN OF LIFE

Once we get some habitable planets, we must ask what fraction will be inhabited by some form of life. (Recall, f_l is the fraction of planets with any sort of life.) Despite common pronouncements to the contrary, we know very little about the origin of life from inorganic matter and/or organic precursors, as well as the development and evolution of complex life. Even so-called simple life requires a dizzying amount of information. In this book we have dealt primarily with the many conditions needed to produce a habitable planet. As far as we know, such a planet is a necessary but nowhere nearly sufficient condition for the existence of complex life. We will avoid making and developing arguments about what else is needed to produce life, since that's a book-length subject in itself.

More important for our purposes is the guiding influence of the Copernican Principle on theorists who do ponder these questions. Concerning the origin and evolution of life, the Copernican Principle takes on a split personality, its explanations refracting into the twin modes of chance and necessity. For some, this principle implies that the origin of life is inevitable, given the right conditions. For others, life is an astronomically improbable fluke. Its bipolar personality thus exposed, we now see that the Copernican Principle is about something more basic than the prevalence of life in the universe.

In the nineteenth century, many thought life at the microscopic level was simple. The nineteenth-century Darwinist Ernst Haeckel, for instance, characterized cells as simple "homogeneous globules of protoplasm." Despite what we now know about the mesmerizing complexity of cells, and the fundamental difference between chemistry and the biological information encoded in chemicals,[25] many still assume that where there's liquid water, there may very well be life. After all, they surmise, if life on Earth emerged very quickly after it was possible to do so, it must be a fairly easy process. (This ignores, among other things, the strong possibility of a selection effect.) Even Don Brownlee and Peter Ward, who argue in *Rare Earth* that complex life is uncommon, nevertheless maintain that "simple" life may be common.[26]

Here one can detect a tension between chance and natural law. Unlike Brownlee, an astronomer, and Ward, a paleontologist, most biologists, who are intimately acquainted with life's mesmerizing and narrowly specified complexity, see life's origin as astronomically improbable. The probabilities for the origination of even fairly "simple" functional proteins through random reactions are dauntingly small. As a result, these biologists argue that we shouldn't expect it to happen more than once.[27] In contrast, some origin-of-life researchers (many outside biology) argue that nature may have certain "self-organizing" laws, enabling life to emerge anywhere given the right chemical conditions.[28]

Because of this fundamental disagreement, there's not even a biased consensus on the chances of life springing up even on an ideally habitable planet. One side seems to assume that any habitable planet will be inhabited. For them, the fraction of habitable planets with life is close to one. The other side is willing to tolerate a universe full of habitable planets, with few or only one actually inhabited. For them, the fraction is close to 0.[29] The one thing both sides seem to agree on is that life shows no signs of purpose or design.

INTELLIGENT LIFE

Finally, of those planets with some form of life, we must determine the fraction of those on which intelligent life evolves (f_i). Recall our distinction between simple, complex, and technological life. Intelligent life would be some smaller subset of complex life. Judging from the Earthly sample, it's probably a tiny fraction of complex life.

It is here, closer to home, that we reach the core of the Copernican Principle. We might sum it up in one simple statement: *We're not special.*

As with the origin of life, so in the debate over the prevalence of intelligent life, we see the tension between chance and necessity, which Charles Darwin had combined in his theory of evolution in the nineteenth century. Darwin argued that random variation (chance) combined with natural selection (necessity or law) would mimic the activities of an intelligent designer and explain the appearance of design in biology. But even those who attribute life's diversity and complexity exclusively to Darwin's mechanism still disagree over which element should have priority.

As a result, the conviction that there's nothing special about *Homo sapiens* can have two quite different meanings. On the one hand, it might mean that once life originates, the evolution of more and more complex and intelligent life forms of some sort is likely if not inevitable. So we have no reason to think we're unique. On the other hand, it might mean that we're the result of a concatenation of chance events, mass extinctions, catastrophes, and dumb luck, and aren't the crowning achievement of some grand cosmic drama. If we're unique, it's not because we're designed and doted on but because we're a fluke—a one-off win at the cosmic slot machine.

As the Copernican Principle trails off in opposite directions, it's easy to lose the scent. Consider, for instance, the debate that raged between the late Stephen J. Gould and Simon Conway Morris. According to former Harvard paleontologist Gould, life's evolution, despite the connotation of the word, is not an ever-increasing, progressive rise in complexity but a meandering and directionless path marked by myriad "contingencies" and mishaps, such as mass extinctions. If we took Earth back to the beginnings of life and let the evolutionary process play itself out again, life would take on completely different forms from those it took on during the first go-round.[30] Darwin's central message, Gould contended, is not one of progress, in which we are the crowning achievement of either creation by divine fiat or a progressive evolutionary process, but purposelessness. In short, according to Gould, the Darwinian revolution bears the same basic message as the Copernican Principle: *We're not special.*[31]

In contrast, Cambridge paleontologist Simon Conway Morris argues that various laws and constraints dramatically narrow the role of chance events in evolution. He points to the many examples of "convergent evolution," such as placental and marsupial mammals, in which similar structural forms appear in creatures that are not closely related.[32] This suggests that some general strictures for life's evolution have been built into nature, so that the chance of intelligent beings emerging is "probably very high,"[33] even if it takes a variety of different pathways. To Conway Morris, this suggests that, in a sense, life may have been built into the cosmos from the beginning, and the history of life may exhibit a larger purpose.[34]

Now, such an idea makes it much more likely that SETI will succeed (assuming there are other planets compatible with life). But its teleological overtones are quite out of keeping with the Copernican Principle. This leaves SETI advocates in a bit of a quandary, because most believe that the existence of extraterrestrial life would confirm the Copernican Principle.

Despite this apparent conflict of interest, the benefits of the idea that life has been built into the cosmos lead most astrobiologists to prefer necessity to chance in the origin and evolution of life.[35] Two interesting exceptions are Peter Ward and Don Brownlee. In *Rare Earth*, they argue, like Gould, that countless contingent events in Earth's history—many very important in the history of life—are highly unlikely to be repeated elsewhere.[36] On Earth, life appeared almost as soon as it possibly could. Intelligent life, however, has only appeared recently, implying that even if so-called simple life is common, intelligent life may not be.

As a generalization, then, many astronomers see the Copernican Principle as suggesting that life, including intelligent life, is common.[37] Many biologists, in contrast, see life, including intelligent life, as a fluke. Yet they don't believe that this contradicts the Copernican Principle but rather confirms it. How can the Copernican Principle produce contrary predictions and still be scientifically useful?[38]

Advanced Civilizations

The last two variables of the Drake Equation, f_c (that fraction of intelligent life that attains radio-communications technology) and f_L (the average lifetime of an advanced civilization) are closely related. (Roughly, f_L is synonymous with our category of technological life.) We can only guess at both, since we have just one data point to work with: us. Most SETI researchers assume that intelligent life will often, if not inevitably, develop science and technology. Avoiding the details of our own such development makes this an easier task. But as we saw in Chapter Eleven, the origin of modern science and technology depended on a precise configuration of economic, cultural, philosophical, and theological precursors, and an unusually long-lasting and stable warm climate. Technology requires dexterity and a level of capacity to communicate that, of the millions of known species of life, only humans possess. It also requires access to an oxygen-rich atmosphere, dry land, and concentrated ores. The laws of physics did not uniquely determine any of these. Until these factors came together, no civilization developed technology advanced enough to harness radio communication. And even on Earth, this has happened only once. What justification do we have for assuming that it's an inevitable result of life, even intelligent life, everywhere?

A different and ironic wrinkle, known as the Doomsday Argument, if correct, would also reduce the chances of successful communication between advanced technological civilizations. The argument goes something like this. The Copernican Principle requires that we assume that we occupy a random rather than privileged sample of space and time (with the Anthropic Principle accounting for exceptions). So we must assume that we don't occupy a special spot along the historical time line of individual observers of the human race. Given the number of generations that have lived, and the fact that the human population has increased dramatically in more recent times, it's unlikely that we now occupy the very early segment of the human race's total history. When applied to advanced technology, the argument is even more depressing, since it would be unlikely that we occupy the very early segment of its existence. Our technological civilization has already survived, say, one hundred years, so there's a good chance that it will not last dramatically longer than that time span into the future. Applying such reasoning generally, this would mean that the average civilization advanced enough to use radio communications would survive only a few hundred years. And what are the chances that two or more will overlap at just the time necessary to communicate with each other?[39]

Apparently, some find this reasoning plausible. But it draws on the doubtful assumptions that we occupy a random sample of time and space and that the future is finite.[40] As a result, it looks to us like a *reductio ad absurdum* of the Copernican Principle, especially since, as we've seen, there are lots of good reasons not to assume these things. The lesson of the Doomsday Argument, it seems, is that we should avoid making facile deductions from the Copernican Principle. Still, the Doomsday Argument is interesting for its irony, since this pessimistic outgrowth of the Copernican Principle implies that SETI, also an inspiration of the Copernican Principle, will be unlikely to succeed. Once again, we see the principle turning on itself, like the snake consuming its own tail.

On these two variables of the Drake Equation—that fraction of intelligent life that attains radio-communications technology, and the fraction of any intelligent, alien race's existence wherein they are an advanced civilization—the correlation between habitability and measurability is a double-edged sword. On the one hand, it gives a ray of hope to SETI researchers. Forget for the sake of argument that the conditions for habitability are extremely rare and that the origin of science depended on many historical contingencies. If intelligent life exists elsewhere in the universe, its local environment probably has many of the same features conducive to scientific discovery that ours does. If any ETIs exist, they probably won't be bereft of a large moon, hemmed in by opaque clouds or restricted to deep, murky water, or surrounded by an otherwise empty Solar System bathed in the dull light of thousands of nearby stars. So they would have similar opportunities for scientific discovery. Moreover, they would have access to information about long-term climate and nearby objects like asteroids and comets, which would enhance their likelihood of surviving for long stretches of time. They're also likely to be in the Galactic Habitable Zone, meaning at least some of them might be nearby. On the other hand, the correlation of habitability and measurability strengthens Fermi's argument, since it diffuses some of the SETI enthusiasts' responses to it. It would give us some reason to suppose that any other intelligent civilizations should be as detectable as we are, and quite possibly able to detect life on Earth and direct their signals and/or their exploration accordingly. So where are they?

Other Galaxies

Drake limited his equation to the Milky Way galaxy because he was concerned only with ETIs that we have some chance of detecting. But there are still billions of other galaxies. In the 1990s, astronomers began training the Hubble Space Telescope deep into space, and resolving thousands of distant galaxies never before seen by human eyes. Some of the light from these Hubble Deep Fields has traveled twelve billion light-years. Such a galactic swarm reminds us that the Milky Way is not the only potential habitat in the universe. In fact, it recently motivated probability theorist Amir Aczel to argue that no matter how small the chances, we can be certain that, somewhere else in the universe, there must be intelligent life.[41]

Unfortunately, Aczel treats galaxies the way early SETI researchers treated stars: one's as good as another. As should be evident from Chapter Eight, this is a serious mistake. Size, age, type, and metallicity all conspire to drastically reduce the number of galaxies capable of harboring not only life but even terrestrial planets. Our Milky Way is among the 2 percent most luminous and, hence, most metal-rich galaxies in the local universe, putting it way ahead on the scale of habitability.

It's difficult to envision, but we have to shake the impression that in the Hubble Deep Fields we are seeing pictures of distant galaxies as they now exist. What we are seeing are ancient images initially transmitted through space billions of years ago during the early stages of the cosmic expansion. Such early galaxies were metal-poor, which means that they probably didn't have enough metals to produce many Earth-size planets; in addition, the radiation levels were more intense then. The entire galactic panorama displayed in the Hubble Deep Fields is therefore an ancient portrait of a time and place almost certainly devoid of habitable planets.

We don't know what the state of those galaxies is "now." In the future, when we know more about galaxy formation, we might be able to determine what percentage of these might have become like the Milky Way, but at the moment, we can only speculate. Perhaps a few now compare nicely with the Milky Way. Perhaps some of these galaxies now have habitable planets with observers looking back at the early Milky Way in their own Hubble Deep Field. The truth is, at the moment we just don't know. But regardless of whether we are literally unique on a cosmic scale, it should begin to dawn on us that, at least in our neighborhood, we might be exceedingly rare. Too much discussion of this issue, however, could distract us from a more fundamental point. For, whether or not it succeeds, SETI, for many, is about something more basic than finding alien races.

SETI AND THE REMAINS OF THE COPERNICAN PRINCIPLE

Complicated and often contradictory, SETI is hard to analyze. It's difficult to miss the semi-religious overtones of its search for higher, albeit natural, intelligence in the heavens, and its desire for a transforming, revelatory message.[42] The fascinating 1997 movie *Contact*, based on Carl Sagan's fictional novel about SETI, typifies this spirit. In the movie's climax, the protagonist and SETI researcher, Ellie Arroway (played by Jodie Foster), is transported to a distant galaxy to communicate directly with an alien representative. She asks him about the purpose of such an interplanetary community, into which humanity has just been inducted. He tells her: "In all our searching, the only thing that makes the emptiness bearable is each other."

Later, after this lone encounter, a congressional panel grills Arroway about her experience. Exasperated by their skepticism, she finally describes her discovery as "a vision that tells us how *tiny* and *insignificant* and *rare* and *precious* we are." In one breath, she sums up the paradoxical charm of SETI. On the one hand, its advocates are following the implications of the Copernican Principle to its bitter end. On the other, they are embarked upon a quasi-religious quest for those who have lost faith in the traditional object of religious belief. The realization that there are many other, alien civilizations might naturally lead one to regard our tiny, relatively primitive civilization as insignificant. But where does she get *rare* and *precious?* She gets them, perhaps, from the god-like and highly advanced alien being who takes the form of her late, loving father. One need not be hyperimaginative here to see this father-like alien as a proxy for a deceased divine father as well, the Father declared dead by nineteenth-century science and philosophy. Every civilization in human history has demanded its god or gods. The SETI enthusiasts apparently are little different.

At the same time, for whatever reason, many SETI researchers tend to have an antireligious streak.[43] They often claim that the discovery of extraterrestrial intelligence would (they usually say *will*) inevitably destroy the traditional religious beliefs of most Earthlings.[44] Lucretius in the ancient world also drew this conclusion, as did SETI researchers in Soviet-era Russia.[45] But the issue is much more subtle than this. Our insignificance no more follows from a universe abundant with life than our individual significance hinges on a scarcely populated planet. The same is true if intelligent life is extremely rare. A universe teeming with life could just as well be purposefully designed and individually valued, as could a universe in which life is rare.[46] Of course, which universe exists—one teeming with life or one where life is rare—is a very interesting and worthwhile empirical question. But the intrinsic interest of both possibilities, along with their theological ambiguity, is often lost on SETI enthusiasts, because the Copernican Principle so thoroughly permeates their work.[47]

In fact, as we've noted, various partisans take both the rarity of life *and* its proliferation as support for the Copernican Principle. Either life is inevitable and hence commonplace or it is rare and hence an accident. But this false dilemma simply presupposes that nature has no design or purpose.

This unstated presupposition against design is apparent in the definition often given to the notion of "contingency." Properly speaking, a contingency is simply something that happens but doesn't have to happen.

Philosophers, therefore, contrast contingent events and necessary events, the latter being events that for some or another reason have to happen. Most scientists see an event as "necessary" if it is determined by the laws of physics. In either the philosophical or the scientific setting, an event can be contingent because it is the result of chance or because it is the result of a free choice. Contingency is the arena of both freedom and accident. The naturalist collapses all contingencies in the natural world into the category of "chance." But it's not the only option, it's just the only option the naturalist is willing to consider. A good way to foreground the disguised presupposition is to avoid the word "contingency." Instead, we should split contingency into its two possible forms, and so speak of chance, design, and necessity.

Astronomer and science historian Steven Dick recently illustrated the point while reflecting on the theological consequences of SETI. He is more theologically astute than many SETI devotees; nevertheless, he concluded that any "monotheistic religions" that wish to "survive extinction" must adjust to new "cosmotheological principles," by which he means the Copernican Principle. The result? If we are to retain any theological language, we must learn to identify "God" with the cosmos itself.[48]

Unintentionally, Dick has summed up the central aspiration of the Copernican Principle: to restrict our gaze to a material universe that, by definition, was not designed. But what if the cosmos has been designed? What if our place in it and its suitability for us as sophisticated observers suggest a purpose? If science involves thinking hard and open-mindedly about the empirical evidence before us, is it really scientific to ignore this evidence because it doesn't fit into some philosophical box? And if we choose not to ignore the evidence, then how do we consider it? To put it differently, if the universe did exist for a purpose, how could we tell?

Chapter 15

A Universe Designed for Discovery

There are more things in heaven and earth, Horatio,
Than are dreamt of in your philosophy.
—William Shakespeare, *Hamlet* Act I Scene V, 185–186

DISCERNING DESIGN

Near the beginning of Arthur C. Clarke and Stanley Kubrick's 1968 sci-fi masterpiece, *2001: A Space Odyssey*, there is a scene that is at once inspiring and terrifying. Human beings have made their first, small steps into space, and the United States has established a colony on the Moon. During their explorations, astronauts unexpectedly uncover a black, domino-like "monolith," buried beneath the lunar surface. When a beam of Sun alights upon the mysterious object for the first time in some four million years, it triggers the transmission of a signal toward Jupiter. For obvious reasons, the United States keeps the discovery secret.

Although no one knows much about the object, one character quickly announces, without doubt or irony, "This is the first evidence of intelligent life off the Earth." We later learn that the monolith has the geometric proportions of one by four by nine (the squares of the first three prime numbers), but we never see any intelligent life other than human beings. In fact, we never learn the purpose of the strange object, even after astronauts discover a giant replica orbiting Jupiter. Still, at no point is there any controversy about the origin of these strange structures. Even initially, no one worries that the lunar monolith might be just a new type of rock. No viewer feels any discomfort that perhaps the characters are jumping to conclusions and that, since they haven't seen any aliens, and know nothing of their intentions, they had best suspend judgment about the origin of the object. No one complains that since they can't falsify the belief that the object is designed, they can't infer that it was. Everyone can tell that it bears the hallmarks of intelligent design—without argument, intricate reasoning, probability calculations, or a forensic investigation. Everyone simply *sees* that the object is designed.[1]

Nor do we need strange monoliths being distributed by standoffish aliens to detect design countless times every day. You're doing it right now, simply by reading this text. Even if you were looking at a text in a language you didn't understand, you would still know it was text and not, say, a pattern formed from decaying paper. We usually don't need to know the meaning, function, or purpose of an object to know it is designed. We rarely even know how we do it. We just do it.

For example, when modern Britons first saw Stonehenge on the Salisbury Plain in Wiltshire, England, they all knew someone had built it. In its discovered state, it was little more than a disheveled circle of large, rough-hewn rocks. (They have since been tidied up a bit.) But almost everyone recognized that it was more than the product of wind and erosion. Research in the last few decades has revealed that Stonehenge is aligned with seasonal events like solstices, and that its builders may have used it to predict astronomical events like eclipses. But even without this knowledge of its purpose, virtually everyone, to put it philosophically, "infers design" when they see it.

Archaeologists commonly use design reasoning, sifting run-of-the-mill stones from ancient tools, arrowheads, and other artifacts. Design plays an important role in a number of other specialized sciences, such as forensics, fraud detection, cryptography (cracking encoded messages), and notably, SETI. Individuals are sentenced to life in prison or execution on the basis of a scientific judgment that a death was the result of criminal design rather than mere accident. And everyone assumes that, at least in principle, SETI researchers will be able to sift out intelligent extraterrestrial radio signals from background radio noise.

In these examples the designer detected is the natural variety—either a human being or some extraterrestrial creature, the latter still natural or creaturely even if novelists and filmmakers enjoy endowing them with semi-divine qualities. But for most of human history, most individuals have inferred design when viewing certain natural objects and nature as a whole. In the West, despite the skeptical arguments of philosophers like David Hume, most have continued to believe that nature is designed, at least in part because of the evidence of the senses. In the English-speaking world, William Paley's famous argument in his *Natural Theology* captured this common intuition. If you were to find a watch lying in a heath, with its intricate and purposeful arrangement of parts, Paley argued, you would reasonably infer that an intelligent agent was responsible for it. As recently as a century ago, his book was still required reading for all undergraduates at Cambridge University.

But times have changed. Charles Darwin is widely touted as having made design explanations superfluous in the biological realm, by proposing that natural selection acting on random variations could mimic the work of an intelligent designer. And the "official" view now among scientists and academics is that the notion of intelligent design is either unscientific or at least superfluous to the practice of all natural science.

If one bases one's judgments on the evidence we have discussed in this book, however, it should come as no surprise that we think that conclusion was premature. With that evidence clearly before us, we are finally situated to consider the broader question: What is the best explanation for the origin and features of the universe we have described?[2] We have argued at length that the correlation between habitability and measurability contradicts the Copernican Principle. But we think it also challenges the assumption that the findings of natural science inevitably confirm naturalism.

SEPARATING THE CHAFF FROM THE WHEAT

In 1967, a year before the release of *2001: A Space Odyssey*, Cambridge University graduate student and radio astronomer Jocelyn Bell detected an extraterrestrial radio transmission that consisted of steadily timed pulses. She and her thesis adviser quickly found four sources of such signals. The signals' period and regularity were dissimilar to those with radio signals from then-known natural sources, suggesting that they had an intelligent origin. The signal and others like it were dubbed, somewhat tongue-in-cheek, LGMs (for Little Green Men). Continued research, however, revealed that the signals had come from spinning neutron stars, the remains of supernovae, which they named pulsars. Although Bell had discovered a new kind of star, she had not detected the signal of an ETI.[3]

Although official SETI research has had a few exciting false alarms, it has not yet had the fortune of detecting an intelligent extraterrestrial signal. But it's easy to imagine what would qualify. In the movie *Contact*, based on Carl Sagan's fictional novel about SETI, researchers receive a repeating signal of beats and pauses, signifying the sequence of prime numbers from 2 to 101. The viewers and the characters know they have found what they were searching for.

In both these cases, researchers trusted their intuitions. But we can't always do this when it comes to inferring design—especially in natural science, where we try to minimize the negative effects of subjective bias. While we shouldn't assume that our intuitive judgments are false, sometimes they are false. We occasionally get it wrong, royally and embarrassingly wrong. Everyone was right to conclude that the black monolith in *2001* wasn't just a rock. We're surely right to believe that Egypt's pyramids did not evolve from sand dunes. And you're right to think that the black scribbles on this page convey the ideas of human authors. But Kepler was wrong when he thought inhabitants had made the Moon's craters. Percival Lowell was wrong to think Martians had constructed canals. UFO enthusiasts are surely wrong to think that some alien race is responsible for the well-known Face-on-Mars, and children are probably wrong to think that clouds really are designed to look like Disney characters. Still, most of the time, in most situations, when our faculties are working properly, we infer design reliably. But we want to avoid "false positives," for although we humans are adept and inveterate pattern detectors, we're also pattern imposers. So how do we separate the chaff from the wheat?

CHANCE, NECESSITY, AND DESIGN

Any event or object submits to a few basic types of causal explanations: chance, necessity, design, or some combination of the three. On these three categories hinge the most important disputes in science and philosophy. Within the world, if an event is regular and repetitive, we are prone to attribute it to natural "necessity"—that is, natural law. Because the repetitive signal Jocelyn Bell received was orderly, she thought it might have come from an intelligent source. But she soon realized that it was the result of a natural, law-like activity. Given that a neutron star has certain physical characteristics, and that it is spinning at a certain rate, it will emit a repetitive, pulsing signal, like a lighthouse with a rotating lamp. Bell didn't need to postulate an intelligent agent to explain her discovery.

As we noted in the previous chapter, if an actual event is not necessary, then it is contingent. But a contingent event can be the result of chance or of intelligent design. If you dump out a box full of Scrabble letters, the law of gravity is responsible for the letters falling to the floor. But gravity doesn't determine that the letters be in any particular order. We would normally attribute the particular configuration of the letters to chance.[4] On the other hand, if you arrange them to spell out a sentence, their order will take a particular configuration because of design. But in both cases, the specific order of the letters on the floor will be a contingency. So how do we separate chance and design?

A common mistake is to assume that the argument for design is merely a matter of calculating probabilities or complexities. The more improbable or complex an event is, many suppose, the less likely that it is the product of chance and the more likely it is the result of intelligent design. For astronomically low probabilities, this might work. But the fact that an event is merely improbable or that a structure is complex gives little justification for inferring design. After all, improbable events—at least when viewed in isolation—happen all the time by chance. The world is a big place. It has vast "probabilistic resources" at its disposal. Lots of stuff happens. Think how improbable it is that you should be reading this book at just this time. What are the chances that the particular tree used to make the pages of this book should be your book and not someone else's? Quite slim. Flip a coin one thousand times, and you've just participated in an enormously improbable event. (If you doubt it, just try to repeat the same sequence of heads and tails.) Nevertheless, there's no reason to suppose that that event was intentionally planned or contrived.

This is why an argument that complex life is rare in the universe does not by itself allow us to conclude that it must be designed. Since the universe has vast opportunities and locations, such rarity might indicate simply that life is a rare, chance occurrence. For this reason, Don Brownlee and Peter Ward can argue in *Rare Earth* that complex life is rare in the universe, without concluding that the universe was designed.[5] The rarity of habitable conditions and complex life itself weighs against the idea that such life flows inevitably from the laws of physics and chemistry. But that rarity alone might suggest chance or design.

Complexity, in some cases a synonym for improbability, is equally ambiguous. If you rake up a pile of leaves, the pile will be enormously complex. If the wind blows it apart, it would be almost impossible for even the most industrious and clever team of engineers to reconstruct the pile as it was before. Similarly, if a chimp spends several hours typing on a laptop, he will produce an immensely complex jumble of letters. But in neither case will intelligent design have played a significant role.

At the same time, complexity, or low probability, usually has something to do with detecting design. It's often a necessary condition, even if it's not a sufficient one. One of the reasons the SETI researchers in *Contact* attribute the string of prime numbers to ETIs is that the sequence is improbable. We know of no natural process that generates such a sequence. And what are the odds of that sequence just happening by chance? But mere improbability is not really the issue, since any non-repetitive pattern of radio signals is improbable. So how did they tell the difference?

A SUITABLE PATTERN

What often seems to be required is the combination of low probability (or complexity) and some sort of suitable pattern, which mathematician and philosopher William Dembski calls a specification.[6] When we correctly infer design, it is often because of the presence of these two properties, improbability (or complexity) and specification. Dembski argues that this joint property—specified complexity—is a hallmark of intelligent agency.

Think of Mount Rushmore. Why do we recognize that the faces of Washington, Jefferson, Theodore Roosevelt, and Lincoln were carved into the side of the mountain by skillful sculptors but do not think the rubble of rocks below them was intentionally assembled? Both are compatible with the laws of physics, but neither is determined by them. Both the faces and the pile of rocks are complex. But only the faces tightly conform to a pattern that we recognize as meaningful—namely, the likenesses of four American presidents. In fact, even if you didn't know what these presidents looked like, you would still recognize that the object was a sculpture and not the product of wind and erosion. It is apparent that some intelligent agent (or agents) chose this particular rock shape from the myriad possibilities available.

But the pattern also has to be sufficiently tight. There is an unusual mountain in the New Mexico Rockies called Hermit's Peak. It juts unexpectedly from the ground, seemingly unrelated to the mountains around it. From a distance, and from just the right angle, it vaguely resembles the profile of Abraham Lincoln as if he were lying on his back, especially to climbers who have been drinking too much beer in the hot sun. But no

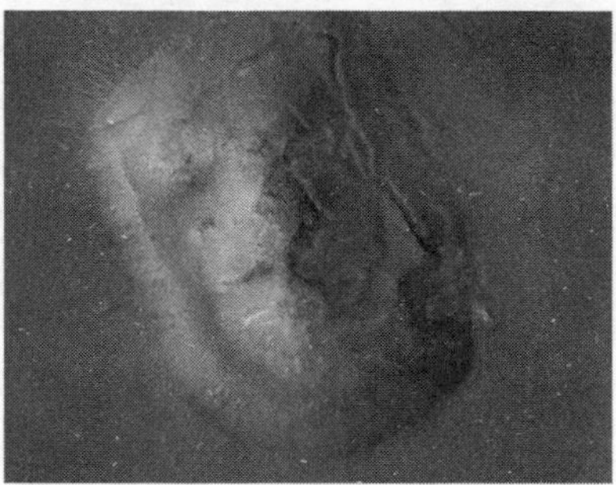

Figure 15.1: The importance of a tight specification. NASA's Viking 1 orbiter took an image of an apparent "Face-on-Mars" in 1976 (left); but the appearance of a face was the result of low resolution and lighting angle. The black dots, one creating a "nostril" on the face, are artifacts of the imaging process. In the much higher resolution image taken by the Mars Global Surveyor in 1998 (right), the structure loses any resemblance to a face.

one seriously entertains the possibility that Hermit's Peak was a secret pork-barrel project of the Parks Service or Department of Interior. (Admittedly, the mountain is suspiciously close to Roswell, but that's another story.) The match between it and Lincoln's face is too loose. The same is true of the famous Face-on-Mars. It looks a little bit like a face from a certain angle, with the right camera resolution and lighting; but in other conditions, it doesn't. It lacks the requisite specificity.

In addition, for a pattern to reliably indicate design, it will need to be relevantly *independent* of the event or structure in question. Otherwise, we could just read a pattern into anything, the way we do with clouds or Rorschach blots. When scientists imaginatively read a pattern into their data, it's called cherry-picking. With a large enough body of data, it's possible to pull out isolated bits that conform to a pattern. But when one looks at the data as a whole, the pattern dissolves. The recent film *A Beautiful Mind* illustrates this point nicely. It tells the story of Nobel laureate John Nash, in his descent into paranoid schizophrenia. As a result of his disease, he begins to see coded Soviet messages embedded in mass print media where none really exist.

Dembski gives another example that illustrates the important distinction between a real pattern and a fake one. If an archer draws a small target on a wall, stands back twenty yards, and puts five arrows into the bull's-eye, we will infer that he is skillful—that is, that the event exhibits masterful design. The bull's-eye is the pattern, and it is relevantly independent of the firing of the arrows. Moreover, the target isn't huge. If it were, the pattern match would not be tight enough, since it would be too easy for the archer to hit the bull's-eye. On the other hand, if the archer shoots one shot at a large, blank wall, and then paints the target around the arrow, the pattern will match. No matter how tight the match, however, it's just a fabrication, since the pattern isn't independent of the event. Therefore we could not determine whether the archer is skillful.

When we infer design, we invariably make use of important background knowledge that enables us to recognize a suitable pattern when we see one. The SETI researchers in *Contact* would not have recognized that the prime number sequence was an intelligent signal if they had not known about prime numbers. The mathematically untutored might never have realized that the transmission wasn't random.

Of course, more can be said about how we detect design. For example, many patterns have additional markers that increase our confidence that they are designed. Philosopher of science Del Ratzsch notes that, historically, arguments for the design of certain natural structures have "almost always involved value," to which we attach meaning, and which is habitually associated with mind and intentions.[7] Such value is difficult to define, but we usually know it when we see it. Art and music have a value that trash and mere noise lack. A functioning car has a value lacking in a pile of scrap metal. Similarly, a living organism, with its interlocking complexity and myriad functions, has an intrinsic value that inanimate objects lack, just as a fine-tuned habitable universe has an intrinsic value that an uninhabitable one would lack. So when we infer design, we often make a qualitative judgment that has a positive and a negative aspect. We judge that chance and blind natural law are unlikely or incapable of producing certain events or objects, and we discern certain features, such as value, that we tend to associate with intelligent agents.

A COSMIC DESIGN

We have to adapt this explanation a bit, since detecting design *within* the natural world can be different from detecting the design of the natural world itself. We normally determine that something is designed by contrasting it with what natural laws and chance can do on their own. Designed objects tend to have what Del Ratzsch calls counterflow. They contrast with the way nature will go if left to operate freely. If events or objects are designed, they will stand out in relief against the background of nature's normal structures and activities. This counterflow was at least partly the reason that the lunar colonists in *2001* identified the monolith as an artifact rather than, say, a geometrically gifted rock. Typical rocks tend to have a more, well, "natural" shape. Unlike the monolith, lunar rocks, and large rocks in general, don't have perfect, geometric angles. We sometimes see perfect geometric shapes in nature at the microscopic level in quartz and other pure minerals. And we see spherical structures like planets and stars at the macroscopic level. A large, black, perfectly geometric rectangle at the "human scale," however, stands out against the more irregular background lunar surface, and exhibits features that we tend to associate with intelligent agents.

But how do we make such a contrast when trying to determine the origin of nature and the natural laws themselves, since we don't have such a background contrast? In fact, as already noted, within the natural world, we tend to *distinguish* between events and objects that are the product of natural necessity and events and objects that are designed. So how can we tell that nature and its laws themselves are designed, are fine-tuned? What is the relevant contrast against which we discern a meaningful pattern?

Whether we realize it or not, in evaluating the apparent fine-tuning of the universe, we are distinguishing between logical and natural "necessity." When scientists talk about natural necessity, they're referring to what will happen given, say, the law of gravity, as long as no agent interferes. For instance, if you lift a ball off the floor, you haven't violated the law of gravity; you've just interfered with its normal course of operation. But there is nothing about the law of gravity that is necessary in the logical sense. No principle of logic requires gravity to be the way it is. The inverse square law does not have the same status as 2 + 3 = 5 or the claim, "All bachelors are unmarried." The law of gravity did not have to obtain, or to have the actual properties it has. Or to put it differently, there could have been objects that obey a different attractive force from gravity as it is in our universe. There are other possible worlds in which gravity (or some counterpart to gravity) has different characteristics.

When we say the laws of physics appear fine-tuned, what we're saying is that in contrast to the many other possible universes in its "universe neighborhood," ours has just the laws that make it habitable. We're contrasting the laws in the actual universe with other, similar (albeit hypothetical) universes with slightly different laws, as well as with the endless sea of chaotic

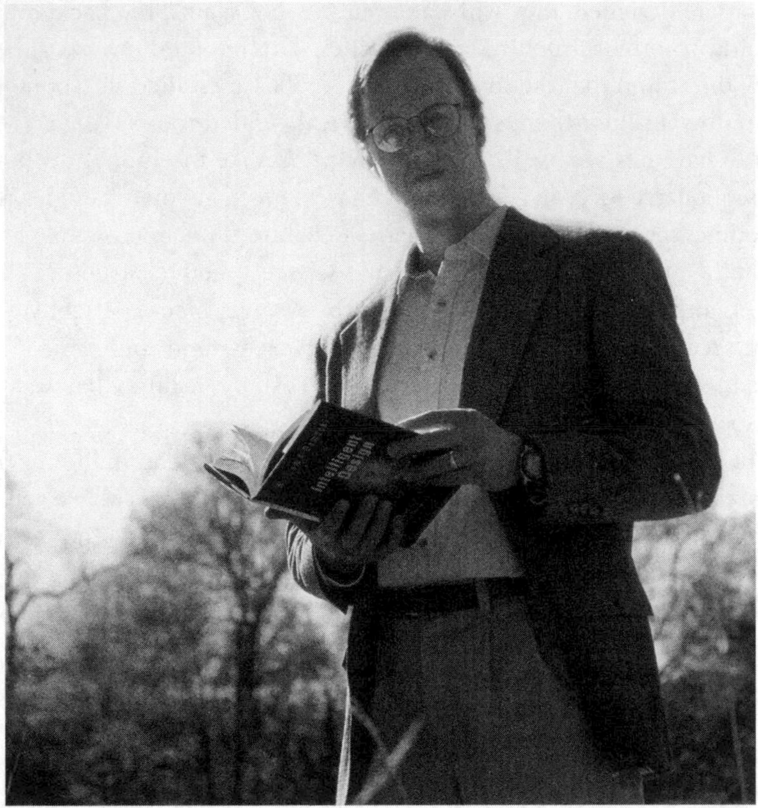

Figure 15.2: Philosopher and mathematician William Dembski has done seminal work in formalizing important aspects of how we detect the activity of intelligent agents.

and disorderly universes that might have existed. We're also recognizing an important distinction between habitable and uninhabitable universes. (We implied this in Chapter Thirteen, but did not make it explicit.) The actual, habitable universe stands out against the many similar but uninhabitable possible ones. The more we learn, the more we realize that if we were just to pick a universe's properties at random, we would almost never stumble across a habitable one. To use Dembski's terms, our universe and our place and time within it appear specified to make possible that most complex of empirical phenomena—technological civilization. Most hypothetical universes in our universe neighborhood do not.

Some objects and events submit to exact probability calculations. Many, including the universe as a whole, might not (unless we restrict the domain of possible universes under consideration). For now, let's assume that we can't determine the exact improbability of the fine-tuning of fundamental laws and constants, and that we can only compare our universe with the range of possible universes in its "neighborhood." That is, we can compare it with other possible universes in which the fundamental constants are just slightly different. This comparison leads to the strong impression that the range of inhospitable universes vastly exceeds the range of hospitable ones in this neighborhood.

A common objection to this argument is that no matter how "improbable" the existence of our actual universe, it's no more improbable than the other possible ones. We can now see why this is not a persuasive objection. It's not a simple improbability that leads us to believe there's something fishy that needs explaining. It's the presence of a telling pattern, a pattern we have some reason to associate with intelligent agency. There is a famous joke about a man who comes home early to find his wife in bed and his sleazy next-door neighbor naked in the bedroom closet. When the husband interrogates him, the neighbor nonchalantly replies, "Everybody's gotta be someplace." Now the reason no one would accept this wily use of probability theory is that there is a meaningful, and in this case highly suspicious, pattern of circumstances.

Related to the crucial role of meaningful patterns, as we mentioned above, is that arguments for design usually involve some tacit judgment of value. When considering universes, everyone recognizes, unless they're trying to avoid a conclusion they find distasteful, that a habitable universe containing intelligent observers has an intrinsic value that an uninhabitable one lacks. Living beings have some value, however difficult to describe, that inanimate objects don't have. That value redounds to a universe that allows for the existence of complex life. Theorists tacitly admit that they share this judgment when they try to explain away the fact that our universe appears fine-tuned for the existence of complex life.

THE CORRELATION AS A MEANINGFUL PATTERN

Of course, this argument is quite abstract: we have to compare the properties of the actual universe with other possible ones. This is not a fatal flaw, but perhaps it's a deficiency, since we have only one actual universe to observe. Our comparisons must be mediated by theoretical judgments about other possible universes. In contrast, the correlation between habitability and measurability is more empirical, since it allows comparisons within the observable universe. Taking the universe as a given, we can compare and contrast the conditions required for habitability and measurability, and can, at least in theory, make more or less specific calculations about their individual probabilities. We have a large universe with many diverse configurations of planets, stars, and galaxies. Highly habitable and measurable environments stand out in contrast with most other regions of the universe.

It's because these conditions are highly improbable, or at least quite rare, that we can see that the correlation forms an interesting pattern. If the universe were perfectly homogenous in this regard, and everywhere compatible with observers and observing, the most we could say is that the universe is generally open to scientific discovery and that it is, in some sense, "rationally transparent."[8] This might be suggestive, but we wouldn't have anything tangible to contrast with our immediate surroundings. As it is, we know that the laws of physics and initial conditions, themselves fine-tuned, are compatible with a wide range of local conditions, only a few of which are habitable. It's intrinsically interesting, and surprising, to find that those few habitable places are also the ones most conducive to diverse types of scientific discovery.

As we noted above, to infer design reliably, normally a pattern must be independent from the event or object in question. We must guard against reading a pattern off of an event, or artificially imposing a pattern on it. For instance, there is a matching pattern between the presence of habitable environments and life. We will only find life in habitable settings. Moreover, these settings are exceedingly rare. Nevertheless, the two elements are obviously dependent, since life will by necessity only exist in areas where it is possible. No one should be surprised to find living beings restricted to environments compatible with their existence, to recall the Weak Anthropic Principle. The situation might be the result of design, but this pattern alone provides inadequate evidence for that conclusion.

In contrast, there's no obvious reason to suppose that habitable environments would also be the ones most conducive to diverse types of scientific discovery. Being habitable and being measurable are distinct properties. We could compile separate lists of the properties that contribute to habitability and those that contribute to measurability. We could analyze one without ever making reference to the other. There is no logically necessary connection between the two.

Moreover, the high degree to which our local environment is congenial to scientific discovery can't be attributed to a selection effect in the same way that habitability alone sometimes can be.[9] While the conditions required for habitability provide for this high degree of measurability, this capacity to measure was itself rarely if ever necessary for our existence. For instance, having perfect solar eclipses or stable polar ice deposits or tree rings, or being able to view the stars or determine their temperature and composition, or having access to the cosmic background radiation was irrelevant to the needs of ancient man. That is, knowledge derived from such phenomena provided no survival advantage to our ancestors.

If we did not know otherwise, in fact, we might even expect that the habitability of an environment would detract from its measurability. For instance, intergalactic space, which is obviously low on the scale for habitability, is "better" for seeing distant galaxies than is the surface of a planet with an atmosphere. We might suspect that this is generally true. An Aristotle or a Ptolemy might reasonably have conjectured that a setting with all the conditions for life probably will hinder our knowledge of the universe. But when we combine the various phenomena that need measuring and observing, it turns out that the opposite is the case.

In a sense, we're *surprised* to discover that these conditions correlate in the actual universe. This discovery calls for an explanation beyond an appeal to blind chance or necessity. It's at least a striking coincidence. As Agatha Christie's detective heroine, Miss Marple, wisely observes, "A coincidence is always worth noticing. You can always discard it later if it *is* just a coincidence."[10] That is, if it's just due to chance. But the correlation between habitability and measurability seems to be the result of more than mere chance. On the contrary, it is a peculiar and telling pattern.

One reason we often prefer an explanation to its competitors is that it resolves our sense of surprise. It offers what philosopher John Leslie calls a "tidy explanation."[11] Design provides just such a tidy explanation here. Think of it this way: If the physical universe were designed so that any observers would find themselves in an environment conducive to many diverse scientific discoveries, then the correlation would be just what they would have expected. Although this evidence might not prove that the cosmos was designed, it would surely confirm it. To put it another way, if we assume that the universe is designed at least in part to allow intelligent observers to make discoveries, the correlation between life and discovery we observe is what we would expect. In contrast, if the cosmos exists by chance and if intelligent observers like human beings are simply a rare and purposeless dross in that indifferent cosmos, we would not expect this. It would be an inexplicable fluke. Whatever the exact probabilities, clearly the correlation is much more likely given design than given chance (which here means "no design"). A universe designed for discovery resolves our surprise. Shrugging our shoulders and chalking it up to chance, or some impersonal and unintended process, does not.

The (undesigned) multiverse scenario is also deficient, since it does not make the existence of any particular kind of universe more likely. In fact, it's utterly indiscriminate. No matter what kind of universe existed—a pure hydrogen universe, a black hole universe, an utterly chaotic universe—it would be equally consistent with the hypothesis of a world ensemble (of course, no one would be around to form the hypothesis, but that's another matter). Even most of its habitable universes would not contain habitable oases for observers that are also the best overall places for observing. Not so with design. We all recognize an intrinsic interest and value in a *discoverable* universe consistent with complex, living observers like ourselves, which would be lacking in these other universes. Like winning the lottery twice in a row, to find those observers just where they can best make diverse discoveries is doubly telling. This is the sort of universe that an intelligent agent would have some interest in designing.[12] It's a fishy pattern, and we know it.

In this case, the pattern we detect has apparently been transmitted through natural laws and initial conditions, although laws and conditions that allow for a significant degree of freedom at the local level. The design, so far as we can tell, is embedded or encoded in the laws and initial conditions themselves. In this case, the artifact of intelligence is the cosmos itself. Although this differs from our usual way of detecting design within the world, there's no reason in principle that we cannot detect design transmitted by laws.

Imagine if, in 2030, the United States and China become bitter enemies, and through a complicated turn of events, complicated as only international relations can be, the countries move to the brink of war. On Christmas Eve, American and Chinese diplomats meet in Jerusalem to begin their final, fragile round of negotiations. At midnight in Jerusalem, however, while diplomats are reaching for another pot of coffee, some Israeli astronomers notice a striking new pattern of craters on the Moon. Focusing their telescopes, they resolve a group of identical craters, which very precisely spells out "Glory to God in the highest, and on Earth, peace among those whom he favors!" in both English and Mandarin. The Chinese and American astronomers confirm that the craters were formed by simultaneous impact from a dense cluster of asteroids moving in normal trajectories through the Solar System. Moreover, so far as they can tell, the pattern was mediated through natural laws and the initial conditions of the Big Bang. The asteroids hadn't been perturbed from their normal courses. Nevertheless, everyone would still recognize a setup. Such highly specified crater configurations are not merely rare but unique, the English and Mandarin languages are clearly independent of the impacts, and the crater pattern matches them very closely. Moreover, the timing of the message's formation fits tightly with world events. To put it somewhat differently, the event is much more likely if we assume a setup.

Similarly, habitable environments are exceedingly rare. The fact that they are also the best overall places for scientific discovery forms a relevantly independent pattern. So we have good reason to suspect that things have been intentionally arranged, even if this came about through the interaction of natural laws and initial conditions.

The correlation not only suggests design but also design that bears a specific purpose. Detecting design is often easier than discerning purpose or meaning. For instance, museums often house artifacts that are clearly designed, although the purposes of the artifacts have been lost to antiquity. Similarly, the American colonists in *2001: A Space Odyssey* knew they were dealing with an alien artifact, even though they had no idea what it was or what it did.

Nevertheless, detecting a purpose usually enhances our confidence that something is designed. For example, a detective may be fairly certain that a husband murdered his wife. Still, discovering that the husband had recently taken out a large life insurance policy on his wife will increase the detective's suspicion.

With the correlation, the pattern we discern is not in an object but a particular situation, but this does not prevent us from seeing the pattern. The first, low-budget season of *Star Trek* contains an episode called "Arena." The story involves an advanced alien race, the Metrons, who capture Captain Kirk as well as the captain of another alien vessel, of the Gorn race. As punishment for invading their space, the Metrons transport Kirk and the Gorn captain to an uninhabited planet to fight to the death, telling them that they—the Metrons—will provide the weapons. But once transported, Kirk and the Gorn captain see no obvious weapons other than large rocks.

The lizard-like Gorn is extremely strong, but Kirk seems somewhat brighter. Before long, Kirk notices a strange abundance of certain minerals such as diamonds, sulfur, potassium nitrate (saltpeter), and coal. "This place is a mineralogist's dream!" he exclaims, dismayed that he can find nothing as simple and useful as a large club.

But eventually he realizes that all the ingredients for gunpowder are present, along with the hardest known substance—diamonds—for projectiles. Rather than supplying him with ready-made weapons, the Metrons have provided the ingredients, which require him to apply his own ingenuity. Once he recognizes a set-up, he begins searching for the rest of the materials he needs. He quickly finds a felicitously shaped hollow tree trunk, which allows him to construct a cannon. With one shot, Kirk wounds the Gorn, but he decides not to kill him. He thus saves himself while impressing the Metrons with his resourcefulness and capacity for mercy.

Although Kirk's intelligence was a prerequisite for success, his surroundings clearly have been set up for this purpose. In fact, the situation is a more fitting challenge than a simple cache of weapons would be, because Kirk must use his intellectual skills to forge a weapon himself. But no amount of genius would have allowed Kirk to make a cannon without the right ingredients. Now, the presence of so many disparate elements, all necessary for making a cannon, is not just a coincidence. We know this in part because, of the countless planets in the universe, few are so well stocked, even those that look strikingly like certain parts of Southern California. We see that the situation is purposefully arranged to allow an inhabitant with enough intelligence to create a cannon. In fact, we begin to realize that the Metrons are a highly intelligent race precisely because of the intriguing fit between the "arena" and Kirk's background knowledge. A less advanced culture would have put Kirk and the Gorn captain in a fighting rink with ready-made weapons, as in the title fight for the World Wrestling Federation. The Metrons have given Kirk and the Gorn captain a real challenge.

We can often discern that a pattern in a situation like this one has been designed, even if we're ignorant of other basic facts. But what about the situation we've described in this book? Is it such a situation? Consider one last illustration. Imagine a mountain climber who decides to climb a high, desolate mountain on the island of Hawaii. Unknown to the climber, it happens to be Mauna Kea. In fact, he knows so little about astronomy and geography that when he reaches the top, he is shocked to see several manmade structures. A closer inspection reveals, among other things, two large telescopes peering out of the openings in two massive white domes. Although he doesn't recognize the famous Keck telescopes, or know anything about how they are built, he recognizes that they are telescopes. But more important for our purposes, he realizes why they are perched atop this mountain and not, say, in downtown Honolulu. Astronomers put telescopes where viewing conditions are best. Telescopes don't exist on mountaintops because of some blind necessity of nature. They're put there for good reason. And appealing to chance in this case is really no explanation at all.

Analogously, our environment has many rare and disparate elements crucial for making scientific discoveries and observations. Those same elements make our environment hospitable to the existence of observers. The rare places with observers are the best overall places for observing. If an otherwise ignorant climber can recognize the purpose in putting telescopes on high mountains, we should be able to see purpose in this striking correlation. Like the telescopes atop Mauna Kea, this isn't just an independent pattern. It's a meaningful one.

INTIMATIONS OF THE PATTERN

But if the correlation is a meaningful pattern, why have so few noticed it? One reason, perhaps, is that most of those acquainted with the relevant evidence have been discouraged from considering design, or of speaking publicly about it. Surely another reason is that much of the evidence necessary for making our argument is of fairly recent vintage. Nevertheless, although no one has developed the argument in any detail, some have caught glimpses of it. Biologist Michael Denton and historian Hans Blumenberg independently noticed the amazing fit between a habitable atmosphere and a relatively transparent and therefore scientifically useful one.[13] In fact, during our work on this book we encountered the following passage from Denton:

> What is so striking is that our cosmos appears to be not just supremely fit for our being and for our biological adaptations, but also for our understanding. Our watery planetary home, with its oxygen-containing environment, the abundance of trees and hence wood and hence fire, is wonderfully fit to assist us in the task of opening nature's door. Moreover, being on the surface of a planet rather than in its interior or in the ocean gives us the privilege to

> gaze farther into the night to distant galaxies and gain knowledge of the overall structure of the cosmos. Were we positioned in the center of a galaxy, we would never look on the beauty of a spiral galaxy nor would we have any idea of the structure of our universe. We might never have seen a supernova or understood the mysterious connection between the stars and our own existence.[14]

He never develops the argument or argues for an astonishingly wide and deep-ranging correlation, but clearly he is sniffing down the same trail. Similarly, physicist John Barrow has discussed the value of a universe with three spatial dimensions for both life and the high fidelity of information transmission so vital to scientific discovery.[15]

Reminiscent of our first chapter, historian of science Stanley Jaki recently argued that the Earth-Moon system contributes not only to Earth's habitability but to scientific discovery as well.[16] Finally, Michael Mendillo and Richard Hart of Boston University anticipated the discovery described in Chapter One. In 1974, they presented a paper called "Total Solar Eclipses, Extraterrestrial Life, and the Existence of God," which was later reported and excerpted in *Physics Today*.[17] They argued, with comic precision, the following:

> **Theorem** An exactly total solar eclipse is a unique phenomenon in the Solar System.
>
> **Lemma** There are observers on Earth to witness the remarkable event of an exactly total solar eclipse.
>
> **Conclusion** A planet/moon system will have exactly total solar eclipses only if there is someone there to observe them. As only Earth meets this requirement, there is no extraterrestrial life in the Solar System.
>
> **Corollary** In a system composed of nine planets and 32 moons, for only Earth with its single moon to have exactly total solar eclipses is too remarkable an occurrence to be due entirely to chance.
>
> Therefore, there is a **God.**

Although their argument is tongue-in-cheek, it exploits our sense that there's something suspicious about finding the best eclipses just where there are observers to enjoy them. What they did not consider is that the same conditions necessary for producing perfect solar eclipses are also important for the existence of observers. And they never imagined that similar correlations would exist in other areas. These facts are highly suggestive, even if they don't pack quite as much theological punch as Hart and Mendillo's conclusion. If the correlation stopped with eclipses, of course, we might chalk it up to coincidence. But as we have argued, eclipses are just the beginning.

We live in a universe with laws and initial conditions finely tuned for the existence of complex life. Although narrowly constrained, they do not inevitably give rise to such life. They are necessary but not nearly sufficient for it. In extremely rare pockets of that universe, conditions are congenial to the existence of beings who can observe the starry heavens above and ponder the meaning of their existence. In at least one of these places, despite struggle and adversity, some came to believe that the world around them was a rational, orderly *uni*verse, accessible not only to rational thought but also to careful investigation. Centuries of study, amplified by technological tools and innovation, have given rise to an unparalleled knowledge of the world around us. The combination of those preliminary discoveries now gives rise to another: The same rare conditions that have sustained our existence also make possible a stunning array of discoveries about the universe.

There is a purposeful value in this. Because of it, and only because of it, can our aspirations for scientific knowledge and discovery be satisfied. Careful investigation, study, and observation of the natural world ultimately succeed. With enough persistence, the natural world discloses itself to us in ways that we do not, and sometimes cannot, anticipate. Once perceived, the thought creeps up quietly but insistently: *The universe, whatever else it is, is designed for discovery*. What better mandate could there be for the scientific pursuit of truth? Scientific discovery enjoys a sort of cosmic prestige, but a prestige apparent only to those open to the possibility that the cosmos exists for a purpose.

CHAPTER 16

THE SKEPTICAL REJOINDER

A fair result can be obtained only by fully stating and balancing the facts and arguments on both sides of each question.
—Charles Darwin[1]

YES, BUT WHAT ABOUT...?

We have offered evidence and suggested what at this moment in history is considered a controversial explanation for that evidence. But even if it weren't controversial, no empirical argument is ever immune to objections. We have drawn on a wide array of scientific disciplines, as well as a great deal of diverse scientific evidence. The argument is not mathematical or deductive. We cannot simply evaluate the premises for contradictions. For our argument to be persuasive, then, it should have the preponderance of evidence in its favor. We think it does. Nevertheless, we can anticipate a number of important objections, and our argument would be incomplete if we did not mention them. (No doubt there are objections other than the ones we consider here.) While we think none of these objections defeats our argument, they do help clarify it, as well as develop its implications.

Objection 1: It's impossible to falsify your argument.
Because our argument is empirical—that is, based on observations—we don't expect perfect agreement between habitability and measurability in every case. In fact, if one considers conditions for measurability in isolation,

it's easy to come up with apparent counterexamples. It's important to emphasize that we have not argued that every condition is *individually* optimized for measurability from Earth's surface. Our claim is that *our habitable environment is an exceptional compromise of diverse conditions for measurability ranging from cosmology and galactic astronomy to geophysics, and that those same conditions are also important for habitability*. This form of the argument accommodates facts that would constitute counterexamples for a less nuanced argument. But precisely because the argument is empirical and carefully nuanced, a single counterexample probably wouldn't refute it. Thus emerges another problem: How could it be falsified?

Recent work in the philosophy of science has revealed the degree to which high level theories tend to resist simple refutation; nevertheless, it's certainly a virtue of scientific proposals to be able to say what evidence would count against it. Indeed, there are a number of possible discoveries that could count against our thesis. Empirical evidence could contradict either element of our argument for a correlation—that is, habitability or measurability. Our presuppositions about habitability are equally open to empirical falsification. The most decisive way to falsify our argument as a whole would be to find a distant and very different environment that, while quite hostile to life, nevertheless offers a superior platform for making as many diverse scientific discoveries as does our local environment. The opposite of this would have the same effect—finding an extremely habitable and inhabited place that was a lousy platform for observation.

Less devastating but still relevant would be discoveries that contradict individual parts of our argument. Most such discoveries would also show that the conditions for habitability of complex life are much wider and more diverse than we claim. For instance, discovering intelligent life inside a gas giant with an opaque atmosphere, near an X-ray-emitting star in the galactic center, or on a planet without a dark night would do it serious damage. Or take a less extreme example. We suggested in Chapter One that conditions that produce perfect solar eclipses also contribute to the habitability of a planetary environment. Thus, if intelligent extraterrestrial beings exist, they probably enjoy good-to-perfect solar eclipses. If we were to find complex, intelligent, indigenous life on a planet without a largish natural satellite, however, this plank in our argument would collapse.

Our argument presupposes that all complex life, at least in this universe, will almost certainly be based on carbon. Find a non-carbon-based life form, and one of our presuppositions collapses. It's clear that a number of discoveries would either directly or indirectly contradict our argument.

Similarly, there are future discoveries that would count in favor of it. Virtually any discovery in astrobiology is likely to bear on our argument one way or the other. If we find still more strict conditions that are important for habitability, this will strengthen our case. In fact, writing Chapter Five was difficult because every extrasolar planet discovery has seemed to bear on our argument. During the roughly thiry-six months we spent in writing, several new requirements for habitability came to light (for example, the planetary spin-orbit core-mantle coupling discussed in Chapter Five). Our argument may be wrong, but it's certainly not unfalsifiable.

Objection 2: It's inevitable. Whatever environment we found ourselves in, we would find examples conducive to its measurability.

There's something to be said for this objection. No doubt it's true that if we were in another environment, with different conditions, then whatever discoveries we made would be made on the basis of *those* conditions. Scientific geniuses would come up with clever ways of testing their hypotheses, no matter what their environment. And we will inevitably base any argument we make on the conditions we know from our actual setting, not the conditions we don't know. So initially it might appear that we are simply under the illusion of a selection effect.

For instance, imagine a loving husband, George, and his wife, Laura, who take the same, highly eccentric walk every morning. For about a mile they walk along the main road, but they always take a shortcut down a side street and through a park. They have no reason to suppose that anyone else ever takes the same route. One Valentine's Day, George is away on business, so Laura decides to invite her somewhat cynical neighbor to join her on the usual walk. On this morning, however, they discover that every hundred feet, a plain, red-paper valentine has been placed on trees, telephone polls, and fence posts. Because the course is so unusual, Laura eventually realizes that her understated husband, knowing the unusual path of their walk, has put up the valentines as a touching romantic gesture.

Her neighbor, however, who has taken one too many philosophy courses, is skeptical. When they arrive back home, she tells Laura that they were probably the victims of a selection effect. After all, they didn't inspect the rest of the town. For all they know, the city might have put up valentines everywhere. They just happened to notice the ones on their walk and would have found the same pattern no matter where they had walked. Disheartened, Laura realizes that her neighbor could be right. Perhaps the entire city is filled with paper valentines, and George has forgotten yet another holiday. Then it dawns on her that she can easily verify the truth by checking the rest of the neighborhood. When she does, she is comforted to find that she was right all along. Not only does the pattern of valentines overlap her route, but the rest of the neighborhood is valentine-free as well.

Notice that as long as Laura couldn't compare the route with the area around it, she couldn't determine whether her husband was responsible for the valentines. She saw that the valentines formed a suitable pattern indicating a very specific design because of the contrast between her route and its immediate surroundings.

We are in a similar situation, since we are able to compare the measurability of our environment with that of other environments. For the discoveries we have made, we can reflect on the conditions necessary for such discoveries, and then compare those conditions with conditions in other settings. For instance, it's unquestionable that a relatively transparent atmosphere is more conducive to astronomical curiosity and discovery than is a murky (translucent) or opaque one. We know that, at least in our Solar System, such an atmosphere is rare, and that it correlates with a habitable atmosphere. We can compare the properties of our atmosphere with the properties of other planetary atmospheres we know about. We can also get some sense of the probability that a planet will bear the properties necessary for a transparent, habitable atmosphere. We can compare the features of our planet that allow us to have a dark night with other planets with different configurations. We can compare the properties of our Sun with other known stars. We can compare the size of our orbit with the orbits of other bodies we know. We can also compare our orbit with other possible orbits of planets and moons. We can compare our local galactic environment with the environments in, say, the galactic center or a globular cluster. We can compare the arrangement and regularity of our Solar System with those of other systems we discover. We can compare the sedimentation and related information-storage processes on Earth with those of other planetary bodies. We can compare solar eclipses as seen from Earth's surface to views from other planets. Theoretically, we can compare our local circumstances in cosmic time with other cosmic times, both past and future. We can also imagine what sorts of experiments we would be able to perform in different environments, and compare their potential effectiveness in discovering the phenomena we've been able to discover from Earth.[2] So we aren't simply captive to our local conditions; we can compare those conditions with other, apparently more common ones.

Notice, moreover, that we can compare our local setting with other similar settings only because of the high capacity for discovery that our setting provides. If we had lived on a world with a murky atmosphere, we would not know of other planets or stars or galaxies, and therefore we would not be able to compare our home with anything beyond its atmosphere. An observer on such a world would have no way of demonstrating that the correlation exists. As we learn to make more and more such comparisons, and comparisons that would have been impossible to make from other environments, this selection-effect objection grows less and less tenable.

Objection 3: Well, then, it's just a selection effect of a different sort. There are phenomena we cannot observe or measure. The argument is biased toward measurable phenomena.

This objection is similar to the previous one and also contains an element of truth. If we are in an environment in which we can't discover something, then we may never know that we can't discover it. Again, however, the argument is overstated.

Contrary to the claims of the anti-realist, who doubts the existence of external truth, scientists aren't locked in a Kantian box where everything we perceive in the universe is primarily the product of our perception. There are many things we have difficulty measuring, and we realize that fact. For instance, we can't determine the distance and properties of some astronomical objects. But we know they exist, since we can detect them either directly or indirectly, and we know that we don't know their distances or many of their intrinsic properties. We can compare the objects in this category with the objects we can both detect and measure, and make generalizations about our ability to measure generally.

Similarly, we are not so bereft of imagination that we can conceive only of those things we directly perceive. If nature is regular in its operation, which we have every reason to believe, then we have some justification for extrapolating what we don't see from what we do see. Theory often predicts the existence of certain objects prior to their discovery, such as additional planets, white dwarfs, black holes, the cosmic background radiation, and neutrinos. For fairly secure theories, we can imagine what conditions would allow us to detect such objects. We can then determine whether our environment allows us to do so and compare it with other settings in the universe. And this has happened numerous times in the past. It is striking how often physicists are able to devise a way to detect entities that are initially predicted for theoretical reasons.

Objection 4: You're cherry-picking. You have used a biased sample to argue for the correlation.

This is always a danger with any general hypothesis like the one we're proposing. When a theorist is looking over a large body of data, it's always possible that he will pick out the pieces that form an intriguing pattern and ignore the pieces that don't. As a result, when the data are considered in their entirety, the pattern dissolves. Any argument involving many different scientific disciplines is especially susceptible to such a danger, since it's impossible to consider every piece of relevant data.

For this reason, we have intentionally chosen important examples from each of the scientific disciplines we've considered. We haven't chosen obscure experiments or conditions of measurability that have little importance for science. For instance, it's difficult to overestimate the importance of a transparent atmosphere and visible stars for astronomy, or sedimentary processes for geology. Any astrophysicist would admit the historical importance of perfect total solar eclipses in the development of stellar physics. No cosmologist would deny the importance of detecting the redshift of distant galaxies, or the cosmic background radiation for our knowledge of the history of the universe. Moreover, as we noted in the previous chapter, other scientists have noticed evidence of the correlation, although none have developed the argument as we have. This makes it less likely that we're creating the correlation out of thin air.

This is an important objection nevertheless, because it would be one way to falsify the claim that there is a correlation between habitability and measurability. If our hypothesis is correct, the correlation will continue to be confirmed not only in the areas we've considered but also in areas we haven't considered. We are convinced that there are still many important discoveries awaiting us—some we can anticipate, some we cannot. At the risk of being wrong, we would be willing to predict that an identifiable subset of gamma ray bursts will one day be found to be useful standard candles.[3] The only reason we have for predicting this is that if the correlation is real, gamma ray bursts would be prime candidates for helping us measure the universe. Perhaps they will allow tomorrow's astronomers to probe even greater redshifts than we can with Type Ia supernovae today.

Another such prediction concerns the evidence for early life. As we mentioned in Chapter Three, Earth's geophysical processes have erased much of the early history of life. If measurability and discoverability are optimized from our vantage point, however, then we might expect that such information will be preserved somewhere accessible to us. The origin of life is a particularly important question. It would be surprising, assuming the correlation, if it could not be investigated. In fact, we might predict that such evidence is available somewhere, if we search diligently enough. It was precisely this prediction that led one of us (Guillermo) to consider the value of lunar exploration for uncovering relatively well-preserved relics of Earthly life from this early period.[4] Finally, we're willing to predict that since carbon and oxygen appear so often among our examples of measurability, they will be central characters in future discoveries as well.

Of course, if we're right about these predictions, this would not prove our position but only further support it. If we're wrong, conversely, it would not destroy our argument but would put a dent in it. But clearly our argument has a predictive dimension. In contrast, the Copernican and Anthropic Principles in their most unrestrained manifestations seem much less useful. Positing the existence of multiple universes, for instance, doesn't offer many fecund research programs within our universe. It looks designed primarily to foreclose certain unwelcome metaphysical possibilities.

Objection 5: Your argument is too speculative. It is based on guesses and a thin empirical base.

Most of the examples we have selected are based on well-understood phenomena, and they are founded on abundant empirical evidence. Examples include the properties of our atmosphere, solar eclipses, sedimentation processes, tectonic processes, the characteristics of the planets in the Solar System, stellar spectra, stellar structure, and our place in the Milky Way galaxy. Some of our other examples have a weaker empirical base, because of the rapid change and recent acquisition of knowledge in certain fields. This new knowledge includes extrasolar planets, additional requirements for habitability, and a host of insights in the field of cosmology. But even in these examples our arguments have a reasonable theoretical basis.

Where our discussions are speculative, we have identified them as such. Thus, our discussion of the Circumstellar Habitable Zone, and all the factors that go into defining it, contain speculative elements, as does our discussion of the Galactic Habitable Zone. While we can't yet estimate the precise boundaries of these habitable zones, present published studies are almost certainly still missing many relevant factors, which, when eventually included, will reduce their sizes, and strengthen our argument. Notice, again, that we're going out on a limb here and making predictions, which makes our argument vulnerable to future discoveries.

Objection 6: Your argument is too subjective. It lacks the quantitative precision necessary to make a convincing case.

This is related to the previous objection. It's true that we have not tried to quantify the correlation between habitability and measurability. We are the first to make this argument; our purpose is not to present a mathematically rigorous treatment. In any case, it's not necessary to quantify habitability and measurability to make the argument.

For example, it's easy to see that Earth's surface is far more habitable than those of other planets in the Solar System, as all astrobiologists now admit, without using a detailed "habitability index." Similarly, we have argued, without benefit of a "measurability index," that Earth's surface offers more opportunity for measurement and scientific discovery than those of the other planets.

But subjectivity also suggests arbitrariness. For instance, a skeptic might claim that we've arbitrarily weighted the values of just the sorts of observations that are feasible from our location, while downplaying the opportunity costs of that location. But any such charge will have to deal with our specific examples. We argue, for instance, that our lack of direct access to the giant black hole in our galaxy's nucleus is a small price to pay in exchange for our ability to discover and measure the cosmic background radiation. Is that arbitrary? Clearly not. We give greater weight to the background radiation because, unlike the nuclear black hole, (1) it is unique, (2) it gives us fairly direct information on the overall properties of the universe, and (3) it gives us a glimpse of the origin of the universe.

Of course, it doesn't follow that a precisely quantified analysis of our argument is impossible. In fact, it would make a very interesting research project.

Objection 7: How can you have a correlation with a sample size of one?

While it's true that Earth is the only example we have of a habitable planet, this does not prevent us from finding a correlation between habitability and measurability. First, our argument is not based merely on the particulars of our home planet and the life we know about. We have argued that life in the universe will almost surely resemble life on Earth, at least at the biochemical level, and a planet very much like ours is probably required for technological life. Starting with these basics, we have used knowledge from a broad range of disciplines to consider a broad range of environments. Discovering a correlation between habitability and measurability, then, is based on our knowledge, not our ignorance.

For example, with knowledge of stellar astrophysics and climatology, we can ask whether a planet around an M dwarf is more or less habitable and offers more or less opportunity for discovery than Earth. Similarly, with our knowledge of galactic astronomy, we can ask how position in the Milky Way affects habitability and the measurability of the local and distant universe.

Objection 8: Since life needs complexity, the correlation is trivial. The greater the complexity, the greater the chance for a correlation between habitability and measurability.

One of our claims is that the conditions for habitability are extremely improbable and, in that sense, complex. As a result, it might appear that the more complex an environment is, the more habitable and measurable it will be. That is, complex life requires complex conditions, including various stellar, planetary, and geological variables. In such a situation, it seems likely that there would be more conditions for measurability available. As a result, measurability will attach to habitability in the same way that inhabitants attach to habitability. Perhaps all our argument amounts to is the observation that our environment is complex, and complexity itself entails that the environment is both habitable and measurable.

There is certainly a grain of truth in this objection, since both complex life and wide-ranging scientific discoveries require a *diverse* environment. But on closer inspection, mere complexity does not clearly correlate with either habitability or measurability. For instance, an opaque, chaotic atmosphere like Jupiter's is not obviously less complex than Earth's. But it is much less habitable. Similarly, a stable planetary system with nine major planets in fairly circular orbits is not clearly more complex than an uninhabitable one with a swarm of diverse and irregular planets in various eccentric orbits. A gas giant planet with several moons of various sizes and orbital parameters is surely more complex than Earth with its single, large Moon. Having a perfect solar eclipse does not require a high degree of complexity compared with other possible systems of moons. A relatively circular galactic orbit, like our own, is less complex than a wildly elliptical orbit that traverses several different galactic regions and is frequently perturbed by the gravitational forces of nearby massive objects. A binary and triple star system is more complex than a singleton like the Sun. The chaos of two, colliding, irregular galaxies is more complex than the orderly spiral of the Milky Way galaxy. And in many other cases we can think of, the less complex condition is not only more habitable but also more measurable.

The difference between the conditions Earth enjoys and those of other, more common ones is not simply that ours is, in many respects, more complex, or even that it is more habitable. Our situation is complex, certainly, but it is also *exhibits a specification, a telling pattern*, in which the rare conditions for habitability and measurability correlate. (Recall that *that* is the pattern on which our argument is based.) This observation, however, won't be of much help to the critic, since that is *our* argument.

Objection 9: There may be separate pathways significantly different from ours leading to equally habitable environments.

This argument would have carried more force thirty years ago, but most discoveries since then work against it. Detailed knowledge of other planets in our Solar System, theoretical understanding of habitable zones, and the discovery of many examples of fine-tuning in physics and astrophysics all make it less likely that an environment very different from ours would be just as habitable. The constellation of evidence we have discussed even suggests that our environment may be near optimal. Recent discoveries about extremophiles notwithstanding, it is beginning to look as though even the most Earth-like locales in the Solar System are too hostile for simple life. We cannot consider a hardy bug in isolation from its Earthly home but must consider it as part of a system (recall the Gaia hypothesis). Even if planets with simple life turn out not to be rare, the restrictions on complex and technological life are much, much higher. We can't rule out the possibility of significantly different habitats elsewhere, but someone will have to offer a realistic example to move this out of the realm of baseless assertion.

Even less likely is the possibility that life might be based on something other than carbon and water. Scientific discoveries have continued to reinforce the unique life-essential properties of water, first employed as a design argument by William Whewell over 150 years ago. This objection is good to keep in mind, though, since finding a form of life significantly different from the water- and carbon-based life we know would compromise our argument.

Objection 10: Your argument is bad for science because it encourages skepticism about cosmology.

Cosmologists often claim that we must assume our location is unexceptional for theoretical reasons, since it allows us to extrapolate from the part of the universe we can see to the parts we can't see. If this were correct, then our arguments against the Copernican Principle might make us doubt our ability to know anything about the cosmos as a whole. Consider Martin Rees's question, mentioned in Chapter Ten, "Why does our universe have the overall uniformity that makes cosmology tractable, while nonetheless allowing the formation of galaxies, clusters and superclusters?" When Rees says cosmology is tractable, he means that it satisfies the following set of assumptions:

1. Space is homogeneous.

2. Space is isotropic (it looks the same from every point, or alternatively, there is no privileged point or direction).

3. The laws of physics are the same everywhere.[5]

Notice how easily his definition conflates the older *cosmological principle* with the Copernican Principle.

As we've seen, the first has required ongoing "qualifications" as our knowledge of the universe has increased. Clearly space isn't homogenous at the size of people, planets, stars, galaxies, or clusters of galaxies. Now it's assumed that space is homogenous above the level of galactic superclusters—that is, clusters of clusters of galaxies. In fact, recent evidence suggests that such superclusters are arranged in a pattern of voids and filaments, somewhat like large bubbles, throughout the cosmos.[6]

The second assumption, at least the way it is usually framed, conflicts with the evidence discussed in this book. Clearly this assumption needs to be reformulated. How the universe looks depends profoundly on one's location in both space and time. In fact, the most overreaching manifestation of the cosmological principle was called the perfect cosmological principle, which stated that every time in the universe is just like any other. The perfect cosmological principle inspired the various Steady State models, and met the same demise.

The third and final assumption—that the laws of physics apply everywhere in the universe—is perhaps the most important. It is the true legacy of the Copernican Revolution. And it does have solid observational support. For example, spectroscopic observations of distant galaxies over a broad range of wavelengths verify that their atoms have the same properties as those measured in Earth laboratories. On the basis of such observations, astronomers and cosmologists feel justified in extrapolating the laws of physics from their tiny laboratories to the entire universe.

Does our argument imply that we can no longer make generalizations about the universe from the parts we can observe? Does it produce a radical skepticism about our ability to learn about the cosmos as a whole? Does it require that we restrict ourselves to observational astronomy and eschew cosmology? Of course not. In fact, a superficially similar, but contrary assumption is needed—namely, that what we see is a good, accurate sample and representation of the universe as a whole. Perhaps we can dub this the Discovery Principle. According to this principle, since so many aspects of our situation have proved to benefit measurability, we can reasonably expect to find other such instances that allow us to gain access to information that reveals the universal laws of physics, as well as the variety of small and large structures in the universe.

In fact, perhaps the most plausible way to put assumptions (1) and (2) above is in the form of a Discovery Principle: *We should be optimistic that the parts of the universe we can't see are relevantly similar to the parts we can see (from our privileged vantage)*. By "relevantly similar" we mean that the sections we can't see don't contain objects or laws fundamentally different from those we can see.

The Discovery Principle even suggests a properly qualified cosmological principle. The cosmological principle has always been attractive to cosmologists because it allows for "simple" solutions to Einstein's equations (though persons not well versed in tensor theory may not find the solutions so simple). This was why mathematical cosmology was such a successful endeavor. The universe we see fits the simplest global expression of the equations of General Relativity. Could the universe have been otherwise? Would a less homogeneous universe be just as habitable? We don't yet know for sure.[7] In any case, the cosmological principle (rightly defined) was not born of the Copernican Principle, and can do just fine without it. And quite apart from demonstrating our mediocrity, it seems to be a prerequisite of a universe designed for discovery for creatures like ourselves, who are restricted to a single, tiny location.

Something like the Discovery Principle is clearly important if we are to seek further evidence about the cosmos as a whole. Discoveries in astrophysics and cosmology don't come cheap, and future investments in such research would seem prodigal without the reasonable optimism that our hypothesis supports. For those scientists weaned on the Copernican Principle, it may seem ironic that to justify conclusions about the universe at large we must have an especially good, even privileged view, not a mediocre one. For instance, probably more planetary matter in the cosmos exists in the form of gas giants than it does in planets like ours. The more mediocre habitat within the depths of a gas giant atmosphere would leave us utterly clueless about the larger universe, so opaque is its atmosphere. We might easily assume that such a heavy atmosphere simply stretched on to eternity.

Similarly, moons around such planets are probably more common than moons around terrestrial planets like our own. And as we argued in Chapter Five, these are quite unlikely to provide the high-quality measuring platform that Earthlings enjoy. What is decidedly not needed is mediocrity. What is needed is the privileged and probably quite rare position of an Earth-like planet. And we don't have merely to assume that we enjoy such a position. It's just what the evidence suggests, at least within the region we know. In contrast, a generalized Copernican Principle would subvert our justification for doing cosmology.[8]

Moreover, notice that the Discovery Principle undercuts the justification for postulating unobservable regions or domains with laws and properties fundamentally different from the observable universe. In other words, cosmology—the study of the cosmos as a whole—is a valid enterprise only if we are correct in assuming that the observable universe to which we have privileged access is a representative sample. Cosmologists, then, have strong motivation for eschewing those many worlds hypotheses that posit radically different "universes" from the one we inhabit.

Of course, it still doesn't follow that we'll learn absolutely every truth about the cosmos. We have reached a moment in the history of scientific exploration where one could reasonably ask whether we have nearly reached the limit of knowledge about the universe. Perhaps we are forever forbidden from peering behind the "curtain" of the cosmic background radiation. After all, how could we see beyond the epoch when matter decoupled from radiation?

The limits imposed on our vision by the background radiation are somewhat like the view an observer would have on the surface of a planet with a thick translucent atmosphere, where one hemisphere continuously faced its host star. The cloud deck on the planet's sunny side would appear uniformly bright, but for all that light, our cloud-befuddled observer could not learn about the distant universe, or even if there was anything beyond the bright clouds. Happily, though, assuming our alien observer possessed an advanced degree in physics, he could obtain a spectrum of the light filtering through the clouds and perhaps infer something about its source. Similarly, the angular power spectrum of the background radiation tells us something about events immediately prior to its own formation, even as it seemingly imposes a limit on what we can know.

And yet, again, the structure and laws of our universe provide a few more tools that future scientists may be able to exploit to peer even farther back. Perhaps one day they will map the neutrino or gravity wave background and gain some knowledge about what happened a little bit beyond a redshift of one thousand, less than 300,000 years after the Big Bang, a blink of the eye in cosmic time. So while we may not learn every truth about our universe, we have already learned more than we could ever expect, and we will surely learn a great deal more.

Objection 11: General Relativity appears to be a superfluous law of nature, which is not obviously required for habitability. Yet it is an important part of science. Does this not contradict the correlation?

George F. R. Ellis has argued that the major features of the cosmos might be just as well determined by Newtonian theory alone.[9] In other words, it may be that life does not require a universe where the laws of nature include General Relativity (GR). Of course, we don't know if this is correct. True, GR seems far removed from our everyday affairs, but it does affect element production, coalescence of neutron stars and black holes, and the immediate environment around giant black holes in the nuclei of galaxies. More work is needed to examine the connections between these processes and life. Since life is sensitive to the cooling rate of our universe, perhaps a Newtonian universe would fail to meet this requirement.

Assuming for the moment Ellis's suggestion is correct, what does it imply? It implies that at least one known law of nature is not obviously linked to our existence in an obvious way. But for all we know, GR is necessary in a universe with the same number of space and time dimensions, laws, and/or forces as ours. If and when we discover the more fundamental law or laws that would unite the various forces and reconcile relativity and quantum theory, relativity theory might strike us an inevitable and essential outgrowth of the whole, like the light-gathering leaves of a tree. Furthermore, relativity theory may prove critical in helping us move on to that next and deeper level of discovery, much as Newton's theories of gravity before it did.

Even ignoring these possibilities, however, the implications of relativity's possible superfluity for our hypothesis are ambiguous, since we have not argued that everything in the universe is relevant to habitability or to the correlation between habitability and measurability. There might be facts about the universe that are useful for measurability that aren't directly related to habitability. This doesn't refute the claim that the conditions for both overlap to a suspicious, seemingly specified degree.

Objection 12: The correlation isn't mystical or supernatural, since it's the result of natural processes.

This, of course, is true. Our argument differs from some traditional and contemporary design arguments in that it does not require or presuppose the direct design of specific entities within the physical world. We do not repudiate all such arguments, but we have not argued that the correlation between habitability and measurability is the result of what philosophers call direct "agent causation" within the natural world. We have focused only on the design of the cosmos considered as a whole. We claim that the correlation forms a meaningful pattern, which, while perhaps embedded in nature's laws and initial conditions, still points to purpose and intelligent design in the cosmos. The issue here is not whether there are natural laws, parameters, and initial conditions, but whether evidence of design and purpose could be built into them. We believe it could be and has been, in part through the striking correlation between measurability and habitability.

In the introduction to his *Principia*, Isaac Newton argued that our Solar System is so intricate and life-fitting that it had to be directly fine-tuned. He concluded, "The most beautiful system of the sun, planets, and comets, could only proceed from the counsel and dominion of an intelligent and powerful Being."[10] For Newton the Solar System itself stood out like an artifact, an object that was directly designed, akin to William Paley's watch resting on a heath. Now, this is not our argument, of course. But while Newton's idea is repugnant to many, it's clearly a possibility. It's also something we could discover to be true or at least likely. For instance, imagine that in the future, we gain detailed knowledge about extrasolar planetary systems and their dynamics. As a result, we are able to determine that the probability, even in our fine-tuned universe, of getting just one Earth-like planet, with its Solar System and all its requisite conditions, is less than one chance in 10^{180}.[11] This would mean that, even in a universe with 10^{11} stars per galaxy and 10^{11} galaxies, totaling 10^{22} available attempts, the chances of getting one such system would still be one chance in 10^{158}.

One chance in 10^{158} may not sound like too terribly daunting a quantity to a reader unfamiliar with dealing with enormous numbers represented so compactly. But to even the most sophisticated and imaginative mathematician, it's an unimaginably small probability. Taking habitability as a "specification"—which, in such a scenario, it surely would be if scientists had in fact determined that the possibility of the universe producing a single habitable planet was 10^{158}—we would have a virtually irresistible argument that our Solar System is a giant artifact of intelligence, like a watch or a Boeing 747. Although right now we can only guess what the probability is, there's no reason to assume that scientists could never develop a sufficiently nuanced understanding of the requirements for life to reasonably estimate it. And we shouldn't simply assume that the odds couldn't be that low.

The problem with this argument—that even one habitable planet is statistically impossible on chance—is not that it violates certain philosophical scruples, though it surely does. The problem is that we don't have enough information to evaluate it at the moment. For argument's sake, we have assumed that, given the (fine-tuned) laws of physics, the initial conditions of the Big Bang, and the number of stars and planets available in the universe, there is some non-trivial probability that at least one habitable planet like Earth, or at least its non-biological prerequisites, will form. This is, however, simply a working assumption, though we think it's a reasonable one. We're now beginning to realize that a single habitable planet requires an array of delicately nested systems ranging from the local to the cosmological level. Perhaps this array is improbable enough that it renders the "chance" formation of even one such system unimaginably improbable, or perhaps not.

Moreover, anyone familiar with the dynamics of planetary systems knows how difficult it is to get a system with nine planets like our own to remain stable for several billion years (our Solar System is, in fact, "chaotically stable"). As a result, we suspect true Earth-like habitable planets will be much, much rarer than most SETI advocates assume. But we don't yet know exactly how rare such systems are. For one thing, we don't yet know what percentage of stars have Earth-like planets, surrounded by a life-nurturing planetary system. Fortunately, we don't need such detailed statistical information, or a local fine-tuning argument like Newton's, to recognize design in this case. Our argument doesn't rest on the notion that Earth is unique or that its environment has been directly fine-tuned. Our argument rests on a particular *pattern*—the correlation between habitability and measurability—of which our environment is an exemplar.

Objection 13: You haven't really challenged naturalism. You've just challenged the idea that nature doesn't exhibit purpose or design.

It's possible to be both a naturalist and to admit design in nature. In fact, in the ancient world, both the Aristotelians and the Stoics did just that. Perhaps, for instance, design is somehow an inextricable part of an eternal cosmos, like matter and energy. We can't conclusively rule this out. The problem in our modern setting is that this strategy would require an essentially pantheistic view of nature that most naturalists deny. A cosmos that includes design and purpose—as well as chance, matter, and natural law—is quite different from "nature" as most modern naturalists understand it.

Moreover, current Big Bang cosmology discourages the view that the cosmos is eternal, which is necessary if design is coextensive with matter, time, and natural law (note also that law is not itself a material entity, nor is it a causal agent). A causal agent that somehow transcends the cosmos is a much more natural explanation for the Big Bang and the resulting physical universe we know than are purely immanent patterns of design. But either is a better explanation than the currently popular view that the physical universe is all there is, was, or ever shall be, and that chance and impersonal necessity exclusively explain its existence.[12]

Objection 14: You haven't shown that ETs don't exist.

This is true, but we did not intend to. In fact, ironically, design might even improve the possibility of ETs. SETI folks should take note. If you're looking to stay optimistic about finding extra-terrestrial life, the naturalism bandwagon offers little hope of getting you to the Promised Land. Whereas the existence of ETs seems increasingly improbable if our only resources are blind chance and necessity, the odds of their existence might improve if we allow the possibility of design. For instance, what if the universe were designed to proliferate with habitable planets and life? Well, it would certainly increase the chances of finding ETs.

While we obviously suspect otherwise, whether there are other habitable planets and ETs is still an open question. Our argument does not require that we are the only intelligent life in the universe, nor does it require that we are not.

MOVING FROM DESIGN TO THEOLOGY

A number of popular objections to contemporary design arguments are basically theological objections, and most of these reduce to a single, simple complaint: "*God wouldn't do it that way*." As a response to design arguments, this is a red herring. We must distinguish between an argument for design and an argument for the existence of God. While a successful argument for the design of the cosmos provides support for belief in the existence of God, it doesn't prove that the God of traditional belief exists. The most it establishes is that there is a designer sufficient to design the universe as we see it.

Moreover, we must distinguish between design on the one hand and perfection or optimality on the other. Something can be designed without being perfect according to some criterion or another. The Model T was designed, but it's hardly perfect. And just because someone can build a better mousetrap doesn't mean that our present mousetraps aren't designed. Similarly, something can be designed, like the guillotine, even if it seems immoral.

Since we haven't offered a detailed theological argument, we will resist the temptation to attempt a detailed treatment of this point at the last moment. But we should mention one argument that often rears its head in the literature of those advocating the Copernican Principle: If the universe were designed for intelligent observers like us, many argue, why is it so large, so old, and so largely uninhabitable? Surely, a wise and benevolent creator would not be so wasteful and inefficient.

Notice that this argument simply doesn't bear on the question of design. Something can be wicked, wasteful, and inefficient but still be designed. Think of the guillotine, the Edsel, and Windows 95. Moreover, this argument presupposes a superficial, and hardly unimpeachable, view of the most likely intelligent designer—namely, God. And it doesn't make clear sense, since the concept of efficiency is only relevant to agents limited to finite material resources. A sufficiently powerful being would not be taxed by the most extravagant of universes. We mention this not because of some irrepressible compulsion to inject theology into science but merely to point out that those who oppose design have already done so, and done so rather poorly at that.

Of course, there's nothing wrong with moving from specifically scientific inferences into theological reflection, but we should be aware of what we're doing. Ironically, theological arguments of this sort often come from those who insist that science cannot consider questions of purpose and design. As a mere matter of logic, they can't have it both ways. Specifically theological objections deserve specifically theological responses.

Conclusion

Reading the Book of Nature

Philosophy is written in this grand book—I mean the universe—which stands continually open to our gaze, but it cannot be understood unless one first learns to comprehend the language in which it is written.

—Galileo, 1623

Although most of us remember Apollo 8 as the first manned mission to orbit the Moon, and perhaps as the mission responsible for the inspiring photograph of Earth "rising" above the Moon's horizon, few know of the mission's long-term effects on its three crewmen. For Frank Borman and Jim Lovell, seeing the distant Earth from lunar orbit strengthened their conviction that human beings were at home in the universe, that there was more than mere matter in motion, that they, and we, existed for a purpose. From 250,000 miles away, Lovell told the world on Christmas Eve, 1968: "The earth from here is a grand oasis in the big vastness of space." For Bill Anders, however, who took the *Earthrise* photo reprinted countless times, this same view had the opposite effect. The Earth, so tiny and isolated within the cold emptiness of space, suggested a lonely purposelessness. Having abandoned the Catholic rituals that he had once held dear, he later described his vague impression that had begun in lunar orbit: "We're like ants on a log."[1]

Even in a crew of three, two opposing interpretations of the world emerged. For some, the cosmos exists for no purpose and submits to no explanation more ultimate than itself. For others the cosmos finds its proper interpretation in terms of purpose, design, and intention.

The discoveries described in this book, and the argument developed in light of these discoveries, bring into focus these two different ways of viewing the world. For those who already believe that nature exists for a purpose, perhaps as a result of a more specific set of philosophical or religious views, our argument may be satisfying, even expected. The skeptic may rightly point out that such individuals may be imposing an artificial pattern on the evidence, much like psychiatric patients impose patterns on Rorschach blots. But such skepticism is a double-edged sword, since such "skeptics" may have blinded themselves to the existence of real patterns in the natural world. The reasonable skeptic—as opposed to the hardened skeptic for whom no evidence is sufficient—should at least consider the possibility that nature exists for a purpose. For those open to such a possibility, the correlation between habitability and measurability should be a compelling discovery.

But even for the open-minded, there's still what we might call an "epistemological hindrance," which could prevent us from seeing patterns that are really there. Since the nineteenth century, we have been discouraged from asking whether there could be evidence of purpose or intelligent design in nature. To recognize the correlation as a meaningful pattern, rather than a mere curiosity or coincidence, requires that we re-awaken certain atrophied intellectual abilities. Fortunately, scientists have never really neglected this ability, even if many of them have suppressed its widest application.

Here's what we mean. Human beings know a great deal more than they can explain or describe explicitly. How exactly we do this is a controversial question. But *that* we do it is obvious. Chemist and philosopher Michael Polanyi often described this capacity by saying, "We know more than we can tell."[2] Consider our ability to recognize a friend or a spouse in a large crowd. We see the specific and unique features of individual faces but we don't directly attend to them. No one thinks, "My wife has brown hair with gray highlights, blue eyes, a rounded nose, high cheekbones, and twenty-six freckles. There's someone with all those features. So that person is probably my wife." Recognition is much more subtle and subterranean. Most of us would have a very hard time describing someone's face in such detail that an artist could reproduce it exactly. Nevertheless, we easily distinguish a familiar face from scores of other similar faces.

Likewise, when mature readers read a written text, they don't attend to individual ink marks or letters. Polanyi said that we attend *from* those letters and words *to* their meaning. We focus on the meaning embedded in the arrangement of ink marks on a page, not the marks themselves. The text is more like the vehicle that transmits its meaning. The ability to discern that meaning, however, has to be cultivated. We can't innately read English, for example, but with a little nurturing, most of us become quite adept at such pattern recognition.

Like readers, natural scientists foster the ability to read patterns in nature. They must develop an ability to "see through" the isolated bits of data to the underlying pattern, like we "see through" the letters and words on a page to discern their meaning. Before the nineteenth century, for instance, fossil finds were relatively rare and uninformative. They were treated mostly as curiosities and mantel pieces. Today, paleontologists know where to look, and they discover fossils regularly. Expert fossil hunters often have an uncanny ability to discern that a structure is a fossil of a once living thing rather than a mere rock. To those untrained in fossil hunting, this ability can seem almost mystical.

Astronomer Edwin Hubble likewise learned to "read" an expanding universe when he photographed redshifted distant galaxies. To do so, he had to entertain the possibility that the universe was not in eternal stasis but had changed profoundly over time, that it might even have had a beginning. His discovery led to an explosion of new areas of research and helped secure the legitimacy of cosmology as a scientific discipline. When scientists read nature accurately, nature discloses itself in new and unanticipated ways, like a rich and multifaceted text to the patient interpreter. A proper reading creates new lines of research and exploration. Herein lies a virtue in seeing the correlation between habitability and measurability as the result of purpose rather than mere coincidence: we should expect to find it elsewhere, and we should expect to continue making discoveries because of it. To one who has discerned that the cosmos is designed, this correlation is much like the sublime beauty and the mathematical elegance of the natural world—no longer a troublesome anomaly to be explained away but something simultaneously fitting and wonderful. Viewing it as a mere coincidence, in contrast, is both theoretically and aesthetically sterile.

Of course, to see design and purpose embedded in nature, we may have to do what scientists in so many other respects have learned to do, learn to "read" the relevant patterns, to cultivate the ability to consider the book of nature as a whole, to read *through* nature to its meaning. We have offered a number of explicit arguments and distinctions to justify our claim that the cosmos gives evidence of design. And all of us are already adept at detecting the activities of intelligent agents. Although archaeologists, detectives, cryptographers, and SETI researchers have specialized expertise and well-defined methods for detecting design, every person who reads a text or understands a language has the same well-developed, if rudimentary capacity. But just as a grammar book can't substitute for reading great literature, knowing arguments for design, and knowing the criteria by which we may infer design cannot substitute for developing the capacity to discern design in nature. Undertaking such an exercise requires, above all else, a mind open to an almost forgotten possibility.

Thus does our inquiry reduce to a single question: Is it possible that this immense, symphonic system of atoms, fields, forces, stars, galaxies, and people is the result of a choice, a purpose or intention, rather than simply some inscrutable outworking of blind necessity or an inexplicable accident? If so, then it's surely possible that there could be evidence to suggest such a possibility.

We've seen that scientific progress and discovery depend on nature being more than meaningless matter in motion, even motion that can be generalized with natural laws. It's an exquisite structure that preserves vast stores of information about itself and its past. Our habitable environment provides access to an exceptional and highly sensitive collection of information-recording "devices," accurately embedding information about the natural world. We, in turn, possess the materials, and the physical and intellectual capacities, to create technologies to decode these devices. Technology extends man's creativity and vision. As eyeglasses and the lightbulb have improved our ability to read written texts, so the microscope and telescope have allowed us to read the book of nature more deeply. In the last few centuries, we have done so with unparalleled speed and success. And here we have discovered a striking pattern: The myriad conditions that make a region habitable are also the ones that make the best overall places for discovering the universe in its smallest and largest expressions. This is the central argument, the central wonder of this book.

The central irony is perhaps this: The more we learn about how much must go right to get a single habitable planet, the more the naturalistic mindset behind the Copernican Principle and SETI actually *reduces* the hope of finding intelligent beings elsewhere. Under such a paradigm, that search may slowly wither before the stingy dictates of chance.

In contrast, our argument could satisfy another quite different hope that inspires the search for extraterrestrial intelligence. That same unarticulated hope runs like a red thread through much science fiction. From *2001: A Space Odyssey* and *Star Trek* to *E.T.* and Carl Sagan's *Contact*, man is ennobled by cosmic mentors: demigods from an earlier age transformed into natural, alien beings. In *Contact*, Ellie Arroway is even mentored by an alien who appears as her wise and loving father. In these archetypal stories, we see modern humanity longing for a personal encounter, groping, half blind, toward some modest form of transcendence. Arroway's quest involves a signal from space, an encrypted sequence of prime numbers. The story is exciting fiction. In reality we have found no such signal. And yet as we stand gazing at the heavens beyond our little oasis, we gaze not into a meaningless abyss but into a wondrous arena commensurate with our capacity for discovery. Perhaps we have also been staring past a cosmic signal far more significant than any mere sequence of numbers, a signal revealing a universe so skillfully crafted for life and discovery that it seems to whisper of an extra-terrestrial intelligence immeasurably more vast, more ancient, and more magnificent than anything we've been willing to expect or imagine.

Appendix A

The Revised Drake Equation

It is the mark of an instructed mind to rest satisfied with the degree of precision which the nature of the subject admits and not to seek exactness when only an approximation of the truth is possible.
—Aristotle

Aristotle was a wise fellow. But few of us, especially scientists and intellectuals, follow his axiom consistently. We often seek more precision than a subject allows. So we must always beware of one sure mark of a bad argument; namely, artificial precision. Artificial precision is an especially dangerous temptation when trying to determine the probability of the existence of a habitable planet. The consequences are just too exciting to resist, and it's much more interesting to read and write pronouncements of certainty than to hem and haw with weasel words like "perhaps," "probably," and "possibly." (We would add, however—to quote Homer Simpson—that it's weaseling out of things that separates us from the animals. Except the weasels.)

But precisely because we don't want a bad argument in print, we're loath to offer exact numbers for the various factors required for habitability. It's not that such numbers don't exist, mind you, but that we don't know what they are. Nevertheless, the Drake Equation is in dire need of updating. Besides, we do have a rough idea of the values of some of the relevant factors. So, based on the evidence discussed in this book, here's our best shot at updating the Drake Equation. Don't construe this appendix, however, as a prediction or direct implication of our argument. It's more of an afterthought.

In their book *Rare Earth: Why Complex Life is Uncommon in the Universe*,[1] Donald Brownlee and Peter Ward offer their own, revised Drake Equation. They thus imply that the original Drake Equation was at best an abbreviation of myriad unknown and unstated factors, all of which must be satisfied to get a single, radio communicating civilization. The Rare Earth version of the equation is as follows:

$$N = N_* \times f_p \times f_{pm} \times n_e \times f_g \times f_l \times f_i \times f_c \times f_L \times f_m \times f_j \times f_{me}$$

Obviously, this version is lengthier than the original. Compared to the original Drake Equation, the new terms are: f_g, the fraction of stars in the Galactic Habitable Zone, f_{pm}, the fraction of metal-rich planets, f_m, the fraction of planets with a large moon, f_j, the fraction of systems with Jupiter-size planets, and f_{me}, the fraction of planets with a critically low number of mass extinction events. But even this version of the equation is inadequate.[2]

Based on the material in Chapters One through Ten, we can add several more factors. But as we include additional factors, we risk separating factors that aren't really independent. For example, the Rare Earth factors f_{me} and f_j are related; some of the mass extinction events are due to comet impacts, which, in turn, depend on the presence of giant planets. So, instead of simply tacking on a few more factors at the end of the Rare Earth equation above, we'll propose our own. We have formulated it to minimize such overlap and redundancy. The probabilities are low as it is. There's no need to exaggerate. Moreover, since we're not interested in the number of communicating civilizations, just the number of technological civilizations, we won't require a factor for radio communication.

Here are the terms and their definitions:

N_*: total number of stars in the Milky Way Galaxy

f_{sg}: fraction of stars that are early G dwarfs and at least a few billion years old

f_{ghz}: fraction of remaining stars in the Galactic Habitable Zone

f_{cr}: fraction of remaining stars near the corotation circle and with low eccentricity galactic orbits

f_{spir}: fraction of remaining stars outside spiral arms

f_{chz}: fraction of remaining stars with at least one terrestrial planet in the circumstellar habitable zone

n_p: average number of terrestrial planets in the CHZs of such systems

f_j: fraction of remaining systems with no more than a few giant planets comparable in mass to Jupiter in large, circular orbits

f_{cir}: fraction of remaining systems with terrestrial planets in CHZ with low eccentricities and outside dangerous spin-orbit and giant planet resonances

f_{oxy}: fraction of remaining planets near enough inner edge of CHZ to allow high oxygen and low carbon dioxide concentrations in their atmospheres

f_{mass}: fraction of remaining planets in the right mass range

f_{comp}: fraction of remaining planets with proper concentration of sulfur in their cores

f_{moon}: fraction of remaining planets with a large moon and the right planetary rotation period to avoid chaotic variations in its obliquity

f_{water}: fraction of remaining planets with right amount of water in crust

f_{tect}: fraction of remaining planets with steady plate tectonic cycling

f_{life}: fraction of remaining planets where life appears

f_{imp}: fraction of remaining planets with critically low number of large impacts

f_{rad}: fraction of remaining planets exposed to critically low number of transient radiation events

f_{lcomp}: fraction of remaining planets where complex life appears

f_{ltech}: fraction of remaining planets where technological life appears

f_L: average lifetime of a technological civilization

Even at this level of detail, we've implicitly included some factors within others. For example, we could have listed metallicity as a separate factor, but it's implicit in the definitions of the Galactic Habitable Zone, the fraction of stars with terrestrial planets, the fraction of stars with giant planets, and the fraction of planets with a low number of large impacts. Similarly, a factor dealing with the dynamical effects of a stellar companion is implicit in the factor requiring that terrestrial and giant planets have circular orbits. Although most radiation threats are taken care of with the f_{ghz}

and f_{spir} factors, some remain, such as gamma ray bursts, supernovae between spiral arms, and giant flares from the host star. To take care of these cases, we've included the f_{rad} factor.

We're not ready to plug in precise numbers for all the factors in our equation. We'll leave that as homework for the reader. It's a useful exercise, though, to estimate the better-known factors. Let's start with 10 percent. Ten percent is probably a reasonable estimate for a few of the listed astronomical factors, but most are probably much smaller. It's unlikely that more than one factor has a probability as high as 50 percent. Let's consider a couple of factors in some detail: f_j and f_{moon}.

Astronomers now know the approximate fraction of Sun-like stars with one or more giant planets, between one and ten times Jupiter's mass, and with orbital periods less than ten years. It's around 5 percent. As we noted in Chapters Five and Eight, all (or nearly all) the known planets are too massive, orbit too closely to their host stars, or have orbits that are too eccentric to permit habitable planets to form and maintain circular orbits in their Circumstellar Habitable Zones. Compared to the giant planets being found around other stars, the planets in our Solar System have more circular orbits. Excluding Pluto (which is better classified as a Kuiper Belt Object), the average eccentricity of the planets in the Solar System is 0.06. If we naïvely assume that all planet eccentricities are uniformly distributed between 0 and 0.8 (as they appear to be for Doppler-detected giant planets), then the probability that our Solar System was selected at random from such a distribution is about one in a billion. Since we still have much to learn about how planetary systems form and evolve, it is best not to rely on this calculation. Ironically, discovering that giant planets around other Sun-like stars can take on a variety of orbits has led us to reduce the value of the f_j factor from prior expectations. Thus, f_j is probably no larger than about 0.1 percent.

The fraction of terrestrial planets with a moon large enough to stabilize a planet's spin axis is probably *much* lower than 10 percent. To produce an estimate for f_{moon}, we need to know how the Moon formed. Most astronomers are convinced that a glancing collision between the proto-Earth and a smaller planetary body is the best explanation for the Moon's origin. As we noted in Chapter One, there is still no consensus on the most likely set of impact parameters of the two colliding bodies. To yield a suitable outcome, at least five parameters must be fine-tuned: the timing of the collision, the impact point on the proto-Earth, the direction of spin of the proto-Earth, the plane of the collision relative to the plane of the ecliptic, and the momentum of the impactor. The third factor is equal to one half and the fourth factor is probably about 20 percent, but the other three factors are probably small. Therefore, f_{moon} is probably no larger than about 0.001 percent.

With these probable upper limits for f_i and f_{moon}, an average of 10 percent for the first thirteen factors is almost certainly an overestimate. Most of the remaining six factors are even less well known, but each one is likely to be very small. But let's be generous. If we assign a quite liberal probability of 10 percent to each of the first thirteen (of a total of twenty) factors, we arrive at a total number of habitable planets in the Milky Way Galaxy of 0.01. So, the probability that the Milky Way Galaxy contains even one advanced civilization is likely to be much less than one. This is an interesting result, of course, since we exist.

Does this mean that SETI researchers are wasting their time? Perhaps. In general, any search strategy for a rare type of object or event must be very accurate. For example, suppose you're searching for ants with a high IQ. From published studies you know that only about one in a billion ants are really smart. You have developed a quick, automated search method that is 99.9 percent accurate. Although that sounds accurate, it's not nearly accurate enough to detect such a tiny fraction of ants. Almost every ant you detect will not be a member of the local ant Mensa chapter, since the sample will be swamped by false positives. You'll need other, probably more time intensive, methods to pick out the ants you want.

If our upper limit estimates for the number of extraterrestrial civilizations in the galaxy are anywhere near the truth, then SETI false positives will overwhelm detections unless the method is very nearly 100 percent accurate. This means that the present criteria for identifying an ETI signal—a very narrow bandwidth that is repeatedly detected on more than one telescope—may not be enough to eliminate false signals. In fact, there have been some false alarms over the past few decades, most attributed to artificial satellites in orbit around Earth. That's why SETI researchers have adopted more stringent criteria. Maybe it's time to move on to a detection threshold like the one used in the movie *Contact*.

SETI funding probably would be better spent refining theoretical calculations of the chances that extraterrestrial intelligence exists. Astronomers proposing to use large radio telescopes for more mundane research projects, say pulsar timing, must convince a group of their peers that they have a reasonable chance of success.[3] If SETI researchers want to compete for the same public instruments, then they should at least demonstrate that their proposals have some chance of success. But much SETI research is now privately funded, and this is a free country. So who are we to complain?

It's not clear that including the other galaxies in the observable universe would significantly increase the chances; it would depend on the importance of galactic properties for habitability and on the values of the remaining "life factors," such as the origin of simple and then complex life. Although the equation, even in its current form, is probably incomplete, it's the best we can do at the moment. There are lots of discoveries still waiting to be made, and many of them will bear on this equation. Still, we're willing to bet that, the more we learn, the smaller the final outcome will become. Any takers?

Appendix B

What about Panspermia?

No modern discussion of the basic requirements of life and of its origin on Earth would be complete without considering possible external sources. The idea that life on Earth was "seeded" by microbes from beyond its environs has come to be called panspermia. This hypothesis has been around since the pre-Socratic philosopher Anaxagoras, but scientists didn't consider it seriously until the Swedish Nobel laureate Svante Arrhenius revived it a century ago.[1] The latest generation of scientists has revisited the panspermia hypothesis, calling upon knowledge from several scientific disciplines, such as microbiology, impact physics, galactic dynamics, and planetary dynamics. It has implications for the origin of life on Earth and its prevalence in the galaxy. Discussions of panspermia deal with two quite different scales: interstellar and interplanetary. Let's consider each in turn.

Prior to the 1990s, most planetary scientists doubted that intact fragments of a planet's crust could be blasted off its surface and hurled into space. That skepticism melted with the identification of meteorites from Mars and the Moon only a few years ago. Because Mars is a relatively large planetary body with an atmosphere, it took a fairly powerful impact by an asteroid or comet to send fragments of its crust on a trajectory that eventually delivered them to Earth. In addition to fist-sized and larger rocks, a big impact also generates much dust-size ejecta. The rock-size ejecta fragments orbit the Sun under the influence of gravity from the Sun and the major planets. A small fraction of the Martian meteoroids get flung out of the Solar System upon close encounter with Jupiter or Saturn. This mode of panspermia is called lithopanspermia. Jay Melosh, a planetary astronomer specializing in impacts, estimates that about fifteen Martian meteoroids larger than ten centimeters in diameter are ejected from the Solar System each year.[2] The number of Terran (Earthly) and Venusian meteoroids available for ejection from the Solar System is far smaller, given the larger surface gravities and thicker atmospheres of these planets. It was only in the early history of the Solar System that substantial quantities of fragments from Earth and Venus would have escaped the Solar System. Of course, not all the ejecta fragments will contain viable organisms. Only about 0.2 percent of the ejecta mass comes from the planetary surface, where life resides, but it is the surface rock that is least likely to be melted and strongly shocked.

Gravity and the pressure from the Sun's radiation determine the trajectory of dust grains. If they're small enough, dust grains can actually achieve escape velocity and leave the Solar System. This mode of panspermia is called radiopanspermia. Small bacteria can be ejected from the Solar System in this way, but the Sun's ultraviolet radiation kills them almost immediately. A well-shielded bacterium could survive the radiation, but the excess mass from such shielding would prevent it from leaving the Solar System.[3]

Only when the Sun becomes a luminous red giant star would there be enough light pressure to push shielded bacteria out of the Solar System. But by then Earth would be lifeless. Thus, radiopanspermia is unlikely to succeed in transporting viable organisms to other planetary systems.

With a typical ejection speed of five kilometers per second, a meteoroid ejected from the Solar System will travel about sixteen light-years in a million years. Another planetary system is much more likely to capture it if it contains giant planets in large orbits, like ours. This is interesting, since such planets are not found around all nearby Sun-like stars (see Chapters Five and Eight). Once captured, a rock from our Solar System will dance around the planets in its new home system, eventually colliding with one of them. If there are terrestrial planets in the target system, they will receive a small fraction of the captured rocks. And of that small fraction, only a small fraction will survive entry to the planet's surface; most will enter the planet's atmosphere so fast that they will vaporize before reaching the surface.

Melosh estimates that about one Martian meteoroid is captured by another stellar system every one hundred million years; if we're interested in estimating how many of these meteoroids can deliver organismal matter, we'll have to include other factors, which will shrink that number substantially. Of course, since Mars is effectively sterile, it doesn't really matter how many Martian meteoroids make it to another planetary system. Likewise, planetary fragments leaving other systems will probably be from Mars-size planets that are also likely to be sterile.

On its way to another planetary system, a meteorite containing viable organisms will be subjected to galactic cosmic rays and radiation from the decay of radioactive isotopes within it. Even the hardiest bacteria are not expected to survive in such an environment for more than about ten thousand years. Recognizing this, some advocates of panspermia propose that a few mere fragments of DNA and RNA may be adequate to jump-start life on another planet.[4]

A simple thought experiment demonstrates the implausibility of this suggestion: Take some bacteria, grind them up, break up their DNA into smaller fragments, and sprinkle the resulting bits onto a sterilized petri dish with agar growth medium; you can add a liquid component and even stir it up to aid the reactions. If you're ambitious, try this hundreds or thousands of times. Nothing of interest will result. And if nothing results after many trials, then why should you expect dead bacteria to seed a distant planet around another planetary system, one that would offer far less friendly conditions for life?

Although interstellar panspermia is extremely unlikely, interplanetary panspermia is much more likely.[5] Earth, Mars, and Venus exchanged a lot of material when the Solar System was still young and large impacts were frequent. The shortest transfer times are measured in thousands, not millions, of years. There is little question that substantial quantities of viable organisms landed safely on the surfaces of Mars and Venus. So why didn't life take on these worlds?

NOTES

INTRODUCTION, THE PRIVILEGED PLANET

1 In I. Good, ed., *The Scientist Speculates* (New York: Basic Books, 1962), 15.

2 Although unmanned missions had orbited the Moon and photographed the side facing away from Earth (often called inaccurately the dark side of the Moon), Apollo 8 was the first manned mission to break loose of Earth's gravity and orbit the Moon. For the story of Apollo 8, see Robert Zimmerman's fast-paced and moving account in *Genesis: The Story of Apollo 8: The First Manned Flight to Another World* (New York: Dell, 1998).

3 Most are familiar with the color version of the *Earthrise* picture taken by Bill Anders. It is almost always shown with the lunar landscape situated at the bottom of the picture, with Earth rising above it. Anders actually took the picture with the Moon situated to the right, since he understood them to be orbiting the Moon at its equator. Moments later Frank Borman took a black-and-white photograph with the lunar horizon at the bottom. Popular reprints of the color Anders photo usually turn it ninety degrees so that it is oriented the same as the Borman photo. See Zimmerman, *Genesis*: The Story of Apollo 8, 200, for discussion.

4 Ibid., 234.

5 Carl Sagan, *Pale Blue Dot* (New York: Ballantine Books, 1994), 7.

6 Throughout the remainder of the book, we will often refer to this claim simply as "the correlation."

7 Carl Sagan, *Cosmos* (New York: Ballantine Books, 1993), 4. This is an apt summary of materialism or its close relative, naturalism. Materialism bears close affinities with a defunct philosophical view called positivism. Positivists hoped to purge the world of metaphysics in the name of science, but the movement collapsed of its own internal contradictions decades ago. It had various expressions. Positivists attempted to tie all meaningful language, or at least all "scientific" language, to that which could be verified by the senses. *Logical* positivists allowed logical truths as well as sense data. The program generated a rat's nest of conceptual problems. Perhaps the most severe one was this: Whatever criterion the positivist attempted to establish to get rid of "metaphysics" inevitably violated itself. For instance, the claim that only statements that can be verified by the senses are meaningful cannot itself be verified by the senses, which means that, by its own accounting, it's meaningless, or at least "unscientific." At the same time, any criterion liberal enough to avoid contradiction and accommodate actual scientific practice tended to let metaphysics in as well. Such problems eventually led to the demise of the entire positivist enterprise. The positivists themselves openly admitted this. For instance, in a BBC radio interview, Brian McGee asked A. J. Ayer, the father of logical positivism, what the main defect of positivism was. Ayer replied that the main problem was that it was "nearly all false." In *Men of Ideas*, B. McGee, ed. (London: BBC, 1978), 131.

Nevertheless, the general positivist ethos lives on among the materialistically inclined, despite its lack of philosophical foundation. Some recent philosophers have argued that, ironically, materialism itself undermines the scientific realism that virtually all scientists presuppose. See, for instance, Roger Trigg, *Rationality & Science: Can Science Explain Everything?* (Oxford: Blackwell, 1993), 80–101, Robert Koons's more technical argument in *Realism Regained: An Exact Theory of Causation, Teleology, and the Mind* (New York: Oxford University Press, 2000), 222–232, and Michael C. Rea, *World Without Design: The Ontological Consequences of Naturalism* (Oxford: Clarendon Press, 2002).

8 See discussion in Steven J. Dick, *Life on Other Worlds: The 20th Century Extraterrestrial Life Debate* (Cambridge: Cambridge University Press, 1998), 209–220.

9 Amir Aczel, *Probability 1: Why There Must Be Intelligent Life in the Universe* (New York: Harcourt Brace, 1998).

10 Stuart Ross Taylor, *Destiny or Chance: Our Solar System and Its Place in the Cosmos* (Cambridge: Cambridge University Press, 1998).

11 Peter Ward and Donald Brownlee, *Rare Earth: Why Complex Life Is Uncommon in the Universe* (New York: Copernicus, 2000).

12 Gonzalez, with Ward and Brownlee, continues to pursue technical research confirming this hypothesis. See Gonzalez, Brownlee, and Ward, "The Galactic Habitable Zone: Galactic Chemical Evolution," *Icarus* 152 (2001): 185–200.

13 Stuart Ross Taylor puts it with refreshing bluntness:

The message of this book is clear and unequivocal: so many chance events have happened in the development of the Solar System that any original purpose, if it existed, has been lost. Superimposed on these chance events from the physical world are those of biological evolution, which has managed to produce one highly intelligent species out of tens of billion attempts over the past four billion years. In *Destiny or Chance*, 204.

14 Henry Petroski, *Invention by Design* (Cambridge: Harvard University Press, 1996), 30.

15 Incidentally, we're only considering "observational" sciences like comparative planetary geology, solar physics, and stellar, galactic, and cosmological astronomy rather than strictly laboratory or experimental sciences, which are much less sensitive to location. To minimize the risk of "cherry-picking" only evidence that fits our theory, and ignoring contrary evidence, we've restricted ourselves to examples particularly important in their disciplines. We will consider perfect solar eclipses, ice cores in Greenland and Antarctica, deep sea cores, tree rings, corals, seismology, the transparency of the atmosphere, meteorites, stellar spectra, stellar trigonometric parallax, stars as isotropic emitters of highly specified information, supernovae and Cepheids, our place in the Milky Way and dust extinction, the capacity to observe the maximum diversity of star types and the distant universe, background radiation, and the particle and event horizons of the universe. A similar argument may be possible in some other observational or historical sciences, such as archaeology and paleontology. But we can't include the data from every field in a single book. Perhaps some budding archaeologists and paleontologists will take up the task themselves.

CHAPTER 1, WONDERFUL ECLIPSES

1 Martin Amis, *London Fields* (New York: Vintage Books, 1991), chap. 22.

2 In this book, we will capitalize "Sun" and "Moon" when referring to our own Sun and Moon, to avoid confusing them with other suns and moons.

3 The results of my experiment were published in "Ground-Level Humidity, Pressure and Temperature Measurements During the October 24, 1995, Total Solar Eclipse," *Kodaikanal Observatory Bulletin* 13 (1997): 151–154. The temperature dropped twenty-five degrees Fahrenheit during the middle of the eclipse, relative to the previous day.

4 S. Brunier and J. P. Luminet, *Glorious Eclipses: Their Past, Present, and Future* (Cambridge: Cambridge University Press, 2000), 6.

5 Ibid., 17.

6 J. Laskar et al., "Stabilization of the Earth's Obliquity by the Moon," *Nature* 361 (1993): 615–617. The Moon stabilizes Earth's obliquity by exerting a torque on Earth that reduces its precession period by about a factor of

three, taking it far from a dangerous resonance. For a planet of a given size, certain combinations of rotation period, precession period, and orbital period produce resonances that cause the obliquity to vary chaotically over a wide range. Earth will approach such a resonance in about 1.5 billion years as its rotation period continues to slow because of the action of the tides. Thus, to have a stable obliquity, several factors relating to a planet's physical and orbital properties must be met simultaneously. See O. Neron de Surgy and J. Laskar, "On the Long Term Evolution of the Spin of the Earth," *Astronomy & Astrophysics* 318 (1997): 975–989.

A rotation period of twelve hours would also produce a stable obliquity, with or without the Moon. So it would seem that a stabilizing moon is not needed if a planet has a rapid spin. The problem with this solution is that Earth's originally high spin was probably caused by the collision event that formed the Moon (see below). The Moon has been gradually robbing the high spin it gave Earth, but as a trade, it has maintained its stable obliquity.

7 D. M. Williams and D. Pollard, "Earth-Moon Interactions: Implications for Terrestrial Climate and Life," *Origin of the Earth and Moon*, R. M. Canup and K. Righter, eds. (Tucson: University of Arizona Press, 2000), 513–525.

8 The prediction that wind and tides are the main drivers of global ocean circulation was made by W. H. Munk and C. Wunsch, "Abyssal Recipes II: Energetics of Tidal and Wind Mixing," *Deep-Sea Research* 45 (1998): 1977–2010. Their prediction was confirmed by Topex/Poseidon satellite altimeter data: G. D. Egbert and R. D. Ray, "Significant Dissipation of Tidal Energy in the Deep Ocean Inferred from Satellite Altimeter Data," *Nature* 405 (2000): 775–778.

9 One of the oceanographers who suggested the deep-water tidal theory notes, "It appears that the tides are, surprisingly, an intricate part of the story of climate change, as is the history of the lunar orbit." C. Wunsch, "Moon, Tides, and Climate," *Nature* 405 (2000): 744.

10 Calculations indicate that the Moon may have formed from a glancing impact by an object two to three times the mass of Mars when Earth was only about half-formed; A. G. W. Cameron, "Higher-Resolution Simulations of the Giant Impact," in *Origin of the Earth and Moon*, R. M. Canup and K. Righter, eds. (Tucson: Univ. of Arizona Press, 2000): 133–144. Other simulations indicate that the impact more probably occurred near the end of Earth's formation and with an impactor the mass of Mars; R. M. Canup, and E. Asphaug, "Origin of the Moon in a Giant Impact Near the End of the Earth's Formation," *Nature* 412 (2001): 708–712. This more recent set of simulations makes the formation of the Moon a more probable event than Cameron's simulations imply, because smaller impactors are more common than bigger ones. But caution is appropriate here, since this is a rapidly developing area of research.

11 Recent measurements of the isotopes of tungsten in Earth and primitive meteorites help to establish when Earth's iron core formed, and confirm the Moon's contribution to that event. This is possible because radioactive halfnium present early on (with a half-life of nine million years, after which it decays to tungsten) and tungsten have different affinities for iron. Thus, when the core formed, some tungsten went down with the iron while halfnium remained behind in the mantle and crust. The data indicate that the core formed about thirty million years after the beginning of the Solar System, about the same time the Moon is thought to have formed. See R. Fitzgerald, "Isotope Ratio Measurements Firm Up Knowledge of Earth's Formation," *Physics Today* (January 2003): 16–18, and references cited therein.

12 For an entertaining and informative, though sometimes speculative, account of the many ways the Moon is important for life on Earth, see N. F. Comins, *What If the Moon Didn't Exist?: Voyages to Earths That Might Have Been* (New York: HarperCollins, 1993). See also C. R. Benn, "The Moon and the Origin of Life," *Earth, Moon, and Planets* 85–86 (2001): 61–66.

13 We'll discuss this in Chapter Seven. Also see G. Gonzalez, "Is the Sun Anomalous?" *Astronomy & Geophysics* 40, no. 5 (1999): 5.25–5.29.

14 C. R. Benn has also suggested a connection between the occurrence of total solar eclipses and life on Earth. He notes that the strength of the tides from the Moon or the Sun depends on its angular size raised to the third power. Thus, the tides induced by the Moon and the Sun are comparable in strength as a result of their similar angular sizes. Benn argues that the beating this produces in the tides may have been instrumental in the origin of life, since the long period of the beating would permit slow chemical reactions on the surface. This argument

breaks down, however, when one considers that the Moon was much closer to Earth when life first appeared on Earth. It is not clear why life would benefit from the similarity in solar and lunar tidal strengths at the present.

15 J. C. G. Walker and K. J. Zahnle, "Lunar Nodal Tide and Distance to the Moon During the Precambrian," *Nature* 320 (1986): 600–602. We'll explain just how astronomers know this in Chapter Two.

16 In this eclipse, the Moon's apparent size was only about forty-five arc seconds greater than the Sun's photosphere. This works out to about 2.5 percent of their mean angular size.

17 The oblateness, or flattening, of the lunar profile in Earth's sky is a mere 0.06 percent. See S. K. Runcorn and S. Hofmann, "The Shape of the Moon," in *The Moon*, S. K. Runcorn, H. C. Urey, eds. (Dordrecht-Holland: D. Reidel, 1972), 22. The lunar axis pointing toward Earth, however, is larger than the other two axes by about 5 kilometers. Had the Moon not yet achieved a rotationally synchronized configuration, the longer axis would result in a less round lunar profile. (A planetary body in a rotationally synchronized orbit keeps the same face aimed toward the body it's orbiting; for the Moon, this state probably set in within about a million years of its formation.)

18 Although the Moon may have started with a relatively round profile, it probably attained its present shape during the Late Heavy Bombardment, about 3.8 billion years ago. At that time large impactors were frequent, and they injected sufficient energy into the Moon to reshape it according to the stronger tidal forces from Earth at that time. See V. N. Zharkov, "On the History of the Lunar Orbit," *Solar System Research* 34 (2000): 1–11.

19 Incidentally, only one moon in the Solar System, Titan, has a substantial atmosphere. The pioneering space artist Chesley Bonestell rendered on his canvas a view of a total lunar eclipse as it would appear from the Moon. See C. Bonestell, *Rocket to the Moon* (Chicago: Children's Press, 1961).

20 Note that Figure 1.4 does not include the many small moons discovered around the outer planets in recent years. As of October 2003, the total number of known moons in the Solar System was 136. All these new moons belong to the "irregular" class (with highly eccentric and inclined orbits), are very small, and do not produce total solar eclipses. For the latest listing of the new moons, see http://www.ifa.hawaii.edu/~sheppard/satellites/. Also not included in the figure are mutual eclipses between moons. These events are far less common than eclipses between moons and their host planets. The mutual eclipses among the Galilean Moons, the four large moons of Jupiter that Galileo first saw through this telescope, are observable in amateur telescopes.

21 In the 1930s Bernard Lyot of France invented the coronagraph and employed it to study the Sun's bright inner corona without the benefit of a total solar eclipse. This was not an easy feat. It required an exceptional site (Lyot observed from Pic du Midi Observatory in France) and carefully designed and built optics. Even today, the best coronagraphs on high mountaintops cannot reveal as much detail in the corona as a total solar eclipse.

22 While eclipse observations made it possible to identify hydrogen as a major constituent of prominences, it was not until the 1920s that Cecilia Payne-Geposchkin first demonstrated quantitatively that hydrogen is the main constituent of the Sun and other stars.

23 Cited by J. B. Zirker, *Total Eclipses of the Sun* (New York: Van Nostrand Reinhold, 1995), 18.

24 To understand what is going on here, imagine the Sun's chromosphere as a very thin shell enveloping the entire surface of the Sun. To produce an absorption line spectrum, the continuous spectrum from the underlying photosphere must pass through this shell. But because its glow is so feeble compared with the photosphere, we cannot observe the chromosphere directly in white light. It is only during a total eclipse that the bright photosphere is completely blocked and the cross section of the faint thin shell is seen against a dark background.

25 From experiments in his lab Gustav Kirkhoff developed three laws of spectroscopic analysis, which together account for the three basic types of spectra seen in nature: continuous, emission, and absorption.

26 One could generate an artificial rainbow during an eclipse with a misty spray of water in front of a dark cloth and thus conduct this experiment without the benefit of a natural rainbow.

27 A recent reassessment of the earliest measurements has shown that the systematic uncertainties were probably underestimated at the time. See J. Earman, and C. Glymour, "Relativity and Eclipses: The British Eclipse Expeditions of 1919 and Their Predecessors," *Historical Studies in the Physical Sciences* 11 (1980): 49–85; Zirker, 170–182.

28 Zirker, *Total Eclipses*, 175–179.

29 Ibid., 179–182. Radio observations can be made in the daytime, so no total eclipse is needed for this type of test. But the number of targets available in the radio are fewer than in the optical bands.

30 L. Morrison and R. Stephenson, "The Sands of Time and the Earth's Rotation," *Astronomy & Geophysics* 39:5 (1998): 13.

31 Once the theoretical tidal component is subtracted from the observed rate of spin-down of the last century, however, it is clear that Earth's rotation is slowing at an irregular rate. Because of this, we cannot extrapolate our present knowledge of Earth's rotation more than a few centuries into the past. Even if our estimate of the length of the day is off by just a tiny amount, this leads to large errors over long periods of time, since clock errors accumulate. The insertion of a leap second in our atomic clocks about once every 1.5 years shows the rate that a steady clock errs in reflecting Earth's rate of rotation.

32 For an illustration of this see M. Littmann, K. Willcox, and F. Espenak, *Totality: Eclipses of the Sun* (Oxford: Oxford University Press, 1999), 130–131.

33 This would become important if the Moon's apparent size were at least twice its present value, since the various twentieth-century starlight deflection experiments generally did not employ stars within about two solar radii of the Sun's limb.

34 For a review on what can and cannot be done from space with respect to study of the Sun's corona, see J. M. Pasachoff, "Solar-Eclipse Science: Still Going Strong," *Sky & Telescope* 101, no. 2 (2001): 40–47.

35 J. O. Dickey et al., "Lunar Laser Ranging: A Continuing Legacy of the Apollo Program," *Science* 265 (1994): 482–490.

36 The future lifetime of the biosphere for complex life is probably no more than 500 million years. See S. Franck et al., "Reduction of Biosphere Life Span as a Consequence of Geodynamics," *Tellus* 52B (2000): 94–107.

37 My (GG) discovery that the observability of perfect solar eclipses correlates with habitability was a rather surprising conclusion, and its publication ["Wonderful Eclipses," *Astronomy & Geophysics* 40, no. 3 (1999): 3.18–3.20] generated some interest—including articles in several newspapers ["Eclipse shows signs of Life," *The Daily Telegraph* (June 23, 1999): 16; "Right Distance for a 'Perfect' Total Eclipse," *The Irish Times* (July 12, 1999): 12; "In the Shadow of Brilliance," *Chicago Sun-Times* (June 27, 1999): 32; "Leben wir unter einem einzigartigen Stern?" *Spectrum* (July 17, 1999): 8], a BBC radio interview, which aired on August 11, 1999 (the day of the European total solar eclipse), and a mention in "The Year in Weird Science" section of the January 2000 *Discover* magazine. This last item is the most revealing one—the moniker "weird" implies that the result of my study contradicted the expectations of *Discover*'s editorial staff.

CHAPTER 2, AT HOME ON A DATA RECORDER

1 New American Standard Bible.

2 D. A. Hodgson, N. M. Johnston, A. P. Caulkett, and V. J. Jones, "Paleolimnology of Antarctic Fur Seal *Arctocephalus gazella* Populations and Implications for Antarctic Management," *Biological Conservation* 83, no. 2 (1998): 145–154; Dominic A. Hodgson and Nadine M. Johnston, "Inferring Seal Populations from Lake Sediments," *Nature* 387 (May 1, 1997): 30–31.

3 While strip chart recorders are still in use by scientists in many disciplines, the computer is today the preferred data-recording device. Still, the basic components in the two types of instruments are essentially the same (a time-stamped signal stored for later analysis).

4 Snow on Greenland and Antarctica, sedimentation on the deep-sea floor and lake bottoms, tidalites (sediments formed in shallow water areas, such as beaches and tidal flats), tree rings, and coral and mollusk growth layers are all examples of Earth's constant and astonishingly effective work as a data recorder. For an overview of the diverse array of natural records of the ancient climate, see the special issue of *Science* on paleoclimatology (April 27, 2001).

5 Researchers Kurt Cuffey and Edward Brook write on ice-core paleoclimatology, "The reward for this effort is an astonishing expansion of our knowledge of past environments, remarkable both for its implications and its level of detail." ["Ice Sheets and the Ice-Core Record of Climate Change," *Earth System Science: From Biogeochemical Cycles to Global Change*, M. C. Jacobson, R. J. Charlson, H. Rodhe, and G. H. Orians, eds. (San

Diego: Academic Press, 2000), 466.] This is the opening quote in a section of their article titled "Ice Sheets as Paleoclimate Archives."

6 The most important tools in the paleoclimatologist's tool chest are isotope ratios. Every chemical element is composed of atoms. Every atom contains at least one positively charged proton in a nucleus and one negatively charged electron outside it. In order to maintain charge neutrality, the number of protons must be equal to the number of electrons; some atoms may be missing an electron or two, but at a macroscopic level, charge neutrality is almost exactly maintained in most places in the universe.

We distinguish isotopes of a given chemical element by the number of neutrons in their constituent atoms, while all the atoms belonging to a given element have the same number of protons. If one plots the number of protons (P) versus number of neutrons (N) for all the known isotopes, one finds that stable ones stick pretty close to the N=P line. The farther away you get from this so-called "valley of stability," the less stable the isotope (and hence the shorter its half-life). Most elements have at least two stable isotopes.

The geography of this valley depends on the particular values of the fundamental physical constants and forces (in particular, the strengths of the strong and electro-weak forces) and also on the particular way they were produced. For much of the past fifty years, astrophysicists have been concerned with understanding how these isotopes are produced in the amounts we observe. The best explanation is that they are produced mostly in the hot, dense interiors of stars via nuclear reactions.

Isotopes, both stable and radioactive, are quite useful to geochemists. They employ radioisotopes to date events over an enormous range of time scales, from minutes to billions of years. Moreover, since radioisotopes are sprinkled throughout the periodic table, geochemists can date geologic processes over a broad range of mineral types. They can date virtually any geophysical process since the origin of the universe with at least one radioisotope type.

Chemical reactions alter the distribution of atoms and how they combine with one another while having little or no effect on the ratios of abundances of different isotopes of the same element. Fractionation (or separation) of isotopes does occur during evaporation and condensation between gaseous and solid or liquid phases, and since these are temperature-dependent processes, the isotope ratios often retain a memory of temperature history. Not all isotopes are of equal value for reconstructing climate history, though. The larger the difference in mass between a pair of isotopes of a given element, the greater their sensitivity to fractionation processes. Water molecules containing the light hydrogen isotope will evaporate from the surface of a body of water more readily than water molecules containing the heavy isotope, and the degree of isotope fractionation will be greater for hydrogen than it is for the oxygen isotopes in the water molecules. These varying qualities make the isotopes in water highly informative.

For one, we can relate changes in the oxygen isotope ratios to changes in the global ice volume. Water evaporating from the ocean will contain more of the light oxygen isotope than water remaining behind. As the water vapor makes its way into colder regions and precipitates as snow in the polar regions, the light oxygen isotope concentration increases in the remaining water vapor in the atmosphere. The ice deposits in the interior of Antarctica and Greenland will have a higher concentration of the light oxygen isotope relative to the oceans (and relative to the ocean sediments). The greater the global ice volume, then, the greater the ratio of heavy to light oxygen isotopes left behind in the oceans.

The isotopic composition of water deposited in the polar regions is also sensitive to temperature. Thus, the oxygen isotope ratios in ice provide a mixed record of global ice volume and local temperature. On the other hand, measurements of the ratio of the two hydrogen isotopes in polar ice show that it correlates very well with temperature. The more mixed record of the oxygen isotopes is not worthless, however. Historically, the global ice volume correlates with the global temperature in the sense that colder intervals (the glacials) are characterized by a greater global ice volume.

Jonathan Lunine, a planetary scientist at the University of Arizona, summarizes the four key requirements for isotopes to be useful as tools in climate reconstructions:

1. availability of stable isotopes of the same element whose separation depends on temperature
2. incorporation of the fractionated isotope mixture in some storage medium that is preserved for a long time

3. ability to measure accurately the ratio of the various isotopes
4. a means to date, in an absolute or relative sense, the age of the stored isotope data

In *Earth: Evolution of a Habitable World* (Cambridge: Cambridge University Press, 1999), 54. Rather tellingly, this quote is located in a section of the book titled "The Measurable Planet: Tools to Discern the History of Earth and the Planets." A surprising variety of processes on Earth fulfill these four basic requirements.

7 Specifically, scientists can relate the amount of deuterium (an isotope of hydrogen) in ice to the local temperature when it was deposited, and the isotope ratio in oxygen in trapped bubbles to global ice volume and, more generally, to the hydrological cycle—the global evaporation, transport, and precipitation of water. For a nontechnical, personal account of fieldwork and laboratory measurements of Greenland ice cores, see R. B. Alley, *The Two-Mile Time Machine* (Princeton: Princeton University Press, 2000); R. B. Alley and M. L. Bender, "Greenland Ice Cores: Frozen in Time," *Scientific American* (February 1998): 80–85. Also see the review by R. J. Delmas, "Environmental Information from Ice Cores," *Reviews of Geophysics* 30, no. 1 (1992): 1–21.

8 This property of ice cores impresses Richard Alley, professor of geosciences at Pennsylvania State University. "The ice brings all this information together, providing a surprisingly complete history of climate changes over much of Earth's surface." R. B. Alley and M. L. Bender, "Greenland Ice Cores: Frozen in Time," *Scientific American* (February 1998): 13.

9 As the authors of the Vostok ice core study observed, "The Holocene, which has already lasted 11 kyr [thousand years], is, by far, the longest stable warm period recorded in Antarctica during the past 420 kyr [thousand years]." J. R. Petit et al., "Climate and Atmospheric History of the Past 420,000 Years from the Vostok Ice Core, Antarctica," *Nature* 399 (1999): 429–436.

10 For example, the date of the Mount Mazama eruption (which formed Crater Lake) has recently been determined to be 7627 ± 150 years ago by C. M. Zdanowicz, G. A. Zielinski, and M. S. Germani, "Mount Mazama Eruption: Calendrical Age Verified and Atmospheric Impact Assessed," *Geology* 27:7 (1999): 621–624.

11 Alley and Bender, "Greenland Ice Cores," 160–161.

12 The abundance variations of beryllium-10 in ice and carbon-14 in tree rings have also been used to reconstruct the sunspot cycle to pre-telescope times. See I. G. Usoskin et al., "A Millennium Scale Sunspot Number Reconstruction: Evidence for an Unusually Active Sun Since the 1940's," *Physical Review Letters*, in press. Interestingly, they find that the solar activity during the last sixty years is unique compared with the past 1150 years.

13 The solar signature is also visible in the dust profile of the Greenland ice cores: M. Ram and M. R. Stolz, "Possible Solar Influences on the Dust Profile of the GISP2 Ice Core from Central Greenland," *Geophysical Research Letters* 28, no. 8 (1999): 1043–1046. Strong evidence that the 1,300-year cycle seen in several proxies is caused by the Sun is given by G. Bond et al., "Persistent Solar Influence on North Atlantic Climate During the Holocene," *Science* 294 (2001): 2130–2136. They write: "Earth's climate system is highly sensitive to extremely weak perturbations in the Sun's energy output, not just on the decadal scales that have been investigated previously, but also on the centennial to millennial time scales documented here."

The Sun's signature has also been detected in other types of proxies discussed later in this chapter. For an example of proxies obtained from stalagmites in caves, see U. Neff et al., "Strong Coherence between Solar Variability and the Monsoons in Oman Between 9 and 6 kyr ago," *Nature* 411 (2001): 290–293. For additional discussions of evidences for Sun-Earth climate links, see F. M. Chambers, M. I. Ogle, and J. J. Blackford, "Palaeoenvironmental Evidence for Solar Forcing of Holocene Climate: Linkages to Solar Science," *Progress in Physical Geography* 23 (1999): 181–204; van Geel et al., "The Role of Solar Forcing upon Climate Change," *Quaternary Science Reviews* 18 (1999): 331–338; K. D. Pang and K. K. Yau, "Ancient Observations Link Changes in Sun's Brightness and Earth's Climate," *EOS, Transactions, American Geophysical Union* 83 (2002): 489–490; N. Wang et al., "Evidence for Cold Events in the Early Holocene from the Guliya Ice Core," *Chinese Science Bulletin* 47 (2002): 1422–1427.

14 R. T. Rood, C. L. Sarazin, E. J. Zeller, and B. C. Parker, "X- or ?-rays from Supernovae in Glacial Ice," *Nature* 282 (1979): 701–703.

15 C. P. Burgess and K. Zuber, "Footprints of the Newly Discovered Vela Supernova in Antarctic Ice Cores?" *Astroparticle Physics* 14, no. 1 (2000): 1–6.

16 The following study claims to have found isotopic evidence in marine sediments of a nearby supernova about two million years ago and associated extinctions: N. Benitez, J. Maiz-Apellaniz, and M. Canelles, "Evidence for Nearby Supernova Explosions," *Physical Review Letters* 88 (2002): 081101.

17 See R. C. L. Wilson, S. A. Drury, and J. L. Chapman, *The Great Ice Age* (Routledge: London, 2000): 163–186.

18 Fossilized terrestrial mollusk shells, too, help us reconstruct past climate. O. Moine et al., "Paleoclimatic Reconstruction Using Mutual Climatic Range on Terrestrial Mollusks," *Quaternary Research* 57 (2002): 162–172.

19 Among the most metrically useful isotopes in the biosphere are hydrogen, deuterium, carbon-12, carbon-13, carbon-14, oxygen-16, oxygen-18, calcium-40, and calcium-44. The standard notation for an isotope is nE, where n is the number of neutrons and E is the element designation. Heavier isotopes found in the biosphere (for example, isotopes of iron) tend to be less useful for reconstructing ancient climates, but iron-60 is thought to be a likely indicator of nearby supernovae.

The really useful isotopes are abundant in organisms, are exchanged with the atmosphere and/or hydrosphere, and vary considerably in mass, making them more sensitive than heavier isotopes to various climate processes. The large mass difference between deuterium and hydrogen makes the deuterium-to-hydrogen abundance ratio more sensitive to temperature variations than any other isotope ratio.

20 Wilson, Drury, and Chapman, *The Great Ice Age*, 71–77. Another method makes use of the magnesium/calcium ratio in planktonic forams and its chemical dependence on temperature. Magnesium atoms often replace calcium atoms in the formation of their shells, and the degree to which this occurs is temperature-dependent. When researchers combine this information with the oxygen isotope ratios from the same shells, they can derive both temperature and global ice volume. See D. W. Lea, T. A. Mashiotta, and H. J. Spero, "Controls on Magnesium and Strontium Uptake in Plantonic Foraminifera Determined by Live Culturing," *Geochimica et Cosmochimica Acta* 63 (1999): 2369–2379. Another promising paleothermometer makes use of calcium isotope ratios. See T. F. Nagler, A. Eisenhauer, A. Muller, C. Hemleben, and J. Kramers, "The $_{44}$Ca-Temperature Calibration on Fossil and Cultured *Globigerinoides sacculifer*: New Tool for Reconstruction of Past Sea Surface Temperatures," *Geochemistry, Geophysics, Geosystems* 1 (2000): paper number 2000GC000091.

21 Biogenic silica produced by diatoms serves as a temperature proxy, with cold periods producing more diatomaceous ooze. See A. A. Prokopenko et al., "Biogenic Silica Record of the Lake Baikal Response to Climatic Forcing During the Brunhes," *Quaternary Research* 55 (2001): 123–132.

22 Among the quantities measured in marine sediments are carbon and oxygen isotope ratios in foram skeletons, detritus from melting icebergs, and extraterrestrial helium-3. Marine sediments do suffer from a slight problem—mixing by worms, or bioturbation. Bioturbation effectively decreases the time resolution of marine cores, but some locations are much less affected than most. Ice cores, for obvious reasons, don't suffer from this problem.

23 J. O. Dickey et al., "Lunar Laser Ranging: A Continuing Legacy of the Apollo Program," *Science* 265 (1994): 482–490.

24 J. C. G. Walker and K. J. Zahnle, "Lunar Nodal Tide and Distance to the Moon During the Precambrian," *Nature* 320 (1986): 600–602; K. A. Eriksson and E. L. Simpson, "Quantifying the Oldest Tidal Record: The 3.2 Ga Moodies Group, Barberton Greenstone Belt, South Africa," *Geology* 28 (2000): 831–834.

25 James Walker and Kevin Zahnle give a succinct description of the basic method:

The most important source of long-term information derives from fossil corals and mollusks. In brief, fine laminae are interpreted either as daily growth increments or as the record of the semidiurnal or diurnal tide, in accordance with the general growth habits of their modern descendents. Modulation of the fine banding by the fortnightly or monthly tidal cycles and by the yearly seasonal cycle allows estimates of the number of days per year, the number of days per month, and the number of months per year ("Lunar Nodal Tide," 600).

26 Researchers also assume that the size of Earth's orbit has not changed significantly over the last few billion years. This assumption receives support from long-term numerical simulations of the orbits of the planets in the Solar System.

27 For a specific example of this type of study, see C. P. Sonett and M. A. Chan, "Neoproterozoic Earth-Moon Dynamics: Rework of the 900 Ma Big Cottonwood Canyon Tidal Laminae," *Geophysical Research Letters* 25 (1998): 539–542. More general reviews of paleoastronomy as applied to tidal rhythmites are given by E. P. Kvale et al.,

"Calculating Lunar Retreat Rates Using Tidal Rhythmites," *Journal of Sedimentary Research* 69, no. 6 (1999): 1154–1168, and G. E. Williams, "Geological Constraints on the Precambrian History of Earth's Rotation and the Moon's Orbit," *Reviews of Geophysics* 38, no. 1 (2000): 37–59.

28 The records from old tree trunks, such as those preserved in bogs and lake bottoms, can be combined to form a long, complete record extending beyond living trees. This is possible if there is overlap among several tree trunks. The irregular pattern of ring thickness greatly facilitates such reconstructions. Had rings and ring patterns been uniform, a given set of tree rings would look like any other, making reconstruction ambiguous and uncertain. The University of Arizona maintains an excellent website on dendrochronology at http://www.ltrr.arizona.edu/.

29 For example, Forrest Mimms III, the famous amateur scientist, has told us of his discovery of what he believes is a proxy in tree rings for the UV level reaching the ground. The mechanism involves the transport of tannin from the leaves to the trunk and branches in response to changing UV levels.

30 G. J. Retallack, "A 300-Million-Year Record of Atmospheric Carbon Dioxide from Fossil Plant Cuticles," *Nature* 411 (2001): 287–290. See also W. M. Kurschner, "Leaf Sensors for CO_2 in Deep Time," *Nature* 411 (2001): 247–248; M. Rundgren and O. Bennike, "Century-Scale Changes of Atmospheric CO_2 During the Last Interglacial," *Geology* 30, no. 2 (2002): 187–189; D. J. Beerling, et al., "An Atmospheric pCO_2 Reconstruction across the Cretaceous-Tertiary Boundary from Leaf Megafossils," *Publications of the National Academy of Sciences* 99 (2002): 7836–7840. More accurately, these studies reconstruct the ancient variations in the carbon dioxide partial pressure, which is just the pressure contributed by a given constituent of the atmosphere. The stomatal index, the proportion of leaf cells that are stomata, is insensitive to other environmental changes, such as sunlight and relative humidity.

31 V. J. Polyak et al., "Wetter and Cooler Late Holocene Climate in the Southwestern United States from Mites Preserved in Stalagmites," *Geology* 29, no. 7 (2001): 643–646.

32 This is due to deviations in the background cosmic ray flux and in the interactions of our magnetosphere with the sunspot cycle–modulated interplanetary field.

33 Wilson, Drury, and Chapman, *The Great Ice Age*, 45. The following study estimates that the Great Barrier Reef in Australia began growing about 600,000 years ago: International Consortium for Great Barrier Reef Drilling, "New Constraints on the Origin of the Australian Great Barrier Reef: Results from an International Project to Deep Coring," *Geology* 29 (2001): 483–486.

34 The study of remnant magnetic variations in sediments and rocks is called *magnetostratigraphy*. For more detailed discussion on this topic, see W. Lowrie, *Fundamentals of Geophysics* (Cambridge: Cambridge University Press, 1997): 297–302.

35 These astronomical cycles are named after Milutin Milankovitch, who in the 1920s and 1930s proposed that long-term changes in the sunshine received at high northern latitudes controlled the growth and decay of the northern ice sheets. For additional background on this topic, see W. S. Broecker and G. H. Denton, "What Drives Glacial Cycles?" *Scientific American* (January 1990), 49–56; R. C. L. Wilson, S. A. Drury, and J. L. Chapman, *The Great Ice Age* (London: Routledge, 2000), 61–65; A. Berger and M. F. Loutre, "Astronomical Forcing through Geological Time," in *Orbital Forcing and Cyclic Sequences: Special Publication of the International Association of Sedimentologists*, P. L. De Boer, D. G. Smith, eds. (Oxford: Blackwell Scientific Publications, 1994), 15–24.

36 P. L. De Boer and D. G. Smith review Milankovitch cycles and their effects on Earth processes in "Orbital Forcing and Cyclic Sequences," in *Orbital Forcing and Cyclic Sequences: Special Publication of the International Association of Sedimentologists*, P. L. De Boer, D. G. Smith, eds. (Oxford: Blackwell Scientific Publications, 1994): 1–14. How scientists use the orbital cycles depends on their ultimate goal. If they want to learn about the long-term orbital evolution of the complex dynamics of the Solar System, they need an independent chronometer. One popular option makes use of the astronomical cycles themselves. This may sound circular—how can one use the astronomical cycles to adjust one's data to study the astronomical cycles? Paleoclimatologists can avoid this problem by using a short-period and well-understood astronomical cycle to calibrate a marine or ice core; usually, they use Earth's 41,000-year obliquity cycle for this purpose. Then they compare the derived periods of other cycles to computer simulations to learn about less well-understood orbital dynamics. On the other hand, if they

just want to age-calibrate a core and aren't interested in studying astronomical cycles, then they employ the astronomical cycles as the primary dating tool.

Two recent representative studies of Milankovitch cycles in marine cores are H. Pälike and N. J. Shackleton, "Constraints on Astronomical Parameters from the Geological Record for the Last 25 Myr," *Earth and Planetary Science Letters* 182 (2000): 1–14; J. C. Zachos et al., "Climate Response to Orbital Forcing Across the Oligocene-Miocene Boundary," *Science* 292 (2001): 274–278.

37 For example, they can compare the amount of dust content or oxygen isotope ratios of ice and marine sediment cores.

38 We should specify that these statements are true given the physical conditions that obtain in our universe. We're not saying that this must be true for any conceivable type of life, with any conceivable set of physical laws and parameters in any possible universe.

39 We can distinguish between simple and complex life in a number of ways. We could differentiate between single-celled and multicellular (or metazoan) organisms, eukaryotes and prokaryotes (with and without a nucleus, respectively), aerobic and anaerobic organisms (oxygen-breathing or not, respectively), or microscopic and macroscopic organisms. Yet another possibility is to define simple life as that which dominated Earth over most of its history.

40 Of course, some scientists have also offered "alternative chemistry" proposals for life forms. See Gerald Feinberg and Robert Shapiro, *Life Beyond Earth: The Intelligent Earthling's Guide to Life in the Universe* (New York: William Morrow, 1980).

41 See N. R. Pace, "The Universal Nature of Biochemistry," *Publications of the National Academy of Sciences* 98 (2001): 805–808.

42 Biochemist A. E. Needham comments on the metastability of carbon reactions: "As in so many other respects, carbon seems to have the best of both worlds, in fact, combining stability with lability, momentum with inertia. Most organic compounds are metastable . . . that is to say they are not in complete equilibrium with their environmental conditions and are easily induced to react further." *The Uniqueness of Biological Materials* (Oxford: Pergamon Press, 1965), 30.

43 In contrast, silicon forms a mineral, silica, when oxidized. This results from the tendency of silicon to form single bonds with the oxygen atoms, rather than double bonds as carbon does. The double bonds carbon forms with each oxygen atom do not leave the oxygen atoms free to form bonds with other carbon atoms in its vicinity. However, this is what happens with the single-bonded oxygen atoms attached to silicon—they bond with other silicon atoms, forming a stable crystalline matrix of SiO_2.

44 Biologist Michael Denton argues that the weak chemical bonds that allow large organic molecules to form three-dimensional shapes are also an essential requirement for life:

> Nearly all the biological activities of virtually all the large molecules in the cell are critically dependent on their possessing very precise 3-D shapes. Nature has provided no other glue to hold together the molecular superstructure of the cell. While we cannot have carbon-based life in the cosmos without covalent bonds, as there would be no molecules, just as certainly, we cannot have carbon-based life without these weak noncovalent bonds—because the molecules would not have stable, complex 3-D shapes.
> *Nature's Destiny: How the Laws of Biology Reveal Purpose in the Universe* (New York: The Free Press, 1998), 114.

45 For example, silicones—alternating chains of silicon and oxygen atoms—offer perhaps the best alternative to carbon-based, or organic, chemistry. They fail on several grounds, however. First, there is no good solvent for silicone chemistry. Second, silicone chemistry becomes more biologically useful if organics are available as side chains to attach to the silicone base; but this makes the silicones superfluous. In other words, if the environment had carbon in the first place, building the life form from silicone would be like building golf club heads out of clay when you had titanium on hand. Finally, we do not see silicone chemistry in nature, only in the laboratory, but we do see long-chained carbon molecules in many places, including Earth, meteorites, and the matter between the stars, called the interstellar medium.

46 By chemical compound, we mean a substance composed of molecules containing more than one type of atom. The most abundant molecule in the universe is molecular hydrogen. It is interesting that the water molecule is composed of the two most abundant reactive elements in the universe.

47 *The Fitness of the Environment: An Inquiry into the Biological Significance of the Properties of Matter* (New York: The Macmillan Company, 1913). Henderson's work was a quantitative extension of the nineteenth-century work by William Whewell, *Astronomy and General Physics, Considered with Reference to Natural Theology* (London: William Pickering, 1833), ninth edition published in 1864.

48 See *Nature's Destiny*; J. D. Barrow and F. J. Tipler, *The Anthropic Cosmological Principle* (Oxford: Oxford University Press, 1986), 524–541; *The Uniqueness of Biological Materials*; and P. Ball, *Life's Matrix: A Biography of Water* (New York: Farrar, Straus and Giroux, 1999). Unlike the "Face on Mars," which disappeared after images with greater resolution were obtained with newer orbiters, the apparently high fitness of water for chemical life has only become more impressive as our "resolution" in chemistry has increased since the nineteenth century.

49 Ice actually has higher heat conductivity than liquid water. But, ice forming on water still acts to slow heat loss from the water below. First, considerable heat is given off as ice forms at the base of the ice sheet, slowing the formation of new ice. Second, the ice does not convect, so heat must flow through it only by conduction; as its thickness grows, the insulation increases. Third, snow forming on the surface of the ice provides extra insulation, since snow is a better insulator than ice.

50 The amounts of heat given off by a gram of water as it condenses from gas to liquid and also from liquid to solid are called latent heats. As heat is applied to a body of water, it both raises the temperature of the water (specific heat) and evaporates some of it (latent heat). But the heat energy that goes into evaporating water does not raise its temperature. The very high latent and specific heats of water work together to reduce temperature changes in an environment that is subject to varying input from energy.

51 Needham, *The Uniqueness of Biological Materials*, 13.

52 K. Regenauer-Lieb, D. A. Yuen, and J. Branlund, "The Initiation of Subduction: Criticality by Addition of Water?" *Science* 294 (2001): 578–580.

53 Stillinger, "Water Revisited," *Science* 209 (1980): 451.

54 Denton, *Nature's Destiny*, 40–41.

55 Denton also describes water's role in regulating body temperature in animals. Its value derives from its viscosity, thermal capacity, heat conductivity, and heat of vaporization. As we noted above, some of these physical properties are also key to the regulation of the climate. In addition, he describes several properties of water (some of which we described earlier) that work together to preserve it in a liquid state. (Denton, 42–45.)

56 J. S. Lewis, *Worlds Without End: The Exploration of Planets Known and Unknown* (Reading: Helix Books, 1998), 199.

57 Henderson, *The Fitness of the Environment*, 248.

58 Denton, *Nature's Destiny*, 115–116.

59 There is as much carbon dioxide in a liter of water as there is in a liter of air. Without the high solubility of carbon dioxide in water, creatures such as ourselves could not rid our cells of this product of oxidative metabolism. While doing so, the dissolved carbon dioxide forms a weak acid in the blood, which helps buffer and regulate acidity in organisms (Denton, 132–137). This synergy between carbon and water is also important on the planetary scale. The carbon dioxide dissolved in rainwater forms a weak acid important in the chemical weathering of exposed rocks.

60 Although we arrived at this important point independently, we should point out that it has also been made explicitly by Denton in *Nature's Destiny* and somewhat less so by Henderson in *The Fitness of the Environment*.

61 There's a more mysterious connection between carbon and one of the two components of water, oxygen. The synthesis of carbon and oxygen in stars in comparable amounts requires that the nuclear energy levels in the carbon nucleus be fine-tuned near the one percent level. This is one of many examples of "fine-tuning" in physics, which we will discuss in Chapter Ten. H. Oberhummer, A. Csoto, and H. Schlattl, "Stellar Production Rates of Carbon and Its Abundance in the Universe," *Science* 289 (2000): 88–90. See more background about

carbon production in Fred Hoyles's autobiography, *Home Is Where the Wind Blows* (Oxford: Oxford University Press, 1997).

62 R. E. Davies and R. H. Koch, "All the Observed Universe Has Contributed to Life," *Philosophical Transactions of the Royal Society of London B* 334 (1991): 391–403; V. Trimble, "Origin of the Biologically Important Elements," *Origins of Life and Evolution of the Biosphere* 27 (1997): 3–21.

63 The case of molybdenum is interesting. It is key to the operation of two enzymes, nitrogenase and nitrate reductase, which are involved in nitrogen fixation. Nitrogen in gaseous form is not useful to life, and so its conversion to a chemical form that life can use is essential. The fact that no other transition metal is used by life for nitrogen fixation, even though molybdenum is rare in some parts of the land, implies that there is no substitute for it. Francis H. C. Crick and Leslie E. Orgel cited the scarcity of molybdenum in Earth's crust as possible evidence that Earth was "seeded" by an extraterrestrial civilization (called directed panspermia, or what a skeptic might call the Little-Green-Man-in-the-Gap theory); "Directed Panspermia," *Icarus* 19 (1973): 341–346.

64 Note also that many chemical elements are found in Earth's oceans only because they have been washed away from the continents. On a planetary body with oceans but without continents, many of the life-essential elements might not be available in sufficient concentration.

65 Geothermal heat can have several sources: radioactive decay, leftover heat from planet formation, differentiation of a planet's interior, and tidal stressing. We have included all these sources of energy within the same category because their basic effect is the same—heating of a planet's surface from below.

66 B. M. Jakosky and E. L. Shock, "The Biological Potential of Mars, the Early Earth, and Europa," *Journal of Geophysical Research* 103 (1998): 19359–19364.

67 Denton notes that chemical reactions with fluorine are too energetic for the stability of organic reactions, and the product of hydrogen and fluorine is a very reactive acid.

68 Henderson, *The Fitness of the Environment*, 247–248.

69 "Extremophiles" are microorganisms that can live in extreme environmental conditions, far from the average in temperature, pressure, moisture, salinity, and acidity. See Michael Gross, *Life on the Edge: Amazing Creatures Thriving in Extreme Environments* (Cambridge: Perseus, 2001); L. J. Rothschild and R. L. Mancinelli, "Life in Extreme Environments," *Nature* 409 (2001): 1092–1101.

70 The interdependence of life makes interstellar panspermia less likely. Even if a particularly hardy bacterium survived the long journey to another planetary system, an extremely improbable event, it is unlikely that, by itself, it could "seed" another planet. Panspermia within a given planetary system is more probable, because large numbers of diverse bugs can be exchanged among planets.

71 For more information on the dependence of subsurface life on surface life, see R. A. Kerr, "Deep Life in the Slow, Slow Lane," *Science* 296 (2002): 1056–1058.

72 We put the word *simple* in quotes because even the simplest organisms on Earth are quite complex, requiring lengthy DNA molecules, numerous proteins and diverse chemicals.

73 That's between 294 and 325 degrees Kelvin (21 to 52 degrees Centigrade), with optimum growth at 309 degrees Kelvin (36 degrees Centigrade). Méndez develops an equation of state of life for prokaryotes, in which he relates the growth rate of prokaryotes to the temperature, pressure, and water concentration of their local environment. See "Planetary Habitable Zones: The Spatial Distribution of Life on Planetary Bodies," paper presented at the 32nd Lunar and Planetary Science Conference, March 12–16, 2001.

74 The diversity of vascular plants correlates with the productivity of an ecosystem. See S. M. Schneider and J. M. Rey-Benayas, "Global Patterns of Plant Diversity," *Evolution and Ecology* 8 (1994): 331–347. More recent studies confirm that biodiversity generally correlates with productivity; see R. B. Waide, et al., "The Relationship Between Productivity and Species Richness," *Annual Review of Ecology and Systematics* 30 (2000): 257–300. Productivity, in turn, depends on such factors as temperature and nutrient availability; see A. P. Allen, J. H. Brown, and J. F. Gillooly, "Global Biodiversity, Biochemical Kinetics, and the Energetic-Equivalence Rule," *Science* 297 (2002): 1545–1548.

75 Denton notes that the solubility of oxygen in water drops rapidly as temperature increases, and the metabolic demand for oxygen doubles with every eighteen-degree rise in temperature; these factors alone limit complex life to temperatures below 115 F. Denton, *Nature's Destiny*, 124.

76 Research on thermophiles indicates that these organisms resist high temperature by incorporating certain key amino acids in their protein structure. These alterations (relative to their mesophilic cousins) make the amino acid sequences much more restrictive; they also make the proteins more rigid at lower temperatures. See C. Vielle and G. J. Zeikus, "Hyperthermophilic Enzymes: Sources, Uses, and Molecular Mechanisms for Thermostability," *Microbiology and Molecular Biology Reviews* 65 (2001): 1–43.

77 James Lovelock concurs:

> There are salt-tolerant bacteria, the halophiles, that live precariously in the saline regions of Earth. These bacteria have solved the problem directly by evolving a special membrane structure that is not disrupted by salt. It works, but at a price; for these organisms cannot compete with mainstream bacteria when the salinity is normal. They are limited to their remote and rare niche, and depend upon the rest of life to keep Earth comfortable for them. They are like those eccentrics of our own society whose survival depends upon the sustenance that we can spare but who could barely survive alone.

The Ages of Gaia (New York: W. W. Norton & Company, 1988), 108.

78 The inability of Earth's microbes to transform Mars into a lush planet weighs against panspermia as an effective way of "seeding" a planet that is not just like Earth; Venus, too, was not maintained in a habitable state by Earth's life. These two natural experiments give us clues as to the close connection between life and the geology of a planet. How much more difficult to expect, then, that a single microbe could seed a distant planet orbiting another star?

79 Another example of ancient records that may enhance future survival is the field of paleotempestology—the study of ancient severe storms from their effects on lake sediments and coastal areas. While it is still a very young field of study, there is hope that enough historical hurricanes and severe storms will be found in the sediment record to discern correlations with the global climate. From this, it may be possible to prepare long-term forecasts of hurricane threats. Two recent studies are J. P. Donnelly et al., "Sedimentary Evidence of Intense Hurricane Strikes from New Jersey," *Geology* 29, no. 7 (2001): 615–618; K. B. Liu and M. L. Fearn, "Reconstruction of Prehistoric Landfall Frequencies of Catastrophic Hurricanes in Northwestern Florida from Lake Sediment Records," *Quaternary Research* 54 (2000): 238–245; A. J. Noren et al., "Millennial-Scale Storminess Variability in the Northeastern United States During the Holocene Epoch," *Nature* 419 (2002): 821–824.

80 The paleoclimate records of the past half-million years show that the times of greatest climate stability are the relatively short-lived "peak interglacials," like the one we are presently living in (though ours is already longer-lived). See J. P. Helmke et al., "Sediment-Color Record from the Northeast Atlantic Reveals Patterns of Millennial-Scale Climate Variability During the Past 500,000 Years," *Quaternary Research* 57 (2002): 49–57, and J. F. McManus, et al., "A 0.5 Million-Year Record of Millennial-Scale Climate Variability in the North Atlantic," *Science* 283 (1999): 971–974. Using otoliths (bony structures in fish for acoustics and balance) as temperature proxies, the following study determined climate conditions about 6,000 years ago: C .F. T. Andrus, "Otolith 180 Record of Mid-Holocene Sea Surface Temperatures in Peru," *Science* 295 (2002): 1508–1511. They showed that sea surface temperatures were three to four degrees Centigrade warmer and El Niño events less severe at that time.

81 For instance, they consider such factors as carbon dioxide concentration, global ice volume, atmosphere and ocean circulation, solar variations, and the Milankovitch cycles.

82 In their study "Future Climatic Changes: Are We Entering an Exceptionally Long Interglacial?" *Climatic Change* 46 (2000): 61–90, they stated, "This insolation variation . . . is really exceptional and has very few analogues in the past."

83 If the Milankovitch astronomical cycles are the main driving force of the ice ages, this could help explain the highly anomalous climate stability of the Holocene, the past twelve thousand years. But there's more going on, because Berger and Loutre predict moderate variations up to sixty thousand years into our future. Moreover, the observed range of the Milankovitch variations since the start of the Holocene has been surprisingly small. Their prediction seems well founded, but there seems to be more going on than they realize. Their explanation for the stability of the Holocene is the low amplitude of the Milankovitch cycles from five thousand years ago until sixty thousand

years into the future, compared to the past three million years or so. This seems to be an odd coincidence. But how can you use the predicted changes in the future as part of the explanation for the anomalous stability of the Holocene so far?

84 A concentration below 260 parts per million by volume seems necessary for substantial glaciation prior to 50,000 years from now. Right now it's at 370 parts per million.

85 Some have argued recently that the Atlantic Ocean circulation can be disrupted by the rapid injection of fresh water into the North Atlantic from the discharge of large numbers of icebergs as the polar ice breaks up during a rapidly warming period (as some predict for the next century), supposedly returning us to glacial conditions in short order (Alley and Bender, "Greenland Ice Cores," 181–184). This is a factor Loutre and Berger did not include in their models, but if it occurs, it's likely we can recover from it quickly with a high carbon dioxide level. In any case, support among climatologists for thermohaline circulation as a driver for the ocean currents is losing ground to wind and lunar tides as the main driving forces (as we noted in Chapter One). See C. Uhlmann et al., "Warming of the Tropical Atlantic Ocean and Slow Down of Thermohaline Circulation During the Last Deglaciation," *Nature* 402 (1999): 511–514.

Even within the framework on the thermohaline circulation model, the most recent atmosphere-ocean circulation simulations suggest that the feared shutdown of the ocean circulation from greenhouse warming will not occur. See M. Latif et al., "Tropical Stabilization of the Thermohaline Circulation in a Greenhouse Warming Simulation," *Journal of Climate* 13 (2000): 1809–1813; S. Sun and R. Bleck, "Atlantic Thermohaline Circulation and its Response to Increasing CO_2 in a Coupled Atmosphere-Ocean Model," *Geophysical Research Letters* 28 (2001): 4223–4226.

86 Even during the relatively stable Holocene, the era since the Younger Dryas, temperatures have fluctuated enough to disturb civilizations. Earth is still recovering from a centuries-long cold spell that ended in the early nineteenth century, although the rising carbon dioxide should delay the next cold spell.

87 We must interpret this data properly if we are to avoid costly and even deadly mistakes, especially since the actual effects of our activities may sometimes be counterintuitive. For instance, this evidence suggests that hindering worldwide economic growth for a slight reduction in the global temperature is unlikely to make Earth more habitable for life in general. In fact, it would probably have the opposite effect, by making it much less hospitable for civilization. We agree with Peter Huber [*Hard Green: Saving the Environment from the Environmentalists, A Conservative Manifesto* (New York: Basic Books, 1999)] that a healthy economy leads to a healthy environment. In addition, there is a huge (and rapidly growing) volume of published material demonstrating the many benefits to the biosphere and to civilization of elevated carbon dioxide levels. Most of the benefits would come from the effects of aerial fertilization on plants and trees. For a review of this topic, see C. D. Idso, "Earth's Rising Atmospheric CO_2 Concentration: Impacts on the Biosphere," *Energy & Environment* 12 (2001): 287–310. The increased plant growth resulting from higher carbon dioxide levels will greatly aid in food production in the coming century as population grows. The overall climate would also be more hospitable. For example, the growing season in cold latitudes would be lengthened; the warming is expected to occur mostly in the winter and nights; evaporation and precipitation would increase, probably resulting in more rainfall throughout the world and more snowfall in artic regions.

The moderate increase in global temperature, too, should benefit civilization. This is not just unfounded speculation. The effects of the climate fluctuations of the past millennium support the notion that warm periods are preferable for civilization. The Medieval Warm Period, also called the *Little Climate Optimum*, peaked near A.D. 1200, reaching perhaps half a degree centigrade higher temperatures than the present. Afterwards, the Little Ice Age saw its coldest temperatures in the seventeenth century. There is much historical anecdotal evidence that the Little Ice Age was much harsher on European peoples than the Medieval Warm Period. For a historical treatment, see H. Lamb, *Climate, Change, and the Modern World*, 2nd ed. (London: Routledge, 1995). There is also increasing evidence that the Medieval Warm Period and Little Ice Age were global phenomena. If the many predicted catastrophes resulting from warmer temperatures (such as the shutdown of the Atlantic circulation or the sudden release of methane from the ocean floor) did not already occur during the Medieval Warm Period (or during the even warmer "Holocene maximum," about five thousand years ago), then they are not likely to occur in the coming centuries either.

88 The notion that we are conducting a vast and dangerous experiment on the climate is often repeated in apocalyptic environmentalist literature. The quote "Through his worldwide industrial civilization, Man is unwittingly conducting a vast geophysical experiment" was an early expression of this view. It originated in R. Revelle, W. Broecker, H. Craig, C. D. Keeling, and J. Smagorinsky, "Restoring the Quality of Our Environment," *Report of the Environmental Pollution Panel, President's Science Advisory Committee* (Washington, D.C.: The White House, 1965), 126. People who hold to this view never imagined that the activities of industrialized society could benefit the world's ecosystems.

CHAPTER 3, PEERING DOWN

1 Brownlee and Ward, *Rare Earth* (New York: Copernicus, 2000), 220.

2 The complete lack of earthquakes would not prevent us from learning something of Earth's deep interior. Seismic waves generated by underground nuclear tests have been used to measure the depth of the inner core, for instance. Of course, the severe limitations imposed on the placement and number of underground nuclear detonations prevent them from competing with natural earthquakes as probes of Earth's interior.

3 A seismograph can record the arrival times and amplitudes of the two types of waves that traverse Earth's interior: compressional (or longitudinal) waves and shear (or transverse) waves. Compressional waves are basically sound waves. Shear waves cannot propagate through a liquid medium. Thus, the relative timings of compressional and shear waves and their presence or absence at a given seismograph station tell a geologist much about density variations.

4 P. G. Richards, "Earth's Inner Core—Discoveries and Conjectures," *Astronomy & Geophysics* 41:1 (2000): 20–24. For the status of the debate on this, see X. Song, "Comment on 'The Existence of an Inner Core Super-Rotation Questioned by Teleseismic Doublets,'" *Physics of the Earth and Planetary Interiors* 124 (2001): 269–273; G. Poupinet and A. Souriau, "Reply to Xiadong Song's Comment on 'The Existence of an Inner Core Super-Rotation Questioned by Teleseismic Doublets,'" *Physics of the Earth and Planetary Interiors* 124 (2001): 275–279. Even if the claim of super-rotation for the core eventually fails, the utility of this experiment in placing strong limits on it has already been demonstrated.

5 Moreover, an original magnetic field was needed to get it started, probably from the Sun. Theoretical models suggest that Earth's solid inner core began forming about a billion years ago, making it at least two billion years younger than the geomagnetic field itself. So a solid inner core doesn't appear to be essential, but it probably yields a stronger, more stable field; see P. H. Roberts and G. A. Glatzmaier, "Geodynamo Theory and Simulations," *Reviews of Modern Physics* 72, no. 4 (2000): 1081–1123. A sharp increase in the magnetic field strength 2.7 billion years ago, as evidenced by paleomagnetic evidence, has been interpreted as the onset of the nucleation of the solid inner core, but more data are needed to know for sure; see C. J. Hale, "Paleomagnetic Data Suggest Link Between the Archean-Proterozoic Boundary and Inner-Core Nucleation," *Nature* 329 (1987): 233–237. Just when the inner core forms in a terrestrial planet depends on the concentration of radioactive elements there. The higher their concentration, the earlier it can form; see S. Labrosse, et al., "The Age of the Inner Core," *Earth and Planetary Science Letters* 190 (2001): 111–123.

6 However, some polarity reversal "events" last less than fifty thousand years. See Lowrie, *Fundamentals of Geophysics*, 295–306.

7 A firsthand account of the research leading to the discovery of the sea floor remnant magnetic field is R. Mason, "Stripes on the Sea Floor," in *Plate Tectonics: An Insider's History of the Modern Theory of the Earth*, N. Oreskes, ed. (Westview Press: Boulder, 2001), 31–45.

8 David Sandwell, "Plate Tectonics: A Martian View," in *Plate Tectonics: An Insider's History of the Modern Theory of the Earth*, N. Oreskes, ed. (Westview Press: Boulder, 2001), 342.

9 Sandwell, *Plate Tectonics*, 343.

10 Ibid.

11 The typical accuracy of potassium-argon dating of lava samples limits the dating of magnetic polarity sequences to less than five million years. Older sequences are dated with Milankovitch cycles or paleontological age markers.

12 For a historical introduction to this older theory, see N. Oreskes, "From Continental Drift to Plate Tectonics," in *Plate Tectonics: An Insider's History of the Modern Theory of the Earth*, N. Oreskes, ed. (Westview Press: Boulder, 2001), 3–27.

13 An introduction to this subject as applied to the Atlantic is J. G. Sclater and C. Tapscott, "The History of the Atlantic," *Scientific American* (June 1979): 156–174.

14 As late as 1960, geology textbooks were still confident of the geosyncline theory. For example, Clark and Stearn write in their textbook, "The geosynclinal theory is one of the great unifying principles in geology. In many ways its role in geology is similar to that of the theory of evolution, which serves to integrate the many branches of the biological sciences. . . . Just as the doctrine of evolution is universally accepted among biologists, so also the geosynclinal origin of the major mountain systems is an established principle in geology." T. H. Clark and C. W. Stearn, *The Geological Evolution of North America* (New York: Ronald Press, 1960), 43.

15 The sunspot cycles are quantified by the "Ap" and "aa" indices. A description of the geomagnetic indices can be found in P. N. Mayaud, "The aa Indices: A 100-Year Series Characterizing the Magnetic Activity," *Journal of Geophysical Research* 77 (1972): 6870–6874.

16 In particular, the values of the aa and Ap indices at sunspot minimum correlate very well with the strength of the ensuing sunspot maximum, making them useful predictors of a given sunspot cycle. G. Gonzalez and K. Schatten, "Using Geomagnetic Indices to Forecast the Next Sunspot Maximum," *Solar Physics* 114 (1987): 189–192.

17 The technique of combining data from widely separated radio telescopes pointed at the same astronomical target is called very long baseline interferometry (VLBI). For a review, see O. J. Sovers, J. L. Fanselow, and C. S. Jacobs, "Astrometry and Geodesy with Radio Interferometry: Experiments, Models, and Results," *Reviews of Modern Physics* 70 (1998): 1393–1454.

18 We say "ironically" because earthquakes can cause death and destruction, which might seem to count against the correlation. Here we should note that the ability to detect and measure earthquakes allows advanced civilizations the freedom to prevent such deaths. Many large population centers are located near plate boundaries, where earthquakes are most frequent, but it doesn't have to be that way. We now know where strong earthquakes are likely to happen with a fairly high degree of confidence. They don't just happen haphazardly, though we still cannot predict their timing. Also, better building codes have greatly improved earthquake survivability of buildings in threatened regions, for those people who choose to remain there. We could virtually eliminate deaths from earthquakes if we chose to do so.

19 See Z. Mian, "Understanding Why Earth Is a Planet with Plate Tectonics," *Quarterly Journal of the Royal Astronomical Society* 34 (1993): 441–448; K. Regenauer-Lieb, D. A. Yuen, and J. Branlund, "The Initiation of Subduction: Criticality by Addition of Water?" *Science* 294 (2001): 578–580.

20 The carbon cycle is too complex for us to cover in detail here. For a general review of the carbon cycle and its interaction with the climate, see L. R. Kump, J. F. Kasting, and R. G. Crane, *The Earth System* (New Jersey: Prentice Hall, 1999), 128–151. For a recent summary of the state of the art in modeling carbon dioxide's role in Earth's ancient climate, see T. J. Thomas and R. A. Berner, "CO_2 and Climate Change," *Science* 292 (2001): 870–872.

21 J. C. G. Walker, P. Hays, and J. F. Kasting, "A Negative Feedback Mechanism for the Long-Term Stabilization of Earth's Surface Temperature," *Journal of Geophysical Research* 86 (1981): 9776–9782.

22 Biological productivity in Earth's oceans is limited primarily by the availability of certain key nutrients, especially phosphorous and iron. Most of the needed nutrients flow into the oceans from the rivers. But it seems that even this is not sufficient for all regions of the oceans. A simple oceanographic experiment conducted in 1995 caught many by surprise. (See K. H. Coale et al., "A Massive Phytoplankton Bloom Induced by an Ecosystem-Scale Iron Fertilization Experiment in the Equatorial Pacific Ocean," *Nature* 383 (1996): 495–501.) Small amounts of water-soluble iron were spread in a small patch of the eastern Pacific Ocean known to have a low biological productivity. Quickly, the populations of phytoplankton and zooplankton skyrocketed. Apparently, iron was the key limiting nutrient in these waters. Because of global wind patterns, some regions of the oceans

receive very little dust from the continents, and the eastern equatorial Pacific is one of them. The iron in the dust is in a form that the biota can use. A planet without continents would lack both rivers and dust as sources of iron.

23 M. H. Hart, "The Effect of a Planet's Size on the Evolution of Its Atmosphere," *Southwest Regional Conference on Astronomy & Astrophysics* 7 (1982): 111–126; G. Gonzalez, D. Brownlee, and P. D. Ward, "The Galactic Habitable Zone: Galactic Chemical Evolution," *Icarus* 152 (2001): 185–200.

24 Even though we do not yet know when or from where Earth received most of its water, there are several reasons a larger Earth-twin would probably start out with deeper oceans. First, it would collect a larger fraction of the water from asteroid and comet impacts, because a smaller fraction of the ejecta would be lost to space. Second, the greater gravitational focusing factor would cause more asteroids and comets to impact than a simple geometrical calculation would indicate. Third, as noted in the text, a larger terrestrial planet will have less surface relief.

25 For a discussion on the physics of mountain heights and other interesting physics problems, see V. F. Weisskopf, "Of Atoms, Mountains, and Stars: A Study in Qualitative Physics," *Science* 187 (1975): 605–612.

26 Here, the relevant scaling relation is the fact that the mass of a sphere of uncompressed matter increases with diameter raised to the third power.

27 Smaller terrestrial planets follow the uncompressed scaling relations fairly closely but begin to deviate significantly before they get as big as Earth. Earth's mean density is 5.52 grams per cubic centimeter. Mars, about half the size of Earth, has a mean density of 3.93 grams per cubic centimeter.

28 J. S. Lewis, *Worlds Without End: the Exploration of Planets Known and Unknown* (Reading: Helix Books, 1998), 64–66. If we had ignored compression, we would expect such a planet to be only about eight times more massive, with twice the surface gravity.

29 A larger terrestrial planet will have a larger impact cross section for asteroids and comets. The impact cross section increases with increasing planet mass for two reasons. First, the physical size of the planet is greater. Second, gravitational focusing is greater for more massive planets, irrespective of their sizes.

30 This is due to self-compression effects. We should add that systems that can form giant Earth siblings will also be likely to form other smaller terrestrial planets (G. W. Wetherill, "The Formation and Habitability of Extra-Solar Planets," *Icarus* 119 (1996): 219–238). Therefore, the problems noted above for a very large terrestrial planet in a given system would still not prevent it from having a terrestrial planet in the required mass range (but there are other problems associated with a system that forms massive planets—more on this in Chapter Five). Of course, the same cannot be said of a system that forms only small terrestrial planets.

31 Ground-based optical telescopes achieve such high resolution by partially compensating for distortions of images of astronomical objects caused by Earth's atmosphere. These kinds of corrections work best for high-altitude observatories.

32 Fire is not a typical exothermic reaction. It releases great heat in a slow, nonexplosive manner. This makes it very useful for a wide variety of technological applications. See Michael Denton, *Nature's Destiny* (New York: The Free Press, 1998), 122–123.

33 Certainly, some building materials can be shaped without metal tools. For example, wood can be shaped with stone axes, but many more construction options are available with bronze- and iron-age tools. Fire is also required to run a steam engine and prepare ceramics and durable bricks. The steam engine and the dynamo generator were arguably the most important prerequisites for the transition from basic to high technology. Both require the ability to shape iron.

34 In retrospect, it seems surprising that human beings learned how to purify metals from ores. About seven thousand years ago, people in the Middle East learned how to smelt copper from colorful azurite and malachite ore. Early potters probably discovered copper smelting by accident when they experimented with powdered malachite and azurite for use as pigments in their carbon-fueled and oxygen-starved kilns. This turned out to be just the environment needed for removing oxygen atoms from the copper oxides in the ore to produce pure copper metal. Soon afterward, the early smelters improved their skills at eliminating impurities by adding a "flux"—another mineral that chemically combined with impurities and formed easily removed slag.

Hematite, an iron oxide, was the most popular flux used in the Middle East for copper smelting. The very same conditions that allowed humans to purify the copper ore allowed them to produce iron metal from the hematite flux. This is probably how iron smelting was discovered. See Stephen Sass, *The Substance of Civilization: Material and Human History from the Stone Age to the Age of Silicon* (New York: Arcane Publishing, 1998). Are these just curious historical accidents on the way to high technology? Was this just an unusually fortuitous path to metalworking?

35 Both of these types of environments have been proposed as possible abodes for life. Neither of them is likely to support life. On a true waterworld, it would be very difficult to get nutrients to the surface where they are needed. On Earth, nutrients make their way from the continents into the oceans via rivers and deltas. If the ocean is not deeper than Earth's, then some methanogens might survive in mid-ocean ridges in a waterworld, but large, complex life is unlikely.

While fire is impossible on a waterworld, electronic circuits may not be. Electric eels, for example, demonstrate that electric circuits are possible in a salty aqueous environment, though it would not be an easy task.

No known organism on Earth has its complete life cycle in the atmosphere, making it less likely that such life exists in the atmospheres of gas giants. The basic problem is that solids in an atmosphere will fall downward and cease to be available to life. Fire might not be impossible in such a setting, but it would be difficult to maintain.

36 There is some recent controversy over this claim, with a few scientists arguing that oil has a nonbiological origin, and may even be a renewable resource like wood. See, for example, Tom Clarke, "Fossil Fuel Without the Fossil," *Nature* (online) (August 14, 2002): http://www.nature.com/nsu/020812/020812-3.html. Most geologists remain unconvinced.

37 The modern Industrial Revolution was actually preceded by the industrial revolution of the European Middle Ages, which took place in monasteries. Monks harnessed the power of rivers with waterwheels attached to various types of labor saving machinery. They also greatly increased the use of iron in agriculture and everyday life. Water power requires mountains to catch the rain and provide the gradient needed to make the water flow rapidly. Windmills began to appear in the twelfth century as another practical source of power. Thus, most of the precursors of the machinery of the modern Industrial Revolution were already in place in the Middle Ages; the biggest change is the source of power. For more information on the technological advances of the Middle Ages, see J. Gimpel, *The Medieval Machine: The Industrial Revolution of the Middle Ages* (New York: Holt, Rinehart & Winston, 1976).

38 Of course, with atmospheric carbon dioxide, there may be such a thing as too much of a good thing. Fortunately, Earth's various data recorders should give us the information necessary for assessing that possibility and responding accordingly.

39 George Brimhall, "The Genesis of Ores," *Scientific American* (May 1991): 84.

40 C. M. R. Fowler, *The Solid Earth: An Introduction to Global Geophysics* (Cambridge: Cambridge University Press, 1990), 225.

41 See P. Karuda, *The Origin of Chemical Elements and the Oklo Phenomenon* (New York: Springer-Verlag, 1982); G. A. Cowan, "A Natural Fission Reactor," *Scientific American* (July 1976): 36–47.

42 Methane gas could be used as a fuel, but nonbiological methane is not likely to coexist with oxygen in planetary atmospheres. Some bodies in the Solar System do have methane, but not with oxygen. Methane may have been abundant on Earth for the first couple of billion years following its formation, but it probably became scarce when the atmosphere became oxygen-rich.

43 Denton, *Nature's Destiny*, 117.

Chapter 4, Peering Up

1 Well into writing this book, we discovered that Hans Blumenberg came very close to reflecting on the correlation between habitability and measurability in this introductory sentence to his monumental volume *The Genesis of the Copernican Revolution*, translated by Robert M. Wallace (Cambridge: MIT Press, 1987), 3, originally published as *Die Genesis der kopernikanischen Welt* in 1975. Regrettably for the progress of natural philosophy

and happily for the progress of our personal publishing ambitions, it serves only to introduce his case for the importance of historical contingencies in scientific progress, and he does not develop the idea in detail.

2 Isaac Asimov, *Nightfall and Other Stories* (Greenwich, Conn.: Fawcett Publications, Inc, 1969), 11–43.

3 See Denton, *Nature's Destiny*, 50–61.

4 *Encyclopaedia Britannica*, fifteenth ed. 18 (1994): 203.

5 For a more detailed discussion on this point, see Denton, 47–70, and references cited therein. For a technical discussion, see W. H. Freeman and A. P. Lightman, "Dependence of Microphysical Phenomena on the Values of the Fundamental Constants," *Philosophical Transactions of the Royal Society of London A* 310 (1983): 323–336; J. D. Barrow and F. J. Tipler, *The Anthropic Cosmological Principle* (Oxford: Oxford University Press, 1986), 338. They derive the typical surface temperature of a star (and thus the typical photon energy) and the typical energy of biological reactions from first principles and confirm that the two are surprisingly close. They consider it a genuine coincidence.

6 George Greenstein, *The Symbiotic Universe* (New York: William Morrow, 1988), 96–97.

7 Here's a riddle: How can astronomers produce images of astrophysical sources in gamma rays, which are stopped high in the atmosphere? This sounds impossible. But thanks to a phenomenon called Cerenkov emission, simple light detectors on the ground can locate and determine the energy of individual gamma rays. When a high-energy gamma ray enters the top of the atmosphere, it hits an atom and triggers a nuclear reaction cascade. The first interaction produces a few particles, and these, in turn, interact with other atoms in the atmosphere, and so on. Eventually, all the original energy is used up and the cascade stops. Along the cascade, any charged particles traveling fast enough (faster than the speed of light in the atmosphere) will produce optical Cerenkov radiation, which can be detected on the ground. Modern Cerenkov detectors can produce images of the gamma ray sources on the sky. For a sample of this type of experiment, see A. Kawach et al., "The Optical Reflector System for the CANGAROO—II Imaging Atmospheric Cherenkov Telescope," *Astroparticle Physics* 14 (2001): 261–269. See also the VERITAS website: http://veritas.sao.arizona.edu/. So from the ground we can see the universe in three separate slices of the spectrum—the gamma ray, optical, and radio. Combined, these three make up a sizable swath of the total range. Still, the optical region is by far the most important.

8 Moreover, the peak in emission of a star's spectrum is only partly related to those processes involved with chemical energetics—that is, with reactions involving only the exchange of electrons and their interaction with photons. It depends on a star's surface temperature, which, in turn, depends on the nuclear reactions in its core and the details of how radiation traverses its interior.

9 How far we can see into the atmosphere of a star depends on its opacity—that is, to its resistance to the transmission of light. The opacity of the gas in a star increases quickly at the temperature that molecules and neutral atoms can form—in other words, where chemistry starts to become important. We can see into the Sun's atmosphere just where molecules are able to start forming. (This has implications for the appearance of the solar spectrum, which we will discuss in Chapter Seven.) Molecules dominate the visible surfaces of significantly cooler stars, while few if any molecules are present in hotter ones. Most of the photons from hot stars would be absorbed in Earth's atmosphere and dissociate molecules, while most of the photons from cool stars would be absorbed by molecules and excite low-energy vibrations in them. Thus, the appearance of the solar spectrum is intimately linked to life on Earth and the transparency of Earth's atmosphere in the optical part of the electromagnetic spectrum.

The surface temperature of a star turns out to be very sensitive to the constants of physics. We'll discuss this at length in Chapter Ten. Thus, the discussions in this chapter apply specifically to the particular set of physical constants that exist in our universe.

10 Typically, Earth's atmosphere limits angular resolution to about one second of arc from the surface. Any sources of radiation with angular sizes smaller than this, such as stars, will appear as unresolved points. Some high-quality mountaintop sites often achieve about one-third to one-half a second of arc resolution.

11 It also requires that we not have ever-present, overly bright moons polluting the sky with their light (although a less polluting bright moon is valuable, as we will discuss in Chapter Six). Finally, it requires that the sky not be so filled with nearby stars that we are blind to the distant universe (which we will discuss in Chapter Eight).

12 For very informative accounts of the folklore, art, and science of rainbows, see R. L. Lee, Jr., and A. B. Fraser, *The Rainbow Bridge: Rainbows in Art, Myth, and Science* (Pennsylvania: The Pennsylvania State University, 2001); C. B. Boyer, *The Rainbow: From Myth to Mathematics* (New York: Thomas Yoseloff, 1959).

13 The best rainbows are produced when raindrops between 0.5 and 1.0 millimeters are abundant. Such conditions are best realized in thunderstorms (with rain falling at a rate near one inch per hour). The high surface tension of liquid water maintains the spherical shape of raindrops less than two millimeters in size as they fall to the ground; larger drops become significantly distorted. A spherical shape is necessary to produce the brilliant colors in a rainbow. A liquid substance with a smaller surface tension could not produce such large spherical drops and therefore could not produce colorful rainbows. Of course, one can also see a rainbow at the bottom of a waterfall, but not without direct sunlight.

14 E. Pallé and C. J. Butler, "Sunshine Records from Ireland: Cloud Factors and Possible Links to Solar Activity and Cosmic Rays," *International Journal of Climatology* 21 (2001): 709–729. Despite the article's title, the authors did not restrict themselves to discussing just the skies over Ireland. We don't yet know how cloudiness varies over long periods of time or how it varies with composition of an atmosphere. It does vary by a few percent on decadal timescales, however.

15 See J. Lovelock, *The Ages of Gaia: A Biography of Our Living Earth* (New York: W. W. Norton & Company, 1988), and J. Lovelock, *Gaia: The Practical Science of Planetary Medicine* (Oxford: Oxford University Press, 2000).

16 While we find much to laud in the Gaia hypothesis, we do not agree with all the extrapolations that Lovelock and his supporters draw from it. We would not agree to call the Earth system a "superorganism," or elevate it to a metaphysical status, or consider people to be infectious "disease agents." We also disagree with Lovelock's characterization that the present is in a "fever" state, while the ice ages are "healthy"; the evidence we presented in Chapter Two strongly disconfirms such a view.

17 A. J. Watson and J. E. Lovelock, "Biological Homeostasis of the Global Environment: The Parable of Daisyworld," *Tellus* 35B (1983): 284–289. The most recent study is by T. M. Lenton and J. E. Lovelock, "Daisyworld Revisited: Quantifying Biological Effects on Planetary Self-Regulation," *Tellus* 53B (2001): 288–305.

18 Such flexibility is most effective while both species of daisies exist. Daisyworld's stability requires that the daisies grow best over a specific temperature range; this is reasonable, since real daisies can't grow in arctic or desert climates. Any other organisms on the planet that require similar temperatures will benefit from this regulation.

19 R. A. Berner, "Paleozoic Atmospheric CO2: Importance of Solar Radiation and Plant Evolution," *Science* 26 (1993): 68–70.

20 R. J. Charlson, J. E. Lovelock, M. O. Andrea, and S. G. Warren, "Oceanic Phytoplankton, Atmospheric Sulfur, Cloud Albedo, and Climate," *Nature* 326 (1987): 655–661. The most recent paper is by R. J. Charlson, et al., "Reshaping the Theory of Cloud Formation," *Science* 293 (2001): 2025–2026. If Charlson and collaborators are correct in their assessment of the importance of the cloud feedback, then the warming predicted for the next century may not materialize.

21 Recent studies of dimethyl sulfide (DMS) production by phytoplankton lend strong empirical support to Lovelock's theory. See G. P. Ayers and R. W. Gillett, "DMS and Its Oxidation Products in the Remote Marine Atmosphere: Implications for Climate and Atmospheric Chemistry," *Journal of Sea Research* 43 (2000): 275–286.

22 We can't yet say how cloudy Earth was during the last glacial period. Determining how clear the atmosphere was in the distant past is a little tricky, since there are no known proxies for cloudiness. But there are indirect clues. One could try to relate the content of aerosols in polar ice known to be important in CCN to cloudiness. But this would probably be somewhat ambiguous, since the deposition of aerosols in the ice would also depend on the wind speeds over the oceans surrounding the polar regions. Another approach would employ computer simulations of ancient climates based on the global conditions recorded in the ice cores. Such a study would be beyond our current understanding of all the processes involved in cloud formation.

23 A. A. Pavlov, L. L. Brown, and J. F. Kasting, "UV Shielding of NH_3 and O_2 by Organic Hazes in the Archean Atmosphere," *Journal of Geophysical Research* 106 (2001): 23,267–23,288; A. A. Pavlov et al., "Organic Haze in Earth's Early Atmosphere: Source of Low-13C Late Archean Kerogens?" *Geology* 29 (2001): 1003–1006. According to these studies, the organic haze played a role similar to that of ozone in the present atmosphere in shielding the ground from dangerous ultraviolet radiation.

24 One could argue that explorers of years past benefited from star navigation, but it is unlikely the survival of the human race depended on it.

25 NASA, in collaboration with the U.S. Air Force, is the main supporter of NEO research at the present time (http:/neo.jpl.nasa.gov). The LINEAR project is currently the most productive program (http:/www.ll.mit.edu/LINEAR/). NASA is committed to discovering 90 percent of NEAs greater than one km in diameter by 2008. See also "Sources of the Asteroid Threat," *Sky & Telescope* 100 (December 2000): 32–33.

26 O. B. Toon, et al., "Environmental Perturbations Caused by the Impacts of Asteroids and Comets," *Reviews of Geophysics* 35 (1997): 41–78; C. R. Chapman, "Impact Lethality and Risks in Today's World: Lessons for Interpreting Earth History," in Proceedings of the Conference on Catastrophic Events and Mass Extinctions: Impacts and Beyond, Vienna Snowbird IV Conference, *Geological Society of America Special Paper* (2001).

27 The Tunguska event was a natural experiment in other ways. Most interesting, the forest beneath the explosion recorded the pattern of the pressure waves it generated. The tree fall pattern formed a butterfly shape; each "wing" was twenty-five miles long. Such a pattern enables us to estimate the object's angle of entry and bearing, and its explosion altitude and magnitude.

28 Hale-Bopp came within Earth's orbit on its closest approach to the Sun in April 1997 (but on the opposite side of the Sun), so it could have hit us had its timing been just right. Comet Hyakutake, the brightest comet of 1996, had passed much closer to Earth (a mere 10 percent of the Earth-Sun distance) but was intrinsically smaller.

29 News item, "Yukon Meteorite Bonanza," *Sky & Telescope* 99 (June 2000): 22. This was a fortunate landing, since it's much easier to see dark meteorites against the white backdrop of ice and snow. Deserts and the ice fields of Antarctica have yielded many good-quality specimens. Meteorite hunters prefer ice fields, though, since they can easily pick out the dark space rocks against the blue-white background. What's more, ice flows against mountain ranges concentrate meteorites on the surface, while the cold temperatures and isolation from the atmosphere help to minimize their erosion. For a firsthand account of meteorite collecting in Antarctica, see B. Livermore, "Meteorites on Ice," *Astronomy* 27 (July 1999): 54–58. See also the readable personal account of Guy Consolmagno, *Brother Astronomer: Adventures of a Vatican Scientist* (New York: McGraw-Hill, 2000).

30 Observations of thousands of asteroids have revealed very few bodies occupying the so-called Kirkwood gaps, verifying the theory. A planetary body is in resonance with another when the ratios of their orbital periods are simple integer values. A pair of planetary bodies in resonance perturb each other's orbits because their closest approach occurs in the same parts of their orbits. The smaller the integers, the more significant are the perturbations. Thus, among the Kirkwood gaps 2:1 resonance yields has a greater effect than a 5:2 resonance. For an introductory review of asteroids and how they are affected by orbital resonances, see R. P. Binzel, "A New Century for Asteroids," *Sky & Telescope* 102 (July 2001): 44–51.

31 In addition to giant-planet resonances, the transfer of asteroidal material to the inner planets is facilitated by the Yarkovsky effect. This effect results from the recoil force on a spinning asteroid from the emission of thermal radiation. It can significantly change the mean orbital distance of an asteroid on timescales of about one million years. See J. Spitale and R. Greenberg, "Numerical Evaluation of the General Yarkovky Effect: Effects on Semimajor Axis," *Icarus* 149 (2001): 222–234.

32 T. Hiroi, C. Pieters, and M. Zolensky, "The Tagish Lake Meteorite: A Possible Sample from a D-Type Asteroid," *Science* 293 (2001): 2234–2236.

33 The bulk mineral types include silicon carbide, nanodiamonds, graphite, corundum, and Si_3N_4. Isotopes include carbon, nitrogen, oxygen, aluminum, silicon, and some heavier elements. For a recent, extensive review of the subject of pre-solar grains, see E. Zinner, "Stellar Nucleosynthesis and the Isotopic Composition of Presolar Grains from Primitive Meteorites," *Annual Review of Earth and Planetary Sciences* 26 (1998): 147–188; L. R. Nittler, "Presolar Stardust in Meteorites: Recent Advances and Scientific Frontiers," *Earth and Planetary Science Letters* 209 (2003): 259–273. Recent research suggests, however, that most nanodiamonds found in meteorites were formed in the early solar nebula. See Z. R. Dai et al., "Possible *in situ* Formation of Meteoritic Nanodiamonds in the Early Solar System," *Nature* 418 (2002): 157–159. However, these authors also note that a fraction of the nanodiamonds in their samples are almost certainly pre-solar. They note, "The identification of pre-solar SiC [silicon carbide] grains in meteorites is secure, not only because of their non-solar 13C/12C compositions, but also because astronomical spectral features due to SiC grains are observed in the outflows of evolved carbon-rich

stars. In contrast, unambiguous evidence of nanodiamonds in the interstellar medium has yet to be rigorously established" (158).

34 In particular, they can measure the important carbon-12 to 13 isotope ratio both in the grains and in cool star atmospheres (isotope ratios of oxygen and magnesium can also be measured in the spectra of some stars). Another method compares isotopic ratios with theoretical "stellar nucleosynthesis," which is the production of isotopes via nuclear reactions at the high temperatures and pressures found deep inside stars.

35 J. E. Chambers and G. W. Wetherill, "Planets in the Asteroid Belt," *Meteoritics & Planetary Science* 36 (2001): 381–399.

36 Carbon is only a trace element in the bulk Earth. Had it not been delivered to Earth's surface near the end of its formation by asteroids and comets, there may not have been enough carbon near the surface for life to flourish. Water was also brought to Earth's surface by asteroids. Some asteroids contain large amounts of water chemically bonded in the minerals (called hydration). Once hydrated, the minerals in asteroids can better retain their water than if the water had not been present in hydrated form. The hydration of the minerals with water requires special conditions. One possibility is that the short-lived radioactive isotopes present in the very early history of the solar nebula heated the asteroid building blocks enough for hydration to occur. Another possibility is that shock waves in the gas provided enough heating and water vapor pressure to hydrate the minerals in the dust. Jupiter may have been the source of the shock wave. See F. J. Ciesla et al., "A Nebular Origin for the Chondritic Fine-Grained Phyllosilicates," *Science* 299 (2003): 549–552.

37 The significance of Jupiter for habitability is discussed in the following review paper: J. I. Lunine, "The Occurrence of Jovian Planets and the Habitability of Planetary Systems," *Publications of the National Academy of Sciences* 98, no. 3 (2001): 809–814. The classic study on the relation between Jupiter and Earth's habitability is by G. W. Wetherill, "Possible Consequences of Absence of Jupiters in Planetary Systems," *Astrophysics and Space Science* 212 (1994): 23–32; in this paper, Wetherill examines the protection offered to Earth from cometary impact.

38 This impresses Donald Brownlee, a planetary astronomer and meteoriticist at the University of Washington:

> Common planetary systems may not have Jupiters or asteroid belts and hence these systems will be also highly deficient in meteoriticists. Without our benevolent asteroid belt, meteoritics might be a science based only on Nakhla, Shergotty, Chassigny, and perhaps Layayette and Zagmi [likely Martian meteorites] and even a few lunar meteorites if Earthlings would have ever convinced themselves such exceedingly rare rocks could actually fall from heaven. Unfortunately the most common interplanetary rocks in any Solar System are probably those from comets. Comet rocks do not appear to survive atmospheric entry to become meteorites, so an earth in an asteroid-free Solar System would be a pretty boring place for people with names like Anders, Wasserburg, Walker, Arnold, Wood, Wasson or Urey [famous meteoriticists].

D. Brownlee, "Mysteries of the Asteroid Belt," *Meteoritics & Planetary Science* 36 (2001): 328–329.

CHAPTER 5, THE PALE BLUE DOT IN RELIEF

1 As translated and published by G. E. Hunt and P. Moore in *The Planet Venus* (London: Faber & Faber, 1982), 74–76.

2 Even though the northern Martian polar cap does not preserve annual layers nearly as well as Earth's polar regions, high-resolution orbiter images of large cracks in the ice have revealed layering. The layering pattern resembles simulations of long-term obliquity and eccentricity changes of Mars over the past million years; see J. Laskar, B. Levrard, and J. F. Mustard, "Orbital Forcing of the Martian Polar Layered Deposits," *Nature* 419 (2002): 375–377. They find a mean deposition rate of 0.05 centimeters per year for the first half of the record and half that amount for the second half.

3 Laskar, et al. "Orbital Forcing," 376; J. Laskar, and P. Robutel, "The Chaotic Obliquity of the Planets," *Nature* 361 (1993): 608–612.

4 The possible Martian climatic responses to large tilt (obliquity) variations are discussed in B. M. Jakosky, B. G. Henderson, and M. T. Mellon, "Chaotic Obliquity and the Nature of the Martian Climate," *Journal of Geophysi-*

cal Research 100 (1995): 1579–1584; T. Nakamura and E. Tajika, "Evolution of the Climate System of Mars: Effects of Obliquity Change," *Lunar and Planetary Science* 33 (2002): 1057. The second study concludes that Mars cannot maintain residual ice caps when its tilt goes above forty-five degrees. Mars does have two very small moons, Deimos and Phobos. Close-up images from the Viking orbiters in the late 1970s revealed shapes and surface details very similar to those of asteroids visited by more recent probes. For this reason, planetary astronomers believe they are captured asteroids. Phobos, the larger and closer of the two moons, orbits Mars in a little under eight hours. Since it is well inside the synchronization distance (where its orbital period equals the planet's rotation period), Phobos is slowly spiraling inward from tidal interactions with Mars. It should impact the planet in about forty million years. Thus, not only are Mars's moons unable to produce total solar eclipses, but they cannot stabilize its obliquity, and one of them will commit a final insult to it in short order. Having more moons does not necessarily make a planet friendlier to scientists.

5 Of course, Mars's weak water cycle, in resembling Earth's cycle, shows that it can record historical information and perhaps could support life at least to the extent that water participates in its climate.

6 J. E. P. Connerney et al., "The Global Magnetic Field of Mars and Implications for Crustal Evolution," *Geophysical Research Letters* 28 (2001): 4015–4018.

7 David Sandwell doubts that Martian remnant fields are evidence of ancient tectonic spreading because "one cannot be this lucky twice." See his chapter, "Plate Tectonics: A Martian View," in *Plate Tectonics: An Insider's History of the Modern Theory of the Earth*, N. Oreskes, ed. (Westview Press: Boulder, 2001), 343. By "lucky," he means it's highly improbable that just the right set of circumstances necessary for plate tectonics (which we discussed in Chapter Three) would also occur on the very next planet out from Earth. The following study presents evidence that the Martian banded magnetic anomalies are due to a process similar to the one that produced the North American Cordillera: A. G. Fairen, J. Ruiz, and F. Anguita, "An Origin for the Linear Magnetic Anomalies on Mars through Accretion of Terranes: Implications for Dynamo Timing," *Icarus* 160 (2002): 220–223.

8 Mars has a greater surface area to volume ratio. A smaller planet has more surface area compared with the volume of its interior and, hence, cools faster.

9 S. M. McLennan, "Crustal Heat Production and the Thermal Evolution of Mars," *Geophysical Research Letters* 28 (2001): 4019–4022.

10 To be fair, there is evidence for some past tectonic activity on Mars, but the interpretation of the data is not universally accepted; see M. H. Acuna et al., "Global Distribution of Crustal Magnetization Discovered by the Mars Global Surveyer MAG/ER Experiment," *Science* 284 (1999): 790–793. Even if this is one day established, Mars's past tectonic activity does not appear to have been the full-blown type we see on Earth.

11 C .C. Reese and V. S. Solomatov, "Non-Newtonian Stagnant Lid Convection and Magmatic Resurfacing on Venus," *Icarus* 139 (1999): 67–80.

12 For a short review of the evolution of the climate of Venus, see M. A. Bullock and D. H. Grinspoon, "Global Climate Change on Venus," *Scientific American* (March 1999): 50–57.

13 By resonance here we mean that the main fluid cycle period is related to one of the motions of the solid body of a planet by some small integer multiple. This is analogous to the resonances in the asteroid belt caused by Jupiter. In the case of a planet in resonance with the fluid in its core, a given fluid mass element is accelerated if it is forced in the same way by the torques from the solid planet at the same spot in each circulation cycle.

14 J. Touma and J. Wisdom, "Nonlinear Core-Mantle Coupling," *Astronomical Journal* 122 (2001): 1030–1050.

15 Recent data also suggest the presence of an ocean in Ganymede, the third Galilean Moon of Jupiter, but it is covered by a thicker ice layer than Europa's ocean.

16 For a review of this topic, see C. F. Chyba and C. B. Phillips, "Europa as an Abode of Life," *Origins of Life and Evolution of the Biosphere* 32 (2002): 47–68.

17 The following study reports on biological activity of bacteria at pressures up to 16,000 times the surface pressure on Earth: A. Sharma et al., "Microbial Activity at Gigapascal Pressures," *Science* 295 (2002): 1514–1516. The biological activity at high pressure was observed to be about 1 percent of that at ambient conditions. It would be interesting to see if these results can be reproduced independently.

18 Chyba and Phillips, "Europa as an Abode of Life," 56.

19 J. Lovelock, *The Ages of Gaia: A Biography of Our Living Earth* (New York: W. W. Norton, 1988), 106. Most life on Earth is limited to salt concentrations near 0.8 molar; the salinity of seawater reaches about 0.68 molar. Molar concentration is the number of moles of a solute per liter of water.

20 W. B. McKinnon and E. L. Shock, "Ocean Karma: What Goes Around Comes Around on Europa (Or Does It?)," *Lunar and Planetary Science* 32 (2001): abstr. 2181.

21 Even with all these limitations on life in Europa, some still speculate that a niche may be available a few meters below the surface in the ice cracks.

22 A. Khan, K. Mosegaard, and K. L. Rasmussen, "A New Seismic Velocity Model for the Moon from a Monte Carlo Inversion of the Apollo Lunar Seismic Data," *Geophysical Research Letters* 27 (2000): 1591–1594.

23 For some preliminary simulations showing how Earth's surface temperature variations would change if its eccentricity were increased, see D. M. Williams and D. Pollard, "Earth-like Worlds on Eccentric Orbits: Excursions Beyond the Habitable Zone," *International Journal of Astrobiology* 1 (2002): 61–69.

24 Not even extremophiles would do well on a planet with larger temperature swings, since they can only tolerate a relatively narrow range of environmental conditions. Recall from our discussion in Chapter Two that extremophiles live in relatively stable, albeit extreme, environments, such as arctic ice, a hot spring, or a salt lake.

25 For instance, finding strong peaks at the expected frequencies in a Fourier power spectrum of a marine sediment core counts as a successful detection of the Milankovitch cycles. Fourier analysis techniques break up a signal measured over some time interval into combinations of sine and cosine functions. In effect, they convert a signal measured in time to one in terms of frequency. A plot of the square of the amplitudes of the sine and cosine terms against frequency is called a power spectrum; it shows how much power is present at each frequency. Power spectra are very useful for searching for weak signals of constant frequency in a noisy dataset; they allow us to separate noise from information.

26 Peterson, in *Newton's Clock*, gives a very readable introduction to the science of chaos in the Solar System; a more technical review of the subject is given by Lecar et al., "Chaos in the Solar System," *Annual Reviews of Astronomy & Astrophysics* 39 (2001): 581–631.

27 One need only spend a modest amount of time with a gravity-simulation computer program to see how quickly multiple planet systems grow chaotic and catastrophic. For example, see the website http://www.7stones.com/Homepage/Publisher/grav.html.

28 J. Laskar, "A Numerical Experiment on the Chaotic Behaviour of the Solar System," *Nature* 338 (1989): 237–238.

29 J. Laskar, "Large-scale Chaos in the Solar System," *Astronomy & Astrophysics* 287 (1994): L9–L12.

30 Lecar et al., "Chaos in the Solar System," 47.

31 The following study considers the stability of the primary variations in Earth's orbital parameters and how that relates to their use in paleoclimatology: A. Berger, M. F. Loutre, and J. Laskar, "Stability of the Astronomical Frequencies over the Earth's History for Paleoclimate Studies," *Science* 255 (1992): 560–566.

32 This is the limit of a purely mathematical orbital simulation. J. Laskar, "The Limits of Earth Orbital Calculations for Geological Time-Scale Use," *Philosophical Transactions of the Royal Society of London A* 357 (1999): 1735–1759.

33 For the latest news on extrasolar planet research, see *The Extrasolar Planets Encyclopaedia* website: http://www.obspm.fr/encycl/encycl.html.

34 For more on the Kepler Mission, see http://www.kepler.arc.nasa.gov.

35 The first planet found around another star was 51 Pegasi, with a four-day orbital period. M. Mayor and D. Queloz published their discovery in "A Jupiter Mass Companion to a Solar-Type Star," *Nature* 378 (1995): 355–359.

36 It is important to note that the eccentricities of the orbits most of these seven planets are still relatively uncertain. Some of the observations are too noisy to determine reliably the eccentricity for a given planet; for these cases, the eccentricities are set to zero.

37 See G. Gonzalez and A. D. Vanture, "Parent Stars of Extrasolar Planets III: ρ^1 Cancri Revisited," *Astronomy and Astrophysics* 339 (1998): L29–L32. By "metals" we mean those elements beyond hydrogen and helium. We will discuss stellar metallicity in more detail in Chapters Seven and Eight.

38 M. Noble, Z. E. Musielak, and M. Cuntz, "Orbital Stability of Terrestrial Planets Inside the Habitable Zones of Extrasolar Planetary Systems," *Astrophysical Journal* 572 (2002): 1024–1030, investigated the stability of terrestrial planets in the 47 UMa system for time spans of a few hundred years. They found that the orbit of a terrestrial planet is stable in the inner edge of the habitable zone of this star for the short duration of their simulation. But it is unlikely that the eccentricity of such a planet would remain as small as Earth's over longer periods. What's more, if the true masses of the giant planets are moderately larger than the observed minimum masses, then the system will be less likely to allow terrestrial planets in the habitable zone with circular orbits.

39 But 16 Cyg B is a member of a binary star system. The two stars have highly eccentric orbits with period around 100,000 years. The present high eccentricity of the planet around 16 Cyg B probably resulted from gravitational perturbations from 16 Cyg A; see T. Mazeh, Y. Krylolowski, and G. Rosenfeld, "The High Eccentricity of the Planet Orbiting 16 Cygni B," *Astrophysical Journal Letters* 477 (1997): L103–L106.

40 These are interaction between a planet and density waves set up in the protoplanetary disk by the planet, gravitational scattering of rocky bodies in the vicinity of a planet, or mutual gravitational interactions among two or more planets, which leaves one in a smaller orbit. Such mutual gravitational interactions tend to leave at least one planet in a highly eccentric orbit.

41 There is another consequence of giant planet migration. As a giant planet migrates inward, it not only scatters away any intervening terrestrial planets, it also clears away planetesimals. Without plenty of planetesimals, Earth-size terrestrial planets cannot form. See P. J. Armitage, "A Reduced Efficiency of Terrestrial Planet Formation Following Giant Planet Migration," *Astrophysical Journal Letters* 582 (2003): L47–L50.

42 The details of the formation of terrestrial planets turn out to be very sensitive to the configuration of the giant planets. See H. F. Levison and C. Agnor, "The Role of Giant Planets in Terrestrial Planet Formation," *Astronomical Journal* 125 (2003): 2692–2713.

43 For example, see D. M. Williams, J. F. Kasting, and R. A. Wade, "Habitable Moons around Extrasolar Giant Planets," *Nature* 385 (1997): 234–236.

44 Especially via captures into temporary orbits. Of course, not every planetary system will have the same comet flux as ours, because of different formative histories and giant planet configurations. A system that forms an Earth-size terrestrial planet, however, is almost certainly going to be accompanied by plenty of comets and asteroids and probably one or more gas giants.

45 P. M. Schenk et al., "Cometary Nuclei and Tidal Disruption: The Geologic Record of Crater Chains on Callisto and Ganymede," *Icarus* 121 (1996): 249–274.

46 An Earth size moon need not generate intrinsic magnetic field in its core. The two Galilean moons, Europa and Ganymede, have fields that are thought to be induced in water with dissolved salts in their interiors; they form conducting electrolytes that generate magnetic fields as the moon crosses the giant planet's magnetic field lines. Ganymede's magnetic field strength is only about 10 percent of Earth's, and Europa's is about twenty-five percent of Ganymede's. Thus, while such induced fields offer moderate protection against charged particle radiation, this does not seem to be as effective as a magnetic field generated in an iron core.

47 If Jupiter were orbiting closer to the Sun, say at the Earth-Sun distance, then its radiation belts would be even more intense. So, it's not quite fair to compare the state of Jupiter's radiation belts to that of a hypothetical giant planet in its host planet's habitable zone.

48 For this to be effective as a source of internal heat, other planet-sized moons must also be present in order to maintain in an eccentric state the orbit of our hypothetical habitable moon-planet. A moon in a circular orbit would not experience the variations in tidal force that are needed to heat its interior.

49 R. Greenberg et al., "Habitability of Europa's Crust: The Role of Tidal-Tectonic Processes," *Journal of Geophysical Research* 105 (2000): 17, 551–17, 562.

50 Our point here is that even Jupiter has an eccentricity about three times that of Earth's orbit. In other words, for the purpose of considering climate stability on a hypothetical habitable moon, the eccentricity should not just be as small as Jupiter's, but it should be as small as Earth's. The variation of sunlight on Earth's surface is about 7 percent over the course of a year. If Earth had an eccentricity as large as Jupiter's, the sunlight would vary by about 21 percent.

51 J. W. Barnes and D. P. O'Brien, "Stability of Satellites Around Close-in Extrasolar Giant Planets," *Astrophysical Journal* 575 (2002): 1087–1093.

52 One could argue that because the Galilean Moons exhibit a systematic decline of volatile content, from the rocky Io to the icy Callisto, then so too will Earth-size moons over a range of distances from their host planet; Ganymede, interestingly, is larger but less massive than the planet Mercury. At some intermediate distance, then, there is a chance that a moon will have just the right volatile content for life. However, there is a problem with this scenario. Most of the differences among the Galilean Moons probably resulted from differences in tidal heating, Io having experienced the most and Callisto the least. The intense volcanic activity on Io drove away most of its initial endowment of volatiles. This was possible because its surface gravity was not strong enough to retain its volatiles in a liquid or vapor state at its surface. An Earth-size moon in the same spot would suffer far less volatile loss from tidal heating than Io did.

Since the abundance of volatiles rises steeply from just inside the water condensation boundary to just outside of it, a giant planet would have to form at just the right distance from its host star and then migrate inward into the habitable zone to have moons with the right level of volatiles.

53 Perhaps such moon dwellers would get stuck on the notion that everything in the universe revolved about their giant planet host.

CHAPTER 6, OUR HELPFUL NEIGHBORS

1 Ivars Peterson, *Newton's Clock: Chaos in the Solar System* (New York: W. H. Freeman, 1993), 286.

2 We owe this insight to philosopher of science Robin Collins. Stanley Jaki makes a similar point about the Earth-Moon system in *Maybe Alone in the Universe After All* (Pinckney, MI: Real View Books, 2000), as does Peterson, 286, 293, with regard to the Solar System.

3 The change from circular to eccentric planetary orbits was a more important paradigm shift in science than most. Kepler first tried to fit the orbit of Mars to a circle, but failed. Only after great effort did he finally convince himself that its orbit was an ellipse. He had broken with the long-held view that planets must orbit in perfect circles. The eccentricity of the orbit of Mars is 0.093; Earth's is 0.017.

4 Of course, some of the stars now known to have only one giant planet might one day be found to have other smaller planets, but many stars are probably accompanied by only one or two planets.

5 One such curiosity is Bode's Law, which was first published in 1772 by the German astronomer Johann Bode. It relates the relative spacings of the planets to a simple mathematical series. See P. Lynch, "On the Significance of the Titius-Bode Law for the Distribution of the Planets," *Monthly Notices of the Royal Astronomical Society* 341 (2003): 1174–1178.

6 Peterson, *Newton's Clock*, 293.

7 The high eccentricity of Mercury's orbit complicates matters even more. In particular, during perihelion (closest approach to the Sun) Mercury's rapid orbital motion overtakes its slower rotation, and, for a few Earth days, the Sun would appear to move backwards in the sky as seen by an observer on Mercury's surface!

8 The hierarchical mass distribution in the Solar System also has relevance to the discovery of the laws of planetary motion. Kepler's laws are actually inexact. The planets don't orbit the Sun, but rather they orbit the center of mass. Newton showed that Kepler's Third Law, P2 = K*a3, is slightly different for each planet, given their non-zero masses. In addition, the planets' mutual gravity causes their orbits to depart from simple ellipses. Both of these minor deviations from Kepler's simple laws would have been more severe had the planets been more massive or had they been closer in mass to the Sun, severe enough perhaps to prevent discovery of the Third Law. We will return to the topic of hierarchical clustering in the universe in Chapter Ten.

9 We don't have empirical evidence to support our claim that people would find the Earth platform better for discovering the laws of planetary motion than other places in the Solar System. Perhaps an enterprising planetarium operator can put together an experiment to test it.

10 J. Laskar, F. Joutel, and P. Robutel, "Stabilization of the Earth's Obliquity by the Moon," *Nature* 361 (1993): 615–617.

11 This is the argument, based on computer simulations, of D. M. Williams and D. Pollard, "Earth-Moon Interactions: Implications for Terrestrial Climate and Life," *Origin of the Earth and Moon*, R. M. Canup and K. Righter, eds. (Tucson: University of Arizona Press, 2000), 513–525.

12 A possible correction to this is the recent evidence for very small amounts of water ice in the regolith on the Moon's polar regions.

13 Currently, there are about 150 impact structures known on Earth, with three to five new ones discovered each year.

14 Crater counting from ground-based observatories is a successful endeavor because the lunar maria are almost exclusively on its near side. Unlike the crater-saturated highlands, a mare presents a relatively smooth surface on which a reliable crater density can be measured. Astronomers use crater density estimates from surfaces of different ages to reconstruct the cratering history of the Moon.

15 Herein, we are embarking on a campaign to banish the phrase "impact gardening" from planetary astronomy. Why use the more complex word "gardening" when "tilling" restricts its meaning to the part of gardening that actually occurs when meteorites till the ground, avoiding the implication that the objects are planting a rutabaga patch upon impact?

16 Even without impact tilling, the Moon would still lack the clear sedimentary layering produced in water media on Earth. For a comparison of sedimentation processes on Earth and the Moon, see A. Basu, and E. Molinaroli, "Sediments of the Moon and Earth as End-Members for Comparative Planetology," *Earth, Moon, and Planets* 85–86 (2001): 25–43.

17 J. Armstrong, L. Wells, and G. Gonzalez, "Rummaging through Earth's Attic for Remains of Ancient Life," *Icarus* 160 (2002): 183–196. The Moon collects fragments not only from the early Earth but also from early Mars and Venus. Since nothing remains on Venus from its early days, the Moon probably provides a unique source of information about this planet.

18 One research group has derived an average terrestrial albedo of 0.297 ± 0.005 from observations obtained at the Big Bear Solar Observatory in California. P. R. Goode, et al., "Earthshine Observations of the Earth's Reflectance," *Geophysical Research Letters* 28 (2001): 1671–1674. See also the website http://www.bbso.njit.edu/Research/EarthShine/.

19 See N. J. Woolf et al., "The Spectrum of Earthshine: A Pale Blue Dot Observed from the Ground," *Astrophysical Journal* 574 (2002): 430–433.

20 Jaki, *Maybe Alone in the Universe After All*, 20–26.

21 Consider what an eclipse of a moon would look like from another planet. For example, an eclipse of one of the Galilean Moons would look quite different to inhabitants of Jupiter's cloud tops. Since Jupiter is immensely larger than its moons, the curve of its shadow would be more difficult to discern. The same could be said of the two small moons around Mars. In general, the curve of a planet's shadow is more discernible when its moon is not much smaller than its parent planet.

22 A very informative and up-to-date introduction to distance measurement in astronomy is by S. Webb, *Measuring the Universe: The Cosmological Distance Ladder* (Chichester, UK: Praxis Publishing Ltd, 1999). A less technical and more historical treatment is by Kitty Ferguson, *Measuring the Universe: Our Historic Quest to Chart the Horizons of Space and Time* (New York: Walker, 1999).

23 Webb, *Measuring the Universe*, 28–32.

24 Hipparchus also estimated the distance to the Moon using lunar eclipses. See Webb, *Measuring the Universe*, 32–34.

25 See A. H. Batten, "Aristarchos of Samos," *Journal of the Royal Astronomical Society of Canada* 75:1 (1981): 29–35.

26 It could also be done by comparing the time intervals between first and third quarters versus third and first quarters, but this would require careful consideration of the Moon's noncircular orbit.

27 Such a precise measurement is theoretically just possible with the naked eye, but in practice the irregular surface of the Moon does not permit it. Telescopic measurements, especially when aided with photography, should make it a relatively easy task.

28 See J. I. Lunine, "The Occurrence of Jovian Planets and the Habitability of Planetary Systems," *Publications of the National Academy of Sciences* 98, no. 3 (2001): 809–814. The classic study on the relation between Jupiter and Earth's habitability is G. W. Wetherill, "Possible Consequences of Absence of Jupiters in Planetary Systems," *Astrophysics and Space Science* 212 (1994): 23–32.

29 To be precise, the effective cross section of the Moon is actually a little different from 7 percent, because Earth's gravity adds to the physical cross section. Thus, the ratio of impacts on the Moon to Earth depends on the Moon's distance, which has been increasing since it formed.

30 Another source of data on large impacts that could have occurred on Earth are the impact craters mapped on Venus by the Magellan probe in 1990. Because of its relatively young surface, only impacts spanning the last 500 to 700 million years are preserved on its surface. In addition, its thick atmosphere does not allow small objects to reach its surface, so the crater count is only a lower limit on the total number of impactors.

31 N. Sleep et al., "Annihilation of Ecosystems by Large Asteroid Impacts on the Early Earth," *Nature* 342 (1989): 139–142.

32 C. Mileikowski et al., "Natural Transfer of Viable Microbes in Space: 1. From Mars to Earth and Earth to Mars," *Icarus* 145 (2000): 391–427.

CHAPTER 7, STAR PROBES

1 Cited in *The Book of the Cosmos: Imagining the Universe from Heraclitus to Hawking*, ed. D. R. Danielson (Cambridge: Helix Books, 2000), 319.

2 As Comte said: "I still believe that the average temperature of the stars must forever elude us. . . . Such being the case, I do not think I am limiting astronomy too much by assigning to it the discovery of the laws governing the geometrical and mechanical phenomena presented by the celestial bodies, that, and that only." *The Essential Comte: Selected from Cours de Philosophie Positive*, Margaret Clarke, trans. (London: Croom Helm, 1974), 76.

3 Astronomers also employ finer divisions within each spectral type category. Thus, a G0 star is slightly hotter than the Sun, a G2 star.

4 We can describe the Sun's interior with the simple ideal gas law (along with a few other equations). The compressional (or pressure; also called p-mode) waves on the Sun's surface have a dominant period of five minutes.

5 To extract the maximum amount of useful information from a pulsating white dwarf star, astronomers must observe them continuously for several days. This is possible when telescopes spread across the globe are trained on the same star for several days in a row. It's also possible from the South Pole during its winter, but observing conditions are not so comfortable. There is a research group based in Ames, Iowa, called the Whole Earth Telescope (WET) that coordinates observations of pulsating white dwarfs from around the world.

6 This is because of the diversity of radiative transitions their electrons can partake in (atomic, vibrational, and rotational). The additional atoms present in a molecule allow more degrees of motion compared with an isolated atom. A hydrogen molecule, for instance, can rotate or vibrate. The changes in the vibrational or rotational states are quantized. Thus, the electrons around molecules preferentially interact with photons of specific wavelengths. The rotational states have the smallest energy steps, so the rotational spectral lines are bunched close together in the spectrum.

7 The basic reason for this is that one type of molecule, such as titanium oxide, produces tens of thousands of absorption lines over just a few hundred ångstroms. Therefore, since only a few titanium oxide lines need be measured to determine the sum of the titanium and oxygen abundances, all the other thousands of titanium oxide lines are superfluous and contribute only toward obscuring other lines from other species.

8 Meteorites give us the isotopic abundances of most elements, while they do not give us reliable abundances for the volatiles. The solar spectrum gives us reliable atomic abundances for most elements, including the volatiles, but relatively few isotope abundances.

9 Even if a given star is too far away to detect its tangential motion relative to other stars, we can still measure its radial velocity—its motion along our lines of sight. All that is required to determine a star's radial velocity is to measure the displacement in the wavelengths of its spectrum relative to emission lines produced in a glowing gas-filled tube on Earth's surface. The shift in the spectrum is most reliably measured if it contains sharp fea-

tures, such as absorption lines. The wavelength shifts are translated into radial velocities via application of the Doppler equation.

10 In fact, the observation of stellar Doppler shifts also tells us Earth's orbital velocity around the Sun. Like stellar parallaxes and aberration of starlight, stellar Doppler shifts are observational evidence of the heliocentric model of the Solar System.

11 That the radial motion of a star could be determined from spectra also impressed Huggins: "The foundation of this new method of research, which, transcending the wildest dreams of an earlier time, enabled the astronomer to measure off directly in terrestrial units the invisible motions in the line of sight of the heavenly bodies." Danielson, *The Book of the Cosmos,* 325.

12 For example, the raising of the tides on Earth by the Moon affects the orbital dynamics of the two bodies. Extrapolation of the Moon's recession into the distant past or future requires some knowledge of the interior structure of Earth and how it responds to tidal stresses. The same kind of knowledge would be required of stars if they were much larger or much closer together.

13 A technical discussion of the astrophysical application of pulsars' stability is S. M. Kopeikin, "Millisecond and Binary Pulsars as Nature's Frequency Standards—II. The Effects of Low-Frequency Timing Noise on Residuals and Measured Parameters," *Monthly Notices of the Royal Astronomical Society* 305 (1999): 563–590. A less technical review of pulsar observations is given by D. R. Lorimer, "Binary and Millisecond Pulsars at the New Millennium," *Living Reviews in Relativity* 4 (2001): http://www.livingreviews.org/Articles/Volume4/2001-5lorimer.

14 Bodies only a fraction of an Earth mass have been found orbiting one pulsar. See V. Wolszczan, "Confirmation of Earth-Mass Planets Orbiting the Millisecond Pulsar PSR B1257 1 12," *Science* 264 (1994): 538–542.

15 J. H. Taylor and J. M. Weisberg, "Further Experimental Tests of Relativistic Gravity Using the Binary Pulsar PSR 1913 1 16," *Astrophysical Journal* 345 (1989): 434–450. The strong gravity limit is testable wherever the gravitational field is very strong, such as the surface of a neutron star.

16 S.-S. Huang, "Occurrence of Life in the Universe," *American Scientist* 47 (1959): 397–402; J. S. Shklovsky and C. Sagan, *Intelligent Life in the Universe* (Holden-Day: San Francisco, 1966); M. H. Hart, "Habitable Zones About Main Sequence Stars," *Icarus* 37 (1979): 351–357; J. Kasting, D. P. Whitmire, and R. T. Reynolds, "Habitable Zones Around Main Sequence Stars," *Icarus* 101 (1993): 108–128; S. Franck et al., "Habitable Zone for Earth-like Planets in the Solar System," *Planetary and Space Science* 48 (2000): 1099–1105.

17 The oceans do not literally boil away. Rather, water gets into the stratosphere, where it is photodissociated by solar UV light and its hydrogen is lost to space.

18 It makes sense to include "biological processes" in this list for Earth, but if we are interested in calculating the size of the CHZ for a planet without life, we have to exclude them. Because of the important role they play in chemical weathering, higher plants help to maintain more stable surface temperatures; this, in turn, benefits other complex life forms. However, most CHZ modelers implicitly include life on their hypothetical planets by calibrating their equations with the present Earth.

19 S. A. Franck et al., "Reduction of Biosphere Life Span as a Consequence of Geodynamics," *Tellus* 52B (2000): 94–107.

20 In this and in the following few paragraphs we are restricting our discussion to the Solar System, but there is likely to be considerable variation in the properties of asteroid belts around other stars. This is one area where we encourage further research.

21 Much of what we say about asteroids in this section also applies to comets, but their distribution peaks near 1.8 AUs. So it is Mars that wanders through the most dangerous region of the Solar System as far as comets are concerned.

22 Mars also stirs some up, deflecting asteroids toward the inner planets. Which effect dominates should depend mostly on Mars's mass, but additional research is needed to quantify them. Our guess is that Mars helps more than it hurts us.

23 This is particularly so, since the energy implanted by a small body colliding with a planet increases as the square of its impact velocity.

24 A more speculative impact-related factor concerns "life-reseeding events." The fraction of ejecta that escape from a planet following a large impact depends on the velocity of the impactor. If its velocity is small enough, then it is possible that no ejecta will be lost from a planet. Thus, the fraction of impacts that do not produce ejecta is greater for increasing distance from the Sun (given the smaller encounter speeds); sterilizing impacts can still occur for the larger impactors, though, even when they are slow-moving. Without microbial life riding on impact-generated ejecta from a sterilizing impact, there is no possibility of reseeding a planet (see L. Wells, J. Armstrong, and G. Gonzalez, "Impact Reseeding During the Late Heavy Bombardment," *Icarus* 162 (2003), 38–46.) Including impact sterilization and reseeding into the mix would place additional constraints on the CHZ, though a closer study of our nearest neighbors needs to be made before we can begin to make an educated guess on the importance of these processes.

25 In addition, the total mass of the atmosphere should not be too great. Otherwise, the surface pressure would be too high to allow air breathing land creatures to survive. The amount of energy expended by an animal in breathing would be prohibitive. See Denton, *Nature's Destiny*, 127–128.

26 Earth receives more sunlight per unit area on its surface than a planet at an average location in the CHZ. On the other hand, a planet in the outer region of the CHZ will have more carbon dioxide in its atmosphere. While the additional carbon dioxide enhances biological productivity of autotrophs, it is harmful to higher life forms. In addition to distance to its host star, a planet's size also has a bearing on the total biological productivity. While the distance to the host star determines the surface energy density available to life, planet size, in part, determines the total biological productivity. The total biological productivity will be proportional to a planet's surface area, which, in turn, is proportional to the square of its size. Thus, even if Mars had been at Earth's location and somehow managed to maintain its atmosphere, its maximum total biological productivity would only be about one-quarter of Earth's. This is important, because the ability of a biosphere to support large mobile organisms depends on the total productivity at the lower levels in the food pyramid.

27 The increase in biological productivity with light intensity is not as dramatic as one may think, however, since light is not the limiting factor for regions near the equator.

28 That is, as defined in the literature. Interestingly, the substitution of water by other abundant liquids made of light molecules would result in habitable zones located farther from their host stars. This is because water has anomalously high melting and boiling temperatures given its small molecular weight. A planet with abundant ammonia or methane, for instance, would have to be farther from its host star to maintain liquid on its surface. But this results in less radiant energy available for biological productivity. So chalk up yet another optimal habitability condition to a property of water.

29 When compared with Earth, meteorites and their associated parent asteroids provide some evidence for this. In particular, the carbon and water content of the carbonaceous meteorites and their parent asteroids are much greater than that of the bulk Earth. Carbonaceous chondrite meteorites contain up to 20 percent water and 4 percent carbon, while the bulk Earth contains about 0.1 percent water and 0.05 percent carbon (Ward and Brownlee, *Rare Earth*, 46).

30 Ibid., 47.

31 It is believed that a substantial fraction of a terrestrial planet's core is made up of material that accreted early in its formation. Later on in the formation process, objects contributing to a planet's mass probably come from a larger area in the disk, because of the increasing eccentricities of their orbits.

32 See Lewis, *Worlds Without End*, 57–59.

33 For a recent discussion on how much potassium might be in Earth's core, see C. K. Gessmann and B. J. Wood, "Potassium in Earth's Core," *Earth and Planetary Science Letters* 200 (2002): 63–78. They estimate that potassium can provide at most 20 percent of the energy required to power the geodynamo in the core.

34 See K. R. Rybicki, and C. Denis, "On the Final Destiny of Earth and the Solar System," *Icarus* 151 (2001): 130–137.

35 Today astronomers can calculate the location of the CCHZ for a star of arbitrary mass. For the latest CCHZ estimates, see S. A. Franck et al., "Determination of Habitable Zones in Extrasolar Planetary Systems: Where are Gaia's Sisters?" *Journal of Geophysical Research* 105 (2000): 1651–1658. For a recent review of the subject,

see Franck et al., "Planetary Habitability: Is Earth Commonplace in the Milky Way?" *Naturwissenchaften* 88 (2001), 416–426.

36 As we noted in Chapter Four, asteroids make their way to Earth via complex orbital dynamics involving giant-planet resonances and the Yarkovsky effect. The Yarkovsky effect changes the mean distances of asteroids from the Sun on timescales of millions of years. It is most pronounced for small asteroids closest to the Sun. Yarkovsky-induced asteroid migration will become more important as the Sun continues to brighten over the next several billion years.

37 A rotationally synchronized planet does offer one significant advantage: the strong tidal forces on the planet from the host star keep its tilt stabilized without the need for a large moon.

38 The following study explores the habitability of planets around M dwarf stars with a range of possible atmospheres: M. J. Heath et al., "Habitability of Planets around Red Dwarf Stars," *Origins of Life and Evolution of the Biosphere* 29 (1999): 405–424. The basic result is that a carbon dioxide level much greater than that of Earth's atmosphere is required to maintain equable temperatures and prevent the atmosphere from freezing out. Such an atmosphere would preclude the existence of complex oxygen-breathing life. Even if some significant fraction of the atmosphere were composed of oxygen, say 10 to 20 percent, this would still not allow animal-like life. And it is not clear that a planetary atmosphere can have high carbon dioxide and oxygen levels at the same time. Earth's atmosphere went from being one dominated by methane and carbon dioxide and simple life to one dominated by oxygen and nitrogen and complex life. If an M dwarf planet made the same atmospheric transition, it would quickly cool.

39 Heath et al. argue in "Habitability of Planets around Red Dwarf Stars" that a rotationally synchronized planet with sufficiently vigorous and globally connected ocean circulation can exchange enough heat between the lit and dark sides to keep water liquid over much of its surface. A simple thought experiment will show that this still does not prevent all the water from freezing out on its dark side. All that is required are some continental land masses on the dark side, where water ice can form and keep from melting (somewhat like the Antarctica and Greenland ice masses on Earth). Once ice begins forming on land, the ocean level will begin to drop around the planet, exposing more land where stable ice can form, lowering the water level more, exposing more land, and so on until all the water freezes. In addition, as the water level drops, the oceans will become less efficient at transporting heat between the hot and cold regions, further accelerating the water freeze-out on the cold, dark side. The planet will grow hot and dry on the lit side and cold and dry on the dark side. Some liquid water may remain at the bottom of some of the basins on the dark side, warmed by heat from the planet's interior. Even some elevated land near the terminator (such as mountain ranges) can trigger a runaway planetary freeze-up as long as ice can form at altitude. Even shallow seas on the dark side would allow for ice to accumulate and the sea level to drop as long as the ice can rest on the sea bottom. In short, surface relief will tend to destabilize a rotationally synchronized planet toward the complete freezing out of its water on its dark side. And as we argued elsewhere, a planet with water and little surface relief doesn't allow for a diverse biosphere and may even be totally sterile. Another major destabilization point is reached when the dropping temperature on the dark side allows carbon dioxide to begin freezing out; this would eventually cause freeze-out of the entire atmosphere. One needs a suspiciously contrived planet, with only continents on the lit side and deep oceans everywhere else, to prevent freeze-out of its water. But as soon as continued tectonics and volcanic activity form land on its dark side, the entire planet will be set on its way toward the freeze-out of all its water. Even with an initially thicker carbon dioxide atmosphere, removing carbon dioxide via the deposition of carbonates on the ocean floor would cool the planet.

40 Flares also emit particle radiation. Charged particles so produced eventually reach Earth and hit the upper atmosphere, sometimes after spiraling along its magnetic field lines. The stronger flares produce more energetic protons, which, upon hitting atoms in the upper atmosphere, generate secondary particles. These include muons, which reach the ground and threaten life there. And more important, the secondary electrons produced in the atmosphere by high-energy ionizing radiation will be downgraded to UV photons that can reach the planet's surface.

41 This will be true for flares of all energies—in particular, the rare energetic ones. Although astronomers have been monitoring solar flares continuously for several decades, they have probably not observed the full intensity range

of flares. The strongest flares will occur less often, once per century, once per millennium, etc. If solar flares have posed any degree of threat to Earth life in the past half-billion years or so, then flares on an M dwarf star of the comparable age will pose a substantially greater threat to life on a planet in orbit about it. The strongest flare ever seen on the Sun occurred on November 4, 2003. It was the most powerful explosive event recorded in our Solar System. Luckily for us, its radiation was mostly directed away from Earth.

Of course, some simple life might survive beneath the surface of a planet around such a star, where temperature swings would be less pronounced, flare radiation would not pose a threat, and geothermal heating could maintain liquid water. But this would not be an adequate environment for complex or technological life.

42 See J. Scalo, J. C. Wheeler, and P. Williams, "Intermittent Jolts of Galactic UV Radiation: Mutagenetic Effects," in *Frontiers of Life; 12th Rencontres de Blois*, ed. L. M. Celnikier, in press; D. S. Smith, J. Scalo, and J. C. Wheeler, "Importance of Biologically Active Aurora-like Ultraviolet Emission: Stochastic Irradiation of Earth and Mars by Flares and Explosions," *Origins of Life and Evolution of the Biosphere*, in press.

43 Because an M dwarf star emits a smaller fraction of its light in the blue than does the Sun, a daytime sky of a planet in orbit around an M dwarf will be less bright. This is because blue light is scattered more strongly than red light by the molecules in a planet's atmosphere (called Rayleigh scattering). The daytime sky for a planet in the habitable zone of its M dwarf host star will be about half as bright as Earth's sky—still too bright to see stars when the star is above the horizon. This is not merely a theoretical point, since Earth itself has partial solar eclipses. Even when the Sun is half-eclipsed by the Moon, very few people notice the darker sky and no one can see stars.

44 Wetherill started with several hundred small asteroidal bodies per simulation. See G. W. Wetherill, "The Formation and Habitability of Extra-Solar Planets," *Icarus* 119 (1996): 219–238.

45 Still, as the extrasolar planet discoveries show, some gas giant planets do end up close to their host stars. We are not certain what might have happened to any terrestrial planets in these systems, but if the gas giants migrated to their present positions, it is likely they were scattered out of their systems by the migrating giants. This is why giant-planet migration is not a likely route for terrestrial planets to end up in the CCHZs of low-mass stars.

It is also worth noting that the present Doppler searches are finding relatively few giant planets around K and M dwarfs. Whether this is a real deficit or some sort of a bias cannot yet be decided. See C. Laws et al., "Parent Stars of Extrasolar Planets VII: New Abundance Analyses of 30 Systems," *Astronomical Journal* 125 (2003): 2664-2677.

46 Here we are making use of the WAP as applied to the observable and quantifiable universe. Other, much more speculative, versions of the Anthropic Principle assume the existence of some "multiverse." We will revisit this topic again in Chapter Thirteen.

47 Jay Pasachoff, *Contemporary Astronomy* (Philadelphia: Saunders College Publishing, 1989), 129. To be fair, we should note that more recently some textbook writers have begun to note that the Sun is not an average star. For example, Michael Zeilik writes in the caption to one of his figures showing the relative numbers of stars of each type in a pyramid drawing, "This pyramid, which shows the relative numbers of common stars, illustrates that the sun is not an 'average' or 'typical' star" (*Astronomy: The Evolving Universe*, Eighth Edition (New York: John Wiley & Sons, 1997), 315, Figure 14.18.)

48 See G. Gonzalez, "Is the Sun Anomalous?" *Astronomy & Geophysics* 40, no. 5 (1999): 5.25–5.29, and G. Gonzalez, "Are Stars with Planets Anomalous?" *Monthly Notices of the Royal Astronomical Society* 308 (1999): 447–458; B. Gustafsson, "Is the Sun a Sun-like Star?" *Space Science Reviews* 85 (1998): 419–428.

49 Neither the existence of stars more massive than the Sun, nor of unusual or unique stars, means that the Sun is not anomalous. For example, the star FG Sagittae has displayed behavior over the last century that can be called unique. Astronomers have learned something new about stellar evolution by studying this star. Similarly, we hope to learn something about habitability requirements by studying the ways the Sun's properties are anomalous. Also, note that by "anomalous" we mean "deviating significantly from the average." We are not claiming that the Sun exhibits weird physics or unique phenomena. Finally, applying the WAP to the Sun may not tell us all the ways its specific parameter values are required for habitability, since some of its life-essential properties might be near the average for Sun-like stars.

50 The following studies describe stellar variability for all spectral types: S. J. Adelman, "On the Variability of Stars," *Information Bulletin on Variable Stars* no. 5050 (2001): 1–2; L. Eyer, and M. Grenon, "Photometric Variability in the HR diagram," *ESA SP-402* (July 1997), 467–472. The least variable stars are found to be A0 to K0 spectral types of luminosity classes II through IV; the main sequence stars (luminosity class V) also tend to be stable, but many of them are young, active stars.

51 G. W. Lockwood, B. A. Skiff, and R. R. Radick, "The Photometric Variability of Sun-like Stars: Observations and Results, 1984–1995," *Astrophysical Journal* 485 (1997): 789–811. Some have argued that our special perspective of the Sun, viewing it near its equator from the ecliptic plane, biases our measurement of its light variations. In contrast to our view of the Sun, we observe other stars with randomly oriented rotation axes. This dependence on observer perspective results from the fact that sunspots tend to occur near the equator and the faculae have a higher contrast near the Sun's limb; if we were viewing the Sun over one of its poles, the light variation would be greater. Numerical simulations show that observer viewpoint cannot explain the low brightness variations of the Sun. See R. Knaack et al., "The Influence of an Inclined Rotation Axis on Solar Irradiance Variations," *Astronomy & Astrophysics* 376 (2001): 1080–1089.

52 As we noted in Chapter Two, there is increasing evidence that Earth's climate is strongly linked to the Sun's activity. Such evidence includes the correlation between global temperature and sunspot cycle length during the past century and the correlation between carbon-14 variations in the atmosphere and auroral activity and the "Medieval Maximum" and the "Little Ice Age." For further information on this topic, see T. I. Pulkkinen et al., "The Sun-Earth Connection in Time Scales from Years to Decades and Centuries," *Space Science Reviews* 95 (2001), 625–637. There are other anomalies concerning the sunspot cycle, but more research is needed to confirm them. See D. F. Gray, "Stars and Sun: Treasures and Threats," in *The Tenth Cambridge Workshop on Cool Stars, Stellar Systems and the Sun*, R. A. Donahue and J. A. Bookbinder, eds. (1998), 193–209.

53 Even if a habitable planet could exist at one AU from an M dwarf star, its orbital period would be longer because of the weaker gravity from the star. This follows from Newton's laws of motion.

54 A very readable review of the history of parallax measurements (successful and not) is A. W. Hirshfeld, *Parallax: The Race to Measure the Cosmos* (New York: W. H. Freeman, 2001).

55 The closest star, Proxima Centauri, at a distance of 4.2 light-years, exhibits the largest parallax angle, 0.77 seconds of arc. Bessel's target, 61 Cygni, has a parallax of only one-third of a second of arc. It is no wonder progress in stellar parallax work proceeded as slowly as it did. Earth's atmosphere blurs stellar images to sizes near one second of arc, but a few sites sometimes achieve one half of a second of arc for short periods. Bessel was able to measure a parallax angle smaller than this atmosphere-imposed limit for two reasons. First, it is possible to measure the position of the center of a star's blurred image with a precision about three to five times better than its apparent size. Second, many observations, accumulated over several years, reduce the statistical uncertainty of the final solution. Nevertheless, the early parallax measurements were difficult and just barely possible from Earth's surface.

56 Of course, our proximity gives us an unfair advantage in comparing the Sun's spectrum to other stars because of rotational Doppler broadening. When we observe the spectrum of a distant, unresolved star, the spectral lines are broadened by its rotation; those parts of the star coming toward us are shifted to the blue and those parts moving away are shifted to the red. We could circumvent the rotational broadening effect by observing the center of a star's disk, which has a motion perpendicular to the observer (the Doppler effect only refers to radial motion). Thus, regardless of the rotational velocity of the star we orbit, we can effectively remove the full rotational broadening effect. Rotation is the dominant line-broadening mechanism for O and B stars. But even a more luminous star (of the same spectral type as the Sun) would still have intrinsically broader spectral lines (owing to greater turbulent motions in its atmosphere).

57 The so-called p-mode oscillations have been detected with the Doppler method in three bright Sun-like stars, Alpha Centauri A, Beta Hydri, and Procyon. For a discussion of the detection of the oscillations in Alpha Centauri A, see F. Bouchy and F. Carrier, "P-mode Observations on ? Cen A," *Astronomy & Astrophysics* 374 (2001): L5–L8.

58 The corresponding values for an F0 star, 1.5 times the mass of the Sun, are 3.5 and 2.1 times those of the Sun. We base these estimates on the following study: H. Kjeldsen, and T. R. Bedding, "Amplitudes of Stellar Oscillations: The Implications for Asteroseismology," *Astronomy & Astrophysics* 293 (1995): 87–106.

59 The nuclear reactions involved in the synthesis of boron-8 in the Sun's core produce large numbers of high-energy neutrinos every second, many of which pass through Earth. While the chance of any particular atom interacting with a solar neutrino is extremely small, the chance that any one of many such neutrinos will interact with some atom in a large detector is vastly greater. The larger the detector, the better the chances. So, solar neutrino detectors are big. Beginning in 1965 Raymond Davis led a small team of scientists to set up an experiment to detect neutrinos produced in the Sun.

His detector consists of one hundred thousand gallons of perchloroethylene in the Homestake Gold Mine in South Dakota. This detector is based on the nuclear reaction that results when a chlorine atom in a perchloroethylene molecule absorbs a neutrino, converting it into an argon atom in the process. Davis's experiment was the only one monitoring solar neutrinos between 1968 and 1988. Even with such a large "neutrino telescope," it detected only one solar neutrino about every three days. This was not only because neutrinos interact very weakly with matter, but also because, in Davis's experiment, the chlorine reaction is produced almost exclusively by the high-energy solar boron-8 neutrinos. Davis's experiment also detected a small fraction of the intermediate-energy solar neutrinos produced by beryllium-7, though none of the abundant proton-proton reaction neutrinos. For his tireless efforts, Davis was awarded a Nobel Prize in Physics in 2002, an indication of the value the scientific community places on solar neutrino research.

For a readable review of the history of solar neutrino astronomy, see J. N. Bahcall, and R. Davis, Jr., "The Evolution of Neutrino Astronomy," *Publications of the Astronomical Society of the Pacific* 112 (2000): 429–433.

60 The production of the important boron-8 neutrinos is very temperature-sensitive. Only about 1.5 billion years ago, the Sun's boron-8 neutrino flux was about one-third its present value. See E. Brocato et al., Stars as Galactic Neutrino Sources," *Astronomy & Astrophysics* 333 (1998): 910–917. The production rate depends approximately on the 25th power of the Sun's central temperature. Had the temperature of the Sun's core been only 5 percent less, it would have produced one-quarter of the boron-8 neutrino flux. In fact, because its core temperature has been increasing since it first began fusing hydrogen in its core, the Sun's boron-8 neutrino flux has also been increasing steadily. A star's core temperature increases during its stay on the main sequence, because its mean molecular weight (and hence density) increases as protons combine to form helium nuclei. The Sun also gets bigger, because the layers outside its core expand in response to the increasing luminosity.

61 The fundamental proton-proton (p-p) reaction produces most of the energy (and most of the neutrinos) in the Sun; the temperature sensitive boron-8 side reaction only produces one in ten thousand solar neutrinos (though they are more energetic and easier to detect). This means that the luminosity of a Sun-like star is directly proportional to its p-p neutrino production rate, but the boron-8 neutrino production rate is much more sensitive to it. Thus, because the visible light energy leaving the Sun's surface scales with the p-p neutrino production rate, observers in the CHZ of another Sun-like star of different mass will be exposed to the same p-p neutrino flux. This is because both the neutrino flux and light intensity fall off with the square of the distance. This relation does not hold for stars significantly more massive than the Sun, which produce most of their energy from the CNO reactions.

62 The early success with the Davis neutrino telescope motivated physicists and governments to build larger, more ambitious projects, such as the Super-Kamiokande water and the GALLEX gallium detectors. The latter experiment is the only one sensitive to the abundant p-p neutrinos. All the neutrino telescopes are threshold detectors. That is, they tell us the total neutrino flux above a specific minimum detectable energy. Each type of neutrino telescope has a different threshold energy.

Today, the interest in solar neutrino research is motivated not only by a desire to confirm stellar models, but also by what we can learn about the neutrinos themselves. Every neutrino telescope since Davis's has detected significantly fewer solar neutrinos than predicted by the best solar models; neither improvements in neutrino telescope technology nor in solar models have removed the discrepancy. The most popular solution to this mystery is the transmutation of electron neutrinos (the type detected by nearly all neutrino telescopes) into other neutrino "flavors" (tau and muon) on their way to Earth.

To test this theory, physicists have constructed a new neutrino detector sensitive to neutrinos of all flavors, the Sudbury Neutrino Observatory (SNO). SNO consists of one thousand metric tons of highly purified heavy water (worth $300 million) in a tank six thousand meters underground. (Needless to say, this is not a very easy experiment to set up.) It's surrounded by 9,456 photomultiplier tubes that detect the light emitted when a neutrino interacts with the water. Like all previous neutrino telescopes, the Sudbury Observatory is sensitive to the boron-8 neutrinos, detecting about ten per day. Its first results were reported in June 2001. Combining their results with those of other neutrino detectors, they confirmed that neutrino transmutation is occurring. See Q. R. Ahmad et al., "Measurement of the Rate of $\nu_e + d \rightarrow p + p + e^-$ Interactions Produced by ^{8}B Solar Neutrinos at the Sudbury Neutrino Observatory," *Physical Review Letters* 87 (2001): 071301.

63 For the sake of completeness we should note that neutrino transmutations are also studied using neutrinos produced in Earth's atmosphere by cosmic rays and also by particle accelerators. See T. Kajita and T. Yoji, "Observation of Atmospheric Neutrinos," *Reviews of Modern Physics* 73 (2001): 85–118. However, these other neutrino experiments were motivated in part by the early success of the solar neutrino experiments. It was the discrepancy between the observed neutrino flux from the Sun and the expected value that led physicists to propose the neutrino transmutation model in the first place. The solar neutrino experiments have the advantage that the production rate of neutrinos is very sensitive to the state of the Sun's core, which astronomers can study with the neutrinos. On the other hand, if they have enough confidence in the solar models, then astronomers can use the Sun as a kind of natural accelerator.

CHAPTER 8, OUR GALACTIC HABITAT

1 Michael Denton, *Nature's Destiny* (New York: The Free Press, 1998), 372.

2 Quoted in Ken Croswell, in *The Alchemy of the Heavens* (Oxford: Oxford University Press, 1995), 13.

3 The story of how astronomers came to discover this will be a subject of the next chapter. To avoid confusion, we now call our home galaxy rather than the band the Milky Way, and we give these other "galaxies" different names. The faint patchy, cloud-like glow extending across the sky on clear, dark summer nights we now call the Milky Way band.

4 The specific Hubble type of the Milky Way is SAB(rs)bc according to G. de Vaucoulers, "Five Crucial Tests of the Cosmic Distance Scale Using the Galaxy as Fundamental Standard," *Proceedings of the Astronomical Society of Australia* 4, no. 4 (1982): 320–327.

5 By "zero-age," we mean very young objects (less than a few million years old) relative to age of the Milky Way galaxy. The scale height is the distance one has to go beyond the mid-plane such that the concentration of objects of a given type is reduced to 37 percent of its value in the mid-plane.

6 J. J. Matese et al., "Periodic Modulation of the Oort Cloud Comet Flux by the Adiabatically Changing Galactic Tide," *Icarus* 116 (1995), 255–268.

7 P. C. Frisch, "The Galactic Environment of the Sun," *American Scientist* 88 (2000): 53–54.

8 For a modern plot of the optical zone of avoidance on the sky, see P. A. Woudt and R. C. Kraan-Korteweg, "A Catalogue of Galaxies Behind the Southern Milky Way. II. The Crux and Great Attractor Regions (l~2890 to 3380)," *Astronomy & Astrophysics* 380 (2001), 441–459. The zone of avoidance obscures about 10 percent of the sky in the infrared region of the spectrum.

9 For a recent example of such a study, see W. C. Keel and R. E. White III, "Seeing Galaxies Through Thick and Thin. IV. The Superimposed Spiral Galaxies of NGC 3314," *Astronomical Journal* 122 (2001), 1369–1382.

10 R. Drimmel, A. Cabrera-Lavers, and M. Lopez-Corredoira, "A Three-Dimensional Galactic Extinction Model," *Astronomy & Astrophysics* 409 (2003): 205–215. Even though there is a deficit of dust at the galactic center, it has some dust due to stars that continue to form there.

11 As biologist Michael Denton correctly noted. But Denton incorrectly claims that we are in a spiral arm. The best evidence indicates we are about halfway between two major spiral arms. See J. P. Vallee, "Metastudy of the Spiral Structure of Our Home Galaxy," *Astrophysical Journal* 566 (2002): 261–266.

12 But most of the bulge would not be so enhanced. Besides the standard light fall-off with the inverse square of the distance, the amount of light extinction from interstellar dust makes the light fall with distance even more steeply. Because of the patchiness of dust between us and different parts of the bulge, however, it is difficult to estimate exactly how much brighter the entire bulge would appear from the inner disk.

13 The only exception might be the relatively rare "blue stragglers." Most globulars only have a few. They are believed to result from the merger of two G dwarfs.

14 In *Worlds Without End: The Exploration of Planets Known and Unknown* (Reading: Helix Books, 1998), 167, planetary scientist John S. Lewis describes the view from a globular cluster in the halo:

> But the naked eye observer within the globular cluster would form a rather different image of his surroundings. He would see a grand total of 120,000 stars in the sky (assuming he could see down to an apparent magnitude of 6, which is about the average limit for humans, who see a total of 3000 stars in Earth's skies). Of these, about 16,000 are brighter than magnitude 3 (compared to 152 for Earth), 3700 are brighter than magnitude 1 (versus 13 for Earth), and 300 are brighter than magnitude –1 (only Sirius appears this bright as seen from Earth). The observer would notice that all the stars brighter than –2 magnitude were yellow, and all those brighter than magnitude 1 are orange or yellow. Of all the visible stars, some 20,000 would be yellow G stars (every one in the cluster would be luminous enough and close enough to be visible), 94,000 would be orange K stars (another 56,000 would be too faint and distant to be seen), and only 6700 would be red M stars.

15 Specifically, astronomers have used radio frequencies of carbon monoxide and neutral hydrogen to map a large fraction of our galaxy's spiral structure. At radio frequencies the galaxy is largely transparent.

16 For an introduction to techniques used to map the spiral structure of the Milky Way galaxy, see D. Mihalas and J. Binney, *Galactic Astronomy: Structure and Kinematics* (New York: W. H. Freeman, 1981), 464–568. Here, we focus on radial velocities and not proper motions (which tell us something about the transverse velocity), because radial velocities can be measured to arbitrary distances, but proper motions become progressively more difficult to measure at increasing distance. Furthermore, proper motions are easiest to measure for stars, not interstellar clouds, which lack a well-defined position.

17 Edwin Hubble first classified galaxies according to their morphology, calling ellipticals early type and spirals late type. Some astronomers once believed this Hubble classification sequence formed an evolutionary sequence. Although astronomers no longer believe this, one can find enough examples of galaxies in the local universe to form a nearly continuous sequence from ellipticals to spirals. Several morphological features do change smoothly across the sequence. In particular, the central bulge becomes less prominent than the flattened disk as one goes from ellipticals to spirals. Thus, for spirals like the Milky Way galaxy, the disk dominates over the bulge, meaning that most stars in our galaxy participate in the differential rotation of the disk (as opposed to the less ordered orbits in the bulge) and are amenable to studies of the type described above.

18 The material in this section is based on the following paper: G. Gonzalez, D. Brownlee, and P. D. Ward, "The Galactic Habitable Zone: Galactic Chemical Evolution," *Icarus* 152 (2001): 185–200; a less technical summary is given in G. Gonzalez, D. Brownlee, and P. D. Ward, "The Galactic Habitable Zone," *Scientific American* (October 2001): 60–67.

19 The relative distribution of the masses of stars that form in a given cluster is called the initial mass function. There is some empirical evidence that the initial mass function does depend on metallicity in the sense that higher metallicity results in relatively more low-mass stars. But this is far from a settled issue. See P. Kroupa, "On the Variation of the Initial Mass Function," *Monthly Notices of the Royal Astronomical Society* 322 (2001): 231–246; P. Kroupa, "The Initial Mass Function of Stars: Evidence for Uniformity in Variable Systems," *Science* 295 (2002): 82–91.

20 J. Bally, "Hazards to Planet Formation," *American Astronomical Society Meeting 198* Abstract #18.01 (2001). Particularly illustrative of hazards to planet formation are the detailed images of the Orion Nebula from the Hubble Space Telescope showing stellar disks being evaporated by the ultraviolet radiation from nearby massive stars.

21 They do this in the form of carbon monoxide. Carbon monoxide is a better tracer of relatively dense gas than hydrogen; this includes the gas disks around forming planetary systems.

22 See J. B. Pollack et al., "Formation of the Giant Planets by Concurrent Accretion of Solids and Gas," *Icarus* 124 (1996): 62–85.

23 For a review of this topic, see G. Gonzalez, "Stars, Planets, and Metals," *Reviews of Modern Physics* 75 (2003): 101-120. This minimum threshold for metallicity admits of three possible explanations. First, if the core instability accretion model is correct, then building giant planets might require a certain minimum initial metallicity; we'll call this the "primordial" explanation. Second, planets might tend to migrate in a system that is initially richer in metals, thus making a giant planet easier to detect with the Doppler method; we'll call this the "migration" explanation. Third, some planets might fall into their host stars, increasing the metals in their atmospheres; we'll call this the "self-enrichment" explanation.

Planet migration may result in the accretion of planetesimals and planets by their host star, possibly even enhancing its surface metallicity. While there is still some debate on the significance of this self-enrichment effect, it's unlikely that it can fully account for the observed high average metallicity of the stars with planets. For instance, some stars with planets have very deep outer convection zones and yet have very high metallicities; it would take the accretion of an unreasonable amount of planetary material to increase the surface metallicities of these stars even by a modest amount. Nevertheless, there is some evidence for self-enrichment among stars with planets.

In A. C. Quillen, and M. Holman, "Production of Star-Grazing and Star-Impacting Planetesimals via Orbital Migration of Extrasolar Planets," *Astronomical Journal* 119 (2000): 397–402, the authors consider the migration of gas giant planets via scattering of rocky planetesimals. Since they are composed of metals, the number of planetesimals formed in a system will depend on its initial metallicity. For a short review of the various mechanisms invoked to account for hot-Jupiters, see R. Malhotra, "Migrating Planets," *Scientific American*, (September 1999): 56–63.

24 For instance, a team of astronomers led by Ronald Gilliland of the Space Telescope Science Institute recently attempted to detect light variations in the globular cluster 47 Tucanae (47 Tuc) caused by transits of giant planets orbiting its stars. If giant planets around 47 Tuc's stars occurred at the same rate as in the Sun's neighborhood, then they should have detected about seventeen planets. They found none. Perhaps stars in a very dense globular cluster tend to have close encounters with other stars, perturbing both developing and established planetary systems. Each such disturbance could ultimately destroy or eject a planet from the system. Or perhaps it's because the stars in 47 Tuc are only about 25 percent as abundant in metals as the Sun, below the most metal-poor stars known to have planets. This tends to support the primordial explanation for giant planet formation. See R. L. Gilliland et al., "A Lack of Planets in 47 Tucanae from an HST Search," *Astrophysical Journal Letters* 545 (2000): 47–51. A transit occurs when a planet crosses in front of a star as seen from Earth.

25 Studies of the dynamics of the known extrasolar planetary systems show that the high eccentricities of their planets do indeed reduce the chance that terrestrial planets would remain in circular orbits within their habitable zones. See K. Menou and S. Tabachnik, "Dynamical Habitability of Known Extrasolar Planetary Systems," *Astrophysical Journal* 583 (2003): 473–488.; M. Noble, Z. E. Musielak, and M. Cuntz, "Orbital Stability of Terrestrial Planets Inside the Habitable Zones of Extrasolar Planetary Systems," *Astrophysical Journal* 572 (2002): 1024–1030.

26 There is a possible way that planet-mass bodies can form in an initially metal-poor system. It involves the condensation of metal-enriched gas from the ejecta of a supernova. It seems that a stellar companion needs to be present prior to the supernova event. This process has been invoked to account for the planets found around the pulsar PSR B1257 + 12. See M. Konacki and A. Wolszczan, "Masses and Orbital Inclinations of Planets in the PSR B1257 + 12 System," *Astrophysical Journal* 591 (2003): L147–L150. It is not clear that these low-mass objects should even be called planets, given their very different nature from the planets in our Solar System. In any case, such systems are not habitable. Therefore, we will not consider them further.

27 At present astronomers do not have any observational evidence for terrestrial planets around other stars, so they cannot say for certain how their formation depends on initial metallicity. But if the core instability accretion model is the correct description of the formation of giant planets, then this implies that terrestrial planet formation, too, will depend strongly on metallicity. A rock-ice core must form before a giant planet can grow by accreting hydrogen and helium.

28 Theoretically, an Earth-mass planet could form in a system with a smaller initial metallicity than the Sun. However, it would have to form where water can condense, beyond the so-called "frost line." In our Solar System

that place is in the outer region of the asteroid belt. Such a planet would have a much higher abundance of volatiles, resembling in some ways the gas giants. Needless to say, such a planet would not be habitable, even if it were to migrate into the Circumstellar Habitable Zone.

29 Charles Lineweaver of the University of New South Wales, Australia has explored some of the ways that terrestrial planet formation and giant-planet migration might depend on metallicity ("An Estimate of the Age Distribution of Terrestrial Planets in the Universe: Quantifying Metallicity as a Selection Effect," *Icarus* 151 (2001): 307–313). He assumes that the mass of a terrestrial planet is linearly proportional to the metallicity of its birth cloud and that the probability of giant-planet migration rises steeply with increasing metallicity, reaching 100 percent for values three times the Sun's metallicity. While Lineweaver's calculations are still somewhat tentative, they are probably not far from the truth. Since a migrating giant planet is detrimental to any terrestrial planets in its path, metal-rich systems will be far from optimal habitats. Like us, Lineweaver concludes that a metallicity near the Sun's value may be optimal for the production of Earth-mass terrestrial planets in stable orbits.

30 At the Sun's location the metallicity gradient is about 5 percent per 1,000 light-years (the gradient is approximately linear with radius on a logarithmic abundance scale, about –0.07 "dex" per kiloparsec). Astronomers measure the metallicity gradient in the disk using spectral features in solar type stars, B stars, open cluster stars, nebulae ionized by O and B stars (H II regions), and planetary nebulae. Studies employing different metallicity indicators have converged to give pretty much the same answer only within the last five years or so. Since the H II regions are very luminous, astronomers have been able to use them to measure the gradient in about two dozen nearby galaxies to date. See, for examples, S. J. Smartt et al., "Chemical Abundances in the Inner 5 kpc of the Galactic Disk," *Astronomy & Astrophysics* 367 (2001): 86–105; L. Chen, J. L. Hou, and J. J. Wang, "On the Galactic Disk Metallicity Distribution from Open Clusters. I. New Catalogues and Abundance Gradient," *Astronomical Journal* 125 (2003): 1397–1406.

31 The overlap of the thin and thick disks also yields a declining metallicity with increasing vertical distance from mid-plane. This is the result of diffusion of stars' orbits away from the mid-plane as they age.

32 One could object that since the content of hydrogen and helium varies by several orders of magnitude among the planets in the Solar System, the variation of initial metallicity by tens of percent in the Galactic disk may get "lost in the noise." This objection is not very relevant to this discussion. The planets in the Solar System fall into two very distinct classes: terrestrial and gas giant planets. The terrestrial planets inhabit the inner region of the Solar System and have very little hydrogen and helium; they were probably formed close to their present locations. The gas giant planets that inhabit the outer Solar System are mostly hydrogen and helium. There is not a smooth gradation of hydrogen content among the planets. What counts in the present context is the formation of a terrestrial planet of the right mass in the CHZ.

33 Only a small part of this spread is intrinsic to stars formed at a given distance from the galactic center at a given time. Other bodies have deflected the orbits of many stars from their original birthplaces at different distances from the galactic center. Such gravitational deflection is called orbital dynamical diffusion. It is analogous to the deflection of the trajectory of a planetary probe by a near encounter with a planet. See G. Gonzalez, "Are Stars with Planets Anomalous?" *Monthly Notices of the Royal Astronomical Society* 308 (1999): 447–458.

Stars in the thin disk are born in nearly circular orbits. Over time, gravity from giant molecular clouds, spiral arms, and other stars increasingly distorts their orbits from this initial state. (As we discuss below, this may lead to problems for life on a planet.) Observations of young stars indicate that the intrinsic metallicity spread of stars born in the solar neighborhood is about 15 to 20 percent. E. Gaidos and G. Gonzalez, "Stellar Atmospheres of Nearby Young Solar Analogs," *New Astronomy* 7 (2002): 211–226.

34 That is, its metallicity is increasing at about 8 percent every billion years.

35 As we noted in Chapter Seven with respect to the CHZ, the concentration of sulfur and potassium-40 in Earth's core determines, in part, its geophysics. Changing the ratio of the mass of the nickel-iron core to that of the mantle should profoundly affect a planet's geophysics; such a ratio will depend on the ratio of iron to silicon plus magnesium. Iron, silicon, and magnesium condense out of cooling gas at similar temperatures. Thus, terrestrial planets forming at different locations in a given system should start with similar relative endowments of these elements. But a giant collision can alter the core-to-mantle mass ratio. For example, the large core mass of Mer-

cury is thought to be the result of a giant collision that removed part of its mantle. Similarly, the giant collision that formed the Moon added to Earth's core. Such collisions also affect the volatile inventory of a planet.

36 In this book we will ignore the finer distinctions within the two basic categories of supernovae.

37 Among the most abundant products are oxygen, silicon, magnesium, calcium, and titanium. Also produced are the so-called rapid, or "*r*-process," elements; in the few minutes following the start of the explosion, many free neutrons are liberated, and they are preferentially and quickly absorbed by heavier elements. This leads to the production of isotopes up to thorium and uranium from initial "seeds" as light as iron.

38 This is the maximum mass that can be supported by electron degeneracy pressure, which is caused by the Pauli Exclusion Principle acting on free electrons. Unlike the gas in Earth's atmosphere or in the Sun, which is supported by the motions of the constituent particles, the gas in a white dwarf is supported by a quantum effect called electron degeneracy. The Pauli Exclusion Principle describes the permitted energy states of free electrons. Unlike macroscopic bodies, elementary particles cannot just occupy any energy state. At the densities encountered in white dwarf interiors, the electron degeneracy pressure is much greater than the familiar gas pressure.

39 Unlike those from a Type II supernova, the most abundant "ashes" from a Type Ia supernova are the iron-peak elements (iron, nickel, cobalt). Lacking are the abundant free neutrons (and thus *r*-process elements) produced in a core-collapse event.

40 R. E. Davies and R. H. Koch, "All the Observed Universe Has Contributed to Life," *Philosophical Transactions of the Royal Society of London B* 334 (1991): 391–403.

41 With galactic chemical evolution models, calibrated by observing the chemical compositions of stars of various ages, astronomers reconstruct the evolution of the star formation rate (and thus the supernova and gas infall rates). Measuring oxygen-to-iron ratios in halo and disk stars with well-determined ages tells us how the Type Ia/Type II supernova ratio has changed through time. They tell us that both the supernova rate and the relative numbers of Type II to Type Ia supernovae have been declining since shortly after the Milky Way galaxy began forming. The Type II supernova rate is equal to the instantaneous production rate of the relatively rare massive stars. The Type Ia supernova rate depends on the production rate of more common intermediate mass stars and responds more slowly to changes in the star formation rate. See F. X. Timmes, S. E. Woosley, and T. A. Weaver, "Galactic Chemical Evolution: Hydrogen Through Zinc," *Astrophysical Journal Supplement* 98 (1995): 617–658; L. Potinari, C. Chiosi, and A. Bressan, "Galactic Chemical Enrichment with New Metallicity Dependent Stellar Yields," *Astronomy & Astrophysics* 334 (1998): 505–539; M. Samland, "Modeling the Evolution of Disk Galaxies. II. Yields of Massive Stars," *Astrophysical Journal* 496 (1998): 297–306.

42 We don't know how long plate tectonics will last on an Earth-size planet with 40 percent less internal heat, but it's a question that geophysicists engaged in astrobiology research are beginning to ponder. See discussion of this topic in Gonzalez, Ward, and Brownlee, "The Galactic Habitable Zone."

43 L. Becker et al., "Impact Event at the Permian-Triassic Boundary: Evidence from Extraterrestrial Noble Gases in Fullerenes," *Science* 291 (2001): 1530–1533; see also the follow-up debate, *Science* 293 (2001): 2343.

44 H. J. Habing et al., "Disappearance of Stellar Debris Disks around Main-sequence Stars after 400 Million Years," *Nature* 401 (1999): 456–457.

45 G. J. Melnick, "Discovery of Water Vapour around IRC 1 10216 as Evidence for Comets Orbiting Another Star," *Nature* 412 (2001), 160–163. The researchers argue that the amounts of water vapor they detect are unexpected for this star. The best explanation they can offer is that the recent increase in the star's luminosity is causing comets around it to vaporize.

46 See J. J. Matese and J. J. Lissauer, "Characteristics and Frequency of Weak Stellar Impulses of the Oort Cloud," *Icarus* 157 (2002): 228–240. To be more precise, the comets in the Kuiper Belt are in less eccentric orbits, and they are less affected by the Galactic tides; close passing stars are the major perturbers of the Kuiper Belt comets.

47 To get comets into the Oort cloud after they formed in the Sun's protoplanetary disk, astronomers believe, they were flung from the giant-planet region via close encounters. A planetary system that forms fewer and/or less massive giant planets will place fewer comets into its Oort cloud. Such a system is likely to form from a more metal-poor cloud. Thus, the number of comets in a star's Oort cloud should be highly sensitive to metallicity.

Other factors that affect the formation of the giant planets will also affect the formation of an Oort cloud. Comets in the Kuiper Belt are believed to form *in situ*.

48 Astrophysicists John Scalo and Craig Wheeler of the University of Texas at Austin have argued that high levels of radiation (such as those provided by extraterrestrial sources) accelerate evolution through the generation of beneficial mutations. See J. Scalo, J. C. Wheeler, and P. Williams, "Intermittent Jolts of Galactic UV Radiation: Mutageneic Effects," *Frontiers of Life; 12th Rencontres de Blois*, ed. L. M. Celnikier, in press. They base their case on simulations of evolution and supposed laboratory evidence for induced evolution in bacterial cultures. Their primary empirical supporting evidence for this claim is M. Vulic, R. E. Lenski, and M. Radman, "Mutation, Recombination, and Incipient Speciation of Bacteria in the Laboratory," *Publications of the National Academy of Sciences* 96 (1999): 7348–7351. While this study has come closer than any other in claiming to produce a new species of bacteria, it has not actually done so. The researchers produced a genetic barrier between two identical lines, which they admit is "much smaller than the barrier between such clearly distinct species as E. coli and Salmonella enterica." Thus, even with the highly artificial and extreme selective pressures applied by the scientists to rapidly reproducing bacteria in a laboratory setting, there is still no evidence that random genetic mutations yield evolutionary innovations above the species level. There is only evidence for negative mutations, which produce cripples eliminated from a population by natural selection. Add to this the fact that the early Earth experienced much higher radiation levels than it has now, via nearby supernovae, solar flares, and potassium-40 in the oceans, yet life hardly did anything interesting for about two billion years.

Even if we granted Scalo and Wheeler's premises about biological evolution, we disagree with one aspect of the application of their theory to astrophysical settings: They argue that a higher rate of nonlethal intermittent radiation events will accelerate evolution. In every astrophysical setting we can think of, however, an increase in the rate of the low intensity events is also accompanied by an increase in the rate of the high-intensity events (this is true for supernovae, gamma ray bursts, stellar flares, etc.). Thus, within their model, an accelerated rate of evolution will be accompanied by a higher probability of sterilization. Complete sterilization will be less likely for the simplest life and most likely for the most complex life. Ironically, even within this model, there is still a preferred place for life in the Galaxy, since low supernova rates will slow the evolution of life and high supernova rates will kill off complex life.

49 There are two main scenarios related to damage to the ozone layer. First, damage to the ozone layer will let through more ultraviolet radiation to the ground. This can happen if radiation is able to ionize atoms in Earth's stratosphere and generate ozone-destroying nitrogen oxides. Second, an increase in the particle radiation flux at the upper atmosphere leads to a greater flux of secondary particles at the surface. Radiation flux intense enough to make itself felt at the surface will also damage the ozone layer. Nearby supernovae may be one cause, but the topic has been controversial for several decades. For a discussion of the physical processes involved, see M. A. Ruderman, "Possible Consequences of Nearby Supernova Explosions for Atmospheric Ozone and Terrestrial Life," *Science* 184 (1974): 1079–1081. For the latest discussion on this topic, see N. Benitez, J. Maiz-Apellaniz, and M. Canelles, "Evidence for Nearby Supernova Explosions," *Physical Review Letters* 88 (2002): 081101; N. Gehrels et al., "Ozone Depletion from Nearby Supernovae," *Astrophysical Journal* 585 (2003): 1169–1176. A high neutrino flux from a nearby supernova could also pose a threat to life; see J. I. Collar, "Biological Effects of Stellar Collapse Neutrinos," *Physical Review Letters* 76 (1996): 999–1002.

50 J. N. Clarke, "Extraterrestrial Intelligence and Galactic Nuclear Activity," *Icarus* 46 (1981): 94–96.

51 J. Ellis and D. N. Schramm, "Could a Nearby Supernova Explosion Have Caused a Mass Extinction?" *Proceedings of the National Academy of Sciences* 92 (1995): 235–238; Gehrels et al.

52 By "extragalactic" we don't mean that gamma ray bursts cannot occur in the Milky Way galaxy. What we do mean is that the observed gamma ray bursts are not local phenomena restricted to the Milky Way galaxy. Many mysteries remain about them, such as the amount of "beaming" of their radiation, and the underlying physical causes, such as merging neutron stars, or the explosion of a very massive star.

53 For a discussion of the probability of a nearby gamma ray burst and its likely effects on Earth life, see A. Dar and A. De Rujula, "The Threat to Life from Eta Carina and Gamma-Ray Bursts," in *Astrophysics and Gamma Ray Physics in Space*, A. Moselli and P. Picozza, eds. (Frascati Physics Series, 2002): 513–523; J. Annis, "An Astro-

physical Explanation for the Great Silence," *Journal of the British Interplanetary Society* 52 (1999): 19–22; J. Scalo and J. C. Wheeler, "Astrophysical and Astrobiological Implications of Gamma-Ray Burst Properties," *Astrophysical Journal* 566 (2002): 723–737. Because of the very short duration of a gamma ray burst, only one-half of a planet's surface is in danger of sterilization from a nearby event.

54 According to John Lewis a dust grain traveling that fast "would have a kinetic energy equal to the explosive power of over one hundred thousand times its weight of TNT." Lewis, *Worlds Without End*, 169.

55 Such radial excursions are about two to three kiloparsecs in about one billion years. See J. R. D. Lepine, I. A. Acharova, and Yu. N. Mishurov, "Corotation, Stellar Wandering, and Fine Structure of the Galactic Abundance Pattern," *Astrophysical Journal* 589 (2003): 210–216. The best places for life would be about one kiloparsec on either side of the corotation circle.

56 We've known since the early decades of the twentieth century that we are not at the Galaxy's center, but we've known only for a few years that we are close to the corotation circle. Yu. N. Mishurov and I. A. Zenina, "Yes, the Sun Is Located Near the Corotation Circle," *Astronomy & Astrophysics* 341 (1999): 81–85; D. Fernandez, F. Figueras, and J. Torra, "Kinematics of Young Stars. II. Galactic Spiral Structure," *Astronomy & Astrophysics* 372 (2001): 833–850.

57 J. Maiz-Apellaniz, "The Origin of the Local Bubble," *Astrophysical Journal* 560 (2001): L83–L86.

58 Priscilla Frisch and her colleagues speculate that the present passage of the Sun through a very low-density region of the interstellar medium might be partly responsible for the climate stability that has allowed civilization to flourish (Frisch et al., "Galactic Environment of the Sun and Stars: Interstellar and Interplanetary Material," *Astrophysics of Life*, M. Livio, ed. (Cambridge: Cambridge University Press), in press.

59 Perhaps the small vertical motion of the Sun has the same cause as its small eccentricity in the plane of the disk. That is, perhaps the gravitational deflections that increase the eccentricity of the Sun's orbit in the plane of the Galactic disk are also likely to increase its motion perpendicular to the plane.

60 In particular, continuing studies of giant black holes in galactic nuclei, supernovae in the nearby universe, distant gamma ray bursts, comets in our Solar System and others, and stellar dynamics in our galaxy's disk will help us better understand the myriad threats to the survival of complex life.

61 This is because the average metallicity of a galaxy correlates with its luminosity. Since luminous galaxies contain more stars, galaxies at least as luminous as the Milky Way galaxy contain about 23 percent of the stars. When we compare our Galactic setting with other galaxies, it is not clear which is the more appropriate statistic. If the total luminosity (or mass) of a galaxy is also relevant (and not just the metallicity of a given star), then the smaller statistic is more appropriate. For observational evidence of the correlation between galaxy luminosity and metallicity, see D. R. Garnett, "The Luminosity-Metallicity Relation, Effective Yields, and Metal Loss in Spiral and Irregular Galaxies," *Astrophysical Journal* 581 (2002), 1019–1031.

62 Of course, since a given galaxy will have its own bell curve of stellar metallicities, a galaxy moderately less luminous than the Milky Way galaxy will contain some solar metallicity stars. They will tend to be closer to the dangerous nucleus, however. Thus, there's no sharp dividing line between galaxies with Earth-mass planets and those without. Nevertheless, a galaxy with a mean metallicity less than one-tenth solar is not likely to have any Earth-mass terrestrial planets.

63 We don't mean to imply that stars are likely to physically collide in an elliptical galaxy. Such events are expected to be very rare in galaxies. But close encounters between stars can perturb the orbits of any planets they might have.

CHAPTER 9, OUR PLACE IN COSMIC TIME

1 Michael S. Turner, "A Sober Assessment of Cosmology at the New Millennium," *Publications of the Astronomical Society of the Pacific* 113 (2001): 653.

2 For a very readable historical account of Hubble's discoveries and the events leading up to them, see R. Berendzen, R. Hart, and D. Seeley, *Man Discovers the Galaxies* (New York: Columbia University Press, 1984). See also A. Sandage, "Edwin Hubble 1889–1953," *The Journal of the Royal Astronomical Society of Canada* 83, no. 6 (1989): 351–362. In later decades Hubble moved to the newer, and larger, 200-inch telescope at Mt. Palomar, also in California.

3 Astronomer Vesto Slipher had earlier noticed various redshifts and blueshifts of astronomical objects, but it was left to Hubble to discover the connection between distance and redshift. Also, in 1917 astronomer Willem de Sitter found static solutions to Einstein's equations that exhibited redshifts. But the de Sitter models predicted a quadratic redshift dependence on distance, whereas a simple expanding universe model predicts linear redshift increase with distance. (His solutions also required a universe without matter, suggesting it might not apply to the actual universe, which contains a bit of matter.) The first published observations of Hubble in the 1920s were insufficient to exclude these models. It was not until 1931 that Hubble's observations of galaxies reached sufficient distance to exclude the de Sitter static models. For more on this interesting historical note, see L. M. Lubin and A. Sandage, "The Tolman Surface Brightness Test for the Reality of the Expansion. IV. A Measurement of the Tolman Signal and the Luminosity Evolution of Early-Type Galaxies," *Astronomical Journal* 122 (2001): 1084–1103.

4 C. F. von Weizsäcker, *The Relevance of Science* (New York: Harper & Row, 1964), 151. Weizsäcker was at one time an assistant to the famous German physicist Werner Heisenberg. Quoted in the introduction to *God and Design: The Teleological Argument and Modern Science*, Neil A. Manson, ed. (London: Routledge, 2003), 3.

5 Originally published as "Über die Krummung des Raumes," *Zeitschrift für Physik* 10 (1922): 377–386.

6 Georges Édouard LeMaître, *La Revue des Questions Scientifiques*, 4e serie 20 (1931): 391.

7 If one considers the Cepheid P-L relation as a simple linear equation with a constant and a slope, then the constant is the "zero-point." The zero point is calibrated with observations of Cepheids with known distances.

8 RR Lyrae variables have typical pulsation periods near half a day. Unlike the Cepheids, all RR Lyrae variables have the same mean luminosity, except for a weak dependence on metallicity. They are found in globular clusters and the halo of the galaxy.

9 While their light curves are very distinctive, astronomers in the first half of the twentieth century failed to notice that there were actually two classes of Cepheids: the Population I (or Classical, or Type I) and Population II (or Type II) Cepheids. The names derive mostly from stellar ages, Population I being younger than Population II. Population II stars are found in elliptical galaxies and the halos of spiral galaxies, including their globular clusters. A Classical Cepheid is about four times brighter than a Type II Cepheid of the same pulsation period. Since Classical Cepheids' typical age is around a few million years, they are only found in galaxies with continuing star formation (in the Milky Way and Andromeda Galaxies, Classical Cepheids are more common than Type II Cepheids). The discovery of this difference in the 1950s resulted in a correction to galaxy distances. Classical Cepheids and Type II Cepheids have different P-L relations.

10 Today, astronomers can determine distances to galaxies containing Classical Cepheids to within about 10 percent uncertainty within about twenty megaparsecs—about sixty-five million light-years—while Type Ia supernovae can yield distance measurements to much farther galaxies with comparable precision. The value of this uncertainty includes systematic errors, arising mostly from imperfectly known calibration equations. For example, uncertainties in the distances to the Magellanic Clouds enter into the uncertainty calculations of Cepheid-based distances. For a typical analysis of observational errors of Cepheids in nearby galaxies, see B. F. Madore et al., "The *Hubble Space Telescope* Key Project on the Extragalactic Distance Scale. XV. A Cepheid Distance to the Fornax Cluster and Its Implications," *Astrophysical Journal* 515 (1999): 29–41. The intrinsic uncertainty of the distance to a galaxy based on multicolor SN Ia light curves can be as small as 5 percent. However, since the Classical Cepheids form the basis for the SN Ia distance calibrations, absolute distances based on SN Ia are not smaller. On the other hand, some experiments compare nearby SN Ia to more distant ones, which is less sensitive to the Cepheid calibrations. For more on the usefulness of SN Ia as standard candles, see A. G. Riess, W. H. Press, and R. P. Kirshner, "A Precise Distance Indicator: Type Ia Supernova Multicolor Light-Curve Shapes," *Astrophysical Journal* 473 (1996): 88–109.

11 Astronomers also sometimes employ as standard candles stars that do not vary in brightness. As we already noted, astronomers can observe B stars, with masses comparable to Classical Cepheids, throughout much of the Milky Way galaxy. The problem with nonvariable standard candles is that astronomers need more information, photometric or spectroscopic, to be certain of an identification. If the candidate star is in a distant galaxy, it's fairly easy to confuse it with other unresolved stars; what looks like a single star may instead be several stars

too close together to resolve. A variable star's light amplitude, though, provides a built-in check on possible contaminating stars. Variable stars of a given type pulsate with a certain light amplitude. If a bright star is positioned so close to a variable star that they cannot be resolved, then its light will combine with the light from the variable star. This will cause its light amplitude to appear smaller. A variable star observed to have an anomalously small light amplitude will be immediately tagged as suspect and removed from further consideration.

12 Arthur Eddington, "The End of the World: From the Standpoint of Mathematical Physics," *Nature* 127 (1931): 447.

13 J. D. Barrow and F. J. Tipler, *The Anthropic Cosmological Principle* (Oxford: Oxford University Press, 1986), 380.

14 The measure of this spectrum is called the "CMBR angular power spectrum." Among other quantities, the CMBR power spectrum can help astronomers estimate the following: the curvature of the universe (flat, open, or closed), the matter energy density, the dark energy density (cosmological constant), the neutrino mass, the reionization redshift, the age of the universe, and the Hubble constant. For a detailed analysis of CMBR data, see M. Tegmark and M. Zaldarriaga, "Current Cosmological Constraints from a 10 Parameter Cosmic Microwave Background Analysis," *Astrophysical Journal* 544 (2000): 30–42.

15 Specifically, from the CMBR power spectrum. See Figure 7 of the important study Perlmutter et al., "Measurements of Omega and Lambda from 42 High Redshift Supernovae," *Astrophysical Journal* 517 (1999): 565–586.

16 Note that matter and radiation give us information not only about their immediate local environment but also about global parameters of the universe. For example, the light we observe from supernovae tells us not only about the supernova itself but also about its host galaxy and the bulk properties of the universe.

17 This test predicts that the surface brightness of a galaxy should decrease as $(1 + z)^4$, where z is the redshift. R. C. Tolman, "On the Estimation of Distances in a Curved Universe with a Non-Static Line Element," *Proceedings of the National Academy of Sciences* 16 (1930): 511–520.

18 L. M. Lubin and A. Sandage, "The Tolman Surface Brightness Test for the Reality of the Expansion. IV. A Measurement of the Tolman Signal and the Luminosity Evolution of Early-Type Galaxies," *Astronomical Journal* 122 (2001): 1084–1103.

19 Specifically, General Relativity, which underwrites the Big Bang model, predicted that time dilation should increase as $(1 + z)$. This means that a supernova at a redshift of 1 will take twice as long to go through its brightness variations as a nearby one. The observed trend of increasing time dilation with redshift also contradicts the tired light theory, which predicts no time dilation effect. The most recent application of this cosmological test is G. Goldhaber et al., "Timescale Stretch Parameterization of Type Ia Supernova *B*-band Light Curves," *Astrophysical Journal* 558 (2001): 359–368.

20 The following studies report on measurements of the temperature of the microwave background at large redshift values: P. Molaro, S. A. Levshakov, M. Dessauges-Zavadsky, and S. D'Odorico, "The Cosmic Microwave Background Radiation Temperature at $z_{abs} = 3.025$ toward QSO 0347-3819," *Astronomy & Astrophysics* 381 (2002): L64–L67 and R. Srianand, P. Petitjean, and C. Ledoux, "The Cosmic Microwave Background Radiation Temperature at a Redshift of 2.34," *Nature* 408 (2000), 931–935. Both studies used "fine-structure transition" absorption lines of neutral and ionized carbon seen against the light of background quasars. Commenting on the work of Srianand et al., John Bahcall remarked, "It is almost as if nature planted an abundance of clues in this anonymous cloud in order to allow some lucky researchers to infer the temperature of the CMB when the Universe was young." ("The Big Bang Is Bang On," *Nature* 408 (2000), 916.) Bahcall was impressed that the properties of the cloud studied by Srianand et al. would conspire to permit them to measure the temperature of the background and to eliminate other possible explanations of the observations.

21 See Lubin and Sandage, "The Tolman Surface Brightness Test," 1086.

22 Many gas clouds also contain significant quantities of elements other than hydrogen and helium, which impress other absorption lines on quasar spectra. This permits astronomers to determine the overall composition of the gas clouds.

23 As we noted earlier, some static models of the universe do result in redshifts, but in de Sitter's static model, the redshifts are quadratic with distance. This would result in the spectra of more distant galaxies being less well separated in redshift space.

24 This point was first made by Robert H. Dicke in a very early application of the Anthropic Principle (before it was given that name): "Dirac's Cosmology and Mach's Principle," *Nature* 192 (1961): 440–441. A number of others have echoed the idea that we in some way select our "now," including Martin Rees, John Barrow, and Paul Davies. For further discussion, see Barrow and Tipler, *The Anthropic Cosmological Principle*, 16–17.

25 C. H. Lineweaver, "An Estimate of the Age Distribution of Terrestrial Planets in the Universe: Quantifying Metallicity as a Selection Effect," *Icarus* 151 (2001), 307–313.

26 The changing carbon isotope ratio might be another time-dependent factor over billions of years. The Solar System was formed from a cloud with a carbon-12 to -13 isotope ratio near 80 (that is, one atom of carbon-13 for every 80 atoms of carbon-12). This ratio should decline as red giant stars continue to return processed matter to the interstellar medium, typically with carbon isotope ratios between six and twelve. This would not directly influence life, but it would affect the habitability of a planet via the greenhouse effect. Greenhouse gases such as carbon dioxide (CO_2) and methane (CH_4) transmit sunlight in the optical region of the spectrum but absorb outgoing infrared light emitted by the warm surface and atmosphere. The absorption is not continuous throughout the infrared but occurs in bands. The precise locations of the absorption bands of a given molecule depend on the isotopes that make it up. The number of absorption bands in the infrared region of the spectrum is increased if one of the minor isotopes in a given greenhouse gas is increased to levels comparable with the primary isotope. Thus, molecules of carbon dioxide and methane in an atmosphere with an increased level of carbon-13 would produce a more efficient greenhouse effect. An Earth twin with relatively more carbon-13 would achieve the same surface temperature with less carbon dioxide and methane. Since Earth already has a low carbon dioxide content, such a planet might not be able to support most plant life. To compensate, the inner edge of the Circumstellar Habitable Zone would need to be farther out. As we noted in Chapter Seven, several processes go into defining the Circumstellar Habitable Zone—change one and the net effect may significantly narrow its total extent. But we are not yet prepared to say to what extent habitability will be affected by a changing carbon isotope ratio.

27 See the NASA website on the Wilkinson Microwave Anisotropy Probe (WMAP) mission: http://map.gsfc.nasa.gov/index.html. Preliminary results based on the data collected during the first year of the WMAP mission were released in early February 2003. While an improvement over previous measurements, especially those of the circa 1990 COBE mission, the WMAP results were not quite the quantum leap in quality expected only a few years ago. This is because experiments from high mountaintops and Antarctica have been more successful than previously thought—yet another example of an important type of measurement that can be done from Earth's surface effectively.

28 E. H. Gudmundsson and G. Björnsson, "Dark Energy and the Observable Universe," *Astrophysical Journal* 565 (2002): 1–16. In 1999, astronomers Fred Adams and Greg Laughlin speculated about the state of the universe 10^{150} years into the future! The recent discovery of the cosmological dark energy quickly made their scenario obsolete. See F. Adams and G. Laughlin, *The Five Ages of the Universe: Inside the Physics of Eternity* (New York: The Free Press, 1999).

29 Gudmundsson and Björnsson, "Dark Energy and the Observable Universe," 11.

30 Lawrence Krauss and Glenn Starkman give the following description of objects approaching the event horizon:

> As the light travels from its source to the observer, its wavelength is stretched in proportion to the growth in a(t). Objects therefore appear exponentially redshifted as they approach the horizon. Finally, their apparent brightness declines exponentially, so that the distance of the objects inferred by an observer increases exponentially. While it strictly takes an infinite amount of time for the observer to completely lose causal contact with these receding objects, distant stars, galaxies, and all radiation backgrounds from the big bang will effectively "blink" out of existence in a finite time; as their signals redshift, the timescale for detecting these signals becomes comparable to the age of the universe.

In "Life, the Universe, and Nothing: Life and Death in an Ever-Expanding Universe," *Astrophysical Journal* 531 (2000): 23.

31 Michael Rowan Robinson's entire quote is as follows:

> It was pointed out by Hoyle that the energy densities of the microwave background, of cosmic rays, of the magnetic field in our Galaxy, and of starlight in our Galaxy are all of the same order, ~10-13 W m^{-3}. . . . But the coincidence of these three Galactic energy densities with the energy density of the microwave background, whose spectral shape and isotropy point to a cosmological origin, remains a mystery. Possibly we just have to accept this as a coincidence, as we have to accept the similar apparent sizes of sun and moon.

In *Cosmology* (Oxford: Clarendon Press, 1977), 144–145. Here energy density is simply the amount of energy per unit volume, averaged over a large volume of space.

32 A useful introduction to the various foreground contaminants of the CMBR is M. Lachieze-Rey and E. Gunzig, *The Cosmological Background Radiation* (Cambridge: Cambridge University Press, 1999).

33 Warm dust tends to have little fine structure, mostly distorting the broad, large-angle features of the CMBR. It affects observations made in the infrared more than it does the radio part of the spectrum. Since stars and distant galaxies are small, they distort the fine details in the CMBR power spectrum as well as the infrared and radio. Charged particle emissions affect mostly broad angles in the CMBR power spectrum in the radio. The "Holy Grail" of CMBR research is the determination of its angular power spectrum. It is a measure of the angular sizes on the sky that most characterize the pattern of brightness fluctuations in the CMBR.

34 Relatively nearby galaxies do not recede from us as quickly as expected from the Hubble Law, because their mutual gravitational attractions are strong enough to counteract the cosmological expansion. For example, the Andromeda Galaxy is actually *approaching* the Milky Way galaxy.

35 Krauss and Starkman, "Life, the Universe, and Nothing," 23.

36 Kalogera *et al.*, "The Coalescence Rate of Double Neutron Star Systems," *The Astrophysical Journal* 556 (2001), 340–356. The typical lifetime of binary neutron stars before they merge is several billion years, so these should continue to be available for would-be astronomers considerably longer, but eventually the implacable hand of time will reduce the frequency of these late-blooming phenomena as well.

37 Deuterium, helium-3, and lithium-7 are particularly important, but these isotopes are easily destroyed in stars. (Certain processes also produce lithium, further complicating the reconstruction of its original abundance.) The helium-4 abundance will continue to deviate from its original value as stars steadily cycle interstellar gas through their interiors.

38 Cosmologist Phillip J. E. Peebles observes, "A galaxy at redshift z~1 is expected to have an angular size on the order of one arc second, which coincidentally is comparable to the angular resolution, or seeing, permitted by our atmosphere." *Principles of Physical Cosmology* (Princeton: Princeton University Press, 1993), 326. Here, Peebles is discussing specifically the angular sizes of galaxies comparable in physical size with the Milky Way's luminous disk.

39 Of course we might survive for billions of years into the future and artificially maintain a habitable environment, but then we're talking about our technological expertise built on the finely tuned habitability and measurability foundation of our own present age.

40 Olbers was actually reframing a similar question asked by Jean Philippe Leys de Cheseaux of Lausanne in 1744, "Why is the sky dark?" Astronomers have been asking this question in some form at least since Kepler.

CHAPTER 10, A UNIVERSE FINE-TUNED FOR LIFE AND DISCOVERY

1 Paul Davies, *The Cosmic Blueprint: New Discoveries in Nature's Creative Ability to Order the Universe* (New York: Touchstone Books, 1989), 203.

2 There is some ambiguity as to what exactly "fine-tuned" means for observations of one universe. Moreover, although the word "fine-tuned" seems to imply a fine-tuner—that is, an intelligent agent to do the fine-tuning—many physicists use the word without that intended connotation. We will address some of these concerns in later chapters.

3 The classic tome on the subject is still John D. Barrow and Frank J. Tipler, *The Anthropic Cosmological Principle* (Oxford: Oxford University Press, 1986).

4 Arguably the most important discussions in the nineteenth century of the fine-tuning of the environment for life were the Bridgewater Treatises. The one that had the greatest influence on Henderson was W. Whewell, *Astronomy and General Physics, Considered with Reference to Natural Theology* (London, 1833); the ninth edition was published in 1864.

5 See Michael Denton's discussion in *Nature's Destiny* (New York: The Free Press, 1998) of the unique fitness for life of magnesium, calcium, iron, copper, and molybdenum: 195–208. See also J. J. R. Frauso da Silva and R. J. P. Williams, *The Biological Chemistry of the Elements* (Oxford: Oxford University Press, 1991).

6 We would also have to look for another metaphor, because snails would not exist.

7 Water was one of the four basic elements of Earth, according to the ancient Greeks. Even until the eighteenth century, scientists still thought water was a basic substance, an element. Water was the most important substance in the chemist's laboratory, serving as the universal medium for studying reactions. Motivated by their new fascination with electricity, Henry Cavendish, James Watt, and other scientists of the late eighteenth century passed an electric current through water (a process called electrolysis). From this they obtained two gases, one of them flammable, as the water disappeared. They also produced water by combining these two gases in a spark chamber. The French scientist Antoine Lavoisier later named these two gases hydrogen and oxygen. As a result, water lost its status as an element and chemists learned the difference between an element and a compound. Popular nature writer Rutherford Platt commented on the significance of this discovery: "As a milestone in the progress of probing nature's secrets, the splitting of a molecule of water into two H's and one O is comparable to the splitting of the atom in our time. Prior to this, scientists had no clues to the composition of water... and H_2O led to the recognition of the elements and modern chemistry." In *Water: The Wonder of Life* (Engle wood Cliffs, N.J.: Prentice-Hall, 1971), 16.

8 F. Hoyle, "On Nuclear Reactions Occurring in Very Hot Stars. I. The Synthesis of Elements from Carbon to Nickel," *Astrophysical Journal Supplement* 1 (1954): 121–146. Although we think this is an excellent example of fine-tuning, not all physicists agree. For example, Steven Weinberg recently wrote on the excited carbon-12 resonance, "But recent calculations show that, as has been long expected, without any fine-tuning of the constants of nature one would in any case expect the carbon nucleus to have a state like an unstable molecule, consisting of a helium nucleus and a beryllium nucleus, which would naturally have an energy close to the values necessary for the synthesis of carbon and heavier elements" ("A Universe with No Designer," *Cosmic Questions*, J. B. Miller, ed. (New York: The New York Academy of Sciences, 2001, 171–172); see also the note at the end of his paper. It is notable that he makes this point on the carbon resonance in a paper that is primarily an antireligious diatribe. In disputing this example of fine-tuning in the context of such a diatribe, he implies that real evidence of fine-tuning would be metaphysically unacceptable to him.

Weinberg bases his claim on the following studies: S. H. Hong, and S. J. Lee, "Alpha Chain Structures in ^{12}C," *Journal of Korean Physics* 35 (1999): 46–48, and M. Livio, D. Hollowell, and J. Truran, "The Anthropic Significance of an Excited State of ^{12}C," *Nature* 340 (1989): 281–284. Contrary to these claims, the following study concludes that the relevant excited carbon-12 resonance is a "genuine three-Alpha resonance," and thus not a beryllium-8 + alpha particle state: R. Pichler et al., "Three-alpha Structures in ^{12}C," *Nuclear Physics A* 618 (1997), 55–64; their result was independently found by D. V. Fedorov and A. S. Jensen, "The Tree-Body Continuum Coulomb Problem and the 3α Structure of ^{12}C," *Physics Letters B* 389 (1996): 631–636. See also H. Oberhummer, R. Pichler, and A. Csoto, "The Triple-Alpha Process and Its Anthropic Significance," *Nuclei in the Cosmos V*, N. Prantzos, ed., in press. Heinz Oberhummer notes, "The paper of Hong and Lee only investigates the ground states of ^{8}Be and ^{12}C and the resonant 0 + state in ^{12}C, whereas with our model we reproduce the three low-lying resonances of ^{8}Be and the 11 lowest natural-parity bound states and resonances of ^{12}C. Finally, the values obtained for the binding energies of the alpha, ^{8}Be, and ^{12}C in the model of Hong and Lee are far off their empirical values making their model very questionable" (personal communication, May 16, 2002). So, contrary to Weinberg's hasty dismissal, this example of fine-tuning, which he seems to find metaphysically troublesome, still stands.

In another recent paper, Robert Klee repeats the claim that the carbon resonance isn't all that fine-tuned. Klee depends only on the paper of Livio et al. for his assertion. See Robert Klee, "The Revenge of Pythagoras: How a Mathematical Sharp Practice Undermines the Contemporary Design Argument in Astrophysical Cosmology," *British Journal of the Philosophy of Science* 53 (2002): 331–354. His other criticisms of athropic arguments focus on the numerological obsession of the number 10^{40} by Dirac, Eddington, and Weyl in the early twentieth century and the mathematical sloppiness of some modern authors in dealing with "orders of magnitude" comparisons. None of his arguments has a bearing on the examples we give in this chapter. Like Weinberg's article, Klee's prose drips with an antitheological and almost conspiratorial tone, as if design arguments are intrinsically suspect because of "ulterior theological motives." We respond to some other aspects of Klee's argument in Chapter Thirteen.

9 Specifically, he predicted the resonance at 7.65 million electron volts (MeV) above its ground level. Hoyle's correct prediction is the first practical application of the weak anthropic principle, even though it had not yet been formulated.

10 Had the beryllium-8 nucleus been stable, helium burning would have been too violent to allow stellar evolution to proceed beyond this stage.

11 This confirmation consisted in self-consistent numerical calculations. See H. Oberhummer, A. Csoto, and H. Schlattl, "Stellar Production Rates of Carbon and Is Abundance in the Universe," *Science* 289 (2000): 88–90; A. Csoto, H. Oberhummer, and H. Schlattl, "Fine-Tuning the Basic Forces of Nature Through the Triple-Alpha Process in Red Giant Stars," *Nuclear Physics A* 688 (2001): 560–562; H. Oberhummer, A. Csoto, and H. Schlattl, "Bridging the Mass Gaps at A = 5 and A = 8 in Nucleosynthesis," *Nuclear Physics A* 689 (2001): 269–279.

Their most recent study is H. Schlattl et al., "Sensitivity of the C and O Production on the 3α Rate," *Astrophysics and Space Science*, in press. In it they include additional stages of stars' evolution, which they did not include in their *Science* paper. As a result, they find less sensitivity of the carbon and oxygen abundances to changes in the energy of the carbon resonance. Interestingly, they also find some sensitivity in the minimum initial stellar mass needed to produce a supernova. This is a rich area for additional research.

Although the ratio of carbon to oxygen in the bulk Earth is much smaller than the ratio produced by stars, what counts is the value in the crust. Carbon is much closer in abundance to oxygen in the crust. Their abundance ratio in the crust was probably set by the carbon and oxygen abundance ratio in comets and asteroids, which are closer to the solar ratio.

12 One could get past the A = 5 bottleneck with lithium-5 or helium 5 with a 10 percent increase in the strong nuclear force. In addition, the resonances in the carbon-12 nucleus are spaced such that a change in the strong force by about 10 percent would bring another resonance into play (Robin Collins, private communication). Of course, there would also have to be another resonance at just the right location in the oxygen nucleus.

13 Barrow and Tipler, 326.

14 Note, however, that the elimination of thorium and uranium (with atomic numbers of 90 and 92, respectively) would not necessarily mean that a terrestrial planet could not derive sufficient heating from radioactive decay to drive plate tectonics. They might be replaced by other radioactive elements, which are otherwise stable in our universe. But they may not have comparable half-lives. There are probably certain ranges of the strong nuclear force that don't yield any isotopes with half-lives between one billion and ten billion years.

15 Note, carbon-14 is only a moderate threat to life, because it has a short half-life. Thus, any carbon-14 produced in stars decayed long ago. It is still present in our environment only because it is continuously generated in our atmosphere in small amounts.

16 Although we won't discuss it here, life's requirements also put significant limits on the minimum half-life of the proton itself.

17 There are probably more problems in this general neighborhood. Higher relative abundances of some of the stable isotopes of the light elements might also diminish the habitability of a planet. As we noted in the previous chapter, significantly increasing the ratio of carbon-13 to carbon-12 would make carbon dioxide and methane much more efficient greenhouse gases, which, in turn, would move the Circumstellar Habitable Zone. Significant

increases in the abundances of the hydrogen and oxygen isotopes would have similar effects. The particular set of isotope ratios that went into forming the Solar System depends on the star-formation history of the Milky Way galaxy and the details of the nucleosynthesis of the elements inside stars. Thus, changing any of a number of constants and forces would significantly change the isotope ratios.

18 The two examples we discuss in this paragraph were originally presented by M. J. Rees, "Large Numbers and Ratios in Astrophysics and Cosmology," *Philosophical Transactions of the Royal Society of London A* 310 (1983): 101–112. We have extended his original discussion a bit.

19 For Sun-like stars, luminosity increases approximately as the fourth power of the mass. The mass-luminosity relation affects the habitability of Earth-like planets via planetary dynamics (via the star's mass), heating of planetary surfaces (via the star's luminosity), and the lifetime of the star in the main sequence.

20 Martin Rees, *Just Six Numbers: The Deep Forces That Shape the Universe* (New York: Basic Books, 2001), 30–31.

21 This dividing line depends on the ratio of the electromagnetic to gravitational fine-structure constants raised to the 20th power. See B. Carter, "Large Number Coincidences and the Anthropic Principle in Cosmology," in *Confrontation of Cosmological Theories with Observation*, M. S. Longair, ed. (Reidel, Dordrecht, 1974), 291–298. See also B. J. Carr and M. J. Rees, "The Anthropic Principle and the Structure of the Physical World," *Nature* 278 (1979): 611.

22 Carter gave a speculative argument for the anthropic importance of the convective dividing line: he suggested that stars contracting toward the main sequence must be convective for planets to form around them. This suggestion still lacks either empirical or theoretical support, though it may still be supported by later observations. In any event, a change in the relative strength between gravity and electromagnetism would also change the stellar mass-luminosity relation.

23 Rees, *Just Six Numbers*, 30.

24 This is because of its larger surface-area-to-volume ratio. A planet half the size of Earth would have about one-eighth its mass and one-quarter its surface area, resulting in twice its ratio of surface area to volume.

25 See Figure 6.1 of Rees, *Just Six Numbers*, 87.

26 M. Tegmark, and M. J. Rees, "Why Is the Cosmic Microwave Background Fluctuation Level 10^{-5}?" *Astrophysical Journal* 499 (1998): 526–532.

27 A. Vilenkin, "Cosmological Constant Problems and Their Solutions," based on talks at "The Dark Universe" (Space Telescope Institute) and PASCOS-2001 (hep-th/0106083). Of course, such an "anthropic explanation" only explains why we happen to be living during this most habitable period. It does nothing to explain why the universe would be configured so that it ever had a habitable period. Regrettably, cosmologists frequently blur this distinction. We discuss this in Chapter Thirteen.

28 V. Sahni and A. Starobinsky, "The Case for a Positive Cosmological Λ-Term," *International Journal of Modern Physics* D9 (2000): 373–444.

29 The first two-parameter fine-tuning plot (strong versus weak forces) was produced by P. C. W. Davies in the early 1970s. See Figure 5.7 in *The Anthropic Cosmological Principle*. For an example of the simultaneous variation in the strong and electromagnetic force strengths, see Figure 5 of M. Tegmark, "Is 'the Theory of Everything' Merely the Ultimate Ensemble Theory?" *Annals of Physics* 270 (1998): 1–51.

30 Interestingly, just two parameters, the electromagnetic force strength and the electron-to-proton mass ratio, determine all chemistry.

31 Arguably the most impressive cluster of fine-tuning occurs at the level of chemistry. In fact, chemistry appears to be "overdetermined" in the sense that there are not enough free physical parameters to determine the many chemical processes that must be just so (like those we discussed in Chapter Two). Max Tegmark notes, "Since all of chemistry is essentially determined by only two free parameters, α and β [electromagnetic force constant and electron-to-proton mass ratio], it might thus appear as though there is a solution to an overdetermined problem with more equations (inequalities) than unknowns. This could be taken as support for a religion-based category 2 TOE [Theory Of Everything], with the argument that it would be unlikely in all other TOEs" (Tegmark, 15). Tegmark artificially categorizes TOEs into type 1, "The physical world is completely mathematical," and type

2, "The physical world is not completely mathematical." The second category he considers as motivated by religious belief.

32 Another way of thinking about fine-tuning of the forces is to consider variations at more fundamental levels. For example, the fundamental forces are believed to be different manifestations of the same force. The forces become distinguishable at low energies (it is not clear, though, that gravity can be made to fit this scheme). Thus, separate variations in strengths of a couple of forces (say, weak and electromagnetic) could be described by variations in the strength of a single force at higher energies (say, the electroweak force). The amount of required fine-tuning is not reduced, but it is consolidated into fewer parameters. For an example of this kind of analysis, see V. Agrawal et al., "Anthropic Considerations in Multiple-Domain Theories and the Scale of Electroweak Symmetry Breaking," *Physical Review Letters* 80 (1998), 1822–1825. Their analysis suggests that over the many orders of magnitude in the strength of the electroweak force that they considered, life is possible only within a narrow window; they base this conclusion on arguments like the ones we discussed in the text concerning the stability of the deuteron and di-proton, among other effects. But even if all the forces could be incorporated into a grand unified theory, it would not "explain" the coincidences. Carr and Rees write, "However, even if all apparently anthropic coincidences could be explained in this way, it would still be remarkable that the relationships dictated by physical theory happened also to be those propitious to life" ("The Anthropic Principle," 612).

33 Virginia Trimble, "Cosmology: Man's Place in the Universe," *American Scientist* 65 (1977): 85.

34 J. Gribbon and M. Rees, *Cosmic Coincidences* (New York: Bantam Books, 1989), 269.

35 E. A. Milne, *Relativity, Gravitation, and World-Structure* (Oxford: Clarendon Press, 1935), 37.

36 J. D. Barrow, "Dimensionality," *Philosophical Transactions of the Royal Society of London A* 310 (1983): 337–346. See also Barrow and Tipler, *The Anthropic Cosmological Principle*, 258–276.

37 P. Ehrenfest, "In What Way Does It Become Manifest in the Fundamental Laws of Physics That Space Has Three Dimensions?" *Proceedings of the American Academy of Science* 20 (1917): 683; G. J. Whitrow, "Why Physical Space Has Three Dimensions," *British Journal for the Philosophy of Science* 6 (1955): 13–31.

38 Barrow, "Dimensionality" 341. In *The Anthropic Cosmological Principle*, Barrow and Tipler describe another value of a low-dimensional world like our own that helps theoretical physicists, which they call "the unreasonable effectiveness of dimensional analysis." Dimensional analysis is a mathematical technique that allows physicists to estimate the magnitude of physical quantities without knowing the precise formula. Such a procedure would not be feasible in a universe with a large number of dimensions. See Barrow and Tipler, *The Anthropic Cosmological Principle*, 270–272.

39 M. Tegmark, "On the Dimensionality of Spacetime," *Classical and Quantum Gravity* 14 (1997): L69–L75. Tegmark further considers possible universes with differing numbers of time dimensions.

40 Paul Davies, *The Mind of God: The Scientific Basis for a Rational World* (New York: Simon & Schuster, 1992).

41 Davies, *The Mind of God*, 159.

42 Harlow Shapley, *Of Stars and Men: Human Response to an Expanding Universe* (Boston: Beacon Press, 1958), 94.

43 Davies, *The Mind of God*, 156–157.

44 George F. R. Ellis has made this point (if only in passing), joining the half-dozen or so people who have hinted at the correlation between habitability and measurability. See "The Anthropic Principle: Laws and Environments," in *The Anthropic Principle: Proceedings of the Second Venice Conference on Cosmology and Philosophy*, F. Bertola and U. Curi, eds. (Cambridge: Cambridge University Press), 31.

45 Within the context of the cooling rate of the expanding universe, Hubert Reeves discusses the minimum requirements for growth of complexity. The key requirement is formation of non-equilibrium structures. Too slow, and the atomic and molecular diversity would be lacking: matter would consist entirely of the final product of equilibrium reactions, iron. Too fast, and large structures—such as planets, stars, and galaxies—could not form. Thus, fine-tuning in the cooling rate is required for complexification to be possible. But we do not agree with Reeves's suggestion that a "Principle of Complexity" should replace the Anthropic Principle, since we are more

than merely complex structures—we are human observers, who can talk about fine-tuning coincidences. See "The Growth of Complexity in an Expanding Universe," in *The Anthropic Principle: Proceedings of the Second Venice Conference on Cosmology and Philosophy,* F. Bertola and U. Curi, eds. (Cambridge: Cambridge University Press), 67–84.

46 Some would argue that life flourishes at the boundaries of stability and chaos. Of course, absolute stability of the constants of nature is not required for a habitable universe. At some small level, some non-stability can be tolerated. Indeed, recent observations of absorption lines seen against distant quasars suggest that the fine-structure constant has changed slightly over the history of the universe. See J. K. Webb et al., "Further Evidence for Cosmological Evolution of the Fine Structure Constant," *Physical Review Letters* 87 (2001): 091301. For an analysis of the sensitivity in the Big Bang production of the light elements to changes in the fine-structure constant, see K. M. Nollett and R. E. Lopez, "Primordial Nucleosynthesis with a Varying Fine Structure Constant: An Improved Estimate," *Physics Reviews D* 66 (2002): 063507. For a review of the diverse constraints on possible changes in the fine-structure constant, see G. Fiorentini and B. Ricci, "α: A Constant That Is Not a Constant?" *Proceedings of ESO-CERN-ESA Symposium on Astronomy, Cosmology and Fundamental Physics*, Report No. INFN-FE-07-02 (astro-ph/0207390). In their search for observational tests of a changing fine-structure constant, physicists are uncovering new examples of fine-tuning.

47 Eugene Wigner, "The Unreasonable Effectiveness of Mathematics in the Natural Sciences," in *Symmetries and Reflections* (Bloomington: Indiana University Press, 1967), 222–237.

48 Mark Steiner, *The Applicability of Mathematics as a Philosophical Problem* (Cambridge, MA: Harvard University Press, 1998).

49 Rees, *Just Six Numbers*, 123. We will discuss, and critique, some of Rees' assumptions in Chapter Fifteen.

50 A. Zee, "The Effectiveness of Mathematics in Fundamental Physics," in *Mathematics and Science*, Ronald E. Mickens, ed. (Singapore: World Scientific, 1990.), 20. We must thank Robin Collins for this reference and several examples in his correspondence with us on this issue. Collins refers to this feature of the universe as "separability," that quality of the universe that provides us with bit-sized chunks that often can be accurately analyzed without considering the whole.

51 Collins explains (from private correspondence):

> Newton's law of gravity, $F = Gm_1m_2/r^2$, can be derived from Einstein's equation, $\mathbf{G} = 8\mathrm{B}T$, by taking the so-called *Newtonian limit*: that is, the limit when the gravitational fields are relatively weak and the velocity of the masses under consideration are small compared to the speed of light. Now, Newton's law tells us the force F between particles of mass m_1 and m_2 separated by a distance r, whereas Einstein's equation tells us the amount by which the total mass-energy (in the form of the mass-energy tensor T) curves four-dimensional space-time, as given by the curvature tensor **G**. (Note that the gravitational constant "G" occurring in Newton's law has an entirely different meaning from the "G" occurring in Einstein's equation.)

Again, we thank Robin Collins for these insights, and look forward to his forthcoming book-length treatment of these issues.

52 It's important to distinguish between a law's being simple and it's being easy to discover. Anyone who has tried to follow the movements of the planets across the sky realizes that it is quite tedious. Discovering laws that describe their regularity is even more difficult. But physicists have often used the simplicity of a discovered law as evidence of its truth. To toil away at observing the complicated motions of heavenly bodies for years on end, only to discover a *simple* mathematical formulation that describes them, is suggestive if not downright fishy. It provides a clue to the investigator that he has stumbled across something more than a mere conventional generalization or mental construct. This is the sense of simplicity that we speak of here. It doesn't imply that discovering, say, the inverse square law is so easy that a child could discover it.

53 Stuart Clark, *Life on Other Worlds and How to Find It* (Chichester, UK: Springer-Praxis, 2000), 1.

54 The better the resolution, the easier it is to separate two close sources of light. The smaller the angle, the greater the resolution required to see them as two sources.

55 Michael Denton, *Nature's Destiny*, 243.
56 Ibid.

CHAPTER 11, THE REVISIONIST HISTORY OF THE COPERNICAN REVOLUTION

1 Quoted in John Noble Wilford, "From Distant Galaxies, News of a 'Stop-and-Go Universe,'" *New York Times* (June 3, 2003). This story is about the role of mysterious dark energy in accelerating the expansion of the universe. Ironically, in the sentence just prior to his summoning of Copernicus, Kirshner is quoted as saying, "We are not made of the type of particles that make up most of the matter in the universe, and we have no idea yet how to sense directly the dark energy that determines the fate of the universe." This would seem to contradict the Copernican Principle, since usually our cosmic ordinariness is taken as evidence for the principle. But as we will see, almost everything and its opposite is taken by someone as evidence for the principle.
2 It was also excerpted in *Time* magazine. The article was published one week after the NASA announcement of possible evidence of life in a Martian meteorite on August 7, 1996.
3 The myth that most ancients thought that Earth was flat seems to have been started by Washington Irving in the English-speaking world. For analysis and critique, see Jeffrey Burton Russell, *Inventing the Flat Earth: Columbus and Modern Historians* (New York: Praeger), 1991.
4 Bruce Jakosky, *The Search for Life on Other Planets* (Cambridge: Cambridge University Press, 1998), 299.
5 Stuart Clark, *Life on Other Worlds and How to Find It* (Chichester, UK: Springer-Praxis, 2000), 1. More generally, physicist Steven Weinberg has said, "The more the universe seems comprehensible, the more it also seems pointless." Earlier in the epilogue to the book, he says, "It is almost irresistible for humans to believe that we have some special relation to the universe, that human life is not just a more-or-less farcical outcome of a chain of accidents reaching back to the first three minutes, but that we were somehow built in from the beginning." Steven Weinberg, *The First Three Minutes* (New York: Basic Books, 1977), 150–155.
6 Bertrand Russell, *Religion and Science* (New York: Oxford University Press, 1961), 222. Quoted in Neil Manson, "Why Cosmic Fine-Tuning Needs to Be Explained," Dissertation at Syracuse University (December 1998), 146.
7 Stuart Ross Taylor, *Destiny or Chance: Our Solar System and Its Place in the Cosmos* (Cambridge: Cambridge University Press, 1998), 11.
8 Thomas Kuhn, *The Copernican Revolution: Planetary Astronomy in the Development of Western Thought* (Cambridge: Harvard University Press, 1957, renewed 1985), 5.
9 In his *Physics* and *On the Heavens*, excerpted in Dennis Danielson, *The Book of the Cosmos* (Cambridge: Perseus, Helix Books, 2000), 42. Danielson's anthology brings together many of the central Western texts on cosmology. We make generous use of the primary texts included in this volume.
10 Quoted in Danielson, *The Book of the Cosmos*, 72. The Christian philosopher Boethius (470–525), who "mediate[d] the transition from Roman to Scholastic thinking," made the same point. See quote and discussion in John Barrow and Frank Tipler, *The Anthropic Cosmological Principle* (Oxford: Oxford University Press, 1986), 45.
11 From Chapter 6 of *De Revolutionibus*, quoted in Danielson, *The Book of the Cosmos*, 112. Copernicus says earlier in this text, "So far as our senses can tell, the earth is related to the heavens as a point is to a body and as something finite is to something infinite."
12 Kuhn, *The Copernican Revolution*, 112–113.
13 Quoted in Hans Blumenberg, *The Genesis of the Copernican Revolution*, Robert M. Wallace, trans. (Cambridge: MIT Press, 1987), xv.
14 For all the talk about Earth's removal from the center of the cosmos, it is ironic that the claim of both the infinite Newtonian universe and the finite (even if boundless) universe of Big Bang cosmology is quite different. It's not that we have been displaced from the cosmic center but that there is no cosmic center.
15 Two nineteenth-century books immortalized the warfare metaphor: Andrew Dickson White, *A History of the Warfare of Science with Theology in Christendom* (1895; reprint New York: George Braziller, 1955), and John W. Draper, *History of the Conflict Between Religion and Science* (1875; reprint New York: Appleton, 1928). An excellent introductory overview of the complex relationship between Christianity and modern science is John Hedley

Brooke, *Science and Religion: Some Historical Perspectives* (Cambridge: Cambridge University Press, 1991). The warfare metaphor recently reared its head in the debate over the prevalence of life in the universe. In *Life Everywhere: The Maverick Science of Astrobiology* (New York: Basic Books, 2001), science writer David Darling attacks one of us (GG), attempting to dismiss GG's arguments for life's rarity on the basis of GG's religious views. The book gives the impression that any scientist who holds religious views inconsistent with Darling's own should be disqualified from practicing science. One also gets the impression that Darling doesn't know the difference between an empirical and an *ad hominem* argument.

16 R. Hooykaas, *Religion and the Rise of Modern Science* (Vancouver: Regent College Publishing, 2000; originally published, Edinburgh: Scottish Academic Press, 1972), 162.

17 Blumenberg, *The Genesis of the Copernican Revolution*, 453–487. For detailed discussion, see also Stanley L. Jaki, *The Road of Science and the Ways to God*, The Gifford Lectures, 1974–75 and 1975–76 (Chicago: University of Chicago Press, 1978).

18 Hooykaas, *Religion and the Rise of Modern Science*, 75–96.

19 See Christopher Kaiser, *Creation and the History of Science* (Grand Rapids: Eerdmans, 1991).

20 Ibid., 7–9, 67–69. "Thus, in total contradiction to pagan religion, nature is not a deity to be feared and worshipped, but a work of God to be admired, studied and managed," 9. For a helpful discussion of technological achievements in the Middle Ages, see J. Gimpel, *The Medieval Machine: The Industrial Revolution of the Middle Ages* (New York: Holt, Rinehart & Winston, 1976). The classic source on Christian faith and medieval technology is Lynn White, Jr., *Medieval Technology and Social Change* (Oxford: Oxford University Press, 1962). Perhaps we should note, contrary to popular assertion, that nothing in the biblical traditions justifies treating nature as a mere means to an end with no intrinsic value, or as a passive backdrop, with which humans can do as they please.

21 An important study on the centrality of the doctrine of creation in the development of science is Eugene M. Klaaren, *Religious Origins of Modern Science* (Grand Rapids: Eerdmans, 1977).

22 Ibid., 29–52. The theological and philosophical movement that emphasized the will of God, insisting that it was not necessary for God to create, or to do so in the way that he did, is called voluntarism. Reijer Hooykaas pioneered study of the influence of voluntarism on the rise of early modern science. Subsequent studies of the influence of voluntarism from Augustine to Ockham to Boyle and Newton are many. The most noteworthy include Francis Oakley, *Omnipotence, Covenant, and Order: An Excursion in the History of Ideas from Abelard to Leibniz* (Ithaca: Cornell University Press, 1984); Margaret J. Osler, *Divine Will and the Mechanical Philosophy: Gassendi and Descartes on Contingency and Necessity in the Created World* (Cambridge: Cambridge University Press, 1994); Jan W. Wojcik, *Robert Boyle and the Limits of Reason* (Cambridge: Cambridge University Press, 1997); Edward Bradford Davis, Jr., "Creation, Contingency, and Early Modern Science: The Impact of Voluntaristic Theology on Seventeenth Century Natural Philosophy" (Ph.D. dissertation, University of Indiana, 1984); and William J. Courtenay, "Capacity and Volition: A History of the Distinction of Absolute and Ordained Power," *Quodlibet: Ricerche e strumenti di filosofia medievale*, no. 8 (Bergamo: Pierluigi Lubrina, 1990).

23 We owe this important point to a lecture and personal conversation with nuclear physicist and science historian Peter Hodgson. For background, see Stanley L. Jaki, *Science and Creation* (Edinburgh: Scottish Academic Press, 1986). For detailed discussion, see Thomas F. Torrance, *Divine and Contingent Order* (Oxford: Oxford University Press, 1981). Torrance summarizes his argument in Thomas F. Torrance, "Divine and Contingent Order," *The Sciences and Theology in the Twentieth Century*, A. R. Peacocke, ed. (Notre Dame: University of Notre Dame Press, 1981), 81–97.

24 In his essay "Abstraction in Modern Science," *Across the Frontiers* (New York: Harper & Row, 1974), 83, 87, the great physicist Werner Heisenberg said:

> This juxtaposition of different intuitive pictures and distinct types of force created a problem that science could not evade, because we are persuaded that nature, in the last resort is uniformly ordered, that all phenomena ultimately take place according to nature's unitary laws.
>
> Perhaps it is also permissible here to mention yet another comparison, from the field of history. That abstraction arises from continuing to ask questions, from the striving for unity, can be clearly recognized from one of the most momentous occurrences in the history of religion. The concept of God in

the Jewish religion represents a higher stage of abstraction, when compared with the idea of many different nature gods, whose activity in the world can be experienced directly. Only at this higher stage is it possible to recognize the unity of divine activity.

25 "[F]aith in the possibility of science, generated antecedently to the development of modern scientific theory, is an unconscious derivative from medieval theology." Alfred North Whitehead, *Science and the Modern World* (New York: Macmillan, 1925), 19. Another important but less common argument is that modern science depended on the Renaissance and especially Protestant desire to return to original texts, and to discern the plain meaning of texts. This had an obvious implication in theology and biblical studies, but the logic may have extended to renewed attempts to "read the plain sense" of the book of nature as well. See Peter Harrison, *The Bible, Protestantism, and the Rise of Modern Science* (Cambridge: Cambridge University Press, 1998).

26 Blumenberg, *The Genesis of the Copernican Revolution*, 483–484.

27 Of course, Thomas Aquinas did not uncritically adopt Aristotle, as anyone who reads his *Summa Theologica* will recognize. Any detailed discussion of the Thomist appropriation of Aristotle would, however, take us too far afield of our topic.

28 John Milton reflects this conviction already in *Paradise Lost*, VIII, lines 66–71 (1667):

> For Heav'n
> Is as the Book of God before thee set
> Wherein to read his wondrous works . . .
> . . . whether Heav'n move or Earth
> Imports not. . . .

Quoted in Owen Gingerich, *The Eye of Heaven* (New York: American Institute of Physics, 1993), 284.

29 Barrow and Tipler call this type of design argument eutaxiological, which requires that order must have a planned cause, and teleological, which argues from order to a consequent purpose. See *The Anthropic Cosmological Principle*, 29, 44, 88–89, 144.

30 Contrary to the conventional description, however, the models had not grown appreciably more complicated from the time of Ptolemy to that of Copernicus, since there were so few systematic observations to improve on Ptolemy's. The main important additions were *trepidations* added by Islamic astronomers before A.D. 1000 to preserve the accuracy of the Ptolemaic values for the location of stars. Accurate systematic observations did not occur until *after* Copernicus's time. (See Gingerich, *The Eye of Heaven*, 25–26.) Thus, it's an overstatement to claim, as Thomas Kuhn does [in *The Structure of Scientific Revolutions* (Chicago: University of Chicago Press, 1962), 67–68], that a "crisis" and "scandal" precipitated Copernicus's innovation. See Gingerich, 193–204.

31 See ibid., 6. This contrasts with the description of Copernicus's insight by the positivists, who claim that it was the result of empirical data, and that Copernicus's system was significantly simpler than the older scheme.

32 Earlier Scholastic critics of Aristotelian cosmology—such as Jean Buridan, Nicole Oresme, and other "Nominalists"—also helped lay the groundwork. And Copernicus was not the first to postulate heliocentrism. Aristarchus of Samos (c. 310–230 B.C.) long before had postulated a revolving Earth in a Sun-centered universe with Earth spinning around its axis. For some reason, however, the intellectual climate in Aristarchus's time was such that the idea did not catch on, as it did for Copernicus.

33 His aspirations for natural philosophy, or what we now call science, however, are still admirable. In his preface to *De Revolutionibus*, written to Pope Paul II, he says, "I know that a philosopher's thoughts are beyond the reach of common opinion—for his aim is to search out the truth in all things—so far as human reason, by God's permission, can do that. But I do think that completely false opinions are to be avoided." Quoted in Danielson, *The Book of the Cosmos*, 105.

34 Blumenberg, *The Genesis of the Copernican Revolution*, xiii, 173–185. Copernicus's anthropocentric assumptions give evidence of the influence of the Neo-Stoic strand of the Renaissance.

35 Kuhn, *The Copernican Revolution*, 139.

36 Dennis Danielson, in a public lecture, "Copernicus and the Tale of the Pale Blue Dot," to the *American Scientific Affiliation*, Lakewood, Colorado, July 25–28, 2003.

37 Kuhn, *The Copernican Revolution*, 131. Note that "center" here has taken on a somewhat different meaning. In the medieval scheme, Earth was literally central, but centrality was not an honor, since Earth was actually a cosmic sump. Copernicus, however, boldly reinterprets the center, so that centrality for the Sun is a place of honor. Nevertheless, since Earth reflected the light of the Sun, its status was much improved from the Aristotelian cosmology. This subtlety is almost always overlooked in popular treatments of the Copernican Revolution.

38 Blumenberg, *The Genesis of the Copernican Revolution*, xvii, 230–255.

39 See the discussion of R. G. Collingwood's defense of this point in Eugene M. Klaaren, *Religious Origins of Modern Science*, 11–13.

40 Kuhn, *The Copernican Revolution*, 135.

41 Hans Blumenberg puts it nicely: "Seen in terms of the whole history of his influence, Copernicus triumphed in the end not so much *over* as *through* his opponents" (*The Genesis of the Copernican Revolution*, 353).

42 In some ways, Brahe's proposal, although it was a compromise between the ancient and the Copernican models, was a more drastic departure from the Aristotelian/Ptolemaic scheme, since it made the nested celestial spheres impossible. His model also simplified the observations, perhaps even more than Copernicus's model. So the superiority of Copernicus's proposal did not really become clear until Kepler modified it using elliptical orbits.

43 See Gingerich, *The Eye of Heaven*, 39–51.

44 Kepler's introduction can be considered as a real paradigm shift from the old attachment to circular orbits. See D. Boccaletti, "From the Epicycles of the Greeks to Kepler's Ellipse—The Breakdown of the Circle Paradigm," presented at *Cosmology Through Time—Ancient and Modern Cosmology in the Mediterranean Area* [Monte Catone (Rome), Italy], June 18-21, 2001. xxx.lanl.gov/abs/physics/0107009.

45 In Kepler, *Optics* II, 277:21–29, in *Gesammelte Werke* (Munich: Beck, 1937), Edward Rosen, trans., in *Kepler's Conversation with Galileo's Sidereal Messenger* (New York: Johnson Reprint Corporation, 1965), 45. We must thank Michael Crowe for reading portions of our manuscript, and alerting us to this remarkable insight of Kepler's.

46 Galileo, *Sidereus Nuncius*, quoted in Danielson, *The Book of the Cosmos*, 150.

47 See Kitty Ferguson's balanced treatment of the subject in her *Measuring the Universe: Our Historic Quest to Chart the Horizons of Space and Time* (New York: Walker, 1999), 94–105.

48 We owe this insight to a conversation with Peter Hodgson, as well as private correspondence and his paper "Galileo the Scientist."

49 Blumenberg, *The Genesis of the Copernican Revolution*, 421–30. For an excellent brief critique of the "Galileo Legend," see also Thomas M. Lessl, "The Galileo Legend as Scientific Folklore," *Quarterly Journal of Speech* 85 (1999), 146–168.

50 One need not look far for examples. For instance, the author of the feature article of the June 2001 issue of *Sky & Telescope* discusses the future possibility of the first *visual* discovery of an extrasolar planet around Epsilon Eridani, the third-closest naked-eye star to the Sun. The story is intrinsically interesting, and has nothing to do with Bruno. Nevertheless, it opens with Bruno's famous quote about countless inhabited worlds, and includes a prominent photo of a statue of Bruno, which marks the spot of his execution in Rome. Govert Schilling, "The Race to Epsilon Eridani," 34–41.

51 Hans Blumenberg treats this in detail in a section entitled "Not a Martyr for Copernicanism: Giordano Bruno," *The Genesis of the Copernican Revolution*, 353–385. Note that Bruno's claim that the world is eternal was an Aristotelian proposition, not a Copernican one.

52 Herbert Butterfield, *The Origins of Modern Science* (New York: The Free Press, 1957), 13–28. Also important was the "transition" concept of impulse developed by Nicole Oresme and Jean Buridan in the mid-fourteenth century. This is just one example illustrating that the "Middle Ages" were important to the rise of science. For an introduction to science in ancient, medieval, and Islamic cultures, see David C. Lindberg, *The Beginnings of Western Science: The European Scientific Tradition in Philosophical, Religious, and Institutional Context, 600 B.C. to A.D. 1450* (Chicago: University of Chicago Press, 1992).

53 This description contradicts the more usual presentation of Newton as a semi-deist. For justification of this interpretation, see Edward B. Davis, "Newton's Rejection of the 'Newtonian World View': The Role of Divine Will in Newton's Natural Philosophy," *Science & Christian Belief* 3 (1991): 103–117.

54 In fact, in the General Scholium, within his famous *Principia*, Newton offers just such a design argument. See Isaac Newton, *Mathematical Principles of Natural Philosophy* (Berkeley: University of California Press, 1960), 542–544.

55 For discussion of the growth of Epicurean materialism in the Renaissance, see Benjamin Wiker, *Moral Darwinism: How We Became Hedonists* (Downers Grove: InterVarsity Press, 2002).

56 This may have been made possible, in part, because of ambivalence in the concept of natural law, which was present in Newton's writings. See John Hedley Brooke, "Natural Law in the Natural Sciences: The Origins of Modern Atheism?" *Science and Christian Belief* 4 (1992): 83–103.

57 We are not endorsing the sufficiency of Darwin's theory to explain adaptation and biological complexity, nor are we endorsing the "Copenhagen" interpretation of quantum phenomena, but only noting that chance, as a scientific explanation, is no longer deemed illicit. "Chance" has a different status in Darwinism than it does in the Copenhagen interpretation of quantum phenomena. For Darwin, variations were the result of chance in the sense that they do not occur with purpose or forethought for the survival or benefit of an organism. In the Copenhagen interpretation, quantum events are more literally chance occurrences in the sense that they are "uncaused," or at least "undetermined."

58 Since we're making harsh generalizations about the textbook treatment of the Copernican Revolution, it's only fair to document it. Almost any introductory astronomy textbook will do. Consider, for instance, Chris Impey and William K. Hartmann, *The Universe Revealed* (Pacific Grove, Cal.: Brooks/Cole, 2000). Chapter Three, "The Copernican Revolution," frames the topic by beginning with Galileo's trial. The authors introduce and explain the topic in this way (53): "The Copernican revolution is one of the most important ideas in history, because it indicates that we are part of a larger cosmic environment, not the masters of nature living in the capital of the universe." No one who understood the history of ideas in the West would speak of Earth in the earlier cosmology as "the capital of the universe." At the same time, astronomy textbooks are a little less sloppy in making this point than books outside the discipline. See, for example, this embarrassing claim by Douglas J. Futuyma, in *Science on Trial* (New York: Pantheon Books, 1983), 195–196: "The geocentric theory of the universe was a theological doctrine developed by St. Clement of Alexandria, Dionysius the Areopagite, the twelfth-century theologian Peter Lombard, and St. Thomas Aquinas. . . . The geocentric theory was unproven; it was untestable; it was held only because it fit Scripture." See also Francisco Ayala's description of Darwinism as the completion of the Copernican Revolution in *Creative Evolution?!*, John H. Campbell and J. William Schopf, eds. (Boston: Jones and Bartlett, 1994), 2–5. Such claims would be embarrassing to almost any historian of science.

59 Dennis R. Danielson, "The Great Copernican Cliché," *American Journal of Physics* 69, no. 10 (October 2001): 1029. This excellent essay does a masterful job of laying to rest the mythology surrounding the Copernican Revolution.

60 From *The Nature of the Gods*, quoted in Danielson, 56. In *Genesis*, Hans Blumenberg notes that, unlike the Stoicism of Cicero, for Christian theology the world was not strictly made for man but rather subjugated to him by God (174). There is a certain irony in the fact that certain Enlightenment figures made "man the measure of all things." In so doing, they exalted man far more than medieval Christianity ever did. Then, in a startling case of historical amnesia and projection, they began to perpetuate the myth that it was the ancient and medieval world picture that erroneously exalted man above his true status. Untangling such a complicated morass, however, would take us too far afield from our chosen topic.

61 Even those who dislike this theme recognize its prevalence. Daniel Quinn identifies this as the fundamental delusion we all tell ourselves—namely, "The world was made for man, and man was made to conquer and rule it." He contends that the biblical creation narratives simply exemplify this more general view. In *Ishmael* (New York: Bantam, 1992), 74. See also Daniel Quinn, *The Story of B* (New York: Bantam, 1996), 129.

CHAPTER 12, THE COPERNICAN PRINCIPLE

1 Carl Sagan, *Pale Blue Dot* (New York: Ballantine Books, 1994), 7.

2 Quoted in Dennis Overbye, "In the beginning ...," *The New York Times* (July 23, 2002): D1.

3 GR neither entails nor implies the cosmological principle. In his highly influential early work in cosmology, E. A. Milne put it this way:

> The cosmological principle is simply used [in Milne's work] as a definition, a principle of exclusion, enumerating the class of systems to be considered and excluding all others. It is in no sense used as a 'law of nature', or principle of compulsion, prescribing what is to happen. Whether, when a system is set up satisfying the cosmological principle, it will continue to satisfy it, is a matter always for investigation.

Relativity, Gravitation, and World-Structure (Oxford: Clarendon Press, 1935), 20.

4 In fact, the early use of the cosmological principle had nothing to do with Copernicus, or with an extension of the Copernican Revolution. Copernicus is rarely if ever mentioned in the writings of early-twentieth-century cosmologists like Milne.

5 Modern writers often simply identify the Copernican Principle with the cosmological principle, by defining the former in terms of the homogeneity of the cosmos. As Ernan McMullin describes it, "A homogeneous large-scaled distribution of matter is (for reasons difficult to specify) more likely to be the actual cosmic state of affairs." In "Indifference Principle and Anthropic Principle in Cosmology," *Studies in the History and Philosophy of Science* 24, no. 3 (February 1993): 359.

6 So J. Richard Gott III argues in "Implications of the Copernican Principle for Our Future Prospects," *Nature* 363, no. 6427 (May 1993): 315:

> The Copernican revolution taught us that it was a mistake to assume, without sufficient reason, that we occupy a privileged position in the Universe. Darwin showed that, in terms of origin, we are not privileged above other species. Our position around an ordinary star in an ordinary galaxy in an ordinary supercluster continues to look less and less special. . . . The Copernican Principle works because, of all the places for intelligent observers to be, there are by definition only a few special places and many nonspecial places, so you are likely to be in a nonspecial place.

7 See discussion in McMullin, "Indifference Principle and Anthropic Principle in Cosmology," 359–367.

8 Of course the modest version of the Copernican Principle could be true and the metaphysical version be false. There is, after all, a difference between a worldview and a world picture. But we might as well admit that they can't be safely quarantined. There's bound to be seepage. If evidence overwhelmingly confirmed that our location is utterly unexceptional, we might suspect that we are unexceptional as well. In any case, practitioners very rarely distinguish between these two senses as we have.

In one of the clearest expositions of the subject, Ernan McMullin (in "Indifference Principle") actually identifies the Copernican Principle with the denial of purpose and "privilege." He explains:

> The Copernican principle has to be understood in terms of what it rejects, namely, older teleological beliefs about the uniqueness of the human and the likelihood that the human abode would be singled out in some special way, e.g., by being at the cosmic center. In practice, the principle reduces simply to asserting that human life is equally likely to be located in any (large) region of cosmic space or (less persuasively) at any point on the cosmic time-line (373).

9 Mario Livio, *The Accelerating Universe: Infinite Expansion, the Cosmological Constant, and the Beauty of the Cosmos* (New York: John Wiley & Sons, Inc., 2000), 263.

10 From Percival Lowell, *Mars* (Boston, 1895), excerpted in Dennis Danielson, *The Book of the Cosmos* (Cambridge: Perseus Pub., Helix Books, 2000), 341. Speculation about extraterrestrial life in the Solar System was usually restricted to the other planets, although William Herschel thought the Sun itself might be habitable. See Steven Kawaler and J. Veverka, "The Habitable Sun: One of William Herschel's Stranger Ideas," *Journal of the Royal Astronomical Society of Canada* 75, no. 1 (1981): 46–55. Today we tend to consider such an idea as quaint, but we should notice that it seems to follow from the Copernican Principle.

11 In John Noble Wilford, *Mars Beckons* (New York: Knopf, 1990), 35, quoted in Danielson, *The Book of the Cosmos*, 341.

12 Quoted in Ken Croswell, *Planet Quest: The Epic Discovery of Alien Solar Systems* (New York: The Free Press, 1998), 185–186. Mayor and Queloz published their discovery in "A Jupiter Mass Companion to a Solar-Type Star," *Nature* 378 (1995), 355–359.

13 The work was originally reported in "The Arecibo Message of November, 1974," *Icarus* 26 (1975): 462–466, authored by "The staff at the National Astronomy and Ionosphere Center," which included Carl Sagan and Frank Drake. Apparently, the signal was transmitted as part of the ceremonies to dedicate the upgraded Arecibo radio telescope on November 16, 1974. See discussion in Steven J. Dick, *Life on Other Worlds: The 20th Century Extraterrestrial Life Debate* (Cambridge: Cambridge University Press, 1998), 217.

14 Shapley overestimated the size of the Milky Way and put us 65,000 light-years out. Astronomers now estimate that we are approximately 27,000 light-years from the galactic center.

15 Ken Croswell, *The Alchemy of the Heavens* (Oxford: Oxford University Press, 1995), 24. Croswell is normally quite astute at challenging conventional wisdom on such matters, and he does not draw any conclusions from Shapley's discovery in favor of the Copernican Principle.

16 S. P. Goodwin, J. Gribbin, and M.A. Hendry, "The Relative Size of the Milky Way," *The Observatory* 118 (August 1998): 201, 207. This paper is shot through with "Copernican" reasoning. In the opening paragraph of the introduction, they say:

> Before the confirmation by Hubble that many 'nebulae' are, in fact, external galaxies, the Milky Way was thought to by many to be the entire Universe. Even after the identification of external galaxies, it seemed at first that these were much smaller objects than the Milky Way, relatively close to our Galaxy. Successive revisions of the cosmic distance scale have placed external galaxies further away from us with the implication that they are correspondingly bigger, reducing the perceived importance of the Milky Way in the Universe (201).

Since Goodwin et al. consider the Milky Way's smaller size as confirming the Copernican Principle, then, to be consistent with their "logic," they should consider its larger mass as contradicting the Copernican Principle.

17 See Table 1 of P. Hodge, "A Comparison of the Andromeda and Milky Way Galaxies," in *The Milky Way Galaxy*, H. van Woerden et al., eds. (Dordrecht: D. Reidel, 1985), 423–430. The face-on blue luminosities of the two galaxies are equivalent within the measurement uncertainties.

18 S. T. Gottesman, J. H. Hunter, Jr., and V. Boonyasait, "On the Mass of M31," *Monthly Notices of the Royal Astronomical Society* 337 (2002): 34–40.

CHAPTER 13, THE ANTHROPIC DISCLAIMER

1 Fred Adams and Greg Laughlin, *Five Ages of the Universe: Inside the Physics of Eternity* (New York: The Free Press, 1999), 201. Emphasis in original.

2 The contemporary philosopher who has done the most to develop this type of argument is William Lane Craig. See especially *The Kalam Cosmological Argument* (Eugene, Ore.: Wipf & Stock, 1979) and William Lane Craig and Quentin Smith, *Theism, Atheism, and Big Bang Cosmology* (Oxford: Clarendon, 1993).

3 Sir Arthur Eddington, who didn't like the implications of the Big Bang model, said, "The beginning seems to present insuperable difficulties unless we agree to look on it as frankly supernatural." In *The Expanding Universe* (New York: Macmillan, 1933), 124.

4 Fred Hoyle, "A New Model for an Expanding Universe," *Monthly Notices of the Royal Astronomical Society* 108, no. 5 (1948): 372–382; Hermann Bondi and Thomas Gold, "The Steady-State Theory of the Expanding Universe," *Monthly Notices of the Royal Astronomical Society* 108, no. 3 (1948): 252–270.

5 Denis Sciama, *The Unity of the Universe* (New York: Doubleday, 1961), 70. Quoted in McMullin, *Indifference Principle*, 368.

6 Stephen Meyer, "The Return of the God Hypothesis," *The Journal of Interdisciplinary Studies* XI, no. 1/2 (1999): 9. The articles substantiating this claim are Alan Guth and Marc Sher, "The Impossibility of a Bouncing Universe," *Nature* 302 (April 7, 1983): 505–507; J. E. Phillip Peebles, *Principles of Physical Cosmology* (Princeton: Princeton University Press, 1993); and Peter Coles and George Ellis, "The Case for an Open Universe," *Nature* 370

(August 25, 1994): 609–613. Of course this doesn't mean that cosmologists have given up trying to get rid of an initial singularity, or beginning, with its repugnant philosophical implications. See, for example, Charles Seife, "Eternal-Universe Idea Comes Full Circle," *Science* 296 (2002): 639.

7 In his widely known book, *A Brief History of Time* (New York: Bantam Books, 1988), Stephen Hawking attempts to accept a finite past while nevertheless avoiding a cosmic beginning (and its presumably distasteful theological implications). He does so with a number of questionable and highly controversial maneuvers, including imaginary time and Richard Feynman's sum-over-all-histories interpretation of quantum theory. We find his strategy not only unconvincing but desperate in the extreme. We point the reader to William Lane Craig's devastating critique of Hawking's argument in *Theism, Atheism, and Big Bang Cosmology*, 279–300.

Tellingly, in Hawking's recent book, *The Universe in a Nutshell* (New York: Bantam Books, 2001), he continues to employ similar arguments, without indicating any awareness of the published critiques of them. He also explicitly endorses the positivism that is easy to detect, at least as a residue, in *Brief History*. This allows him, albeit inconsistently, to skirt the issue of whether he intends his use of imaginary time to accurately represent reality. If it works mathematically, he claims, such questions of correspondence with reality are superfluous. He doesn't seem to realize that this immunization strategy undercuts his desire to do away with a beginning, since we are still left with the question of whether the *actual* universe began to exist. Just because a theoretical physicist can concoct a mathematical justification for avoiding the question doesn't mean anyone is obliged to follow him, especially when he refuses to argue that his conjecture represents reality. Moreover, once again, he seems blithely unaware of the trenchant critiques of positivism that have led virtually all philosophers of science to abandon it.

8 This point is often misunderstood. See, for instance, Anthony Aguirre, "The Cold Big-Bang Cosmology as a Counter-example to Several Anthropic Arguments," *Physical Review D*, 64, Issue 8 (Oct. 15, 2001), 083508. The argument does not require that there be only one set of physical constants compatible with complex life. It only requires that among the range of alternative universes considered in the "universe neighborhood," the range of life-permitting universes is quite narrow.

9 Paul Davies, *The Cosmic Blueprint* (New York: Simon & Schuster, 1988), 203.

10 Fred Hoyle, "The Universe: Past and Present Reflections," *Annual Review of Astronomy and Astrophysics* 20 (1982), 16.

11 Stephen Hawking, *A Brief History of Time* (New York: Bantam Doubleday Dell, 1998, tenth anniversary edition).

12 For an excellent presentation of this argument, see Timothy McGrew, Lydia McGrew, and Eric Vestrup, "Probabilities and the Fine-Tuning Argument: A Skeptical View," *Mind* 110 (October 2001). Philosopher Robin Collins has responded to their argument in "Fine-Tuning and Comparison Range: An Answer to McGrew, Vestrup, and Manson" at http://www.messiah.edu/hpages/facstaff/rcollins/crange.htm.

13 Actually, this example needs a nuance, the purpose of which might not be immediately obvious. Explaining it, however, makes it a bit tedious and complicated. (That's why it's in this endnote.) It's not quite correct to say that there is only one set of constants (in the example, one numeric combination) compatible with a habitable universe. After all, even the force of gravity relative to the electromagnetic force is "only" fine-tuned out to the fortieth decimal place. So the illustration requires a *range* of combinations rather than a single number. Imagine in this story that Q has discovered that he can add any number to the right of the sequence—that is, past the eighth decimal place—without affecting the outcome. But any other change to the combination above that level of resolution spells disaster. Here's the idea: Q determined that this number, 91215225.79141425, was required by the Universe-Creating Machine to generate a habitable universe, but that the combination tolerated variations in the number past the eighth decimal point. So combinations such as 91215225.79141425*1*, 91215225.79141425*2*, and so on still permit a habitable universe, but none that he can find without the initial number sequence 91215225.79141425. This would suggest that every number *between* 91215225.79141425 and 91215225.79141426 would permit a habitable universe. And of course, mathematically speaking, there is a potential infinity of numbers between any two numbers. Nevertheless, despite the fact that Q had found a potentially infinite range of combinations at that level of resolution that generated a habitable universe, he would still not doubt that the original combination had been set. At least one reason

is because a single specific piece of knowledge, which profoundly restricts the range of possibilities, would be necessary to find the "set" of combinations that would work. One might even say the combination to the Universe-Creating Machine is still singular, since it can be expressed thus: "The code for generating a habitable universe is any natural number from 91215225.79141425 to 91215225.79141426, excluding 91215225.79141426." No one would buy the argument that we should remain agnostic about whether the combination had been initially set, because there are a potentially infinite number of combinations that can produce a habitable universe.

14 Brandon Carter initiated the contemporary discussion of the Anthropic Principle with a formulation similar to this: "What we can expect to observe must be restricted by the conditions necessary for our presence as observers." In "Large Number Coincidences and the Anthropic Principle in Cosmology," in M. S. Longair, ed., *Confrontation of Cosmological Theory with Astronomical Data* (Dordrect: Reidel, 1974), 291–298. Reprinted in John Leslie, ed., *Physical Cosmology and Philosophy* (New York: Macmillan, 1990). In addition to the Weak Anthropic Principle, Carter proposed a Strong Anthropic Principle. There is an exegetical debate surrounding the meaning of the Strong Anthropic Principle, and in particular, the meaning of the word "must" in Carter's formulation. Leslie interprets Carter in this way: The Weak sense refers to our location in time and place. The Strong sense refers to the universe as a whole. SAP means, "Any universe with observers in it must be observer-permitting." See Leslie's "Introduction" in *Modern Cosmology and Philosophy*, John Leslie, ed., (Amherst, NY: Prometheus, 1998), 2. Whether or not this is a correct interpretation, the Strong Anthropic Principle has been taken in all sorts of ways, and often does not conform to Carter's initial formulation. We are following this definition of Weak and Strong Anthropic Principles, without judging whether this is consistent with Carter's initial formulation.

15 Carter seems to have understood it in just this way. In "Large Number Coincidences" (reprinted in *Physical Cosmology and Philosophy*, 131), he says:

> [I]t consists basically of a reaction against exaggerated subservience to the "Copernican principle." Copernicus taught us the very sound lesson that we must not assume gratuitously that we occupy a privileged *central* position in the Universe. Unfortunately there has been a strong (not always subconscious) tendency to extend this to a most questionable dogma to the effect that our situation cannot be privileged in any sense.

16 Konrad Rudnicki, "The Anthropic Principle as a Cosmological Principle," 123.

17 We thank Paul Nelson for this story.

18 As we explain in endnote 14, there is a confusing variety of different definitions of the Strong Anthropic Principle in the literature. Often it is defined as meaning that the laws and constants somehow had to be what they are, or that they are necessarily what they are. Others define WAP as a selection effect and SAP as implying that the fine-tuning is the result of intentional design. This diversity is confusing, so we've decided to go with the definitions that are easiest to grasp. Nothing crucial in our argument depends on this way of defining WAP and SAP.

19 This is derived from the popular story told by philosophers Richard Swinburne, John Leslie, and others. See John Leslie, *Universes* (London: Routledge, 1989), 13–15, 108.

20 Alan H. Guth, *The Inflationary Universe* (Reading, Mass.: Perseus Books, 1997). Still more speculative proposals are Andrei Linde, "The Universe: Inflation Out of Chaos," in *Modern Cosmology and Philosophy*, and "The Self-Reproducing Inflationary Universe," *Scientific American* 271 (November 1994): 48–55.

21 For one highly speculative proposal along these lines, see Lee Smolin, *The Life of the Cosmos* (New York: Oxford University Press, 1997). We sympathize with those who object to using the term "universe" in this way, since "universe" implies singularity. Nevertheless, cosmologists now frequently talk about multiple worlds or universes, so we have decided to follow convention on this point.

22 In their classic tome on the subject, *The Anthropic Cosmological Principle* (Oxford: Oxford University Press, 1986), 4, John D. Barrow and Frank J. Tipler observe, "In a sense, the Weak Anthropic Principle may be regarded as the culmination of the Copernican Principle, because the former shows how to separate those features of

the Universe whose appearance depends on anthropocentric selection, from those features which are genuinely determined by the action of physical laws."

23 The arguments on this point are somewhat complicated. For details, see Craig, *The Kalam Cosmological Argument,* and Craig and Smith, *Theism, Atheism, and Big Bang Cosmology,* 3–191. There is some confusion about actual infinities in the scientific literature. For instance, some physicists and cosmologists assume that if the universe is open and will continue to expand indefinitely, then the universe contains or will contain an infinite number of planets, stars, and galaxies. This doesn't follow. Even if the matter in a continually expanding universe continued to congeal into stars, planets, and galaxies (which we have no reason to assume), it would never reach an actual infinite set of such objects. It would continually be in process of doing so, and would be part of a potentially infinite set, not an actually infinite one. No amount of theoretical modeling can change this logical point. For examples of authors who make this mistake, see Jonathan I. Lunine, "The Frequency of Planetary Systems in the Galaxy," in Ben Zuckerman and Michael H. Hart, eds., *Extraterrestrials: Where Are They?* (Cambridge: Cambridge University Press, 1995), 223–224; and G. F. R. Ellis and G. B. Brundrit, "Life in an Infinite Universe," *Quarterly Journal of the Royal Astronomical Society* 20 (1979): 128–135.

24 As astronomer and science historian Steven Dick explains, "Such a claim [that many universes exist] has emerged over the last few years after pondering the fact that our universe seems to be finely tuned for life." Steven J. Dick, "Extraterrestrial Life and Our World View at the End of the Millenium," *Dibner Library Lecture* (Washington, DC: Smithsonian Institution Series, 2000), 29.

25 F. Adams and G. Laughlin, *Five Ages of the Universe: Inside the Physics of Eternity,* 199, 201. Emphasis in original. Not surprisingly, Adams and Laughlin do not mention even the possibility of design. Similarly, in *Our Cosmic Future: Humanity's Fate in the Universe* (Cambridge: Cambridge University Press, 2000), 239, Nikos Prantzos notes that, despite "the impossibility of any physical contact among the bubble universes," "chaotic inflation is one of the most attractive from a philosophical point of view." One wonders what philosophical perspective he has in mind. Martin Rees uses similar reasoning. The Prologue to his book *Our Cosmic Habitat* (Princeton: Princeton University Press, 2001) is entitled: "Could God Have Made the World Any Differently?" He answers "yes," and proposes that what we call laws of nature are "no more than local bylaws—the outcome of historical accidents during the initial instants after our own particular Big Bang" (xvii). But although he answers "yes" to the question, he doesn't really mean it. In fact, the question seems to be a red herring, since he appeals to multiple universes:

> In this book I argue that the multiverse concept is already part of empirical science: we may already have intimations of other universes, and we could even draw inferences about them and about the recipes that led to them. In an infinite ensemble, the existence of some universes that are seemingly fine-tuned to harbor life would occasion no surprise; our own cosmic habitat would plainly belong to this unusual subset. Our entire universe is a fertile oasis within the multiverse.

But clearly Rees is posing a false dilemma. How exactly has he ruled out the possibility that our universe looks fine-tuned because it *is* fine-tuned? He hasn't ruled it out. He's just failed to consider it. Despite Rees's claim that the multiverse concept has empirical consequences, it's very difficult to see what function it serves other than to avoid the bothersome implications of the evidence of fine-tuning.

26 An infinite World Ensemble actually combines the virtues of chance and necessity, hence its charm. The fact that we find ourselves in a habitable universe, according to this explanation, is not due to any purpose. In that sense, it's a result of chance. Unless the existence of that ensemble is contingent, however (which the many worlds theorist is likely to deny), then the fact that there are habitable universes is due to necessity, since the Ensemble is *infinite.* Reality in this scheme is like a cosmic lottery. Individual universes are the tickets. Since every possible universe exists, the chance of a universe like ours existing is exactly one.

27 As philosopher of science Robin Collins has argued in "The Fine-Tuning Design Argument: A Scientific Argument for the Existence of God," *Reason for the Hope Within,* Michael Murray, ed., (Grand Rapids: Eerdmans, 1999), 61:

> In all currently worked out proposals for what this universe generator could be . . . the generator itself is governed by a complex set of laws that allow it to produce universes. It stands to reason, therefore,

that if these laws were slightly different the generator probably would not be able to produce any universes that could sustain life.

28 Obviously a great deal more remains to be said. But a detailed critique of various many worlds hypotheses would take us too far afield from our chosen topic. An excellent general treatment of many worlds hypotheses is John Leslie's *Universes.* For other arguments against many worlds hypotheses, see Roger White, "Fine-Tuning and Multiple Universes," *Nous* 34, no. 2 (2000): 260–276; Robin Collins, "The Fine-Tuning Design Argument"; William Lane Craig, "Barrow and Tipler on the Anthropic Principle v. Divine Design," *British Journal for the Philosophy of Science* 38 (1988): 389–395; Jay W. Richards, "Many Worlds Hypotheses: A Naturalistic Alternative to Design," *Perspectives on Science and Christian Faith* 49, no. 4 (1997): 218–27; William Lane Craig, "Cosmos and Creator," *Origins & Design* 20, no. 2 (Spring 1996): 18–28; Richard Swinburne, " Argument from the Fine Tuning of the Universe," in *Modern Cosmology and Philosophy,* John Leslie, ed., 160–179. An excellent and brief summary of some of these arguments is Stephen Meyer, "The Return of the God Hypothesis," 7–17.

29 This conclusion will surprise many astronomers and cosmologists. For instance, in *Our Cosmic Future: Humanity's Fate in the Universe,* 205, astrophysicist Nikos Prantzos recently asserted:

> The spatial version of this principle [that there is nothing extraordinary about us as intelligent observers in the Universe], usually associated with the name of Copernicus, has proven correct every time it has been confronted with observational evidence. Indeed, we live on a small planet, orbiting a very ordinary star among myriad others of the same kind, within an unexceptional galaxy among hundreds of billions of other galaxies in the observable Universe.

He is right provided one arbitrarily and illogically limits considerations of specialness to geometrical position in the cosmos. In other words, as long as one ignores the myriad ways that our position benefits life and observation, our position as living, sophisticated observers in the universe doesn't seem special at all. This is sort of like saying Michael Jordan is not being a special basketball player at a particular instant because he's not in the center of the basketball court but rather ten feet in and a few feet to the side of center; never mind that, at the moment he is backpedaling away from a game-winning buzzer shot from twenty-five feet that he made under double coverage, or that he's, well, Michael Jordan.

30 This is his definition of the "cosmological principle," which is a synonym for the Copernican Principle. Stuart Clark, *Stars and Atoms: From the Big Bang to the Solar System* (New York: Oxford University Press, 1995), 23. Konrad Rudnicki also defines the "Generalized Copernican Principle" as the view that "the Universe looks roughly the same for any observer located on any other planet." "The Anthropic Principle as a Cosmological Principle," *The Astronomical Quarterly* 7 (1990): 121. Physicist Richard Feynman also describes the principle, which he calls "the grand hypothesis that nearly every cosmologist makes," in terms of the uniformity of the appearance of the universe from every location. In *Feynman Lectures on Gravitation* (Reading, Mass.: Addison-Wesley, 1995), 166. He also recognizes its tenuousness:

> It is a completely arbitrary hypothesis, as far as I understand it. . . . I suspect that the assumption of uniformity of the universe reflects a prejudice born of a sequence of overthrows of geocentric ideas. . . . It would be embarrassing to find, after stating that we live in an ordinary planet about an ordinary star in an ordinary galaxy, that our place in the universe is extraordinary, either being the center or the place of smallest density, and so forth. To avoid this embarrassment we cling to the hypothesis of uniformity.

31 A quark is about 10^{-16} m in diameter. The adult human being is about 2 meters tall. Earth is about 10^{7} m in diameter. The distance between Earth and the nearest stars is about 10^{17} m. The observable universe is about 10^{26} m. See "Scale of the Universe" in Clark, *Stars and Atoms,* 76–77.

CHAPTER 14, SETI AND THE UNRAVELING OF THE COPERNICAN PRINCIPLE

1 For a good survey, see Steven J. Dick, *Plurality of Worlds: The Origins of the Extraterrestrial Life Debate from Democritus to Kant* (Cambridge: Cambridge University Press, 1982). For a detailed study of the growth of the idea

from Kant to the twentieth century, see Michael J. Crowe, *The Extraterrestrial Life Debate 1750–1900: The Idea of a Plurality of Worlds from Kant to Lowell* (Cambridge: Cambridge University Press, 1986).

2 Steven J. Dick, *Life on Other Worlds: The 20th-Century Extraterrestrial Life Debate* (Cambridge: Cambridge University Press, 1998), 54. This book is an abridgment of a much longer book defending the same thesis, entitled *The Biological Universe* (Cambridge: Cambridge University Press, 1996).

3 Douglas Adams, *The Hitchhiker's Guide to the Galaxy* (New York: Pocket Books, 1979), 76.

4 According to a recent Gallup poll (Darren K. Carlson, "Life on Mars?" *Gallup News Service*, Feb. 27, 2001), 61 percent of those surveyed "said they thought that other forms of life existed in the universe, and 33 percent did not. When asked about the specific form of extraterrestrial life, 41 percent of Americans said they thought there could be 'people somewhat like ourselves living on other planets in the universe' while 54 percent did not think so."

5 See Dick, *Life on Other Worlds*, 164–165.

6 Dick, *Life on Other Worlds*, 3, 261–266.

7 Dick, *Life on Other Worlds*, 218–221.

8 Frank Tipler, "A Brief History of the Extraterrestrial Intelligence Concept," *Quarterly Journal of the Royal Astronomical Society* 22 (1981): 133–145; Frank Tipler, "Additional Remarks on Extraterrestrial Intelligence," *Quarterly Journal of the Royal Astronomical Society* 22 (1981): 271–291.

9 But we do think it's a strong argument, nonetheless. See G. Gonzalez, "Extraterrestrials: A Modern View," *Society* 35, no. 5 (July/August 1998): 14–20. For an interesting survey of the debate, see the essays in B. Zuckermann and M. H. Hart, ed., *Extraterrestrials: Where Are They?* (Cambridge: Cambridge University Press, second edition, 1995). Note that most of the skeptics are biologists. The British journal *Astronomy and Geophysics* is one of the few scientific journals that have remained balanced in their treatment of the subject, publishing articles from both advocates and critics. (Too many others have become partisan campaigners.) See I. A. Crawford, "How Common Are Technological Civilizations?" *Astronomy and Geophysics* 38, no. 4 (1997): 24–26; T. Joseph, W. Lazio and James M. Cordes, "The Number of Civilizations in our Galaxy," *Astronomy and Geophysics* 38, no. 6 (1997): 16–18; I. A. Crawford, "Galactic Civilizations: A Reply," *Astronomy and Geophysics* 38, no. 6 (1997): 19.

10 This response is unlikely to solve Fermi's Paradox given the evidence we present in this book. If habitable planets like Earth are as rare as we argue, then Earth would have been near the top of any colonizers' target lists. In addition, any spacefaring civilization would have the technology to survey any nearby stars for the presence of giant and terrestrial planets. And since Earth has been inhabited by relatively simple life over most of its history, there would be little motivation for quarantining it. So, it's unlikely Earth was passed by either by oversight or by intention by a wave of colonization.

11 For an overview of such responses, see David Brin, "Mystery of the Great Silence," in *Are We Alone in the Universe?: The Search for Alien Contact in the New Millennium* (New York: ibooks, 1999), 139–157. A serious problem with most of these responses is that they are only plausible if relatively few alien civilizations exist. But this contradicts the view of most believers in ETI that such civilizations are abundant. For discussion, see Ian Crawford, "Where Are They?: Maybe We Are Alone in the Galaxy After All," *Scientific American* (July 2000), 29–33.

12 Few have recognized this point explicitly, but planetary scientist Christopher P. McKay hints at it in "Astrobiology: The Search for Life Beyond the Earth," *Many Worlds: The New Universe, Extraterrestrial Life & the Theological Implications*, Steven J. Dick, ed. (Philadelphia: Templeton Foundation Press, 2000), 56:

> [I]t is clear that the search for life beyond the Earth is based on the assumption that our experience of life on Earth is typical for the cosmos—the Copernican principle. . . . The Copernican principle is well established only for stars and elements. . . . [It] is not established with respect to biology, culture, or ethics. In this context, the question "Are we alone?" is a deep philosophical question.

13 Tom Clarke, "SETI Sees the Light: Alien Hunters Look for Light Signals from Other Worlds," *Nature News* (July 25, 2001).

14 Public funding for SETI research waxes and wanes. The federal government backed SETI research for years through NASA, but such funding was shut off in 1992. At the moment, most direct SETI work, such as that done

by the SETI Institute, is supported from private sources. Nevertheless, NASA continues to support astrobiology programs at a number of institutions, all of which bear on the general question of extraterrestrial life. And in 2003, the SETI Institute received a major grant from the federally funded NASA Astrobiology Institute.

15 The meeting took place at the National Radio Astronomy Observatory (NRAO) in Green Bank, West Virginia. For discussion of the equation, and Drake's more recent reflections on the topic, see Frank Drake and D. Sobel, *Is Anybody Out There?: The Scientific Search for Extraterrestrial Intelligence* (New York: Delacorte Press, 1992). Drake had already performed an early search at NRAO called Project OZMA. Independently, Cornell physicists Philip Morrison and Giuseppe Cocconi published a seminal paper "Searching for Interstellar Communications," *Nature* 184 (1959): 844–846, making the same calculations as Drake. The literature on the subject is quite large. For a brief, sympathetic description of the historical details, see Dick, *Life on Other Worlds*, 200–235. For a description of the variety of estimates that different astronomers have assigned to the Drake Equation, see Dick, *Life on Other Worlds*, 217, and Dick, *The Biological Universe*, 441.

16 The history of the equation is somewhat complicated. Drake's original equation was slightly different from the one here. He used *R* instead of N_* and *L* instead of f_L. *R* stood for the rate of star formation in the galaxy, and *L* stood for the average number of years that a communicating civilization survives. See http://www.activemind.com/Mysterious/Topics/SETI/drake_equation.html for a calculator using our version, and http://www.seti.org/science/drake-bg.html for the SETI Institute description.

17 Although SETI seeks empirical evidence, one shouldn't suppose that SETI researchers are withholding their conclusions until the evidence is in. On the contrary, most of them are certain that ETI exists. They're just trying to prove it. Frank Drake, for instance, sees it as a certainty, as do most of the SETI enthusiasts. See Drake's own discussion of the optimism of the original Green Bank meeting participants in Drake and Sobel, *Is Anybody Out There?*, xi, 45–64. See also David Koerner and Simon LeVay in *Here Be Dragons: The Scientific Quest for Extraterrestrial Life* (Oxford: Oxford University Press, 2000), 162.

18 D. Schwartzman and L. J. Rickard, "Being Optimistic About SETI," *Bioastronomy: The Next Steps*, George Marx, ed. (Dordrecht: Kluwer Academic Publishers, 1988), 305. For a recent example of an unrealistically optimistic evaluation of the Drake Equation, see "The Drake Equation," *Astronomy* (September 2002), 47.

19 Quoted in Dick, *Life on Other Worlds*, 217. Bruce Jakosky says similarly that the "Drake equation is just a mathematical way of saying, who knows?" In *The Search for Life on Other Planets* (Cambridge: Cambridge University Press, 1998), 285.

20 Sagan makes clear his debt to the Copernican Principle in his book, coauthored with Soviet astrophysicist I. S. Shklovskii, *Intelligent Life in the Universe* (San Francisco: Holden-Day, 1966). In Chapter 25, entitled "The assumption of mediocrity," (357), they say:

> The idea that we are not unique has proved to be one of the most fruitful of modern science. . . . Let us assume that the Earth is an average planet, and that the Sun is an average star. Then the diameter of the Earth, its distance from the Sun, and its albedo, or reflectivity, should be characteristic of planets in general.

21 Dick, *Life on Other Worlds*, 102. It's also long been a favorite for disciples of the Copernican Principle. In 1943, Henry Norris Russell wrote an article about a supposed planet around 61 Cygni called "Anthropocentrism's Demise." Ibid., 82.

22 Quoted by Fred Heeren in "Home Alone in the Universe?" *First Things* 120 (March 2002): 38–46.

23 Y. Tsapras et al., "Constraints on Jupiters from Observations of Galactic Bulge Microlensing Events During 2000," in *Monthly Notices of the Royal Astronomical Society* 337 (2002): 41–48.

24 See Serge Tabachnik and Scott Tremaine, "Maximum Likelihood Method for Estimating the Mass and Period Distributions of Extra-solar Planets," *Monthly Notices of the Royal Astronomical Society* 335 (2002): 151–158. More specifically, these researchers estimate that the true occurrence rate of planets with masses between one and ten times Jupiter's mass and orbital periods between two days and ten years is 3 percent. This underestimates the total occurrence rate of giant planets. First, the estimate does not include giant planets less massive than Jupiter, which suffer from larger observational selection effects. Second, planets with long orbital periods are

more difficult to detect (though Tabachnik and Tremaine try to correct for this bias). This estimate assumes that giant planets occur with equal probability among main-sequence stars of any mass. But the present data hint at a declining incidence of planets among low-mass stars [see Laws et al., "Parent Stars of Extrasolar Planets. VII. New Abundance Analyses of 30 Systems," *Astronomical Journal* 125 (2003): 2664–2677], which we have taken into account in our estimate of 4 percent.

25 On the importance of information in biology, see Hubert Yockey, *Information Theory and Molecular Biology* (Cambridge: Cambridge University Press, 1992); Bernd-Olaf Küppers, *Information and the Origin of Life* (Cambridge: MIT Press, 1990); Bernd-Olaf Küppers, *Molecular Theory of Evolution* (Heidelberg: Springer, 1983); W. Loewenstein, *The Touchstone of Life* (New York: Oxford University Press, 1998). On the difference between biological information and the chemical structures that carry that information, see Michael Polanyi, "Life's Irreducible Structure," *Science* 160 (1968): 1308, and Michael Polanyi, "Life Transcending Physics and Chemistry," *Chemical and Engineering News* (Aug. 21, 1967), 54–66.

26 "Life arose here almost as soon as it theoretically could. Unless this occurred utterly by chance, the implication is that nascent life itself forms—is synthesized from nonliving matter—rather easily. Perhaps life may originate on any planet as soon as temperatures cool to the point where amino acids and proteins can form and adhere to one another through stable chemical bonds. Life at this level may not be rare at all." Brownlee and Ward, *Rare Earth: Why Complex Life is Uncommon in the Universe* (New York: Copernicus, 2000), xix. See also 1–14. Stuart Ross Taylor, also a skeptic about the pervasiveness of intelligent life, similarly argues that the origin of life is basically a matter of getting the right chemical conditions. *Destiny or Chance: Our Solar System and Its Place in the Cosmos* (Cambridge: Cambridge University Press, 1998), 187.

27 See, for instance, Jacques Monod, *Chance and Necessity* (New York: Knopf, 1971). Astrophysicist Fred Hoyle famously joined biologists in this assessment. See his *The Intelligent Universe* (London: Michael Joseph, 1983).

28 See, for instance, Stuart Kauffman, *At Home in the Universe: The Search for the Laws of Self-Organization and Complexity* (New York: Oxford University Press, 1995), and Christian De Duve, *Vital Dust: A Cosmic Imperative* (New York: Basic Books, 1996).

29 Since we have not dealt with such issues in this book, we will not enter into a detailed discussion of them here. In general, we're skeptical of the assumption that since life appeared quickly on Earth, it will probably evolve on any habitable planet. This ignores several other possibilities—for example, that we are the victims of a selection effect. Moreover, self-organizational scenarios generally suffer from a basic conceptual problem. The self-organizing systems known in nature create repetitive ordered patterns. But this is not what needs explaining in biology. In fact, specified biological information is nonrepetitive and has more in common with the seemingly random structure of a book's letters written in an unknown language and alphabet than with repetitively ordered ones. For these reasons, we suspect that the likelihood of life originating, given only the known laws of physics and chemistry, and even on an ideally hospitable planet, is much closer to zero. For discussion, see Paul Davies, *The Fifth Miracle: The Search for the Origin and Meaning of Life* (New York: Simon & Schuster, 1999). For a strong critique of the viability of self-organizational scenarios for explaining the origin of life, see William A. Dembski, *No Free Lunch: Why Specified Complexity Cannot be Purchased Without Intelligence* (Lanham: Rowman & Littlefield Publishers, 2002), 179–237.

The longer the origin-of-life experiments continue without producing the basic informational molecules of life in plausible settings without undue experimenter intervention, the smaller the chances for the naturalistic origin of life. There's a similar argument for SETI: the longer they go without detection, the smaller the chances for ETIs in the Milky Way.

30 See Gould's *Wonderful Life: The Burgess Shale and the Nature of History* (New York: W.W. Norton, 1989).

31 See discussion in Koerner and LeVay, *Here Be Dragons*, 140–153. The title of Chapter 6 is "What Happens in Evolution?: Chance and Necessity in the Origin of Biological Complexity." Gould elsewhere criticizes what he calls the "anthropic principle," by which he means the idea that our environment has been designed by a "Supermind." See his "Mind and Supermind," in *Modern Cosmology and Philosophy*, John Leslie, ed. (Amherst, NY: Prometheus, 1998). Harvard seems to breed commitment to chance in evolution. Harvard's George Gaylord Simpson argued similarly several decades ago in "The Nonprevalence of Humanoids," *Science* 143 (1964): 769,

as has Ernst Mayr, Emeritus Professor of Zoology at Harvard. See, for instance, his "The Probability of Extraterrestrial Life," in *Extraterrestrials: Science and Alien Intelligence*, Edward Regis, Jr., ed. (Cambridge: Cambridge University Press, 1985), 23–30.

32 See Simon Conway Morris, *The Crucible of Creation: The Burgess Shale and the Rise of Animals* (Oxford: Oxford University Press, 1998). Although many SETI researchers and astrobiologists are encouraged by his argument, so far as we know, Morris himself does not seem committed to the Copernican Principle.

33 This is a quote from Simon Conway Morris in Koerner and LeVay, *Here Be Dragons*, 146.

34 See his important recent book, *Life's Solution: Inevitable Humans in a Lonely Universe* (Cambridge: Cambridge University Press, 2003).

35 Dick, *Life on Other Worlds*, 169, 195. Stuart Clark, following Paul Davies, argues that astronomers are here assuming the Copernican Principle and the Principle of Plenitude, which means, basically, that whatever can happen, will. We are less confident that the Principle of Plenitude qualifies as a common methodological principle in the same way that the Copernican Principle does. Stuart Clark, *Life on Other Worlds and How to Find It* (Chichester, UK: Springer-Praxis, 2000), 2–6.

36 Peter D. Ward and Donald Brownlee, *Rare Earth*, 83–242.

37 "My feeling is that the ultimate vindication of the Copernican Principle would be if Earth, as a planet, were not even privileged by the emergence of life. If Copernican ideas can be extended to biology, then there is no reason to assume that life on Earth is a cosmic one-off." Clark, *Life on Other Worlds*, 3. Some SETI advocates manage to blend all these elements into an uneasy amalgam. For instance, in *The Search for Life on Other Planets*, 301, Bruce Jakosky says:

> Finding non-terrestrial life would be the final act in the change in our view of how life on Earth fits into the larger perspective of the universe [notice the stereotypical history of the Copernican Revolution]. We would have to realize that life on Earth was not a special occurrence, that the universe and all of the events within it were natural consequences of physical and chemical laws, and that humans are the result of a long series of random events.

38 One final prediction of the Copernican Principle is: *There's nothing special about water- and carbon-based life forms. In other environments, life probably evolves according to different chemistries and energy needs.* Although this is difficult to falsify, we argued in Chapter Two that there is a great deal of evidence that carbon's diverse capacity for chemical bonding makes it especially, if not uniquely, fitted for the job of bearing biological information. Silicon is a distant second, and no other elements are plausible competitors. And water is by far the best solvent for chemical reactions. For these reasons, we should restrict habitability to those places with plenty of liquid water, and to expect that, if extraterrestrial life exists, it will be carbon-based.

39 Brandon Carter first suggested this idea. Philosopher John Leslie developed it in *The End of the World: The Ethics and Science of Human Extinction* (London: Routledge, 1996). For detailed analysis, see Nick Bostrom, *Anthropic Bias: Observation Selection Effects in Science and Philosophy* (London: Routledge, 2002). J. Richard Gott III defends this type of reasoning in "Implications of the Copernican Principle for Our Future Prospects," *Nature* 363 (May 1993): 315–319. Their various arguments render different judgments and timescales, which we do not discuss here. For a critique of Gott's argument, see Carlton M. Caves, "Predicting Future Duration from Present Age: A Critical Assessment," *Contemporary Physics* 41, no. 33 (2000): 143–153. The Doomsday argument is often combined with additional suggestions that advanced civilizations may have a propensity to destroy themselves, either through weapons of mass destruction or environmental degradation. It may be these politically useful aspects of the argument, rather than its intrinsic plausibility, that lead intellectuals to take it seriously.

40 All such temporal applications of the Copernican Principle assume that the future is *finite*. If, in contrast, the future is (potentially) infinite, the argument fails, since observers will always find themselves near the "beginning" of cosmic time.

41 Amir D. Aczel, *Probability 1: Why There Must Be Intelligent Life in the Universe* (New York: Harcourt Brace, 1998), 208–214.

42 See Steven J. Dick, "Cosmotheology: Theological Implications of the New Universe," *Many Worlds*, 202.

43 David Darling's recent defense of SETI optimism, *Life Everywhere: The Maverick Science of Astrobiology* (New York: Basic Books, 2001), is a prime example. But we know of one exception. See Michael D. Papagiannis, "May There Be an Ultimate Goal to the Cosmic Evolution?" *Bioastronomy: The Search for Extraterrestrial Life—The Exploration Broadens*, J. Heidmann and M. J. Klein, eds. (Berlin: Springer-Verlag, 1990).

44 In 2001, Seth Shostak of the SETI Institute gave a public lecture as part of a Templeton Foundation lecture series entitled "The Search for Extraterrestrials and the End of Traditional Religion." In a similar vein, see the essay by Jill Cornell Tarter (also of the SETI Institute), "SETI and the Religions of the Universe," *Many Worlds*, 143–149.

45 Koerner and LeVay in *Here Be Dragons*, 173.

46 Our own views about the existence of ETI have changed over the years, but for empirical rather than theological reasons. There are interesting philosophical and theological issues here, however. Some scientists, in fact, argue that the existence of ETIs would enhance the prospects for design. Physicist Paul Davies, for instance, argues, "If life is widespread in the universe, it gives us more, not less, reason to believe in cosmic design." "Biological Determinism, Information Theory, and the Origin of Life," *Many Worlds*, 15. In *The Face That Is in the Orb of the Moon*, Plutarch argued similarly for life on the Moon on the basis of intelligent design. See Lewis White Beck, "Extraterrestrial Intelligent Life," in Regis, *Extraterrestrials: Science and Alien Intelligence*, 3–6. Of course, we're not arguing that the existence of ETI would have no specific theological consequences. It obviously would. For some discussion, see Ernan McMullin, "Life and Intelligence Far from Earth: Formulating Theological Issues," in *Many Worlds*, 151–175. Historically, theological views have influenced astronomical research in a surprising diversity of ways. See Michael J. Crowe, "Astronomy and Religion: Some Historical Interactions Regarding Belief in Extraterrestrial Intelligent Life," *Osiris* 16 (2001): 209–226.

47 We find the approach of Christian author and scholar C. S. Lewis agreeable. He made no theological pronouncements on the ETI question, though he included extraterrestrials in his fictional works, particularly in *Out of the Silent Planet* (New York: Scribner's, 1996). Speaking as a theologian, he argued, decades before the discovery of an extrasolar planet, that the existence of ETI is, and must remain, an open question:

> We know that God has visited and redeemed His people, and that tells us just as much about the general character of the creation as a dose given to one sick hen on a big farm tells us about the general character of farming in England. . . . It is, of course, the essence of Christianity that God loves man and for his sake became man and died. But that does not prove that man is the sole end of nature. In the parable, it was one lost sheep that the shepherd went in search of: it was not the only sheep in the flock, and we are not told that it was the most valuable—save insofar as the most desperately in need has, while the need lasts, a peculiar value in the eyes of Love. The doctrine of the Incarnation would conflict with what we know of this vast universe only if we knew also there were other rational species in it who had, like us, fallen, and who needed redemption in the same mode, and they had not been vouchsafed it. But we know of none of these things.

Lewis dryly commented on atheists' attempts to use both sides of the ETI debate as a weapon against Christianity:

> If we discover other bodies, they must be habitable or uninhabitable: and the odd thing is that both these hypotheses are used as grounds for rejecting Christianity. If the universe is teeming with life, this, we are told, reduces to absurdity the Christian claim—or what is thought to be the Christian claim—that man is unique, and the Christian doctrine that to this one planet God came down and was incarnate for us men and our salvation. If, on the other hand, the earth is really unique, then that proves that life is only an accidental by-product in the universe, and so again disproves our religion. Really, we are hard to please.

In C. S. Lewis, "Dogma and the Universe," in *The Grand Miracle and Other Essays on Theology and Ethics from 'God in the Dock,'* W. Hooper, ed. (New York: Ballantine Books, 1990), 14.

48 In Steven J. Dick, "Cosmotheology: Theological Implications of the New Universe," *Many Worlds*, 200–205. In "Extraterrestrial Life and Our World View at the End of the Millenium," *Dibner Library Lecture* (Washington, DC: Smithsonian Institution Series, 2000), 14–25, Dick argues similarly:

> I propose that . . . there are two, and only two, "chief world systems," in the sense of overarching cosmologies that affect, or should affect, our world view at a lower level of the hierarchy such as philosophical or religious. . . . These two mutually exclusive world views, stemming from the concept of cosmic evolution, are as follows: that cosmic evolution commonly ends in planets, stars, and galaxies, or that it commonly ends in life, mind, and intelligence.

Obviously he is committing one of the oldest and best-known logical fallacies, the description of which has been a staple of freshman logic textbooks ever since such things as freshmen and textbooks have existed. The false dilemma doesn't seem like a fallacy from inside a dogma. From outside the dogma, the fallacy and its speaker can sound almost surreal in their disregard of other possibilities.

CHAPTER 15, A UNIVERSE DESIGNED FOR DISCOVERY

1 We thank Gershon Robinson and Mordechai Steinman for popularizing this example. For an intriguing argument that *2001: A Space Odyssey* makes a somewhat unconscious argument for design, see their *The Obvious Proof* (CIS Publishers, 1993) as well as their interesting website, "The 2001 Principle," http://www.2001principle.net/.

2 We are alluding here to a common method of reasoning in the historical and origins sciences, often called abductive reasoning or "inference to the best explanation." For discussion, see Stephen C. Meyer, "The Scientific Status of Intelligent Design," 176–192, 213–228, in *Science and Evidence for Design in the Universe* (San Francisco: Ignatius Press, 2000), and Peter Lipton, *Inference to the Best Explanation* (London: Routledge, 1991). For a recent collection of articles about the modern design argument, see Neil A. Manson,ed., *God and Design: The Teleological Argument and Modern Science* (London: Routledge, 2003).

3 See her personal account of the episode in S. Jocelyn Bell Burnell, "Little Green Men, White Dwarfs or Pulsars?" *Annals of the New York Academy of Science* 302 (Dec. 1977): 685–689.

4 Of course, if the universe were deterministic, then nothing within it would be the result of chance. Even among those who allow chance explanations, many restrict true chance explanations to the realm of quantum indeterminacy. Protons might decay by chance, but falling Scrabble letters do not. So they might insist that if we had sufficient knowledge of the initial conditions, we could trace a determining cause for every falling Scrabble piece. We don't intend to take a position on such controversies. Here we're simply describing the way we normally make causal explanations in everyday life. In this example, our central point is that the law of gravity alone does not determine the specific configuration of the Scrabble letters.

5 Peter Ward and Don Brownlee, *Rare Earth: Why Complex Life Is Uncommon in the Universe* (New York: Copernicus, 2000).

6 He first developed and articulated this concept in William A. Dembski, *The Design Inference* (Cambridge: Cambridge University Press, 1998). He developed it for the general reader in *Intelligent Design: The Bridge Between Science and Theology* (Downers Grove: InterVarsity Press, 1999). He has continued to refine the concept in *No Free Lunch: Why Specified Complexity Cannot Be Purchased Without Intelligence* (Lanham, Md.: Rowman & Littlefield, 2001). There are many complicated philosophical issues surrounding Dembski's argument, which we have avoided here. Although our account differs in certain respects from Dembski's, we owe a great deal of our analysis here to his seminal insights.

7 Ratzsch, *Nature, Design, and Science*, 72–73.

8 This phrase is from a public speech by Sir John Polkinghorne at a conference entitled *Cosmos and Creator*, held in Seattle, Washington, on April 27, 2001.

9 There is another type of selection effect, however, that could be relevant. One might object that we will find that our environment makes possible those discoveries we have made because we can only make those discoveries that our environment does make possible. We will discuss that in the next chapter.

10 As with attempts to explain away the appearance of fine-tuning, we suspect that there will be attempts to explain away the correlation. Ironically, however, such attempts may be a tacit admission that the correlation, if it really exists, suggests design.

11 See John Leslie, *Universes* (London: Routledge, 1989). This type of argument is somewhat subjective if considered in isolation. As a supplement to the previous analysis, however, it seems to capture one of the reasons we often infer design.

12 On both these points, see the excellent brief article by Neil Manson, "Cosmic Fine-Tuning, 'Many Universe' Theories and the Goodness of Life," in *Is Nature Ever Evil?*, William Drees, ed. (London: Routledge, 2003), 139–146.

13 Hans Blumenberg, *The Genesis of the Copernican Revolution*, Robert M. Wallace, trans. (Cambridge: MIT Press, 1987), 3, originally published as *Die Genesis der kopernikanischen Welt* in 1975; M. J. Denton, *Nature's Destiny* (The Free Press, New York, 1998), 50–61.

14 *Nature's Destiny*, 262.

15 John Barrow and Frank Tipler, *The Anthropic Cosmological Principle* (Oxford: Oxford University Press, 1986), 258–276.

16 S. L. Jaki, *Maybe Alone in the Universe After All* (Pinckney, Mich.: Real View Books, 2000).

17 In "Resonances," *Physics Today* 27, no. 2 (1974): 73.

CHAPTER 16, THE SKEPTICAL REJOINDER

1 Charles Darwin, *The Origin of Species*, edited and with an introduction by Gillian Beer (Oxford: Oxford University Press, 1996), 4 (originally published in 1859).

2 Indeed, skeptics of our argument are welcome to do just this. Every clever and credible observational technique they invent that would only work in a seemingly less life-friendly environment would chip away at our argument's credibility. No one such idea would destroy our argument, because it would take many, many such ideas to show that Earth was less than optimal over a range of observational situations than some uninhabitable place. Such a slew of credible insights is possible, though, and is one more example of how the correlation is falsifiable.

3 For the sake of full disclosure, we note that we made this prediction shortly after we began writing this book, in February or March 2001. Some discussions of the possibility of using gamma ray bursts as standard candles had already appeared in press. As of 2003, astronomers are still debating the issue.

4 See J. Armstrong, L. Wells, and G. Gonzalez, "Rummaging Through Earth's Attic for Remains of Ancient Life," *Icarus* 160 (2002): 183–196. See also the News and Views piece on the study: C. R. Chapman, "Earth's Lunar Attic," *Nature* 419 (2002): 791–794.

5 In *The Cosmological Background Radiation*, 7, M. Lachieze-Rey and E. Gunzig mention four such assumptions:

- Homogeneity of space applies. Thus it is assumed that all points of space are equivalent and the properties associated with each point are the same, that is, the laws of physics are the same everywhere. This homogeneity is taken to apply to spatial scales greater than those of galaxies, clusters of galaxies and even superclusters, i.e. greater than a few hundred Mpc.
- Isotropy of space applies. This means that there is no privileged direction in space. (Again this refers to large scales.) These two assumptions make up what is called the *cosmological principle*.
- That the matter in the universe can be described very simply in terms of what is called a perfect fluid. In this case its properties are completely given by its density ρ and its pressure p.
- That the laws of physics are the same everywhere.

Notice that they repeat the assumption about the laws of physics in the first and fourth item. But homogeneity of space is clearly a separate issue from the uniformity and universality of the laws of physics.

6 See S. Nadis, "Size Matters," *Astronomy* 30 (March 2002): 28–32.

7 The cosmologist P. J. E. Peebles notes that inflation theory violates Einstein's cosmological principle, but the boundary of the universe we can see is distant from the wider chaotic domains. *Principles of Physical Cosmology* (Princeton: Princeton University Press, 1993), 15.

8 John Barrow makes a related comment in his *Impossibility: The Limits of Science and the Science of Limits* (Oxford: Oxford University Press, 1998), 169–70:

> Before the possibility of inflation was discovered, it was generally assumed that the Universe should look pretty much the same beyond our horizon as it does inside it. To assume otherwise would have been tantamount to assuming that we occupied a special place in the Universe—a temptation that Copernicus taught us to resist. . . . The general character of inflationary universes reveals that we must expect the Universe to be far more exotically structured in both space and time than we had previously expected.

Barrow seems to imply that chaotic inflationary models contradict the Copernican Principle. Actually, these theories, which posit multiple universes that differ in their constants and initial conditions, seem to us an example of the remarkable plasticity—and essentially metaphysical character—of the Copernican Principle. Formerly we were told that the Copernican Principle required us to assume the uniformity of laws and constants throughout the universe. When such an assumption led to a universe that seemed disturbingly fine-tuned for the existence of complex life, it suddenly required that we posit a multiverse with varying properties between individual universes.

9 "The Anthropic Principle: Laws and Environments," in *The Anthropic Principle: Proceedings of the Second Venice Conference on Cosmology and Philosophy*, F. Bertola and U. Curi, eds. (Cambridge: Cambridge University Press), 29.

10 Isaac Newton, *Mathematical Principles of Natural Philosophy* (Berkeley: University of California Press, 1960), 542–544.

11 William Dembski argues that we can decisively rule out chance for an event if, given all the "probabilistic resources" available, the chance for an event is still one in 10^{150} (10^{-150}). He arrives at this number by combining the number of elementary particles in the observable universe, the duration of the observable universe until heat death, plus the Planck time, which is the smallest unit of time for a physical event. The idea is to treat every particle in the universe as part of giant computer. One in 10^{158} would obviously exceed this "universal probability bound."

Nevertheless, we often infer design quite well even when probabilities are much higher than Dembski's very stringent universal probability bound. Moreover, in many cases, we know little about the actual probability of producing a structure. How would we make such a determination, for instance, for Mount Rushmore? See Dembski, *The Design Inference* (Cambridge: Cambridge University Press, 1998), 203–214.

12 As we mentioned in Chapter Thirteen, there is also a powerful philosophical argument against the possibility of a universe with an infinite past. See, for example, William Lane Craig, *The Kalam Cosmological Argument* (Eugene, Ore.: Wipf & Stock, 1979), and William Lane Craig and Quentin Smith, *Theism, Atheism, and Big Bang Cosmology* (Oxford: Clarendon, 1993). Craig's most recent article defending this argument is "In Defense of the *Kalam* Cosmological Argument," *Faith and Philosophy* 14 (1997): 236–247.

CONCLUSION, READING THE BOOK OF NATURE

1 Robert Zimmerman, *Genesis: Apollo 8: The First Manned Flight to Another World* (New York: Dell Publishing, 1998), 242, 293.

2 The most complete articulation of Polanyi's view is his *Personal Knowledge: Towards a Post-Critical Philosophy* (Chicago: University of Chicago Press, 1958).

APPENDIX A, THE REVISED DRAKE EQUATION

1 Donald Brownlee and Peter Ward, *Rare Earth* (New York: Copernicus, 2000).

2 In their version of the Drake equation, Ward and Brownlee give n_g, the number of stars in the Galactic Habitable Zone, but it should be f_g, as we present it in the text.

3 For a remarkably cogent analysis and critique of the very "pursuit-worthiness" of SETI, see André Kukla, "SETI: On the Prospects and Pursuitworthiness of the Search for Extraterrestrial Intelligence," *Studies in the History and Philosophy of Science* 32, no. 1 (2001): 31–67.

APPENDIX B, WHAT ABOUT PANSPERMIA?

1 Svante Arrhenius, *Worlds in the Making* (New York: Harper & Row, 1908).

2 Jay Melosh, "Exchange of Meteoritic Material Between Stellar Systems," *Lunar and Planetary Science XXXII* (2001): 2022.

3 J. Secker, J. Lepock, and P. Wesson, "Damage Due to Ultraviolet and Ionizing Radiation During the Ejection of Shielded Microorganisms from the Vicinity of 1 M_{sun} Main-Sequence and Red Giant Stars," *Astrophysics and Space Science* 219 (1994): 1–28; J. Secker, P. Wesson, and J. Lepock, "Astrophysical and Biological Constraints on Radiopanspermia," *Journal of the Royal Astronomical Society of Canada* 90 (1996): 184–192.

4 Secker, et al. "Damage Due," "Astrophysical and Biological Constraints" (1994, 1996).

5 C. Mileikowsky, et al., "Natural Transfer of Viable Microbes in Space: 1. From Mars to Earth and Earth to Mars," *Icarus* 145 (2000): 391–427.

Acknowledgments

Many people deserve credit for their help with this book—so many, in fact, that we fear we may leave some unmentioned. Several individuals were kind enough to review parts or all of our manuscript. Others contributed needed information, crucial distinctions, and insightful arguments. We are especially grateful to Jonathan Witt and Andrew Sperling. Jonathan provided countless editorial suggestions to an early version of our manuscript, and was bold enough to encourage us to cut some unneeded text. Andrew provided indispensable help in creating many of our line drawings. Without his help, the text would have been much more sparsely and much less impressively illustrated. We would also like to thank Sam Fleishman for seeing the potential in this project when it was little more than an idea with a few examples.

Ben Wiker and Mike Keas gave us excellent advice on the manuscript, especially on our historical sections. Bruce Nichols provided keen editorial advice in the very early stages of our book. Thanks also to Kerry Magruder, Dennis Danielson, Michael Crowe, and Ted Davis for reading our historical chapter (and for saving us from making several mistakes). We also benefited from suggestions by Nancy Pearcey, Allan Sandage, Kyler Kuehn, Bill Dembski, Peter Hodgson, Josh Gilder, Phil Skell, Daniel Bakken, David Snoke, Scott Minnich, and Ginny Richards.

Robin Collins and Del Ratzsch offered some very helpful technical advice. Robin and Del were also participants in the symposium held at Notre Dame, "The Mathematics of the Fine-Tuning Argument," in April

2003, along with Neil Manson, Tim and Lydia McGrew, Nick Bostrom, Roger White, Alex Pruss, and Brian Pitts. We were delighted to be included in this meeting, and benefited greatly from it.

Thanks to John Armstrong, Don Brownlee, Tom Quinn, and Forrest Mims for providing scientific data and illustrations, and to Miriam Moore at Regnery, for her careful editing of our final manuscript.

We would like to thank the John Templeton Foundation, which generously provided financial support for one of us (GG) for research on this book within the context of its "Cosmology & Fine-Tuning Research Program" (Grant ID #938-COS187). Finally, our thanks to the Discovery Institute, and especially Bruce Chapman, Steve Meyer, John West, and Bob Cihak, for their unflagging support and encouragement while we have worked on this project.

An acknowledgement should not be interpreted as implying that an individual agrees with us. And any errors remaining are, of course, our own.

Figure Credits

Many of the line drawings were created with the valuable help of Andrew Sperling at Discovery Institute. Of these, most are original, but some were drawn, with certain changes, after other published line drawings. For many of these, complete bibliographic information is included in the endnotes of the relevant chapters. Figures 1.1, 1.3, and 1.7 are drawn after Littmann et al., p. 8, p. 131, and p. 88, respectively. Figure 2.4 is after Figure 1 of Zachos et al., "Trends Rhythms, and Aberrations in Global Climate 65 Ma to Present," *Science* 292 (2001): 686–693. Figure 3.1 is after Figure 17.10 of Tarbuck and Lutgens (1999). Figure 3.2 is after Figure 19.28 of Tarbuck and Lutgens (1999). Figure 3.5 is after P. Cloud, *Oasis in Space: Earth History from the Beginning* (New York: Norton), 1988. Figure 3.6 is after Van der Voo (1990). Figure 8.2 is after Vallée (2002). Figure 8.7 is after Figure 39 of Timmes et al. (1995) and Figure 1 of Samland (1998). Figure 9.3 is after Figure 7 of Perlmutter et al. (1999). Figure 9.5 is after Figure 17-13, Neil F. Comins and William J. Kaufmann III, *Discovering the Universe*, sixth edition (New York: W.H. Freeman and Company), 2003. Figure 10.2 is after Figure 7.15 of Rolfs and Rodney (1988). Figure 10.3 is after Figure 4 of Tegmark (1998). Plate 15 is after Figure 9-15 of Michael A. Seeds, *Stars and Galaxies*, 2nd edition (Pacific Grove, CA: Brooks/Cole, 2001).

Figures 5.6, 12.2, and Plates 1, 16, are courtesy NASA. Figure 5.3 and Plates 2 and 10 are courtesy NASA/JPL/Cal Tech. Figure 15.1 is courtesy NASA, Viking Project and Malin Space Science Systems. Plate 12 is courtesy R. Evans, J. Trauger, H. Hammel, the HST Comet Science Team and NASA. Plate 18 is courtesy Robert Williams, Hubble Deep Field Team, and NASA. Plate 19 is courtesy NASA/WMAP Science Team.

Figures 1.5, 6.1 (Einstein portrait), 9.2, and 9.8 are courtesy US Naval Observatory Library. Figure 1.6 is courtesy Lick Observatory. Figure 2.2 is courtesy Jeremy Young and the Natural History Museum in London. Figure 2.3 is courtesy NOAA. Plate 6 is courtesy the National Geophysical Data Center's Marine Geology & Geophysics Division/NOAA. Figure 3.3 is from A. D. Raff and R. G. Mason, "Magnetic Survey off the West Coast of North America, 40°N Latitude to 52°N Latitude," *Geological Society of America Bulletin* 72 (1961): 1267–1270, courtesy the Geological Society of America.

Figures 4.2 and 8.5 are courtesy Donald Brownlee. Figures 5.5, 6.2, 8.1 (left), 8.4, 8.9, 8.10, and Plates 3 (top), 8, 9, 13, and 14, are by Guillermo Gonzalez. Figure 8.3 is courtesy Ronald Drimmel. Figure 9.1 is courtesy the Observatories of the Carnegie Institution of Washington. Figure 14.1 is from University of Chicago, courtesy AIP Emilio Segrè Visual Archives.

Figure 15.2 is courtesy William Dembski. Plate 3 (bottom) is courtesy Mike Reynolds. The original photo used in Plate 4 is courtesy Jagdev Singh of the Indian Institute of Astrophysics, Bangalore, India. Plate 11 is courtesy Robin Canup and Bill Ward.

Figure 2.1 is based on data described in Petit et al. (1999). Figure 2.6 is based on data from the 1.9 mile deep Greenland summit ice core obtained as part of the Greenland Ice Sheet Project 2 (GISP2), and data from the Quaternary Isotope Laboratory at the University of Washington. In Figure 5.2, data for Earth are courtesy Thomas Quinn, and data for Mars are courtesy John Armstrong. In Figure 7.5, the comet data are from JPL's DASTCOM database as of June 21, 2003. Asteroid data are from the Minor Planet Center and include all objects in their database larger than about half a kilometer, as of June 21, 2003. Both sets of data are restricted to asteroids and comets with known orbital elements. Figure 8.6 is after Figure 4 of Laws et al. (2003).

The historical images in Figures 6.1, 6.3, 11.1, 11.2, 11.4–11.14, and 13.3 are courtesy University of Oklahoma, History of Science Collections, and some of the information in the captions for these figures is courtesy Kerry Magruder, History of Science Collections, University of Oklahoma.

INDEX

absorption lines
 and chromosphere, 350n24
 and CMBR, 389n20
 discovery, 12
 fine-structure constant, 396n46
 and quasars, 179–181
 in solar spectrum, 15, 123–125
 titanium oxide, 374n7
active galactic nucleus (AGN) outbursts, 162
Aczel, Amir, 289
Adams, Douglas, 152
Adams, Fred, 268–269
age, of galaxies, 258
agency, 263–264, 327–330, 333–334, 406–407n27
 see also God; purpose; rationality
Airy, George, 13
albedo, 69–71, 109, 373n18
alpha particles, 198–199
Alpher, Ralph, 175
Anaxagoras, 343
ancients
 cosmology, 111–114, 225–227, 393n3, 399n30
 and distance ladder, 111–113
 on human condition, 218, 245
 naturalistic views, 248–249
 on nature, 329
 and panspermia, 343
 and time, 228
Anders, Bill, 331, 347n3, Plate 1
Andromeda, 150, 167, 173, 258
annular eclipses, 18
Antarctica, 23–24, 37
anthropic coincidence, 197–203
 see also fine-tuning
Anthropic Principle, 136–138, 164–166, 405n14
 and Copernican Principle, 265–268, 272, 405–406n22
anthropocentrism, 251
Apian, Peter, 111(fig.)
Apollo 8 mission, 331, 347n2
Aquinas, Saint Thomas, 230
Arctic, 37
Aristarchus of Samos, 112–114, 399n32
Aristotle
 cosmology, 225–227, 229–231, 238, 239, 242, 260
 and distance ladder, 111–113
 on human condition, 218
 on nature, 329
 projectile motion, 242
Arrhenius, Svante, 343
Asimov, Isaac, 65
asteroid belt, 77–79, 83(fig.), 368n38
 distance from, 128–131
 orbits, 93
 and planet formation, 384–385n28
asteroids, 73–79
 hydration, 368n36
 impact rates, 131
 Kirkwood Gaps, 367n30

asteroids (*continued*)
and metals, 156–157, 161
parallax measure, 114
and star luminosity, 132
and Sun, 133(fig.), 377n36
see also Yarkovsky effect
astrology, 103
astronomy
ancient, 111–113, 225–227
and continental drift, 54–55
and cosmology, 104–106
early twentieth century, 120
nineteenth century, 113–114, 120, 140
and solar eclipses, 12–15
twentieth century, 183–185
last decade, 284
see also specific astronomers
asymptotic giant branch (AGB) stars, 76–77
atmosphere
and breathing, 376n25
CO_2 level, 130
lack of, 90, 91
of Mars, 84
and moons, 350n19
of outer planets, 91
and planet size, 7
pressure, 39, 355n30
and star-planet distance, 130
of stars, 365n9
ultraviolet role, 135
of Venus, 86, 107
see also atmosphere, of Earth
atmosphere, of Earth
clouds, 69–71
early components, 72–73
and glacial periods, 72, 366n22
hydrological cycle, 73, 91
and radiation, 66, 68
and technological life, 64
and telescopes, 60, Plate 8
transparency
benefits, 66–68, 72–75, 316
and observability, 90–91, 365n10
oxygen effect, 209
see also carbon dioxide; greenhouse gases; methane; nitrogen; oxygen; water vapor
atoms, 126, 201–202, 380n59
Augustine, Saint, 228
autotrophs, 37, 40
energy from, 130

B stars, 139–140, 146
bacteria
and panspermia, 344–345
and pressure, 369n17
salt-tolerance, 359n77
and uranium, 63
Barrow, John, 176, 209, 310
***A Beautiful Mind*, 299**
Bell, Jocelyn, 295
Berger, A., 41–42
beryllium isotopes, 25, 353n12
beryllium-8, 199, 393n10
Bessel, Friedrich, 140
bias, 317–319
in nature, 332
of SETI research, 281
see also Anthropic Principle
Big Bang(s)
causal agent, 329
chemical elements, 152–153, 202
and gravity, 197
series of, 261
Big Bang theory, 172, 174–178, 188
and CMBR, 273
biodiversity, 92, 358n74, 376n16, 376n27
biomineralization, 63
black holes
and galactic interactions, 167
and quasars, 179
and radiation, 162
and Relativity Theory, 326
Blumenberg, Hans, 309, 364n1
Bode's Law, 372n5
body temperature, 34, 357n55
bolides, 75
Bondi, Hermann, 174, 260
Borman, Frank, 331, Plate 1
boron-8, 380nn59–61
Boyle, Robert, 231
Brahe, Tycho, 104, 213–214, 235–238, 400n42
Bridgewater Treatises, 392n4
Brimhall, George, 62, 64
Brownlee, Don, 285, 287, 297, 338
Brunier, Sergei, 3, 4
Bruno, Giordano, 222, 223, 242

calcium, 354n20
calendars, 17
Callisto, 97, 372n52
capillary action, 34

carbon
and biological information, 411n38
and complex life, 32–33, 314–315, 322
in Earth, 156–157, 368n36, 393n11
and fine-tuning, 198–199
and fire, 217–218
in hydrocarbons, 37, 72–73
and marine sediment, 354n22
in meteorites, 376n29
and oxygen, 357–358n61, 393n11
stability, 356n42
and stars, 154, 209
carbon-12
carbon-13 ratio, 209, 368n34, 390n26, 393–394n17
resonance, 198–199, 392n8, 393n12
carbon-14
dating, 29–30, 209
presence, 393n15
and sunspot cycles, 25, 353n12
carbon cycle, 55–57
see also greenhouse effect
carbon dioxide
and climate, 41–42, 72
exchangeability, 33
high levels, 57, 360n87, 364n38, 376n26
and ice, 23–24, 41–43, 360n85
and leaves, 29
on Mars, 82, 84
and M dwarf stars, 377n38
as oxidation product, 37
partial pressure, 355n30
and planetary mass, 59
and stellar distance, 130
and water, 357n59
see also greenhouse gases
carbon monoxide, 209, 382n15, 382n21
carbonaceous chondrites, 76
Carter, Brandon, 204
causal agent. *see* agency; purpose
Cepheids
Classical, 122, 172–174, 189, 190, 388nn9–10
Type II, 190, 388n7, 388n9
Cerenkov emission, 365n7
chance
atoms and neutrinos, 380n59
and contingency, 292
for Darwin, 401n57
vs. design, 296–298
chance (*continued*)
vs. fine-tuning, 196–197
and Jupiter moons, 372n52
vs. necessity, 286–287, 296, 413n4
and many worlds, 406n26
and orbits, 383n25
and purpose, 348n13
and quantum phenomena, 401n57
ruling out, 415n11
and SETI, 334
see also coincidence(s); contingency
Chandrasekhar limit, 157
chaos, 93–94, 105–106, 211, 370n27
chemical compounds, 357n46, 392n7
hydrocarbons, 37, 72–73
chemical elements
in cosmic time, 181–182
heavy elements, 182, 200–201
and nuclear force, 201
origin, 178
r-process, 385n37
and Relativity, 326
and supernovae, 152–153, 157–158
see also specific elements
chemical energy, 37, 67–68
chemical weathering, 34–35, 55–57, 375n18
chemistry
fine-tuning, 197–198, 394–395n31
key factors, 394n30
chlorine, 25
chromosphere, 13, 15, 18, 350n24
Cicero, 245
Circumstellar Habitable Zone (CHZ), Plate 17
carbon-12: carbon-13, 319–320n17
Continuously Habitable (CCHZ), 129(fig.), 132–134
description, 6–7, 127–132
and giant planets, 97
speculativeness, 319–320
civilization(s)
and Drake Equation, 288–289
extraterrestrial, 408nn10–11
probability of, 338–342
and purpose, 302
and temperature, 360n86
Clark, Stuart, 215, 224, 273
climate
and axial tilt, 92
change data, 22–29
and clouds, 41, 69–70, 72, 366n20
and Earth orbit, 94

climate (*continued*)
  feedback cycle, 35–36
  forecasting, 40–41
  and isotopes, 352–353n6
  of Mars, 82
  and Moon, 108–109
  Moon effect, 5–6, 107
  and orbital locking, 7
  and planet size, 93–94
  present, 41, 42
  and reflecting surfaces, 69–71
  simulations, 41
  and solar activity, 379n52
  stable periods, 359n80
  storms, 359n79
  and technological life, 61
  and tides, 349n9
  water effect, 23, 33–34
cloud condensation nuclei (CCN), 72
clouds
  and climate, 41, 69–70, 72, 366n20
  in early Earth, 366n22
  of gases, 181, 389n22
  giant molecular, 160
  intergalactic, 209
  interstellar (H II), 145–146, 153–154
  Lyman-alpha, 178, 179–181, 190
  Magellanic, 172, 190
  and outer planets, 91
  see also Oort Cloud
coincidence(s)
  and Copernican Principle, 284–285
  and evolution, 286–287
  large number, 197–203, 394n21
  see also fine-tuning
Collins, Robin, 214, 215
colors, 12, 140, 204
  Rayleigh scattering, 378n43
  see also redshift
comets
  and Copernican Principle, 256
  as data source, 78–79
  disintegration of, 213, 368n38
  and Earth atmosphere, 73–75
  Hale-Bopp, Plate 9
  and metals, 156–157
  and planet size, 97
  reservoirs of, 160
  shields from, 114–115
  Shoemaker-Levy, 91, 97
compasses, 52–53
complexity
  and design, 297–298
  and habitability, 321–322
  layers, 211, 376n26
  and measurability, 321–322
  principle of, 395–396n45
compression waves, 361n3, 374n4
Comte, Auguste, 120
condensation, 30, 99–100, 352n6
conduction, of heat, 357n49
constancy, 211, 370n24, 396n46
Contact, 290, 296, 298, 300, 335, 341
continental drift, 53–55
continents
  and carbon cycle, 57
  lack of, 36, 358n64
  before separation, 47, 52
  see also plate tectonics
contingency(ies), 229, 287, 291–292, 297, 406n25
convection
  in Earth mantle, 57, 63
  on Mars, 85
  and potassium, 131
  stagnant lid, 86
  and stars, 203–204
  on Venus, 86
Conway Morris, Simon, 286–287
Copernican Principle
  and Anthropic Principle, 265–268
  basic principle, 292, 402n8
  and cosmological principle, 248–250, 402n5, 407n30
  and extraterrestrial life, 281, 287, 290–292, 408n12, 411n37
  and future, 411n40
  and human beings, 402n6, 405n15, 407n29
  implicit predictions, 250–265
    carbon-based life, 411–412n38
    in cosmology, 259–261
    major, 251–257
    in physics, 261–265
  and intelligent life, 284, 286
  and laws of physics, 323
  and life origination, 284
  and many worlds, 268–271, 306, 319, 415n8
  and measurability, 273
  and Milky Way, 258
  and recent findings, 271–274
  and theology, 330

Copernicus
  and ancients, 226, 235
  critique, 242, 244, 245
  intellectual context, 231–235
  and Kepler, 400n42
  and neo-Platonism, 232–234
  official story, 222–224
  in textbooks, 401n58
copper, 64, 85–86, 363–364n34
coral, 28, 30, 55–56, 354n25
core instability accretion, 154, 383n27
coronagraph, 350n21
corotation circle, 165, 387nn55–56
Cosmic Habitable Age (CHA), 181–183, 187
cosmic microwave background radiation (CMBR)
  angular power spectrum, 389n14, 391n33
  and Big Bang *vs.* Steady State, 273
  contaminants, 187–188
  discovery, 174–175
  and dust, 391n33
  energy density, 31
  and expanding universe, 175, 179
  and gravity, 204
  and redshift, 179
  and subjectivity, 320
cosmic rays, 187, 391n31
cosmological constant, 186, 205–206
cosmological principle, 323–324, 402nn3–5, 407n30
  defined, 248
cosmology
  of ancients, 111–114, 225–227, 393n3, 399n30
  of Aristotle, 230–231, 238, 239, 242, 260
  and astronomy, 104–106
  of Christianity, 227–231, 244–245
  and Copernican Principle, 259–261
  of Copernicus, 231–232, 399n30
  geocentrism, 244–245, 400n37, 401n58
  heliocentrism, 222–225, 232–233
  and Kepler, 104–106
  many worlds, 268–271, 306, 319, 325
  in Middle Ages, 400n37
  oscillating-universe, 261
  skepticism, 322–326
  see also universe
Cottingham, Edwin, 15–17
craters, 108, 160, 373n14
Crommelin, Andrew, 16
crust
  of Earth, 62–63, 156–157, 393n11
crust (*continued*)
  of Mars, 85
  of Venus, 86
Curie point, 48
cycles, life on Earth, 36–37

Danielson, Dennis, 233, 244
Dante, 227–228
dark energy, 176–177, 186, 390n28
  vs. gravity, 205
dark matter problem, 142
dark sky, 68, 191–193
Darwin, Charles, 286, 402n6
Davidson, Charles, 16
Davies, Paul, 210–211, 263
days
  length
    changes in, 28
    and Earth resonance, 87
    and Earth tilt, 5
    leap seconds, 351n31
    Moon effect, 107
    on outer planet moons, 100
    on Venus, 86
  temperature, 7, 68, 97–98, 107
Dead Sea, 37
decoupling, radiation and matter, 179–181
Deism, 243–244
Dembski, William, 298, 299–300
density
  matter-energy, 176–178, 205
  and seismographs, 361n3
  vacuum energy, 205
Denton, Michael, 34–35, 64, 217, 309–310
Descartes, René, 69, 248
design
  and agent, 327 (*see also* agency)
  vs. chance, 415n11 (*see also* chance)
  cosmic, 300–306
  counterflow, 300–301
  and ETI, 412n46
  green dot metaphor, 262–263
  inference of, 293–300, 322
  vs. optimality, 330
  and probability, 197, 297–298, 302–303, 415n11
  purposeful, 244, 307, 327, 332–334, 348n13
  and theology, 330
  see also rationality
details, discernment limit, 217
deuterium, 178, 190, 353n7, 354n19, 391n37

diamonds, 367–368n33
Dick, Steven, 292
differential rotation, 151–152
dimensional analysis, 395n38
dimethyl sulfide, 72
discoverability, 208–218
  and design, 294
  and habitability, 304–307, 334
    objections, 313–329
  incrementalism, 214–216, 311, 326
  and infinite universe, 192(fig.)
  and Judeo-Christian tradition, 229
  and measurability, 272–273
  of nature, 333
  separability, 396n50
  vs. simplicity, 210, 396n52
  and solar eclipses, 18–19
Discovery Principle, 323–326
distance ladders, 111–113, 173–174
distances
  from asteroid belt, 128–129
  asteroids, from Sun, 133(fig.), 377n36
  and Classical Cepheids, 388n10
  Moon, from Earth
    ancient calculations, 112
    and impact events, 374n29
    and observability, 68
    recession rate, 18, 27
    size factor, 7
    and tides, 18, 27–28
  and orbital speed, 129
  to other stars, 125, 138–140
  planet from host star
    and asteroids, 129–131, 137(fig.)
    and eclipses, 9–10
    gas giants, 378n45
    and habitability, 6–7, 130–131, 133
    and parallax, 140
    and surface energy density, 376n26
  and redshift, 173
  and solar eclipses, 9–10
  from Sun
    of Earth, 113, 129, 140
    of impact events, 376n24
  and supernovae, 388n10, 389n19
  trigonometric parallax, 138–140
  of visibility, 184(fig.), 186–190
DNA, 33, 345, 386n48
Doppler effect
  and p-mode oscillations, 379n57
  and planet detection, 94
Doppler effect (*continued*)
  and rotational broadening, 379n56
  as velocity marker, 125, 374n9, 375n10
  see also redshift
Drake Equation, 279–281, 409n16
  revised, 337–342
Drake, Frank, 278
dust
  and CMBR, 391n33
  from comets, 78–79
  and Earth atmosphere, 72–73
  interstellar reddening, 140, 147–148
  and iron, 362–363n22
  and kinetic energy, 387n54
  and M dwarf stars, 136
  on Mars, 82, Plate 10
  in Milky Way, 147–148
  and panspermia, 344
  prestellar, 189
  radioactive, 187
dust disks, 284
dwarf stars. *see* M dwarf stars
dynamos
  description, 48
  on Mars, 85
  and potassium, 131
  and technological life, 363n33

Earth
  atmosphere of (*see* atmosphere, of Earth)
  axial tilt, 4–5, 30, 83, 85, 92, 107, 284
  and CHZ, 130
  climate history, 22–26
  core, 6, 86–87, 131, 349n11, 361n5, 384n35
  cratering, 108, 160
  crust, 62–63, 156–157, 393n11
  distance from Sun, 140
  extinction events, 159–164
  formation, 152–153
  magnetic field (*see* magnetic field)
  mantle, 57, 63, 131
  in Milky Way, 188
  Moon effect, 4–6, 107, 284, 348–349n6
  NEO impacts, 73–79, 159–161 (*see also* impact events)
  orbit (*see under* orbits)
  radioactive heat source, 63
  reflecting surfaces, 69, 109
  resonance period, 87
  rotation
    rates, 17–18, 351n31

Earth
rotation *(continued)*
vs. revolution, 106
size, 59, 112
temperature, 34
and universe, 190–191
see also geocentrism
earthquakes, 45–48, 60, 361n2, 362n18
see also plate tectonics
Earthrise, 347n3, Plate 1
earthshine, 108–109, 239–240
Eddington, Arthur, 15–17, 174
Ehrenfest, P., 209
Einstein, Albert, 15–17, 171, 248, 260
and Newton, 396n51
electricity
from magnetism, 48
in oceans, 364n35
electromagnetic coupling constant, 207(fig.)
electromagnetic spectrum, 12, Plate 8
electromagnetism
and gravity, 197, 203–204, 207–208, 394nn21–22, 404n13
and nuclear force, 201
electron degeneracy pressure, 385n38
electron-to-proton mass ratio, 207(fig.)
elliptical galaxies, 151
Ellis, George F. R., 326
Emerson, Ralph Waldo, 65
emission lines, 12–15, 123–125
energy
fire, 217–218
fuels, 61–62, 64
from impact events, 375n23
for life, 36–37, 67–68, 130
and light, 67
in Middle Ages, 364n37
nuclear resonance, 198–199
and stars, 132, 134, 139, 203–204
and Sun, 25, 70, 203–204, 380n61
for technological life, 61–62
and temperature, 365n5
see also vacuum energy
energy density, 391n31
Eratosthenes of Cyrene, 112
Eros (asteroid), 114
ETI. *see* extraterrestrial intelligence
Europa, 88–90, 98, 252, 371n46
evaporation
and isotopes, 352n6
and Milankovitch cycles, 30
evaporation *(continued)*
ponds and lakes, 34
and salinity, 90
event horizon, 186–187, 390n30
evolution, 67, 286–287
see also Darwin, Charles
extinction events, 159–164
extraterrestrial intelligence (ETI)
and design, 412n46
and Drake Equation, 341
public opinion, 275–278, 408n4
relevance, 329
search for (SETI), 278–290
civilizations, 288–289, 408nn10–11
and Copernican Principle, 281, 287, 290–292
Drake Equation, 279–284
extragalactic, 289–291
funding, 408–409n14
intelligent life, 286–288
life origination, 284–286
extraterrestrial life
assumptions, 32
and Copernican Principle, 281, 287, 290–292, 408n12, 411n37
implications, 411n37
from material exchange, 39
origination, 410n29
support, 36–37
extreme environments, 37–40
extremophiles, 37–40, 322, 359n76, 369n17
defined, 358n69
and temperature swings, 370n24
eyes, 67

feedback
life and measurability, 40
life with Earth, 36–37
light reflection, 69–71
Fermi, Enrico, 141, 276–277
Fermi's Paradox, 276–278, 289, 408n10
fine structure constant, 207(fig.)
fine-tuning
carbon and oxygen, 357–358n61
in chemistry, 197–203
dimensions, 209–210, 395n38
for discovery, 208–218
of forces, 395n32
gravity, 197, 203–205
of human beings, 215–218
mass-luminosity relation, 202
multituning, 205–208

fine-tuning (*continued*)
  and necessity, 301
  Newton view, 327
  nuclear physics, 201–202
  and probability, 196–197, 303
  and Unified Grand Theory, 264
  vacuum energy, 205–206
fire
  heat release, 363n32
  and human size, 217–218
  on Mars, 86
  and technological life, 60–61, 363n33
flash spectrum, 15
fluorine, 358n67
food chain, 37, 376n26
forams, 26, 354n22
forces, 69–71, 395n32
  see also strong nuclear force; weak force; *specific forces*
fossil magnetic field, 48–49, 85
fossils
  of coral and mollusks, 354n25
  of cosmic history, 181
  of mites, 29
  and patterns, 333
  tidalites, 28, 107
  of tree leaves, 29
Fourier analysis, 93, 370n25
Fraunhofer lines, 12–13
friction, 87
Friedmann, Aleksandr, 171–172
fuels, 61–62, 64
future, prediction of, 40–41

Gaia hypothesis, 70–71, 322, 366n16
Galactic Habitable Zone (GHZ), 152–164, Plate 17
  and Drake Equation, 282
  vs. galactic center, 257
  speculativeness, 319
  trends, 167
  width, 167
galaxies
  cannibalization, 190
  classification, 382n17
  clusters, 167–168, 273, 382n19
  distance to, 388n10
  in early stages, 179–181 (*see also* quasars)
  habitability, 258 (*see also* Galactic Habitable Zone)
  interactions between, 167–168
galaxies (*continued*)
  Local Group, 150, 167, 273
  metal content, 167, 258
  recession of, 391n34
  shapes
    and CMBR, 188–189
    elliptical, 151, 168, 188–189
    in Local Group, 167
    spiral, 144, 146–150, 258
  surface brightness, 389n17
  terminology, 381n3
  and time, 175–176, 185–186, 290
  and universe expansion, 189
  see also black holes; Milky Way
Galilean Moons, 9–10
  see also under Jupiter
Galileo, 222, 239–242
gamma ray bursts
  description, 162–164
  discovery, 284, 318
  in Drake Equation, 340
  threat, 161–162, 386–387n53
gamma rays, 12, 182
  images in, 365n7
Gamow, George, 175
Ganymede, 97, 371n46
gas clouds, 181, 389n22
gas giants
  atmospheric life, 364n35
  formation, 154
  and host stars, 378n45
  and M dwarf stars, 134
  and magnetic field, 114
  moons, 97–98, 114
  Pegasi B, 254
  view from, 325
  see also Jupiter; Saturn
General Relativity. *see* **Relativity, Theory of**
genetics, 386n48
geocentrism, 244–245, 400n37, 401n58
geosyncline theory, 53–54, 362n14
geothermal heat, 358n65
giant molecular clouds (GMCs), 160
giant planets
  and asteroids, 377n36
  and CHZ, 97
  migration, 371n41
  occurrence rate, 409–410n24
  and terrestrial planets, 371n42
  see also gas giants

glacial periods, 72, 366n22
glaciers, 17, 41–43
global albedo, 69–71, 109, 373n18
global warming, 360n87, 366n20
globular clusters, 164, 382n14, 383n24
God
 biblical view, 229, 398–399n24
 Christian view, 412n47
 for Copernicus, 233–234
 in Middle Ages, 228
 for Newton, 242–244, 327
Gold, Thomas, 174, 260
Goodwin, S. P., 258
Gould, Stephen J., 286
Grand Unified Theory, 263–264
gravity
 and Copernican Principle, 250
 deflection, 384n33
 and electromagnetism, 197, 203–204, 207–208, 394nn21–22, 404n13
 and fine-tuning, 197, 203–205
 in interacting systems, 93
 in interstellar nebulae, 126
 and light, 15–17, 88
 and natural necessity, 301
 and nearby galaxies, 391n34
 Newtonian limit, 396n51
 and orbits, 371n40
 and planet size, 59–60
 and rotation, 87
 and stars, 203
 vs. vacuum energy, 186, 205
greenhouse effect, 56–57, 360n85
greenhouse gases, 35–36, 390n26, 393n17
 and Venus, 86
 see also carbon dioxide; methane; water vapor
Greenland, 24–26
Greenstein, George, 67
Gribbon, John, 206–207, 258
Guth, Alan, 261

H-R diagrams, 120–121
habitability
 and asteroids, 78
 autotroph effect, 40
 and axial tilt, 92
 and carbon isotopes, 390n26
 and Copernican predictions, 250–258
 and discoverability, 304–307, 334
  objections, 313–315
habitability (*continued*)
 and Drake Equation, 279–281, 409n16
  revision, 337–342
 of Europa, 88–90
 of extrasolar planets, 96–97
 in extreme environments, 37–40
 fine-tuning, 196–197 (*see also* fine-tuning)
 and host star distance, 6–7, 130–131, 133
 and infinite universe, 192(fig.)
 and M dwarf stars, 133–134, 377n38
 and measurability (*see under* measurability)
 and metals, 61–64, 153–156, 258
 of moons, 97–101
 negative feedbacks, 69–71
 of non-terrestrial planets, 87–88
 pathways to, 322–333
 and planet mass, 59–60
 and plate tectonics, 55–57
 prerequisites, 31–36, 271–272
 and solar eclipses, 7, 18–19, 314
 and stars, 127
 of terrestrial planets, 82–88
 timeline, 180(fig.)
 and water, 127–128, 322, 411n38
 see also Circumstellar Habitable Zone; Cosmic Habitable Age; Galactic Habitable Zone
Hadean period, 91
Haeckel, Ernst, 285
Hale-Bopp comet, Plate 9
Hart, Richard, 310
Hartmann, Georg, 52
Hawking, Stephen, 263–264, 404n7
heat
 conduction, 357n49
 geothermal, 358n65
 release
  in fire, 363n32
  and gas giant moons, 98
  on Mars, 85
  from radioactive decay, 158
  on Venus, 86
  from water, 357n50
heavy elements, 182, 200–201
heliocentrism, 222–225, 232–233, 399n32
helioseismology, 122
helium
 and beryllium, 393n10
 discovery, 13
 isotopes, 391n37, 393n12
 and marine sediments, 354n22

helium (*continued*)
and metallicity, 384n32
and planet formation, 153–154
stars, burning, 202–203
in universe, 178, 202
helium-shell burning, 198–199
Henderson, Lawrence, 33, 37, 197–198, 199
Henderson, Thomas, 140
Hendry, M. A., 258
Herman, Robert, 175
Herschel, John, 13
Herschel, William, 275
Hertzsprung, Ejnar, 120
heterotrophs, 37
hierarchical clustering, 212–214
hierarchical simplicity, 214–216
Hipparchus of Nicaea, 112
Hodgson, Dominic, 21
Hooykaas, Reijer, 228
host stars, 132–136
distance from (*see under* distances)
Hot Big Bang model, 172
hot-Jupiter systems, 94–99
Hoyle, Fred, 171, 174, 187, 198–199, 217, 260, 263
Hubble constant, 176, 178, 273
Hubble Deep Fields, 175–176, 185–186, 289–290, Plate 18
Hubble, Edwin, 169–174, 260, 333
Hubble Space Telescope, 126, 162, Plate 18, Plate 19
Huggins, William, 120
human beings
detail discernment, 217
evolution, 286–287
knowledge, 332–333
position
ancient views, 218, 222, 245
biblical view, 228–229, 244–245, 273–274
and Copernican Principle, 224–228, 402n6, 405n15, 407n29
and Einstein principle, 248
Enlightenment view, 401n60
and purposeful design, 245, 248–249, 401n61
on size scale, 215–217, 407n31
hurricanes, 359n79
hydrocarbons, 37, 72–73
hydrogen
abundance, 357n46
and carbon, 37, 72–73
and CMB photons, 175

hydrogen (*continued*)
in Earth
atmosphere, 72
bulk Earth, 156–157
and fluorine, 358n67
isotopes, 352n6, 354n19 (*see also* deuterium)
and metallicity, 384n32
and Milky Way mapping, 382n15
molecule rotation, 374n6
and planet formation, 153–155
and quasars, 179–181
and stars
burning, 202–203
life cycle, 132
temperature, 120–121
in Sun, 13, 350n22
in universe, 178
hydrological cycle, 40, 73, 91
Hyperion, 93, 106

ice
and climate, 33, 69, 83–84
cores, 22–26, 41
glacial periods, 72, 366n22
glaciers, 17, 41–43
global volume, 26
heat conductivity, 357n49
life forms, 37
on Mars, 82–85, 368n2, 368n4
Moon effect, 107
in outer planet moons, 99–100
and oxygen isotopes, 352n6
and plankton, 354n20
polar breakup, 360n85
and water density, 33
see also glaciers
impact events
energy from, 375n23
extinctions, 159–164
extrasolar planets, 344
Mars, 115, 359n78
Moon formation, 6, 340–341, 349n10
NEOs, 73–79, 159–161 (*see also* near Earth objects)
and oceans, 363n24
and planet rotation, 86–87
protection from, 114–115
rate, 373n13
seeding life (*see* panspermia)
Tunguska event, 74, 367n27

impact events (*continued*)
  and Venusian craters, 374n30
Independence Day, 278
Industrial Revolution, 43, 61–62
industrialization, 360n87, 361n88, 364n37
inertia, 242
infinity, 192(fig.), 260–261, 406n23
inflation theory, 414n7, 415n8
infrared light, 188, 390n26
initial conditions, 306–307
initial mass function, 382n19
intelligence. *see* agency; design; extraterrestrial intelligence; rationality
intelligent life, 285–291
  see also extraterrestrial life; technological life
interpretation, 136–138, 164–166, 265–271
interstellar grains, 76–77
Io, 372n52
iron
  and anthropic coincidence, 201
  in Earth core, 6, 131, 349n11, 384–385n35
  on Mars, 85
  in newer planets, 157, 159
  in oceans, 362–363n22
  smelting, 363–364n34
isotopes
  beryllium, 25, 199, 353n12, 393n10
  boron, 380nn59–61
  carbon, 354n22, 368n34, 390n26 (*see also* carbon-12; carbon-14)
  chlorine, 25
  and climate reconstruction, 352–353n6
  and condensation, 352n6
  description, 352n6
  in greenhouse gases, 390n26
  helium, 391n37, 393n12
  hydrogen, 352n6, 354n19 (*see also* deuterium)
  lithium, 391n37, 393n12
  in marine sediments, 26
  and meteorites, 209, 374n8
  notation, 354n19
  and nuclear force, 202
  oxygen, 26, 352n6, 354n22
  potassium, 131, 201–202, 384n35
  as proxies, 23
  radioactive (*see* radioisotopes)
  ratios, 393–394n17
  and stellar nucleosynthesis, 190
  and sunspot cycles, 25
  thorium, 30
isotopes (*continued*)
  tungsten, 349n11
  of uranium, 30, 63
  valley of stability, 352n6

Jaki, Stanley, 310
Jakosky, Bruce, 222–223
Janssen, Pierre J. C., 13
Judeo-Christian tradition
  Christian theology, 229–231, 244–245, 401n60, 412n47
  creation *vs.* Creator, 228–229
  God, 229, 398–399n20
  human role, 244–245
  Psalm, 274
Jupiter
  atmosphere, 91
  and Earth formation, 78
  as Earth protector, 114–115, Plate 12
  hot-Jupiter systems, 94–99
  and meteorites, 75
  moons (*see also specific moons*)
    and eclipses, 9–10, 373n21
    habitability, 88–90, 97–101
    magnetic fields, 371n46
    tidal heating, 372n52
  radiation belt, 371n47
  and Shoemaker-Levy 9 comet, Plate 12
  see also hot-Jupiter systems

Kant, Immanuel, 169
Kepler, Johannes
  and Brahe, 235–238
  and Copernicus, 400n42
  correction, 372n8
  laws, 104–106, 113–114, 210
  and Moon craters, 251, 275, 296
  and Newton, 104, 213–215, 372n8
  and orbit shape, 372n3
  and orbital speed, 129
Kirchoff, Gustav, Plate 5
Kirchhoff laws, 15
knowledge, 332–333
Kuhn, Thomas, 232, 235
Kuiper Belt, 160–161, 385–386n47

lakes, 34
land, 69, 70
  see also continents; plate tectonics
Landau, Lev, 247

large number coincidences, 197–203, 394n21
  see also fine-tuning
Laughlin, Greg, 268–269
laws
  applicability, 323–324
  for Deists, 243–244
  and design, 296–297, 306–307
  discoverability, 210–216
  as local by-laws, 406n25
  natural necessity, 301
  for Newton, 242–243
  and origin of life, 285
  of physics, 127, 210–211, 323–324
  simplicity *vs.* discoverability, 396n52
  see also Relativity, Theory of
Leavitt, Henrietta, 172
Lemaître, Georges, 171–172
Leslie, John, 305
Lewis, Thomas, 35
life
  biodiversity, 92, 358n74, 376n16, 376n27
  breathing, 376n25
  and carbon, 314–315
  extraterrestrial (*see* extraterrestrial life)
  feedback loops, 36–37, 40
  and gravity, 204
  and impact events (*see* panspermia)
  intelligent, 285–287
  minimum complex, 32
  and NEO impacts, 115
  origination, 284–286
    data sources, 318–319
    eclipse role, 349–350n14
    on other planets, 410n26, 410n29
    panspermia, 343–345
  prerequisites, 32–37, 356n44
  and pressure, 39
  simple and complex, 39, 211, 356n39, 376n26
  in Solar System, 251–252
  technological, 60–64 (*see also* technological life)
  and temperature, 39
light
  and atmosphere, 66
  bending, 15–17, 88
  and chemical reactions, 67–68
  and electromagnetic spectrum, 12
  and M dwarf stars, 378n43
  in nebulae, 126
  and photosynthesis, 135
light (*continued*)
  from quasars, 179–181
  reflection, 69–71
  see also luminosity; tired light theory
light spectra, 12, 66–67, 378n43
  and eyes, 67
  see also redshift
limits, 326
lithium, 178
  isotopes, 391n37, 393n12
lithopanspermia, 344
Livio, Mario, 249–250
Local Group, 150, 167, 273
Lockyer, Joseph N., 13
logical positivism, 347n7
Loutre, M., 41–42
Lovell, Jim, 331
Lovelock, James, 36, 70
Lowell, Percival, 251, 296
luminosity
  calculation, 139
  and gravity, 203
  and helium, 202
  of M dwarf stars, 134
  in main sequence stars, 132
  and mass, 202, 394n19
  and metallicity, 387nn61–62
  and oscillation amplitude, 141
  in other galaxies, 167
  period-luminosity relation, 172–173, 190, 394n22
  standard candles, 139–140, 172–174
  of Sun, and CHZ, 128
  and temperature, 120–121
lunar eclipses, 9, Plate 13
Lyman-alpha clouds, 178, 179–181, 190

Magellanic Clouds, 172, 190
magnesium
  in planets, 157, 159
  in shells, 354n20
magnetic field(s)
  dynamo effect, 48, 85
  and Earth core, 6, 131, 361n5
  and energy density, 391n31
  fossil, 48–49, 85
  in future planets, 158
  and gas giants, 97, 114
  and geologic past, 48–50
  and ice cores, 25

magnetic field(s) (*continued*)
interplanetary, 54
on Mars, 85
and Moon, 6
of moons, 371n46
and oceans, 49–51, Plate 6
and planet mass, 59
and plate tectonics, 57–58
in pulsars, 127
reversals, 30, 49–50, 54
rock records, 52–53
of Sun, 54
synchrotron emission, 187
three-D compasses, 52
and Venus, 86
magnetostratigraphy, 355n34
main sequence stars, 132, 136–137, 204
and Drake Equation, 282
many worlds hypothesis (MWH), 268–271, 306, 319, 325, 406n26
Margulis, Lynn, 70
marine life, 37–38, 56, 354nn20–21
forams, 26, 354n22
tidalites, 28, 107
see also plankton
marine sediments
bioturbation, 354n22
as data source, 26–28
and Earth orbits, 94
and helium, 354n22
measurements, 354n22
Moon effect, 107
and radioactive decay, 30
radioisotopes, 26–28
Mars
M dwarf stars, 133–134, 138, 377n38, 378n43
and asteroid belt, 128
crust, 85
as Earth protector, 115, 375n22
face on, 299(fig.)
ice layers, 82–85, 368n2, 368n4
and Kepler, 104
life on, 39–40, 251
panspermia, 115, 343–345, 359n78
metallicity, 85–86
and meteorites, 345
and obliquity, Plate 11
orbit, 94
and oxygen, 86
and parallax measurement, 114
Mars (*continued*)
plate tectonics, 85, 369n7, 369n10
rotation axis, 83, 84–85
surface: volume, 369n8
water, 82, 369n5
mass-luminosity ratio, 202, 394n19, 394n22
mass-luminosity relation, 202, 394n19
materialism, 347n7
mathematics, 212
matter
clustering, 212–214
dark, 142
Einstein cosmological principle, 248
Judeo-Christian tradition, 228–229
nuclear density, 127
and radiation, 179–181
in Steady State model, 260
matter-energy density, 176–178
Mayor, Michel, 254
measurability
and Copernican Principle, 273
and habitability
con arguments, 313–315, 321–322, 327–328
and Copernican Principle, 221, 295
feedback, 40
and naturalism, 295
pro arguments, 31–32, 304–305, 328–329
and purpose, 333
SETI, 289
universe size, 272
and hierarchical clustering, 212
vs. observability, 18–19
Melosh, Jay, 344, 345
membranes, 34
Méndez, Abel, 39
Mendillo, Michael, 310
Mercury
core, 384–385n35
habitability, 87–88
orbit, 94, 372n7
planetary laws, 106
rotations and orbits, 106
metabolism, 37, 64
metallicity
and dust, 189
in Earth
core, 131
crust, 62–63, 156–157
in galaxies, 167, 258
in gases, 383n26

metallicity (*continued*)
and habitability, 61–64, 153–156, 258
and helium, 384n32
and hydrogen, 384n32
and luminosity, 387nn61–62
on Mars, 85–86
and planet formation, 153–156
in revised Drake Equation, 339
smelting, 363–364n34
and Sun, 384n30
threshold, 383n23
see also copper; iron; nickel
meteorites, 75–79
carbon content, 376n28
from Earth, 109, 344
and isotopes, 209, 374n8
and Mars, 85, 115, 251, 343–345
from Moon, 108
and Venus, 344
methane
and climate, 35–36
in Earth atmosphere, 72, 364n42
exchangeability, 33
and global warming, 360–361n87
and ice cores, 23–24
and planetary mass, 59
see also greenhouse gases
Meyer, Stephen, 261
microbes, 39, 63, 344–345
Middle Ages, 227–228, 360n87, 364n37, 400n37, 400n52
Milankovitch cycles
description, 30–31
detection, 370n25
and Holocene, 359–360n83
uses, 41–42, 355–356n36, 362n11
Milky Way
and Copernican Principle, 258, 403n16
discovery, 134(fig.), 143
dust, 147–148
early materials, 76
Earth position in, 188, 403n14
gravitational measurements, 126
habitable zone, 152–164, 167, 257
heavy elements, 182
intergalactic event, 167
location, 144
mapping, 150–151, 382nn15–16
metallicity, 153–156
nucleus, 188
Milky Way (*continued*)
observation from, 146–151
regions, 146, 148–150
satellite galaxies, 172, 190
Solar System position in, 257
vs. universe, 169–170
zone of avoidance, 147, 381n8
millisecond pulsars, 189–190
Milne, E. A., 208
minerals
in asteroids, 368n36
biomineralization, 63
distribution of, 34–35, 55–56
in Earth core, 131
in interstellar grains, 76, 367n33
in Martian crust, 85
in oceans, 36
and technological life, 62–64
mites, 29
molecules
absorption and emission lines, 12, 123–124
absorption bands, 390n26
and complex life, 32
in cool or hot stars, 365n9
light absorption, 67
most abundant, 357n46
rotation *vs.* vibration, 374n6
mollusks, 28, 354n18, 354n25
molybdenum, 36, 37, 201, 358n63
Moon
and atmosphere transparency, 68
and cosmology, 104, 106–108
craters, 108, 373n14
cross section, 374n29
and Earth
distance between (*see under* distances)
early life relics, 319
effect on, 4–6, 107, 284, 348–349n6
as protector, 115
resonance period, 87
Earthrise picture, 347n3, Plate 1
and eclipses, 10
interior, 92
and Newtonian physics, 104
orbit, 28, 93, 349n9
origination, 6–7, 340–341, 349n10
seismic events, 92
shape, 9, 350n16;17
size, 7, 10, 340, 351n33
and Solar System history, 108–109

Moon (*continued*)
and Sun, 11(fig.)
as telescope, 109–110
moons
with atmosphere, 350n19
and Copernican Principle, 256
as habitats, 97–101
of Jupiter (*see under* Jupiter)
magnetic fields, 114, 371n46
of Mars, 368–369n4
of Neptune, 11
orbits, 11
of outer planets, 10
of Saturn, 10, 11, 97–101, 252 (*see also* Hyperion)
shapes, 9
size, 7, 10, 11
mountains, 60
movies, 276, 335
see also specific movies
Myrhvold, Nathan, 222

naturalism, 248, 274, 292, 295
challenge to, 328–329
natural necessity, 301
natural selection, 67, 286
nature
vs. design, 300–301
disclosure of, 333
Judeo-Christian view, 229, 398–399n20
and naturalism, 329
and purpose, 248–249 (*see also* rationality)
purpose in, 332
see also laws
near Earth objects (NEOs), 73–79, 159–161, 367n25
see also asteroids; comets; impact events; meteorites
nebulae, 126, 145–146, 169–170
see also galaxies
neo-Platonism, 232–234
Neptune, moons of, 11
Nereid, 11
Nernst, Walther, 171
neutrinos
boron-8, 380nn59–61
dark matter problem, 141–142
and Sun temperature, 122–123
transmutations, 380nn62–63
and weak-force, 202
neutron stars. *see* pulsars
neutrons, 203
Newton, Isaac
and Einstein, 396n51
on fine-tuning, 327
and God, 243–244
and infinite universe, 260
and Kepler laws, 104, 213–215, 372n8
and Relativity Theory, 215
and spectroscopy, 69
nickel, 384–385n35
nights
dark sky, 68, 191–193
temperature, 7, 68, 97–98, 107
nitrogen, 36, 37, 156–157, 358n63
nonlinear interactions, 211
nuclear density, 127
nuclear force, 201–202
nuclear fusion, 139
nuclear resonance, 198–199
nucleosynthesis, 76–77, 178, 190

O stars, 139–140, 146
obliquity, 348–349n6, 355–356n36, Plate 11
observability
and Copernican Principle, 402n6
detail discernment, 217
and galaxy shape, 188–189
from globular cluster, 382n14
and habitability, 7
limits, 326
of solar eclipses, 7, 18–19, 310–311, 351n37
vantage point, 323
observatories, 90–91, 251
observer selection effect, 136–138, 164–166, 405n14, 405n18
and Copernican Principle, 265–268, 272, 405–406n22
oceans
and asteroids, 161
circulation, 349n8, 360n85 (*see also* tides)
on Europa, 88–90
and impact events, 363n24
in large Earth twin, 363n24
and light spectrum, 67
and magnetism, 49–51
nutrients, 36, 362–363n22, 364n35
and oxygen isotopes, 26
polarity reversals, 30
as reflecting surface, 69, 70
and rotational synchronization, 377n39
salinity, 90, 370n19

oceans (*continued*)
and silicate rocks, 55–56
spreading ridges, 47, 49–51, 54, 89–90
surface temperature, 72
thermal vents, 37–38
and winds, 349n8, 362n22
see also tides
oil, 364n36
Olbers, Heinrich W. M., 191–193
Oort cloud, 160–161, 385–386n47
optical spectra, 12–15, 123–125, 350n25, 390n26, Plate 7
see also infrared light; ultraviolet light
optimality, 330
orbital dynamical diffusion, 384n33
orbit(s)
around pulsars, 127
and chaos, 93–94, 105–106
and comets, 160
of Earth
as baseline ruler, 138
and climate, 92–93
earlier, 94
Kepler laws, 104
and Milankovitch cycles, 30
size, 354n26
and sunlight, 371n50
velocity, 375n10
in elliptical galaxies, 151
of gas giant moons, 88, 98–99
and gravity, 371n40
of Mars, 94
of Mercury, 94, 372n7
of Moon, 93, 349n9
of planets, 83(fig.), 372n8
outer planets, 88
size factor, 93–94, 97, 340
of Pluto, 93
and rotation direction, 87
shape
angular size ratio, 11
and CHZ, 128
discovery, 104, 238, 372n3
and Drake Equation, 339, 340
of extrasolar planets, 383n25
of Solar System planets, 340
speed, 129, 238
of stars, 384n33
extrasolar, 94–96

orbit(s) (*continued*)
of terrestrial planets, 135–136, 371n38
of Venus, 87, 94
see also resonance(s)
Orion Nebula, 284
oscillating-universe, 261
oscillations
stars, 379n57
on Sun, 141
in universe, 261
oxidation, 37, 135
oxygen
and carbon, 357–358n61, 393n11
in Earth
atmosphere, 72–73, 209
crust, 393n11
and extremophiles, 36–37
and fine-tuning, 198–199
and fire, 86, 217–218
isotopes, 26, 352n6, 354n22
and life, 32, 37
and Mars, 86
and methane, 364n42
and newer planets, 157
and ores, 64
and silicon, 356n43, 356n43, 356n45
solubility, 358n75
and stars, 154, 204
ozone layer, 161, 386n49

p-mode oscillations, 141, 379n57
paleoclimatology, 94, 351n5, 352n6
paleotempestology, 359n79
Pangaea, 47
panspermia, 343–345
and impact velocity, 376n24
interstellar, 358n70
Mars seeding, 115, 359n78
and molybdenum, 358n63
probability, 358n70
pantheism, 329
parallax effect
angle measurement, 379n55
definition, 112
and early astronomers, 112–114
and Milky Way, 150
stellar trigonometric, 138–140
particle horizon, 184(fig.), 186–190
particle radiation, 188–189, 377n40, 386n49

Pasachoff, Jay, 136
patterns, 298–310, 322, 333–334
  resistance to, 332
Pauli Exclusion Principle, 385n38
Pauli, Wolfgang, 141
Pegasi B, 253
pendulum, double, 93
Penzias, Arno, 174
perfect solar eclipses, 7–9, 17–18
perfection, 330
period-luminosity relation, 172–173, 190, 388n7, 394n22
periodic table, 38(fig.), 201–202
perturbation theory, 213–214
Peterson, Ivars, 105
Phobos, 368–369n4
photometric color indices, 120
photons
  and absorption lines, 123–124
  and chemical reactions, 67–68
  cosmic microwave (CMB), 175, 176
  in pulsars, 127
  and temperature, 365n5
  tired light hypothesis, 178–179
photosynthesis, 55, 67, 135
physics
  laws of, 127, 210–211, 323–324 (*see also* Kepler, Johannes; Relativity, Theory of)
    and Copernican Principle, 261–265
phytoplankton, 72, 366n21
planets
  axial tilt, 7
  core fluid resonance, 369n13
  distance from Sun/host star (*see under* distances)
  earliest, in universe, 182
  extrasolar, 94–97, 282–283, 370n33
  formation, 153–156, 181–183, 384–385n28
  giant (*see* gas giants; giant planets)
  gravity fine-tuning, 204
  magnetic fields, 6, 59, 158
  material exchange, 39
  migration, 371n41, 383n23
  non-terrestrial, 87–88, 97, 114 (*see also* gas giants)
  number of, 115
  orbits (*see under* orbits)
  relative distances, 113–114, 372n5
  size of
    and biosphere, 376n26
    and gravity, 59–60
planets
  size of (*continued*)
    for life support, 7
    and moons, 10
    and orbits, 93–94, 97
    and pressure, 59
  see also terrestrial planets; *specific planets*
plankton, 26–27, 56, 354n20
  phytoplankton, 72, 366n21
plant life, 70, 358n74, 375n18
  see also phytoplankton; trees
plate tectonics
  and carbon cycle, 55–57
  and erosion, 57
  in future planets, 158–159
  heat for, 393n14
  and impact event, 6
  and internal heat, 385n42
  and magnetic field, 57–58
  on Mars, 85, 369n7, 369n10
  and potassium isotope, 201–202
  and radioisotopes, 183
  size factor, 7
  theories, 52–54
  variability, 128
  and Venus, 86
  and water, 34
  see also earthquakes
Plato, 229
  neo-Platonism, 232–234
Pluto, 93, 160
Polanyi, Michael, 332–333
pollen, 30
positivism, 347–348n7, 404n7
potassium
  in Earth
    core, 131, 376n33, 384n35
    crust, 63, 131
    mantle, 131
  in interstellar medium, 158
  isotopes, 131, 201–202, 384n35
  on Mars, 85
power. *see* energy
power spectra, 370n25
  CMBR, 389n14
precision, 337
pressure
  in atmosphere, 39, 355n30
  electron degeneracy, 385n38

pressure (*continued*)
on Europa, 88–89
and life, 39, 88–89, 369n70
and planet size, 59
at star center, 139
see also partial pressure
Principle of Indifference. *see* Copernican Principle
Principle of Mediocrity. *see* Copernican Principle
Principle of Plenitude, 411n35
probability
chance *vs.* design, 197, 297–298, 302–303, 415n11
of Earth-like planet, 327–328
of extraterrestrial civilizations, 338–342, 408nn10–11
of giant planets, 409–410n24
probes, stars as, 125–126
prokaryotes, 39
Prometheus, 10, 11
prominences, 13
protons, 202, 207(fig.), 380n61
proxies, 23
Proxima Centauri, 379n55
Psalm, 274
Ptolemy, 222, 225, 226, 230, 399n30
pulsars
signals, 295
trends, 189–190, 391n36
uses, 127
purpose
and human beings, 245, 248–249, 401n61
in universe, 307, 327, 332–334, 348n13 (*see also* design)

quantum indeterminacy, 211, 413n4
quantum phenomena (Copenhagen), 401n57
quantum theory, 326
quarks, 211
quasars, 179–181, 189
and continental drift, 54
Queloz, Didier, 253

radiation
background, 187–189 (*see also* cosmic microwave background radiation)
from Big Bang, 174–176
cosmic *vs.* galactic, 187
in early universe, 181–182
and Earth atmosphere, 66, 68
and evolution rate, 386n48
radiation (*continued*)
extraterrestrial bursts, 161–164
and gas giants, 97
and M dwarfs, 134–135
and matter, 179–181
and stars, 65–68, 203–204
supernovae, 162, 188–189
and temperature, 67–68, 179
see also particle radiation; *specific types*
radioactive decay, 30, 158
see also carbon-14
radioisotopes
and anthropic coincidence, 201
and asteroids, 161
carbon-14 dating, 29–30
data combinations, 362n17
in Earth crust, 63, 131
in interstellar medium, 158, 182–183
and plate tectonics, 159
uses, 352n6
radio signals
and pulsars, 295
and SETI, 279, 290–291, 296, 341
radio telescopes, 127, 341–342, Plate 8
radio waves, 12, Plate 7
radiopanspermia, 344
rainbows, 15, 68–69, 366n13
artificial, 350n26
rainfall, 5–6, 55–56
rationality
abductive reasoning, 413n2
bias against, 332
nested complexity, 211–212
for Newton, 327
and SETI, 290–292
superintellect, 262–263
see also agency; design; God
Ratzsch, Del, 300
Rayleigh scattering, 378n43
red giant stars, 390n26
redshift
and angular resolution, 391n38
and Big Bang theory, 178
and carbon, 209
and CMBR, 179, 389n20
de Sitter models, 388n3, 389n23
and distance, 173, 390n30
Hubble Constant, 176
and nebulae, 169–170
and photons, 175

redshift (*continued*)
- tired light hypothesis, 178–179, 389nn19–20
- and universe expansion, 175, 181, 186–187

Rees, Martin, 203, 204, 206–207, 212, 323
reflection, 69–71, 75–76, 109
reflection nebulae, 145–146
Relativity, Theory of
- and cosmological principle, 248
- and expanding universe, 171
- and Hubble, 260
- and Kepler laws, 104
- and Newton, 214–215, 242–244
- and pulsars, 127
- and quantum theory, 326
- tests, 15–17, 88, 249–250
- and time, 389n19

religion
- causative agent, 327–330
- and Galileo, 241–242
- Judeo-Christian tradition, 228–229, 244–245, 398–399n20
- pantheism, 329
- Psalm, 274
- *vs.* science, 224, 228–229, 397–398n15, 399n25
- and SETI, 291
- and UFOs, 276
- voluntarism, 398n22
- *see also* theology

Renaissance, 230–231
resonance(s)
- with core fluid, 369n13
- and Earth, 87, 348–349n6
- and Europa, 90
- excited carbon-12, 392–393n8
- and host star distance, 131
- and NEOs, 367nn30–31
- nuclear, 198–199

respiration, 376n25, 377n38
reversing layer, 15
RNA, 33, 345
rocks
- chemical weathering, 34–35, 55–57
- and magnetic field, 52–53

rotation
- around M dwarf stars, 133
- differential, 151–152
- direction, 87
- of Earth, 17–18, 351n31
- and impact events, 86–87
- of Milky Way disk, 150–151

rotation (*continued*)
- of molecules, 374n6
- *vs.* revolution, 106
- and stable obliquity, 348–349n6
- of stars, 379n56
  - pulsars, 127

rotational synchronization, 7, 97–98, 377n37, 377n39
rotation axis, 83, 84–85
Rowan-Robinson, Michael, 187
RR Lyrae, 172, 388n8
Russell, Bertrand, 224
Russell, Henry Norris, 120–121

Sagan, Carl, 251, 280, 290
Sagittarius, 190
salinity, 90, 370n19
- Great Salt Lake, 37

salt tolerance, 359n77
Sandage, Allan, 178–179
Sandwell, David, 50
satellites, 54
- *see also* moons

Saturn
- atmosphere, 91
- as Earth protector, 114–115
- moons of, 10, 11, 97–101, 252 (*see also* Hyperion)

scale, of sizes, 215–217, 407n31
Schiaparelli, Giovanni, 251, 275
Schrödinger Equation, 212
Sciama, Denis, 260–261
science
- and Copernican Principle, 248–249, 256 (*see also* Copernican Principle)
- and Middle Ages, 400n52
- myth of, 244–245
- *vs.* religion, 224, 228–229, 397–398n15, 399n25
- *see also* discoverability; measurability; observability

science fiction, 152, 275–276
Secchi, Angelo, 13
sedimentation
- fossilized remains, 28
- and magnetism, 48–50
- Milankovitch cycles, 30
- Moon effect, 107
- polarity reversals, 30
- and seal hairs, 21–22
- *see also* ice cores; marine sediments

seismic waves, 91, 361n2, 361n3

seismographs
as data source, 45–48
of Moon, 92
of Sun, 122
and wave types, 361n3
SETI. *see* extraterrestrial intelligence
Shapley, Harlow, 172–173, 210, 245, 257
shear waves, 361n3
shellfish, 56, 354n20
Shoemaker-Levy comet, 91, 97
silicate rocks, 55–56
silicon
and biological information, 411n38
in newer planets, 157, 159
and oxygen, 356n43, 356n45
silicon carbide, 76
silicones, 356n45
simulations
of chaotic orbit, 93
of CHZ, 128
of climate, 41
of early gravity, 78
of hot-Jupiters, 98–99
of interacting star gravities, 126
of Mercury orbit, 94
of Milky Way, 283
of Moon origination, 349n10
of other solar systems, 253
of terrestrial planet formation, 135–136
size scale, 215–217, 407n31
solar eclipses, Plates 3–4
and ancients, 112
annular *vs.* total, 18
description, 1–4
and design inference, 310
discoveries from, 12–15
and Earth rotation, 17–18
and General Relativity, 15–17
and habitability, 7, 18–19, 314
and host star distance, 9–10
and life origins, 349n14
and Rayleigh scattering, 378n43
super-eclipses, 7, 17–18
solar flares, 134, 377nn40–41
solar spectrum, 15, 123–125
Solar System, 83(fig.)
chaos in, 93–94
and Copernican Principle, 254–258
early gravity simulation, 78
early materials, 75–76, 99, 109
Solar System (*continued*)
extrasolar counterpart, 96
fine-tuning, 327–328
formation, 390n26
history of, 108–109
intelligent life in, 251–252
mass distribution, 372n8
metallicity, 157
orbit eccentricity, 340
relative distances, 111–114
solutions, aqueous, 198
sound waves, 122, 361n3
space, 323
specification, 298, 322
spectra, 66–67, 123–125, 350n25, Plate 5, Plate 14
emissions, 12–15, 365n8
see also absorption lines; light spectra; power spectra; redshift
spectrohelioscope, 13
spectroscopy, 15, 69, 75–76
spiral galaxies, 144, 146–150, 258
spiral nebulae, 169–170
stabilizing forces, 69–71
stagnant lid convection, 86
standard candles, 139–140, 172–174, 190, 388n11
gamma ray bursts as, 318
star nurseries, 147
stars
asymptotic giant branch (AGB), 76–77
atmosphere, 365n9
and chemical elements, 181–182
color of, 204
companion, 339
distance from, 125, 129–131
and Drake Equation, 282, 340
exploding, and ice cores, 25–26
formation, 203, 385n41
and gravity, 203
helium- *vs.* hydrogen-burning, 198–199, 202–203
host stars, 132–136
internal structure, 121–123
life-support roles, 132, 134
luminosity, 120, 132
mass:luminosity ratio, 202, 394n19, 394n22
main sequence, 132, 136–137, 204, 282
most Sun-like, 96
nucleosynthesis, 76–77, 178, 190
orbits, 94–96, 384n33
planets around, 94–97, 202

stars (*continued*)
as probes, 125–126
radiation, 65–68, 203–204
red giants, 390n26
size
and life cycle, 132
and luminosity, 136–137
and neutrinos, 141
and planetary orbits, 202
spectral emissions, 123–125, 365n8, Plate 14
temperature, 120, 124
velocity, 125
radial, 374n9
see also B stars; Cepheids; O stars; pulsars; quasars; supernovae
Star Trek
movie, 276, 335
series, 307–308
Steady State models, 174, 175, 260–261
and CMBR, 273
Steiner, Mark, 212
stellar candles, 139–140, 172–174
stellar trigonometric parallax, 138
Stillinger, Frank H., 34
Stoics, 329
storms, 359n79
Strong Anthropic Principle, 266–267, 405n14, 405n18
strong nuclear force, 201–202, 393n12
Struve, Wilhelm, 140
subduction zones, 46–47
sulfur, 131, 384n35
Sun
for ancients, 112–113, 225–226
anomaly, 378n49
and chemical reactions, 67–68
and climate, 379n52
and Copernican Principle, 253
for Copernicus, 232–234
corona, 12, 18
distance, from Earth, 113, 129, 140
energy
movement within, 203–204
output, 25, 70
source, 380n61
as hot/cool star, 141
hydrogen content, 13, 350n22
interior, 122, 123(fig.), 141, 374n4
knowledge of, 12–15
light spectra, 66–67, 122–123
luminosity, 128, 132, 136–137, 379n51
Sun (*continued*)
magnetic reversals, 54
metallicity, 384n30
and moon sizes, 11
and neutrinos, 141–142
oscillations, 141
and panspermia, 344
signature proxies, 353n13
size, 9–10, 18, 137
temperature, 67–68
wobbling, 213
see also heliocentrism; light
sunspot cycles
energy output, 70
light output, 137
and magnetic field, 54
quantification, 362nn15–16
reconstruction, 25, 353n12
super-eclipses, 7, 17–18
supernovae
and Big Bang theory, 179
and chemical elements, 152–153, 157–158
as data source, 389n16
and distance(s), 388n10, 389n19
in early universe, 182, 188–189
and ice cores, 25–26
and marine sediment, 354n16
and matter-energy density, 177–178
and neutrinos, 142
and radiation, 162, 188–189
as standard candles, 173
Type Ia and Type II, 157, 179, 385n39, 385n41
and universe expansion, 179, 186, 189
surface of last scatter, 175
surface:volume ratio, 213, 369n8, 394n24
synchrotron emission, 187

Taylor, Stuart Ross, 224–225, 348n13
technological life
defined, 60–61
and design, 301–302
and fuels, 61–62, 64
and human size, 217–218
and Mars, 85–86
and Middle Ages, 364n37
and ores, 62–64
prerequisites, 363n33
tools, 61, 85–86
see also Drake Equation; extraterrestrial intelligence
teleology, 230, 242–244, 248

telescope(s)
and design inference, 308–309
ground-based, 60, 363n31
Hubble, 126, 162
and human size, 217
Moon as, 109–110
neutrino, 380n62
radio-, 127, 341–342
temperature
and biodiversity, 358n74
and carbon cycle, 56–57
and civilizations, 360n86
of days and nights, 7, 68, 97–98, 107
of Earth
core, 131
surface, 34
and energy, 365n5
global warming, 360n87, 366n20
and hydrogen isotopes, 354n19
and luminosity, 139
and magnetism, 48
and marine life, 354nn20–21
Moon effect, 107
of ocean surface, 72
and oxygen solubility, 358n75
and prokaryotes, 39
and radiation, 67–68
and redshift, 389n20
and rotational synchronization, 97–98
and star-planet distance, 129–130
of stars, 120–121, 124
of Sun, 66–67, 122–123
and surface energy, 365n5
and surface:volume ratio, 369n8
on Venus, 86
and water, 34, 352n6, 357n50
wide swings, 370n24
see also body temperature; climate; heat
terminator, 7, 133–134
terrestrial planets
cores, 376n31
and Drake Equation, 282–284
extrasolar, 382n24
formation, 135–136, 154–158, 371n42
habitability, 82–88
impact threats, 114–115, 129–130
and metallicity, 387n62
and metals, 258, 383nn27–28, 384n29
orbits, 135–136, 371n38, 383n24
in other galaxies, 167
terrestrial planets (*continued*)
rotation, 86–87
size, 363nn25–26, 363nn29–30
spacing, 135–136
with stabilizing moon, 340
see also Earth; Mars; Mercury; Venus
theology
Christian, 227–231, 244–245, 401n60, 412n47
and design, 330
and ETI, 292, 412n46
and solar eclipses, 310–311
theories, beautiful, 250
thermal vents, 37–38
thermophiles, 359n76
thorium
in crust, 63
initial seeds, 385
interstellar, 158
isotopes, 30
in Mars, 85
and plate tectonics, 393n14
three-dimensional tomography, 46–47
tidalites, 28, 107
tides
and climate, 349n9
and Earth rotation, 17
galactic, and comets, 160
host star effect, 7
and life origins, 349–350n14
and M dwarf stars, 133
and Moon
distance from, 18, 27–28
effects, 6, 107
seismic events, 92
and moons, of gas giants, 98
and planet stability, 7
on Venus, 87
time
and Big Bang theory, 260
and CMBR, 189
dilation, 179
dimension(s), 210
estimates, 354n25
and fine-tuning, 205–208
and galaxies, 175–176, 185–186, 290
habitability
timeline, 180(fig.)
window for, 181–183
infinity, 192(fig.), 260–261, 406n23
and Judeo-Christian tradition, 228

time (*continued*)
linear *vs.* cyclical, 228
and Relativity Theory, 389n19
Tipler, Frank, 176
tired light theory, 178–179, 389nn19–20
titanium oxide, 374n7
Tolman, Richard, 178–179
tomography, three-dimensional, 46–47
tools, 61, 85–86
tree rings, 22, 28–30, 353n12, 355nn28–29
trees, 29, 30, 70
trigonometric parallax, 138
Trimble, Virginia, 206, 283
tungsten, 349n11
Tunguska event, 74, 367n27
***2001: A Space Odyssey*, 276, 293, 295, 301, 307, 335**

UFOs, 276
ultraviolet light, 179–180, 181, 355n29
ultraviolet radiation
and gravity, 204
and panspermia, 344
protection, 366n23
role of, 134–135
universe(s)
ancient views, 225–227
Aristotle, 230–231, 238, 239, 242
center of, 226
dimensions, 209, 305n38
for discoverability, 313–315 (*see also* discoverability)
and Earth surface, 191
expansion(s)
and CMBR, 175–176, 179
with contractions, 261
cooling rate, 395–396n45
and galaxies, 189
Hubble work, 170–171, 176
and infinity, 406n23
and redshift, 175, 181
tests, 178–179
and visible distance, 184(fig.), 186
and gravity, 204
habitability, 181–183, 196–197, 301–302
history, 185(fig.)
human role (*see under* human beings)
infinite, 192(fig.), 260–261, 406n23
multiple, 268–271, 306, 319, 325, 406n26
origination, 404n7 (*see also* Big Bang theory)
oscillating, 261

universe(s) (*continued*)
rational underpinning, 211–212, 229–231, 242–244 (*see also* design; rationality)
size scales, 216(fig.)
timeline, 180(fig.)
see also design; fine-tuning; habitability; purpose
uranium, 63, 158, 385n37, 393n14
on Mars, 85
uranium-234, 30
uranium-235, 63
Urey, Harold, 26
47 Ursa Majoris, 96

vacuum energy, 176–177, 186, 390n28
vs. gravity, 205
value, 303
vaporization, of water, 34
see also water vapor
Venus
atmosphere, 86, 107
craters, 374n30
as Earth protector, 115
meteorites from, 344
orbit, 87, 94
and panspermia, 343, 345, 359n78
planetary laws, 106
very long baseline interferometry (VLBI), 60
voluntarism, 398n22
Von Weizsäcker, C. F., 171
Vostok ice cores, 23–24

Ward, Peter, 285, 287, 297, 338
water
in asteroids, 161, 368n36
and body temperature, 357n55
capillary action, 34
and carbon dioxide, 357n59
and chemistry of life, 33–35
and climate, 23, 33–34
and clouds, 72
condensation boundary, 99–100
density, 33
element *vs.* compound, 392n7
on Europa, 89
and fine-tuning, 198
and habitability, 127–128, 322, 411n38
heat effect, 357n50
hydrological cycle, 40
isotopic composition, 352n6
on Mars, 82, 369n5

water (*continued*)
and M dwarf stars, 134
and Milankovitch cycles, 30
and mineral ores, 63
as oxidation product, 37
oxygen solubility, 358n75
and planetary mass, 59
substitutions for, 376n28
temperature, 34, 352n6, 357n50
water vapor
in atmosphere, 66–67, 68–69
and climate, 35–36
condensation effect, 99–100
and heat, 34
waterworld, 364n35
Watson, Andrew, 70–71
Watterson, Bill, 278
Weak Anthropic Principle (WAP), 136–138, 164–166, 405n18
and Copernican Principle, 265–266, 272, 405–406n22
and design inference, 304
weak force, 202–203, 395n32
weathering, of rocks, 34–35, 55–55
Wegener, Alfred, 53–54
Weizsäcker, C. F. von, 171
Wells, H. G., 276
Wetherill, George, 135–136
Whewell, William, 322
white dwarfs. *see* Cepheids, Classical
Whitrow, G. J., 209
Whole Earth Telescope (WET), 374n5
Wigner, Eugene, 212
Wilkinson Microwave Anisotropy Probe (WMAP), 390n27
Wilson, Robert, 174
wind(s)
on Mars, 82
in Middle Ages, 364n37
and oceans, 349n8, 362n22
worldviews, 331–335

X-rays, 12
and early universe, 182
and globular clusters, 164
and M dwarf stars, 134

Yarkovsky effect, 367n31, 377n36
Yellowstone springs, 37
Young, Charles A., 14–15
Younger Dryas, 25, 41

Zee, A., 214
zone of avoidance, 147, 381n8